T0213723

Lecture Notes in Computer Science   9866

Commenced Publication in 1973
Founding and Former Series Editors:
Gerhard Goos, Juris Hartmanis, and Jan van Leeuwen

Matt Bishop · Anderson C.A. Nascimento (Eds.)

Information Security

19th International Conference, ISC 2016
Honolulu, HI, USA, September 3–6, 2016
Proceedings

 Springer

Editors
Matt Bishop
University of California
Davies, CA
USA

Anderson C.A. Nascimento
University of Washington
Tacoma, WA
USA

ISSN 0302-9743   ISSN 1611-3349   (electronic)
Lecture Notes in Computer Science
ISBN 978-3-319-45870-0   ISBN 978-3-319-45871-7   (eBook)
DOI 10.1007/978-3-319-45871-7

Library of Congress Control Number: 2016949632

LNCS Sublibrary: SL4 – Security and Cryptology

This Springer imprint is published by Springer Nature
The registered company is Springer International Publishing AG Switzerland

Preface

This volume contains the papers presented at the 19th Annual International Conference on Information Security (ISC 2016) held on September 3–6, 2016, in Honolulu, Hawaii, USA.

Held annually, the Information Security Conference is an international conference covering research in the theory and applications of information security. It started as a workshop in 1997 and became a conference in 2001. It has been held on five different continents.

Rui Zhang from the University of Hawaii, USA and Masahiro Mambo, from Kanazawa University, Japan organized ISC 2016, in cooperation with the International Association for Cryptologic Research.

Submissions to ISC 2016 were required to be anonymous. The 76 submitted papers were reviewed using a double-blind process. Selecting the final accepted papers was a difficult task that demanded many rounds of interaction. As a result, the Program Committee accepted 19 full papers (resulting in an acceptance rate of 25 %) and 7 short papers. We would like to thank the Program Committee for their valuable work.

We also thank Rui Zhang and Masahiro Mambo for the superb work they did organizing the conference. We are grateful to the ISC Steering Committee for their support, to Depeng Li for handling the local arrangements, and the creators and maintainers of Easychair, the superb platform we used to manage the review process.

Finally, we would like to thank all the authors who submitted papers to ISC 2016. They are the main reason for the success of this conference.

September 2016

Matt Bishop
Anderson C.A. Nascimento

Organization

Information Security Conference 2016, Honolulu, Hawaii, USA, September 7–9, 2016

General Chairs

Masahiro Mambo	Kanazawa University, Japan
Rui Zhang	University of Hawaii, USA

Program Chairs

Matt Bishop	University of California, Davis, USA
Anderson C.A. Nascimento	University of Washington, Tacoma, USA

Local Arrangements

Depeng Li	University of Hawaii, USA

Steering Committee

Colin Boyd	Norwegian University of Science and Technology, Norway
Ed Dawson	Queensland University of Technology, Australia
Javier Lopez	University of Malaga, Spain
Masahiro Mambo	Kanazawa University, Japan
Eiji Okamoto	University of Tsukuba, Japan
Susanne Wetzel	Stevens Institute of Technology, USA
Rui Zhang	University of Hawaii, USA
Yuliang Zheng	University of Alabama at Birmingham, USA

Program Committee

Gail-Joon Ahn	Arizona State University, USA
Luis Antunes	University of Porto, Portugal
Diego Aranha	State University of Campinas, Brazil
Claudio Ardagna	Università degli Studi di Milano, Italy
Elias Athanasopoulos	Vrije Universiteit Amsterdam, The Netherlands
Nuttapong Attrapadung	AIST, Japan
Tuomas Aura	Aalto University, Finland
Elisa Bertino	Purdue University, USA
Alex Biryukov	University of Luxembourg, Luxembourg
Matt Bishop	UC Davis, USA
Liqun Chen	Hewlett Packard Labs, USA
Sherman S.M. Chow	Chinese University of Hong Kong, Hong Kong, China

Junjie Zhang	Wright State University, USA
Rui Zhang	University of Hawaii, USA
Ziming Zhao	Arizona State University, USA
Jianying Zhou	Institute for Infocomm Research, Singapore

Contents

Encryption, Signatures and Fundamentals

Cryptanalysis

Truncated and Multiple Differential Cryptanalysis of Reduced Round Midori128

Mohamed Tolba, Ahmed Abdelkhalek, and Amr M. Youssef[(✉)]

Concordia Institute for Information Systems Engineering,
Concordia University, Montréal, Quebéc, Canada
youssef@ciise.concordia.ca

Abstract. Midori is a family of SPN-based lightweight block ciphers designed to optimize the hardware energy consumption per bit during the encryption and decryption operations. At ASIACRYPT 2015, two variants of the cipher, namely Midori128 and Midori64, which support a 128-bit secret key and a 64/128-bit block, respectively, were proposed. Recently, a meet-in-the-middle attack and an invariant subspace attack were presented against Midori64 but both attacks cannot be applied to Midori128. In this paper, we present truncated and multiple differential cryptanalysis of round reduced Midori128. Our analysis utilizes the special structure of the S-boxes and binary linear transformation layer in order to minimize the number of active S-boxes. In particular, we consider differentials that contain only single bit differences in the input and output of the active S-boxes. To keep this single bit per S-box patterns after the *MixColumn* operation, we restrict the bit differences of the output of the active S-boxes, which lie in the same column after the shuffle operation, to be in the same position. Using these restrictions, we were able to find 10-round differential which holds with probability 2^{-118}. By adding two rounds above and one round below this differential, we obtain a 13 round truncated differential and use it to perform a key recovery attack on the 13-round reduced Midori128. The time and data complexities of the 13-round attack are 2^{119} encryptions and 2^{119} chosen plaintext, respectively. We also present a multiple differential attack on the 13-round Midori128, with time and data complexities of $2^{125.7}$ encryptions and $2^{115.7}$ chosen plaintext, respectively.

Keywords: Truncated differential cryptanalysis · Midori128 · Multiple differential cryptanalysis

1 Introduction

Over the past few years, many lightweight block ciphers such as HIGHT [6], mCrypton [10], DESL/DESXL [9], PRESENT [2], KATAN/KTANTAN [3], Piccolo [12] and PRINTcipher [7] were proposed. On the other hand, there has been little work that focuses on determining the design choices that lead to the most energy efficient architecture. While power and energy are clearly correlated, optimizing the power consumption for block ciphers does not necessarily lead to the

© Springer International Publishing Switzerland 2016
M. Bishop and A.C.A. Nascimento (Eds.): ISC 2016, LNCS 9866, pp. 3–17, 2016.
DOI: 10.1007/978-3-319-45871-7_1

most energy efficient designs since a low power optimized cipher may have high latency, i.e., it takes longer to perform the encryption and decryption operations and hence the required energy increases. In other words, there is no guarantee that low power block cipher designs would lead to low energy designs and vice versa.

By identifying some design choices that are energy efficient and by choosing components specifically tailored to meet the requirements of low energy design, at ASIACRYPT 2015, Banik *et al.* [1] proposed an SPN-based lightweight block cipher, Midori. In particular, Midori is designed to optimize the hardware energy consumption per bit during the encryption and decryption operations. Two variants of the cipher, namely, Midori128 and Midori64 which support a 128-bit secret key and a 64/128-bit block, respectively, were proposed. The linear and non-linear operations of both versions were selected to optimize this objective.

The state in Midori is represented as 4×4 matrix, where the size of each cell depends on the version of cipher, e.g., the cell size in Midori128 is 8 bits. Midori uses 4×4 almost MDS binary matrix because, compared to other MDS matrices, this almost MDS matrix is more efficient in terms of area and signal-delay. To compensate for the low branch number of the almost MDS matrix (4 as compared to 5 in the case of MDS), the designers utilized an optimal cell-permutation layer in order to improve the diffusion speed and increase the number of active S-boxes.

Recently, Midori64 has been analyzed by two different techniques. The first is a meet-in-the-middle with differential enumeration and key dependent sieving [11] which attacks 11 rounds (resp. 12 rounds) with time, memory, and data complexities of 2^{122} (resp. $2^{125.5}$) encryptions, $2^{89.2}$ (resp. 2^{106}) 64-bit blocks, and 2^{53} (resp. $2^{55.5}$) chosen plaintext. The second is an invariant subspace attack [5] against the full cipher. The latter attack proves that the security margin of Midori64 is 96 bits instead of 128 bits. Both of these attacks are not applicable to Midori128.

In this paper, we present truncated differential cryptanalysis of round reduced Midori128. Our attack utilizes the following two observations. First, Midori128 uses four different 8-bit S-boxes, namely SSb_0, SSb_1, SSb_2 and SSb_3, where each one is composed of two 4-bit S-boxes Sb_1 in addition to input and output bit permutation. Consequently, in order to minimize the number of active S-boxes, we consider only single bit differences (i.e., $1, 2, 4, 8, 16, 32, 64, 128$) in the input and output of the 8-bit S-boxes. Second, given the binary nature of the almost MDS transformation, and the fact that the Hamming weight of each row is 3, it follows that the active bytes in a column after the *MixColumn* operation have a single bit difference if and only if the active bytes in the input column are all equal and each one has a single active bit. Hence, to maintain the pattern of single bit differences after the *MixColumn* operation, we restrict the bit differences of the output of the active S-boxes, which lie in the same column after the shuffle operation, to be in the same position.

Based on these observations, we are able to find a 10-round differential that holds with probability 2^{-118}. Then, we added two rounds above and one round below this 10-round differential to obtain a 13-round truncated differential [8] that holds with probability 2^{-230}. Using this truncated differential, we can recover the master key of the 13-round reduced cipher with time and data complexities of 2^{119} encryptions, and 2^{119} chosen plaintext, respectively. We also present a multiple differential attack [4] on the 13-round reduced cipher with time and data complexities of $2^{125.7}$ encryptions and $2^{115.7}$ chosen plaintext, respectively.

The rest of the paper is organized as follows. In Sect. 2, we provide the notations used throughout the paper and a brief description of Midori128. In Sect. 3, we describe and analyze the algorithm we use to efficiently search for long differentials with small number of active S-boxes. Details of our truncated differential attack on Midori128 reduced to 13 rounds are presented in Sect. 4. Section 5 presents our multiple differential attack. Finally, the paper is concluded in Sect. 6.

2   Specifications of Midori128

2.1   Notations

The following notations are used throughout the rest of the paper:

- a^t: Transposition of the vector or the matrix a.
- K: The master key.
- RK_i: The 128-bit round key used in round i.
- WK: The 128-bit whitening key.
- x_i: The 128-bit input to the *SubCell* operation at round i.
- y_i: The 128-bit input to the *ShuffleCell* operation at round i.
- z_i: The 128-bit input to the *MixColumn* operation at round i.
- w_i: The 128-bit input to the *KeyAdd* operation at round i.
- $x_i[j]$: The jth byte of x_i, where $0 \leq j < 16$.
- $x_i[j \cdots l]$: The bytes from j to l of x_i, where $j < l$.
- $x_i[j, l]$: The bytes j and l of x_i.
- $\Delta x_i, \Delta x_i[j]$: The difference at state x_i and byte $x_i[j]$, respectively.

2.2   Specifications

Midori128 can be considered as a variant of Substitution Permutation Networks (SPNs). The state in Midori128 is represented as a 4×4 array of bytes as follow:

$$S = \begin{pmatrix} s_0 & s_4 & s_8 & s_{12} \\ s_1 & s_5 & s_9 & s_{13} \\ s_2 & s_6 & s_{10} & s_{14} \\ s_3 & s_7 & s_{11} & s_{15} \end{pmatrix}$$

Midori128 iterates over 20 rounds. Each round, except the last one, has 3 layers: S-layer (*SubCell*) which maps $\{0,1\}^{128} \to \{0,1\}^{128}$, P-layer (*ShuffleCell* and *MixColumn*) which maps $\{0,1\}^{128} \to \{0,1\}^{128}$ and a key-addition layer (*KeyAdd*) which maps $\{0,1\}^{128} \times \{0,1\}^{128} \to \{0,1\}^{128}$. The last round contains only the S-layer. Moreover, before the first and after the last rounds, prewhitening and postwhitening are performed using *WK*. In what follows, we show how these operations update the 128-bit state S:

- *SubCell*: A nonlinear layer applies 4 8-bit S-boxes, namely SSb_0, SSb_1, SSb_2, and SSb_3, on each byte of the state S in parallel, where $s_i \leftarrow SSb_{(i \bmod 4)}[s_i]$ and $0 \le i \le 15$. As shown in Fig. 1, each 8-bit S-box SSb_i is composed of input and output bit permutations and 2 4-bit S-box Sb_1, where Sb_1 is 4-bit S-box (see Table 1).
- *ShuffleCell*: The bytes of the state S is permuted as follow: $(s_0, s_1, \cdots, s_{15}) \leftarrow (s_0, s_{10}, s_5, s_{15}, s_{14}, s_4, s_{11}, s_1, s_9, s_3, s_{12}, s_6, s_7, s_{13}, s_2, s_8)$
- *MixColumn*: Each column in the internal state is multiplied by a binary matrix M, where

$$M = \begin{pmatrix} 0 & 1 & 1 & 1 \\ 1 & 0 & 1 & 1 \\ 1 & 1 & 0 & 1 \\ 1 & 1 & 1 & 0 \end{pmatrix}$$

Hence, the internal state is updated as follows:

$$(s_i, s_{i+1}, s_{i+2}, s_{i+3})^t \leftarrow M(s_i, s_{i+1}, s_{i+2}, s_{i+3})^t, i = 0, 4, 8, 12.$$

- *KeyAdd*: Where 128-bit round key RK_i is XORed with the state S.

Table 1. 4-bit bijective S-box Sb_1 in hexadecimal form [1]

x	0	1	2	3	4	5	6	7	8	9	a	b	c	d	e	f
$Sb_1[x]$	1	0	5	3	e	2	f	7	d	a	9	b	c	8	4	6

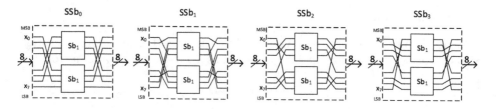

Fig. 1. SSb_0, SSb_1, SSb_2, and SSb_3 [1]

The data encryption procedure of Midori128 is illustrated in Algorithm 1 where $R = 20$ denotes the number of rounds.

Algorithm 1. Data Encryption Algorithm 1

Data: $X, WK, RK_0, ..., RK_{R-2}$
Result: Y
$S \leftarrow KeyAdd(X, WK)$;
for $i \leftarrow 0$ *to* $R - 2$ **do**
$\quad$ $S \leftarrow SubCell(S)$;
$\quad$ $S \leftarrow ShuffleCell(S)$;
$\quad$ $S \leftarrow MixColumn(S)$;
$\quad$ $S \leftarrow KeyAdd(S, RK_i)$;
$S \leftarrow SubCell(S)$;
$Y \leftarrow KeyAdd(S, WK)$;

The prewhitening and postwhitening key WK in Midori128 are equal to the master key K, the rounds keys $RK_i = K \oplus \beta_i$, $0 \le i \le 18$, where β_i is a constant, X is the plaintext, and Y is the ciphertext. Throughout our analysis, we measure the time complexity of our attack in terms of the equivalent number of reduced-round Midori128 encryptions. For further details about the design rational of the cipher, the reader is referred to [1].

3 A 10-round Differential of Midori128

As mentioned above, in order to minimize the number of active S-boxes, we consider differentials that have only single bit differences in the input and output of the active 8-bit S-boxes. In this section, we describe and analyze the algorithm we use to efficiently find such differentials whose probabilities are greater than 2^{-128}. In each round, we have 4 operations. The operations that can disturb the single bit difference propagation patterns are the $SubCell$ and $MixColumn$ operations. From the structure of the S-boxes, it follows that for any active S-box and for a given 1-bit input difference, there are at most 4 possible output differences of 1-bit difference because only 1 4-bit S-box will be active. Furthermore, from the properties of the binary almost MDS matrix, preserving single bit differences propagation patterns requires that we restrict the bit differences of the output of the active S-boxes, which lie in the same column after the shuffle operation, to be in the same position. As a result, the active bytes in each column after the $MixColumn$ operation have the same value. Therefore, at each round we have at most $(8 \times 15)^4 \approx 2^{28}$ possible input differences. The term 8 in the previous formula denotes the number of possible values of the difference in each column $(1, 2, 4, \cdots, 128)$. The term 15 denotes the total number of combinations for active bytes within each column and the exponent 4 denotes the total number of columns. As noted above, for a given input difference ΔS_i at the beginning of each round, after the S-box layer the values of the active bytes, which lie in the same column after the shuffle operation, should be equal. Therefore, for each input difference i, we have a set Ω_i which contains at most 4^4 possible output differences (in each column we have at most four 1-bit differences, and we can have at most four active columns.)

Algorithm 2 describes the procedure used to find the maximum number of rounds r, such that a differential with the above S-box propagation patterns exists and its differential probability is greater than 2^{-128}. The algorithm run time is upper bounded by $2^{28} \times 4^4 \times r$. It utilizes four tables where each table has 2^{28} entries corresponding to the possible input differences at each round. In what follows we describe the use of each table:

1. Each entry i in the table *InputDiffProb* indicates the probability to reach the difference i at the beginning of the considered round.
2. Each entry i in the table *OutputDiffProb* indicates the probability to reach the difference i at the end of the considered round.
3. Each entry i in the table *InputParent* indicates the input difference used at the beginning of round 0 to reach difference i at the beginning of the considered round. The value -1 is used to indicate that the difference i cannot be reached from any input difference.
4. Each entry i in the table *OutputParent* indicates the input difference used at the beginning of round 0 to reach difference i at the end of the considered round. The value -1 indicates that the difference i cannot be reached from any input difference.

As explained in Algorithm 2, we iterate over the input differences that have differential probability > 0 round by round, i.e., at each round we propagate all the input differences and begin with the obtained differences as input to the next round. In our implementation, the 28-bit index i encodes the state difference ΔS as follows: we use 7 bits for each one of the four columns where these 7 bits are divided into two parts. The first 3 bits represent the value of the difference of the active bytes in the column and the remaining 4 bits represent the different 15 combinations of the active bytes within the column. After applying the 3 operations: *SubCell*, *ShuffleCell*, and *MixColumn*, we have 3 cases: (i) this difference did not appear before (*Outparent*[j] $= -1$). In this case we set its probability and parent, (ii) this difference appeared previously and its previous parent is the same as the new parent. Therefore, we add the probabilities, and (iii) this difference appears previously but with a different parent. In this case, we choose the parent with the higher probability. At the end of each round, we copy the *OutputDiffProb* and *OutParent* into *InputDiffProb* and *InputParent*, respectively, and initialize *OutputDiffProb* and *OutParent* to begin another round.

Using a PC with Intel(R) Xeon(R) CPU E3-1280 V2 @ 3.6 GHz and 32 GB RAM, a non-optimized implementation of Algorithm 2 terminates in about 3 hours and outputs $r = 10$ rounds. A 10-round differential which holds with probability 2^{-118} is illustrated in Fig. 2. This 10-round differential has 2554 characteristics that are distributed as shown in Table 2. The characteristic that holds with probability 2^{-123} is detailed in Table 3.

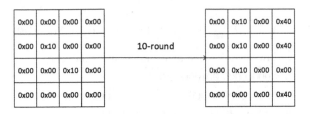

Fig. 2. A 10 rounds differential of Midori128

Table 2. The characteristics distribution of the 10-round differential

# of characteristics	Probability of each characteristic
1	2^{-123}
7	2^{-124}
23	2^{-125}
50	2^{-126}
83	2^{-127}
2390	$\leq 2^{-128}$

4 13-round Truncated Differential Cryptanalysis of Midori128

In this section, we show how we can extend the differential obtained in the previous section by two rounds above, and one round below, to obtain a truncated differential over 13 rounds (see Fig. 3). Then, we present a key recovery attack on Midori128 reduced to 13 rounds using this truncated differential.

The total probability of the 13-round differential can be calculated from its three parts. First, the differential probability of the top two rounds which is 2^{-112} and can be computed as follows: (i) $4 \rightarrow 2, 3 \rightarrow 1, 4 \rightarrow 2$, and $3 \rightarrow 1$ transitions over $MixColumn$ ($z_0 \rightarrow w_0$) which happens with probability $2^{-16} \times 2^{-16} \times 2^{-16} \times 2^{-16} = 2^{-64}$, (ii) $3 \rightarrow 1$ and $3 \rightarrow 1$ transitions over $MixColumn$ ($z_1 \rightarrow w_1$) which happens with probability $2^{-16} \times 2^{-16} = 2^{-32}$ and (iii) $w_1[5]$

and $w_1[10]$ equal difference 16 which happens with probability $2^{-8} \times 2^{-8} = 2^{-16}$. Second, the differential probability of 10-round is 2^{-118}, calculated by Algorithm 2. Third, the bottom 1 round has a differential probability equals to 1. Therefore, the differential probability of the 13-round is 2^{-230}.

Table 3. The 2^{-123} 10-round characteristic of Midori128

i	Input difference at round i (in hexadecimal)
0	00000000 00100000 00001000 00000000
1	00000000 00000000 00000000 00080800
2	40004040 00000000 00101010 00000000
3	40400040 40400000 08080008 08080000
4	00400040 00000040 08000000 08000800
5	40000040 00404000 08000008 00000000
6	40404000 00400040 08080008 08000800
7	00400000 40004000 00080000 08000800
8	00000000 00400000 00000800 00000000
9	00000000 00000000 00000000 00080800
10	40004040 00000000 00101010 00000000

In what follows, we show how we can perform our key recovery attack on Midori128 reduced to 13 rounds using the above differential. The attack is decomposed of two steps. The first step is the *data collection* in which we collect many pairs of messages to guarantee that at least one of them confirms to the 13-round truncated differential in Fig. 3. The second step is the *key recovery* in which the collected data pairs are used to identify the key candidates.

Proposition 1. *(Differential Property of bijective S-boxes) Given two non-zero differences, Δi and Δo, in $\mathbb{F}256$, the equation: $S(x) + S(x + \Delta i) = \Delta o$ has one solution on average. This property also applies to S^{-1}.*

Data Collection. To reduce the number of required chosen plaintext and get enough pairs to launch the attack, we use the structure technique. Here, our structure takes all the possible values in all the bytes except bytes 2 and 11. These bytes take fixed value. Therefore, one structure generates $2^{14 \times 8} \times (2^{14 \times 8} - 1)/2 \approx 2^{223}$ possible pairs. We need to collect 2^{230} message pairs because the total probability of the 13-round differential is 2^{-230}. Since each structure contains 2^{223} message pairs, we need to collect 2^7 structures to find the right pair. Therefore, we ask the encryption oracle for the encryption of 2^{119} messages.

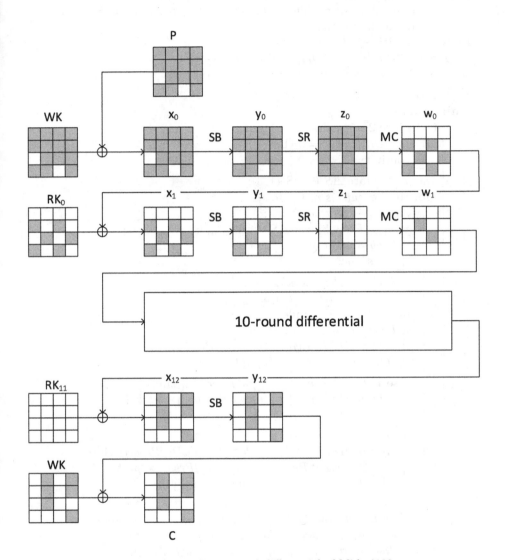

Fig. 3. 13-round truncated differential of Midori128

Algorithm 2. Find the maximum number of rounds, r, that has a differential which holds with probability $> 2^{-128}$ considering only the single bit difference for the active S-boxes

Result: X: Input difference, Y : Output difference, *Probability:* The probability of the differential, r: The number of rounds covered by the differential

for $i \leftarrow 0$ *to* $2^{28} - 1$ **do**
    | $InputDiffProb[i] \leftarrow 1$, $OutputDiffProb[i] \leftarrow 0$;
    | $InputParent[i] \leftarrow i$, $OutputParent[i] \leftarrow -1$;

$r \leftarrow 0$, *valid*$\leftarrow$ *true*;

while *valid* **do**
    for $i \leftarrow 0$ *to* $2^{28} - 1$ **do**
        if $InputDiffProb[i] = 0$ **then**
        | continue;

        map i to state ΔS (see Sect. 3);
        forall the *entries in Ω_i* **do**
            $\Delta S_{temp} \leftarrow nextvalue(\Omega_i)$;
            $prob \leftarrow$ Probability $(\Delta S \rightarrow \Delta S_{temp})$;
            if $prob = 0$ **then**
            | continue;

            $\Delta S_{temp} \leftarrow ShuffleCell(\Delta S_{temp})$;
            $\Delta S_{temp} \leftarrow MixColumn(\Delta S_{temp})$;
            map ΔS_{temp} into index j;
            if $OutParent[j] = -1$ **then**
            | $OutputDiffProb[j] = InputDiffProb[i] \times prob$;
            | $OutputParent[j] = InputParent[i]$;

            else if $OutParent[j] = InputParent[i]$ **then**
            | $OutputDiffProb[j] = InputDiffProb[i] \times prob +$
            | $OutputDiffProb[j]$;

            else
               if $InputDiffProb[i] \times prob > OutputDiffProb[j]$ **then**
               | $OutputDiffProb[j] = InputDiffProb[i] \times prob$;
               | $OutputParent[j] = InputParent[i]$;

    $InputDiffProb \leftarrow OutputDiffProb$, $InputParent \leftarrow OutParent$;
    initialize $OutputDiffProb$ to 0 , initialize $OutputParent$ to -1;
    get the index l of the maximum entry in $InputDiffProb$;
    if $InputDiffProb[l] <= 2^{-128}$ **then**
    | valid $\leftarrow$ false;
    $r \leftarrow r + 1$;

Get the index l of the maximum entry in $InputDiffProb$;
$X \leftarrow InputParent[l]$, $Y \leftarrow l$, $Probability \leftarrow InputDiffProb[l]$;

Key Recovery. In this step, we try to identify the key candidates that confirm to the 10-round differential. First, we try to identify the number of key suggestions of 26 bytes $WK[0, 1, 3 \cdots 10, 12 \cdots 15], RK_0[1, 3, 6, 9, 11, 14]$, and $WK[4, 5, 6 , 12, 13, 15]$ that correspond to each pair of messages. This can be achieved as follows: to deduce the values of the 6 bytes $WK[4, 5, 6, 12, 13, 15]$, we know that the 6 bytes $\Delta x_{12}[4, 5, 6, 12, 13, 15]$ only take one value, the same difference of the end of the 10-round differential since the key add layer does not change the difference. The knowledge of the ciphertext allows to compute $\Delta y_{12}[4, 5, 6, 12, 13, 15]$. Using the differential property of the S-box, we can evaluate $y_{12}[4, 5, 6, 12, 13, 15]$. The knowledge of the ciphertext with $y_{12}[4, 5, 6, 12, 13, 15]$ allows us to deduce the values of the 6 bytes $WK[4, 5, 6, 12, 13, 15]$. From the other side, we know the difference ΔW_1 since it is the same difference at the beginning of the 10-round differential. Then we propagate this difference linearly trough *MixColumn* and *InvShuffleCell* to get the difference Δy_1. Δy_1 has only 6 active bytes and each, after the *SubCell* operation, has only 6 possible differences. Therefore, we have $6^6 \approx 2^{15.6}$ possible differences at Δx_1. Then after the *MixColumn* and *InvShuffleCell*, we have only $2^{15.6}$ possible differences at Δy_0. The knowledge of the plaintext allows us to compute the difference at Δx_0. Then, guessing the $2^{15.6}$ possible differences of Δy_0 and using the S-box proposition, we get the value of $x_0[0, 1, 3 \cdots 10, 12 \cdots 15]$. From the knowledge of the plaintext we can drive the value of $WK[0, 1, 3 \cdots 10, 12 \cdots 15]$. As a result we have $2^{15.6}$ key candidates for $WK[0, 1, 3 \cdots 10, 12 \cdots 15]$ and $WK[4, 5, 6, 12, 13, 15]$ but we have 6 bytes common; therefore, we have only $2^{15.6-48} = 2^{-32.4}$ key candidates. To derive the 6 bytes value $RK_0[1, 3, 6, 9, 11, 14]$, we know that we have only one difference at Δy_1. Then we get one key candidate for $RK_0[1, 3, 6, 9, 11, 14]$ but also we have 5 bytes filter between $WK[0, 1, 3 \cdots 10 , 12 \cdots 15]$ and $RK0[1, 3, 6, 9, 11, 14]$, from the key schedule as the whitening key is the master key K and the round keys $RK_i = K \oplus \beta_i$ and β_i is constant. Therefore, we have only $2^{-32.4} \times 2^{-40} = 2^{-72.4}$ key candidates for each message pair. To identify the remaining key candidates for all the message pairs, we should identify the remaining message pairs after the ciphertext filter. The ciphertext has a filter probability of $2^{-112.3}$ and is computed as follows: we have 10 bytes of zero difference which have probability of 2^{-80} and the remaining 6 bytes have probability of $2^{15.7-48} = 2^{-32.3}$ since we know that each byte of the 6 active bytes in Δx_{12} has only one difference and after the S-box layer 5 bytes out of these 6 active bytes, each, has 6 possible difference and the remaining byte has 7 possible differences. Hence, we have $6^5 \times 7 = 2^{15.7}$ possible differences at Δy_{12} which also the same differences in the 6 bytes in the ciphertext. Therefore, after the ciphertext filter, we have $2^{230} \times 2^{-112.3} = 2^{117.7}$ remaining message pairs to identify the key candidates. As a result, we have $2^{117.7} \times 2^{-72.4} = 2^{45.3}$ remaining key candidates for 15 bytes of the master key K.

In order to determine efficiently the remaining key candidates for all the remaining message pairs after the ciphertext filter, we perform the following steps:

1. From the ciphertext side, we have only one value for the active bytes of Δx_{12}. Then, using the S-box proposition, we can derive the 6 bytes $WK[4, 5, 6, 12, 13, 15]$. Therefore, we have $2^{117.7}$ key candidates for $WK[4, 5, 6, 12, 13, 15] \equiv K[4, 5, 6, 12, 13, 15]$.

2. From the plaintext side, by guessing the 6 possible differences of $\Delta w_0[14]$ and propagating them backward through the linear operations $MixColumn$ and $InvShuffleCell$ we can know the values of $\Delta y_0[7, 8, 13]$. Hence, we can use the S-box proposition to derive the values of $x_0[7, 8, 13]$. Then, knowing the plaintext and $x_0[7, 8, 13]$ allows to derive $WK[7, 8, 13] \equiv K[7, 8, 13]$. Using the one value of the difference $\Delta y_1[14]$, we can derive the value of $x_1[14]$ using the S-box proposition. Then, we can derive $RK_0[14] \equiv K[14]$. At this stage, we have $2^{117.7}$ key candidates and we guess 6 values to derive $K[7, 8, 13, 14]$ but we have one filter of $K[13]$ between this step and the previous step. Therefore, in total, we have $2^{117.7} \times 6 \times 2^{-8} = 2^{112.3}$ remaining key candidates for $K[4 \cdots 8, 12 \cdots 15]$.

3. By guessing the 6 possible differences of $\Delta w_0[6]$ and propagating them backward through the linear operations $MixColumn$ and $InvShuffleCell$, we can determine the values of $\Delta y_0[1, 4, 14]$. Hence, we can use the S-box proposition to derive the values of $x_0[1, 4, 14]$. Then, knowing the plaintext and $x_0[1, 4, 14]$ allows us to derive $WK[1, 4, 14] \equiv K[1, 4, 14]$. Using the one value of the difference $\Delta y_1[6]$, we can derive the value of $x_1[6]$ using the S-box proposition. Then, we can derive $RK_0[6] \equiv K[6]$. At this stage we have $2^{112.3}$ key candidates and we guess 6 values to derive $K[1, 4, 6, 14]$ but we have 3 filters of $K[4, 6, 14]$ between this step and the previous step. Therefore, in total we have $2^{112.3} \times 6 \times 2^{-24} = 2^{90.9}$ remaining key candidates for $K[1, 4 \cdots 8, 12 \cdots 15]$.

4. By guessing the 6^2 possible differences of $\Delta w_0[1, 3]$ and propagating them backward through the linear operations $MixColumn$ and $InvShuffleCell$ we can determine the values of $\Delta y_0[0, 5, 10, 15]$. Hence, we can use the S-box proposition to derive the values of $x_0[0, 5, 10, 15]$. Then, knowing the plaintext and $x_0[0, 5, 10, 15]$ allows us to derive $WK[0, 5, 10, 15] \equiv K[0, 5, 10, 15]$. Using the one value of the difference $\Delta y_1[1, 3]$, we can derive the value of $x_1[1, 3]$ using the S-box proposition. Then, we can derive $RK_0[1, 3] \equiv K[1, 3]$. At this stage we have $2^{90.9}$ key candidates and we guess 6^2 values to derive $K[0, 1, 3, 5, 10, 15]$ but we have 3 filters of $K[1, 5, 15]$ between this step and the previous step. Therefore, in total we have $2^{90.9} \times 6^2 \times 2^{-24} = 2^{72.1}$ remaining key candidates for $K[0, 1, 3 \cdots 8, 10, 12 \cdots 15]$.

5. By guessing the 6^2 possible differences of $\Delta w_0[9, 11]$ and propagating them backward through the linear operations $MixColumn$ and $InvShuffleCell$ we can determine the values of $\Delta y_0[3, 6, 9, 12]$. Hence, we can use the S-box proposition to derive the values of $x_0[3, 6, 9, 12]$. Then, knowing the plaintext and $x_0[3, 6, 9, 12]$ allows us to derive $WK[3, 6, 9, 12] \equiv K[3, 6, 9, 12]$. Using the one value of the difference $\Delta y_1[9, 11]$, we can derive the value of $x_1[9, 11]$ using the S-box proposition. Then, we can derive $RK_0[9, 11] \equiv K[9, 11]$. At this stage we have $2^{72.1}$ key candidates and we guess 6^2 values to derive $K[3, 6, 9, 11, 12]$ but we have one filter in this step of $K[9]$ and 3 of $K[3, 6, 12]$ between this step and the previous step. Therefore, in total we have $2^{72.1} \times 6^2 \times 2^{-32} = 2^{45.3}$ remaining key candidates for $K[0, 1, 3 \cdots 15]$.

Attack Complexity. The time complexity of the key recovery phase can be derived from the previous steps as follows: step 1 needs $2 \times 2 \times 2^{117.7}/(4 \times 13) = 2^{114}$ encryptions, step 2 needs $2 \times 2^{117.7} \times 6/(4 \times 13) = 2^{115.6}$ encryptions, step 3 needs $2 \times 2^{112.3} \times 6/(4 \times 13) = 2^{110.2}$ encryptions, step 4 needs $2 \times 2^{90.9} \times 6^2/(4 \times 13) = 2^{91.4}$ encryptions, step 5 needs $2 \times 2^{72.1} \times 6^2/(4 \times 13) = 2^{72.6}$ encryptions. Therefore, the time complexity to find $2^{45.3}$ key candidates for $K[0, 1, 3 \cdots 15]$ is $2^{114} + 2^{115.6} + 2^{110.2} + 2^{91.4} + 2^{72.6} \approx 2^{115.6}$. To retrieve the master key we make an exhaustive search for the remaining key candidates with $K[2]$ which needs $2^8 \times 2^{45.3} = 2^{53.3}$ encryptions. Therefore, the time complexity of the attack is dominated by the time needed to build the required structures which is 2^{119} encryptions. The data complexity of the attack is 2^{119} chosen plaintext.

5   Multiple Differential Cryptanalysis of Midori128

In this section, we describe a multiple differential attack that offers some time-data trade-off compared to the previous attack. Using Algorithm 2 and by enumerating all the 10-round differentials that hold with probability $> 2^{-124}$, we found 700 such differentials. In here, we show how to exploit 16 of them to launch a multiple differential attack on Midori128. These 16 differentials are shown in Table 4. As shown in the table, all these differentials have the same input difference, but have different output differences that are all active in the same bytes. The first differential in Table 4 is the differential that is used in the attack described in the previous section. Therefore, the previous attack can be applied with multiple differentials with small modifications.

In this attack, we retrieve the same key bytes as in the previous attack. The total differential probability of the 16 differential is $2^{-114.7}$. Therefore, the total differential probability of the 13-round differential is $2^{-112} \times 2^{-114.7} = 2^{-226.7}$. Consequently, we need $2^{226.7-223} = 2^{3.7}$ structures with $2^{3.7} \times 2^{112} = 2^{115.7}$ chosen plaintext. As shown in Table 4, $\Delta x_{12}[4, 5, 6]$ have the following possible differences $\{4, 8, 16, 32\}$ and $\Delta x_{12}[12, 13, 15]$ have the following possible differences $\{8, 16, 32, 64\}$. Therefore, after the S-box layer, we have the following: each one of $\Delta y_{12}[4, 6, 15]$ has 19 possible differences, $\Delta y_{12}[5]$ has 15 possible differences, $\Delta y_{12}[12]$ has 17 possible differences, and $\Delta y_{12}[13]$ has 20 possible differences. As a result, we have $19^3 \times 15 \times 17 \times 20 = 2^{25.1}$ possible differences at the ciphertext. Consequently, we have $2^{226.7} \times 2^{-80} \times 2^{25.1-48} = 2^{123.8}$ remaining message pairs after the ciphertext filter. The remaining key candidates for each pair are $16 \times 2^{15.6} \times 2^{-88} = 2^{-68.4}$. Therefore, the number of remaining key candidates for all the remaining message pairs, after the ciphertext filter, is given by $2^{123.8} \times 2^{-68.4} = 2^{55.4}$ which can be exhaustively searched with the remaining key byte. We can use the same steps that are used in the previous attack to determine theses $2^{55.4}$ remaining key candidates. The time complexity of the attack is dominated by step 2. At the beginning of step 2, we have $2^{123.8} \times 16 = 2^{127.8}$ key candidates for $K[4, 5, 6, 12, 13, 15]$. Therefore, the time complexity of step 2 is $2 \times 2^{127.8} \times 6/(4 \times 13) = 2^{125.7}$ encryptions. The data complexity is $2^{115.7}$ chosen plaintext.

Table 4. 10-round differentials of Midori128

Input difference (in hexadecimal)	Output difference (in hexadecimal)	Probability
00000000 00100000 00001000 00000000	40004040 00000000 00101010 00000000	$2^{-118.09}$
00000000 00100000 00001000 00000000	08000808 00000000 00080808 00000000	$2^{-118.12}$
00000000 00100000 00001000 00000000	08000808 00000000 00202020 00000000	$2^{-118.12}$
00000000 00100000 00001000 00000000	40004040 00000000 00080808 00000000	$2^{-118.18}$
00000000 00100000 00001000 00000000	40004040 00000000 00202020 00000000	$2^{-118.18}$
00000000 00100000 00001000 00000000	10001010 00000000 00080808 00000000	$2^{-118.36}$
00000000 00100000 00001000 00000000	10001010 00000000 00202020 00000000	$2^{-118.36}$
00000000 00100000 00001000 00000000	10001010 00000000 00101010 00000000	$2^{-118.43}$
00000000 00100000 00001000 00000000	10001010 00000000 00040404 00000000	$2^{-118.8}$
00000000 00100000 00001000 00000000	40004040 00000000 00040404 00000000	$2^{-118.8}$
00000000 00100000 00001000 00000000	20002020 00000000 00101010 00000000	$2^{-118.81}$
00000000 00100000 00001000 00000000	08000808 00000000 00040404 00000000	$2^{-119.29}$
00000000 00100000 00001000 00000000	20002020 00000000 00080808 00000000	$2^{-119.81}$
00000000 00100000 00001000 00000000	20002020 00000000 00202020 00000000	$2^{-119.81}$
00000000 00100000 00001000 00000000	08000808 00000000 00101010 00000000	$2^{-120.32}$
00000000 00100000 00001000 00000000	20002020 00000000 00040404 00000000	$2^{-120.43}$

6 Conclusion

We showed how to exploit the structure of the S-boxes and the *MixColumn* operations of Midori128 in order to obtain long differentials that use single bit difference for the inputs and outputs of the active S-boxes. Then, we developed an algorithm that can be used to efficiently enumerate all such differentials for a given number of rounds. Using this algorithm, we obtained a 10-round differential that holds with probability 2^{-118}. By appending 2 rounds above and one round below this 10-round differential, we obtained a 13 round truncated differential and used it to launch a key recovery attack attack on 13-round reduced Midori128. The time and data complexities of the attack are 2^{119} encryptions and 2^{119} chosen plaintext. Moreover, we presented a multiple differential attack on the 13-round reduced cipher with time and data complexities of $2^{125.7}$ encryptions and $2^{115.7}$ chosen plaintext, respectively.

References

1. Banik, S., Bogdanov, A., Isobe, T., Shibutani, K., Hiwatari, H., Akishita, T., Regazzoni, F.: Midori: a block cipher for low energy. In: Iwata, T., Cheon, J.H. (eds.) ASIACRYPT 2015. LNCS, vol. 9453, pp. 411–436. Springer, Heidelberg (2015). doi:10.1007/978-3-662-48800-3_17
2. Bogdanov, A.A., Knudsen, L.R., Leander, G., Paar, C., Poschmann, A., Robshaw, M., Seurin, Y., Vikkelsoe, C.: PRESENT: an ultra-lightweight block cipher. In: Paillier, P., Verbauwhede, I. (eds.) CHES 2007. LNCS, vol. 4727, pp. 450–466. Springer, Heidelberg (2007)

3. De Cannière, C., Dunkelman, O., Knežević, M.: KATAN and KTANTAN — a family of small and efficient hardware-oriented block ciphers. In: Clavier, C., Gaj, K. (eds.) CHES 2009. LNCS, vol. 5747, pp. 272–288. Springer, Heidelberg (2009)
4. Canteaut, A., Fuhr, T., Gilbert, H., Naya-Plasencia, M., Reinhard, J.-R.: Multiple differential cryptanalysis of round-reduced PRINCE. In: Cid, C., Rechberger, C. (eds.) FSE 2014. LNCS, vol. 8540, pp. 591–610. Springer, Heidelberg (2015)
5. Guo, J., Jean, J., Nikolić, I., Qiao, K., Sasaki, Y., Sim, S.M.: Invariant Subspace Attack Against Full Midori64. IACR Cryptology ePrint Archive, 2015/1189 (2015). https://eprint.iacr.org/2015/1189.pdf
6. Hong, D., Sung, J., Hong, S.H., Lim, J.-I., Lee, S.-J., Koo, B.-S., Lee, C.-H., Chang, D., Lee, J., Jeong, K., Kim, H., Kim, J.-S., Chee, S.: HIGHT: a new block cipher suitable for low-resource device. In: Goubin, L., Matsui, M. (eds.) CHES 2006. LNCS, vol. 4249, pp. 46–59. Springer, Heidelberg (2006)
7. Knudsen, L., Leander, G., Poschmann, A., Robshaw, M.J.B.: PRINTCIPHER: a block cipher for IC-printing. In: Mangard, S., Standaert, F.-X. (eds.) CHES 2010. LNCS, vol. 6225, pp. 16–32. Springer, Heidelberg (2010)
8. Knudsen, L.R.: Truncated and higher order differentials. In: Preneel, B. (ed.) FSE 1994. LNCS, vol. 1008, pp. 196–211. Springer, Heidelberg (1995)
9. Leander, G., Paar, C., Poschmann, A., Schramm, K.: New lightweight DES variants. In: Biryukov, A. (ed.) FSE 2007. LNCS, vol. 4593, pp. 196–210. Springer, Heidelberg (2007)
10. Lim, C.H., Korkishko, T.: mCrypton – a lightweight block cipher for security of low-cost RFID tags and sensors. In: Song, J.-S., Kwon, T., Yung, M. (eds.) WISA 2005. LNCS, vol. 3786, pp. 243–258. Springer, Heidelberg (2006)
11. Lin, L., Wu, W.: Meet-in-the-Middle Attacks on Reduced-Round Midori-64. IACR Cryptology ePrint Archive, 2015/1165 (2015). https://eprint.iacr.org/2015/1165.pdf
12. Shibutani, K., Isobe, T., Hiwatari, H., Mitsuda, A., Akishita, T., Shirai, T.: *Piccolo*: an ultra-lightweight blockcipher. In: Preneel, B., Takagi, T. (eds.) CHES 2011. LNCS, vol. 6917, pp. 342–357. Springer, Heidelberg (2011)

Improved Linear Cryptanalysis
of Round-Reduced ARIA

Ahmed Abdelkhalek, Mohamed Tolba, and Amr M. Youssef[✉]

Concordia Institute for Information Systems Engineering,
Concordia University, Montréal, Québec, Canada
youssef@ciise.concordia.ca

Abstract. ARIA is an iterated SPN block cipher developed by a group
of Korean cryptographers in 2003, established as a Korean standard in
2004 and added to the Transport Layer Security (TLS) supported cipher
suites in 2011. It encrypts 128-bit blocks with either 128, 192, or 256-
bit key. In this paper, we revisit the security of round-reduced ARIA
against linear cryptanalysis and present a 5-round linear hull using the
correlation matrix approach to launch the first 8-round key recovery
attack on ARIA-128 and improve the 9 and 11-round attacks on ARIA-
192/256, respectively, by including the post whitening key. Furthermore,
sin all our attacks, we manage to recover the secret master key. The
(data in known plaintexts, time in round-reduced encryption operations,
memory in 128-bit blocks) complexities of our attacks are $(2^{122.61}, 2^{123.48},
2^{119.94})$, $(2^{122.99}, 2^{154.83}, 2^{159.94})$, and $(2^{123.53}, 2^{238.13}, 2^{239.95})$ for ARIA-
128, ARIA-192, and ARIA-256, respectively.

Keywords: Block cipher · Cryptanalysis · Linear cryptanalysis ·
ARIA · Key recovery · Linear hull · Correlation matrix

1 Introduction

ARIA is an iterated Substitution Permutation Network (SPN) block cipher that
operates on 128-bit blocks with 128, 192 or 256-bit key. It was designed by a
group of Korean cryptographers and published in ICISC 2003 [11]. When ARIA
was published in ICISC, it had 10/12/14 rounds for key sizes of 128/192/256
bits, respectively, and used 4 distinct S-boxes. In 2004, it was adopted by the
Korean Agency for Technology and Standards (KATS) as the Korean 128-bit
block encryption algorithm standard after increasing the number of rounds to
12/14/16 and introducing some modifications in the key scheduling algorithm.
The life span of ARIA has been extended since then and the latest extension was
in December 2014 where its life span was extended for another 5 years (KS X
1213-1:2014) [9]. Since 2011, ARIA is also one of the ciphers that are supported
in the Transport Layer Security (TLS) protocol [10].

Since its introduction, the security of ARIA was scrutinized by several cryptographers. After the initial analysis of ARIA by its designers, Biryukov *et al.* [4]

© Springer International Publishing Switzerland 2016
M. Bishop and A.C.A. Nascimento (Eds.): ISC 2016, LNCS 9866, pp. 18–34, 2016.
DOI: 10.1007/978-3-319-45871-7_2

evaluated the security of ARIA against many cryptanalytic techniques. The best attack they developed was based on a 7-round truncated differential. They have also put forward dedicated linear attacks on 7-round ARIA-128 and 10-round ARIA-192/256 in the weak-key setting, i.e., these attacks succeed for a limited number of weak keys. Apart from the cipher designers, Wu et al. [24] were the first to evaluate the security of ARIA against impossible differential cryptanalysis. They have proved, in contrast to the designers' expectations, that 4-round impossible differentials do exist and they can be used to mount a 6-round attack on ARIA. The impossible differential attack proposed by Wu et al. was independently enhanced by Li and Song [15] and Li et al. [21], and then it was extended to 7-round ARIA-256 by Du and Chen [7]. Li et al. [14] presented 3-round integral distinguishers that can be used to attack 4/5-round ARIA and 6-round ARIA-192/256. Afterwards, these 3-round integral distinguishers were modified by Li et al. [16] to 4-round integral distinguishers which improved the complexity of the 6-round integral attack and extended it to 7-round attack on ARIA-256. Boomerang attacks on 5/6-round ARIA and 7-round ARIA-256 were presented by Fleischmann [8]. Meet-in-the-Middle (MitM) attacks were applied to ARIA for the first time by Tang et al. [23], where they presented 5 & 6/7/8 MitM attacks on ARIA-128/192/256, respectively. The complexities of these MitM attacks were further improved by Bai and Yu [3] which enabled them to extend the MitM attacks to 7-round ARIA-128 and 9-round ARIA-256. The complexities of the 7/8-round MitM attacks on ARIA-192/256 were also enhanced by Akshima et al. [2] and they presented the first master key recovery attacks on ARIA. Although the designers of ARIA did not expect the existence of effective attacks on 8 or more rounds of ARIA with any key size using linear cryptanalysis, Liu et al. [17] managed to attack 7/9/11-round ARIA-128/192/256, respectively, by presenting a special kind of linear characteristics exploiting the diffusion layer employed in ARIA. However, the attacked rounds by Liu et al. [17] did not include the post whitening key. This means that if the post whitening key is considered, then the number of the reported rounds in their attacks will be reduced by one for all versions of ARIA. Finally, after the introduction of the Biclique cryptanalysis, it was applied on the full-round ARIA-256 [25], however, this class of attacks is considered as an optimized exhaustive search.

Linear cryptanalysis is one of the major cryptanalysis techniques used against symmetric-key ciphers. It was applied for the first time to FEAL and then to DES by Matsui [18,19]. In linear cryptanalysis, which is a known plaintext attack, the adversary tries to find a linear approximation between some bits from the plaintext, ciphertext and the secret key which can be used as a statistical distinguisher over several rounds of the cipher. Such linear distinguishers are then extended to key-recovery attacks on a few additional rounds using partial decryption and/or encryption. Subkeys of the appended rounds are guessed and the ciphertext is decrypted and/or plaintext is encrypted using these subkeys to calculate intermediate state value at the ends of the distinguisher. If the subkeys are correctly guessed then the distinguisher should hold and it fails, otherwise. After the introduction of linear cryptanalysis, many extensions and improvements have been proposed. One particular improvement that we use in this paper is the introduction of the notion of linear hull by Nyberg [20]. A linear hull is a set of linear approximations that involve the same bits in the plaintext and ciphertext and each one involves different intermediate state bits. An equally important framework for the description and understanding of the mechanisms of linear cryptanalysis is the concept of correlation matrices of boolean functions which was introduced by Daemen *et al.* [5]. The elements of the correlation matrices of a boolean function F are all the correlation coefficients between linear combinations of input bits and that of output bits of F.

In this paper, we revisit the security of ARIA against linear cryptanalysis. Inspired by the work of Liu *et al.* [17], we first explore all the iterative patterns across ARIA's diffusion layer which have 8 active S-boxes in 2 rounds such as 3-5-3 and 4-4-4. Then, in order to have a good balance between the complexity of the analysis rounds and the number of S-boxes involved in the distinguisher, we focus our attention on the patterns that involve 4 S-boxes in each round, i.e., 4-4-4. Among these patterns, we found 2 patterns that involve only 2 distinct S-boxes (out of the 4 possible distinct S-boxes used in ARIA) in both the even and odd rounds. Then, to simplify our analysis, we focus on these 2 patterns and build their correlation potential matrices to estimate their linear hull effect. In a correlation potential matrix, every element of the correlation matrix is squared. One of these patterns provide a new 5-round linear hull distinguisher with correlation $2^{-114.93}$ which gives us one more round as compared to [17]. Based on this 5-round linear hull, we append 3/4/6 analysis rounds which enables us to mount the first attack on 8-round ARIA-128 and improve the 9 and 11-round attacks on ARIA-192/256, respectively, to include the post whitening key. Further, we use the recovered bytes of information from the round keys to recover the master key. Our results and all previous attacks are summarized in Table 1.

The rest of the paper is organized as follows. Section 2 provides a description of ARIA and the notations adopted in the paper. In Sect. 3, we briefly give the concepts required for the linear cryptanalysis of ARIA. In Sect. 4, we use the correlation potential matrix to establish a linear hull of ARIA and present our 8, 9 and 11-round attacks on ARIA-128/192/256. We also show how the master key can be recovered. Finally, we conclude the paper in Sect. 5.

2   Specification of ARIA

ARIA [12] is an iterative 128-bit block cipher that follows the SPN structure. It can be used with 3 different key lengths, i.e., 128, 192 and 256 bits. The number of rounds in ARIA differs by the key length, i.e., 12 rounds for ARIA-128, 14 rounds for ARIA-192 and 16 rounds for ARIA-256. Similar to AES, the internal state of ARIA can be represented as a 4×4 matrix, where each byte of the matrix is an element in $GF(2^8)$. An ARIA round applies the following three transformations to the state matrix:

- Add Key (AK): XORing a 128-bit round key with the internal state. The round keys are deduced from the master key via the key scheduling algorithm which is described later in this section.

Table 1. Summary of attacks on ARIA

Key size	Rounds	Attack type	Data	Time	Memory	Reference
128/192/256	4	IC	2^{25} CP	2^{25}	*	[14]
	5	IDC	$2^{71.3}$ CP	$2^{71.6}$	$2^{72\dagger}$	[21]
	5	IC	$2^{27.2}$ CP	$2^{76.7}$	$2^{27.5\dagger}$	[14]
	5	MitM	25 CP	$2^{65.4}$	2^{121}	[23]
	5	BA	2^{109} ACPC	2^{110}	2^{57}	[8]
	6	IDC	2^{121} CP	2^{112}	$2^{121\dagger}$	[24]
	6	IDC	2^{120} CP	2^{96}	*	[15]
	6	IDC	$2^{120.5}$ CP	$2^{104.5}$	$2^{121\dagger}$	[21]
	6	IDC	2^{113} CP	$2^{121.6}$	$2^{113\dagger}$	[21]
	6	MitM	2^{56} CP	$2^{121.5}$	2^{121}	[23]
	6	IC	$2^{99.2}$ CP	$2^{71.4}$	*	[16]
	6	BA	2^{128} KP	2^{108}	2^{56}	[8]
	7	TDC	2^{81} CP	2^{81}	2^{80}	[4]
	7	TDC	2^{100} CP	2^{100}	2^{51}	[4]
	$7\ddagger$	LC	$2^{105.8}$ KP	$2^{100.99}$	$2^{79.73}$	[17]
	7	MitM	2^{121} CP	$2^{125.7}$	2^{122}	[3]
192/256	6	IC	$2^{124.4}$ CP	$2^{172.4}$	$2^{124.4\dagger}$	[14]
	7	MitM	2^{113} CP	2^{132}	2^{130}	[23]
	7	MitM	2^{96} CP	$2^{161.3}$	2^{185}	[23]
	$9\ddagger$	LC	$2^{108.3}$ KP	$2^{154.83}$	$2^{159.77}$	[17]
	$10^{\ wk}$	LC	2^{119} KP	2^{119}	2^{63}	[4]
128	$7^{\ wk}$	LC	2^{77} KP	2^{88}	2^{61}	[4]
	8^{mk}	LC	$2^{122.61}$ KP	$2^{123.48}$	$2^{119.94}$	This paper
192	7	MitM	2^{113} CP	$2^{135.1}$	2^{130}	[2]
	9^{mk}	LC	$2^{122.99}$ KP	$2^{154.83}$	$2^{159.94}$	This paper
256	7	IC	$2^{100.6}$ CP	$2^{225.8}$	*	[16]
	7	IDC	2^{125} CP	2^{238}	*	[7]
	7	BA	2^{128} KP	2^{236}	2^{184}	[8]
	7^{mk}	MitM	2^{115} CP	$2^{136.1}$	2^{130}	[2]
	8	MitM	2^{56} CP	$2^{251.6}$	2^{250}	[23]
	8	MitM	2^{113} CP	$2^{244.61}$	2^{130}	[3]
	8^{mk}	MitM	2^{56} CP	$2^{251.6}$	2^{252}	[2]
	8^{mk}	MitM	2^{113} CP	$2^{245.9}$	2^{138}	[2]
	9	MitM	2^{121} CP	$2^{253.37}$	2^{250}	[3]
	$11\ddagger$	LC	$2^{110.3}$ KP	$2^{218.54}$	$2^{239.8}$	[17]
	11^{mk}	LC	$2^{123.53}$ KP	$2^{238.13}$	$2^{239.95}$	This paper
	16^{mk}	BC	2^{80} CP	$2^{255.2}$	*	[25]

Time in round-reduced ARIA encryptions and memory in 128-bit blocks

BA: Boomerang Attack

BC: Biclique Cryptanalysis

IC: Integral Cryptanalysis

IDC: Impossible Differential Cryptanalysis

LC: Linear Cryptanalysis

MitM: Meet-in-the-Middle

TDC: Truncated Differential Cryptanalysis

ACPC: Adaptive Chosen Plaintexts and Ciphertext

CP: Chosen Plaintext   KP: Known Plaintext

mk: Recovers the master key wk: Weak-key setting

*: Not given in the related paper   †: Estimated in [8]

‡: Without post whitening key

- SubBytes (SB): Applying non-linear invertible 8-bit to 8-bit S-box to each byte of the state. ARIA employs 4 distinct S-boxes, namely, S_1, S_2 and their inverses S_1^{-1}, S_2^{-1}. Moreover, the order in which the S-boxes are applied to the internal state differs between odd and even rounds. In the odd rounds, the S-boxes are applied, column-wise, in the order: $(S_1, S_2, S_1^{-1}, S_2^{-1})$ while in the even rounds, the order, for each column, is: $(S_1^{-1}, S_2^{-1}, S_1, S_2)$. Figure 1 depicts the order in which the S-boxes are applied in both odd (X_1) and even (X_2) rounds.
- MixState (MS): Multiplication of the internal state by an involutional binary matrix that has a branch number of 8. Given an input state Y, the output state Z of the MS operation is computed as:

$$
\begin{pmatrix} Z[0] \\ Z[1] \\ Z[2] \\ Z[3] \\ Z[4] \\ Z[5] \\ Z[6] \\ Z[7] \\ Z[8] \\ Z[9] \\ Z[10] \\ Z[11] \\ Z[12] \\ Z[13] \\ Z[14] \\ Z[15] \end{pmatrix}
=
\begin{pmatrix}
0 & 0 & 0 & 1 & 1 & 0 & 1 & 0 & 1 & 1 & 0 & 0 & 0 & 1 & 1 & 0 \\
0 & 0 & 1 & 0 & 0 & 1 & 0 & 1 & 1 & 1 & 0 & 0 & 1 & 0 & 0 & 1 \\
0 & 1 & 0 & 0 & 1 & 0 & 1 & 0 & 0 & 0 & 1 & 1 & 1 & 0 & 0 & 1 \\
1 & 0 & 0 & 0 & 0 & 1 & 0 & 1 & 0 & 0 & 1 & 1 & 0 & 1 & 1 & 0 \\
1 & 0 & 1 & 0 & 0 & 1 & 0 & 0 & 1 & 0 & 0 & 1 & 0 & 0 & 1 & 1 \\
0 & 1 & 0 & 1 & 1 & 0 & 0 & 0 & 0 & 1 & 1 & 0 & 0 & 0 & 1 & 1 \\
1 & 0 & 1 & 0 & 0 & 0 & 0 & 1 & 0 & 1 & 1 & 0 & 1 & 1 & 0 & 0 \\
0 & 1 & 0 & 1 & 0 & 0 & 1 & 0 & 1 & 0 & 0 & 1 & 1 & 1 & 0 & 0 \\
1 & 1 & 0 & 0 & 1 & 0 & 0 & 1 & 0 & 0 & 1 & 0 & 0 & 1 & 0 & 1 \\
1 & 1 & 0 & 0 & 0 & 1 & 1 & 0 & 0 & 0 & 0 & 1 & 1 & 0 & 1 & 0 \\
0 & 0 & 1 & 1 & 0 & 1 & 1 & 0 & 1 & 0 & 0 & 0 & 0 & 1 & 0 & 1 \\
0 & 0 & 1 & 1 & 1 & 0 & 0 & 1 & 0 & 1 & 0 & 0 & 1 & 0 & 1 & 0 \\
0 & 1 & 1 & 0 & 0 & 0 & 1 & 1 & 0 & 1 & 0 & 1 & 1 & 0 & 0 & 0 \\
1 & 0 & 0 & 1 & 0 & 0 & 1 & 1 & 1 & 0 & 1 & 0 & 0 & 1 & 0 & 0 \\
1 & 0 & 0 & 1 & 1 & 1 & 0 & 0 & 0 & 1 & 0 & 1 & 0 & 0 & 1 & 0 \\
0 & 1 & 1 & 0 & 1 & 1 & 0 & 0 & 1 & 0 & 1 & 0 & 0 & 0 & 0 & 1
\end{pmatrix}
\begin{pmatrix} Y[0] \\ Y[1] \\ Y[2] \\ Y[3] \\ Y[4] \\ Y[5] \\ Y[6] \\ Y[7] \\ Y[8] \\ Y[9] \\ Y[10] \\ Y[11] \\ Y[12] \\ Y[13] \\ Y[14] \\ Y[15] \end{pmatrix}
$$

In the last round of ARIA, the MS linear transformation is replaced by an AK operation, which is referred to as the post whitening key. The full encryption function of an r-round ARIA is given in Fig. 1, where the ciphertext C is computed from the plaintext P via r rounds using $r+1$ round keys.

Key Schedule. The key schedule algorithm of ARIA takes the master key and outputs 13, 15, or 17 128-bit round keys for ARIA-128/192/256, respectively. First, the master key is divided into 2 128-bit values KL and KR, where KL is the leftmost 128-bits of the master key and KR is the remaining bits, if any, of the master key, right-padded with zeros to a 128-bit value. Then, a 3-round, 256-bit Feistel structure, as shown in Fig. 2, is used to compute 4 128-bits words ($W0, W1, W2$, and $W3$), where F_o and F_e denote ARIA odd and even round functions replacing the AK operation with pre-defined constants addition. The

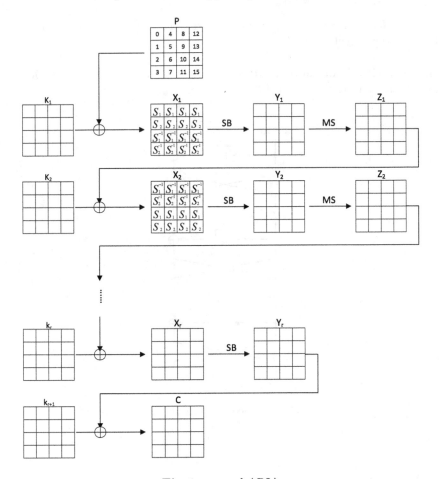

Fig. 1. r-round ARIA

round keys are deduced from $W0, W1, W2,$ and $W3$ as follows:

$$K_1 = W0 \oplus (W1 \ggg 19),$$
$$K_3 = W2 \oplus (W3 \ggg 19),$$
$$K_5 = W0 \oplus (W1 \ggg 31),$$
$$K_7 = W2 \oplus (W3 \ggg 31),$$
$$K_9 = W0 \oplus (W1 \lll 61),$$
$$K_{11} = W2 \oplus (W3 \lll 61),$$
$$K_{13} = W0 \oplus (W1 \lll 31),$$
$$K_{15} = W2 \oplus (W3 \lll 31),$$
$$K_{17} = W0 \oplus (W1 \lll 19),$$

$$K_2 = W1 \oplus (W2 \ggg 19),$$
$$K_4 = (W0 \ggg 19) \oplus W3,$$
$$K_6 = W1 \oplus (W2 \ggg 31),$$
$$K_8 = (W0 \ggg 31) \oplus W3,$$
$$K_{10} = W1 \oplus (W2 \lll 61),$$
$$K_{12} = (W0 \lll 61) \oplus W3,$$
$$K_{14} = W1 \oplus (W2 \lll 31),$$
$$K_{16} = (W0 \lll 31) \oplus W3,$$

where $a \lll b$ and $a \ggg b$ denote that a is circularly rotated by b bit to the left and right, respectively.

For more detailed information regarding the S-boxes and the key schedule algorithm, the reader is referred to [12].

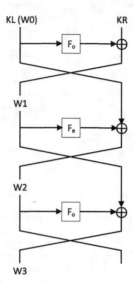

Fig. 2. ARIA key schedule - Initialization phase

2.1   Notations

The following notations are used throughout the rest of this paper:

- I_i: State value at the input of round i, where I_1 is the plaintext P.
- X_i: State value after the AK operation of round i, where X_{R+1} is the ciphertext C and R is 12 for ARIA-128, 14 for ARIA-192, and 16 for ARIA-256.
- Y_i: State value after the SB operation of round i.
- Z_i: State value after the MS operation of round i.
- O_i: State value at the output of round i, i.e., $O_i = I_{i+1}$.
- $S_i[j]$: The $(j+1)^{th}$ byte of state S at round i, where $0 \leq j \leq 15$, as numbered in P in Fig. 1.
- $S_i^k[j]$: The $(j + 1)^{th}$ byte of state S^k at round i which corresponds to the plaintext/ciphertext pair (P^k, C^k).
- $K_i^{\{a,b,c,d\}}$: The XOR of 4 bytes of K_i, i.e., $K_i[a] \oplus K_i[b] \oplus K_i[c] \oplus K_i[d]$.

3   Linear Cryptanalysis

As mentioned above, linear cryptanalysis [18,19] is a known plaintext crypt-analysis technique, in which the adversary attempts to construct linear approx-imations for each round of a block cipher E, such that the output mask of a round equals the input mask of the next round. The concatenation of these lin-ear approximations creates a linear trail (Ω) whose correlation is computed by multiplying the correlations of each round linear approximation. This results in a linear distinguisher covering several rounds of E that can be used to distinguish it from a random permutation. A linear approximation of a block cipher E is typically given by a plaintext mask α and a ciphertext mask β, such that the corresponding correlation $CO_E(\alpha, \beta)$ is non-negligible:

$$CO_E(\alpha, \beta) = |2 \times Pr[\alpha \bullet P \oplus \beta \bullet C = \gamma \bullet K] - 1| \gg 2^{-n/2},$$

where α, β, and γ denote the masks of the plaintext, ciphertext, and key, respectively, n denotes the block length of the cipher and $a \bullet b$ denotes the bitwise inner product of a and b. To distinguish E, the adversary gathers $N = \mathcal{O}(1/CO_E^2(\alpha, \beta))$ plaintexts and their corresponding ciphertexts and com-putes the empirical correlation $\hat{CO}_E(\alpha, \beta)$:

$$\hat{CO}_E(\alpha, \beta) = |2 \times \#\{i : \alpha \bullet P^i \oplus \beta \bullet C^i = 0\}/N - 1|.$$

The computed empirical correlation is close to $CO_E(\alpha, \beta)$ for the attacked block cipher E, and smaller than $1/\sqrt{N}$, with high probability, for a random permu-tation [13]. By adding more rounds, the so-called analysis rounds, at the bottom and/or the top of such linear distinguisher, it can be used to perform a key recov-ery attack using partial decryption and/or encryption. The attack proceeds by guessing the round keys used in the appended rounds, and computing an inter-mediate state value(s) from the guessed round keys, ciphertext and/or plaintext. The distinguisher is then applied to the deduced intermediate state value(s): if the round keys guess is correct, the distinguisher is expected to hold, and fail for wrong key guesses.

Linear Hulls. The notion of linear hulls was introduced by Nyberg [20], where an r-round linear hull of a block cipher E is a set of all linear trails having the same input mask α, output mask β and can differ in the intermediate masks. If we denote the square of a correlation by correlation potential, then the average correlation potential of a linear hull over r rounds of a key-alternating block cipher, averaged over all values of the expanded key (i.e. the concatenation of all round keys), is the sum of the correlation potentials of all individual trails that compose that linear hull, assuming independent round keys (Theorem 7.9.1 in [6]).

Correlation Matrices. High-probable linear hulls can be found by creating a correlation matrix, or rather a correlation potential matrix, a notion that was introduced by Daemen et al. [5]. For a key-alternating cipher of n-bit block length, a correlation potential matrix M is an $2^n \times 2^n$ matrix where the element M_{ij} in row i and column j of the matrix corresponds to the correlation potential of an input mask α_i and an output mask β_i. Computing M^r gives the correlation potential after r rounds [1]. Constructing the correlation potential matrix for modern block ciphers is infeasible as n is quite large. An alternative approach, then, is to construct a submatrix of the correlation potential matrix that enables us to obtain a lower bound on the average correlation potential of a linear hull.

4   Linear Cryptanalysis of ARIA

Liu *et al.* [17] have proposed a special kind of linear characteristics for byte-oriented SPN block ciphers and applied it on ARIA. Their proposal exploited the MS linear transformation in ARIA by finding a linear relation between 4 bytes of its input and 4 bytes of its output. Then, the linear approximation over one round is formed by applying an input mask α and an output mask β to the XOR of these input/output bytes, i.e.,

$$\alpha \bullet \oplus_{i \in V} I_r[i] = \beta \bullet \oplus_{i \in V} O_r[i],$$

where V is the set of the input/output bytes positions. For example, in their attack $V = \{0, 3, 12, 15\}$.

Inspired by their work, we have first explored the space of all iterative patterns that have 8 active S-boxes in 2 rounds such as 3-5-3 and 4-4-4. We have found that, for 5-round distinguisher and 3 analysis rounds, there is a trade-off between the number of S-boxes involved in the linear characteristic, or rather the linear hull, and the number of key bytes to be guessed in the analysis rounds. On the one hand, the more S-boxes involved in the linear hull, the smaller the correlation potential of the linear hull will be and thus the higher data complexity of the attack will be. On the other hand, the more key bytes to be guessed in the analysis rounds, the higher time complexity will be. Therefore, such a trade-off can be thought of as a trade-off between the data complexity and the time complexity. As an example, in a 3-5-3-5-3-5-3-5 pattern, its first 5-round linear hull involves a total of 19 S-boxes and in its last three analysis rounds, there are 13 key bytes to be guessed as will be illustrated in our attacks later. If the same pattern is shifted by one round to be 5-3-5-3-5-3-5-3, then the number of S-boxes involved in the 5-round distinguisher increases to 21 while the number of key bytes to be guessed in the analysis rounds drops to 11 bytes. The pattern that achieves the balance between the number of S-boxes in the distinguisher and the number of guessed key bytes in the analysis rounds is the pattern 4-4-4.

We have automated the search for all the 4-4-4 patterns across ARIA's MS linear transformation and found that there are 204 such patterns. Among all these patterns, there are only 2 patterns that have 2 active distinct S-boxes even though the order of the application of the S-boxes alternates between the odd and even rounds. The first set of these two patterns is $V1 = \{8, 10, 12, 14\}$ which has S_1 and S_1^{-1} as the active S-boxes in both the odd and even rounds (the gray cells in Fig. 3). The other set is $V2 = \{9, 11, 13, 15\}$ which has S_2 and S_2^{-1} as the active S-boxes, once again in both the odd and even rounds (the black-hatched cells in Fig. 3). Based on these two patterns, a 1-round linear trail of round i with input mask α and output mask β can be written as:

$$V1 : \alpha \bullet (I_i[8] \oplus I_i[10] \oplus I_i[12] \oplus I_i[14]) =$$
$$\beta \bullet (O_i[8] \oplus O_i[10] \oplus O_i[12] \oplus O_i[14]),$$
$$V2 : \alpha \bullet (I_i[9] \oplus I_i[11] \oplus I_i[13] \oplus I_i[15]) =$$
$$\beta \bullet (O_i[9] \oplus O_i[11] \oplus O_i[13] \oplus O_i[15]).$$

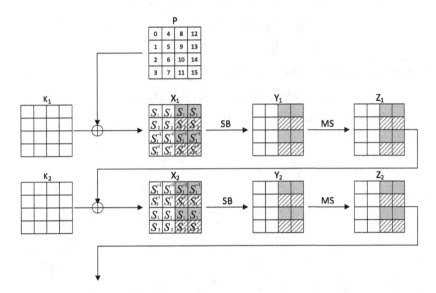

Fig. 3. ARIA 4-4-4 iterative patterns involving 2 distinct S-boxes, each. The gray cells represent pattern $V1$ while the black-hatched cells represent pattern $V2$.

Since both α and $\beta \in GF(2^8)$, the 1-round correlation potential matrix M for each pattern has a size of a $2^8 \times 2^8$ and as the S-boxes involved in these patterns do not change over the odd and even rounds, an M^r correlation potential matrix to get the average correlation potential after r rounds can be constructed by simply raising M to the power r. Such a correlation potential matrix can be regarded as a correlation potential submatrix of ARIA, restricting the inputs and outputs of the matrix to the values that follow our specific patterns. We have automatically constructed the 1-round correlation potential matrix for both patterns. We were not able to go for more than 5 rounds as the highest correlation potential starting M^6 exceeds 2^{-128}. So, for M^5 of pattern $V1$, the highest average correlation potential was found to be $2^{-114.93}$ when the input mask α is $0x09$ and the output mask β is $0x0E$ while for M^5 of pattern $V2$, the highest average correlation potential was found to be $2^{-115.63}$ when the input mask α is $0x24$ and the output mask β is $0xD3$.

4.1 Key Recovery Attacks on ARIA

As the highest correlation potential in $V1$ is greater than the highest one in $V2$, we have opted for using pattern $V1$. In our attacks, we have placed the 5-round linear hull to cover rounds 1–5, hence it is represented as:

$$0x09 \bullet (I_1[8] \oplus I_1[10] \oplus I_1[12] \oplus I_1[14]) =$$
$$0x0E \bullet (O_5[8] \oplus O_5[10] \oplus O_5[12] \oplus O_5[14])$$

and since:

$$O_5[8] \oplus O_5[10] \oplus O_5[12] \oplus O_5[14] =$$
$$X_6[8] \oplus X_6[10] \oplus X_6[12] \oplus X_6[14] \oplus$$
$$K_6[8] \oplus K_6[10] \oplus k_6[12] \oplus K_6[14]$$

the 5-round linear hull can be re-written as:

$$0x09 \bullet (P[8] \oplus P[10] \oplus P[12] \oplus P[14]) =$$
$$0x0E \bullet (X_6[8] \oplus X_6[10] \oplus X_6[12] \oplus X_6[14])$$

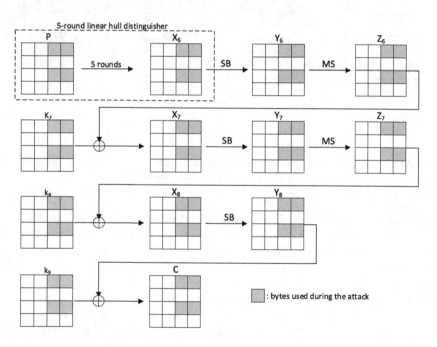

Fig. 4. Attack on 8-round ARIA

ARIA-128. The attack on 8-round ARIA-128 is based on the above 5-round linear hull and adding 3 more rounds at its end, as illustrated in Fig. 4. The attack proceeds as follows:

1. First, we gather N plaintexts and their corresponding ciphertexts (P^i, C^i), where $1 \leq i \leq N$.
2. Next, we initialize 2^{32} counters U_m, where the size of each counter is $\lceil log_2 N \rceil$ bits and $0 \leq m \leq 2^{32} - 1$. Then, for each plaintext/ciphertext pair (P^i, C^i), we increment (resp. decrement) the counter U_m by 1 if the parity of

$$0x09 \bullet (P^i[8] \oplus P^i[10] \oplus P^i[12] \oplus P^i[14])$$

   is 0 (resp. 1) and m equals the value of $C^i[8]\|C^i[10]\|C^i[12]\|C^i[14]$.
3. We initialize 2^{120} counters U_l, where the size of each counter is $\lceil log_2 N \rceil$ bits as well and $0 \leq l \leq 2^{120} - 1$ and l represents the possible value of the 15 bytes of $K_9^{\{8,10,12,14\}}\|K_8^{\{8,10,12,14\}}\|K_7^{\{8,10,12,14\}}\|Y_8[8]\ \|Y_8[10]\|Y_8[12]\|Y_8[14]\|$ $Y_7[8]\|Y_7[10]\|Y_7[12]\|Y_7[14]\|Y_6[8]\|Y_6[10]\|Y_6[12]\|Y_6[14]$.
4. Then, for each possible value of $K_9^{\{8,10,12,14\}}$, $K_8^{\{8,10,12,14\}}$ and $K_7^{\{8,10,12,14\}}$, we do the following:
   (a) For each possible value of the 2^{32} values of m, compute $Y_8^m[8] \oplus Y_8^m[10] \oplus$ $Y_8^m[12] \oplus Y_8^m[14] = m[0] \oplus m[1] \oplus m[2] \oplus m[3] \oplus K_9^{\{8,10,12,14\}}$ and denote this value as t_8^m.
   (b) For any value of the 2^{24} values of $Y_8[8]\|Y_8[10]\|Y_8[12]\|Y_8[14]$ satisfying t_8^m, we deduce $X_8^m[8], X_8^m[10], X_8^m[12], X_8^m[14]$ from the corresponding S-boxes. Then, using the guessed value of $K_8^{\{8,10,12,14\}}$, compute $Y_7^m[8] \oplus$ $Y_7^m[10] \oplus Y_7^m[12] \oplus Y_7^m[14] = Z_7^m[8] \oplus Z_7^m[10] \oplus Z_7^m[12] \oplus Z_7^m[14] = X_8^m[8] \oplus$ $X_8^m[10] \oplus X_8^m[12] \oplus X_8^m[14] \oplus K_8^{\{8,10,12,14\}}$ and denote this value as l_7^m.
   (c) Then, for any value of the 2^{24} values of $Y_7[8]\|Y_7[10]\|Y_7[12]\| Y_7[14]$ satisfying t_7^m, we deduce $X_7^m[8], X_7^m[10], X_7^m[12], X_7^m[14]$ from the corresponding S-boxes. Then, using the guessed value of $K_7^{\{8,10,12,14\}}$, compute $Y_6^m[8] \oplus Y_6^m[10] \oplus Y_6^m[12] \oplus Y_6^m[14] = Z_6^m[8] \oplus Z_6^m[10] \oplus Z_6^m[12] \oplus Z_6^m[14] =$ $X_7^m[8] \oplus X_7^m[10] \oplus X_7^m[12] \oplus X_7^m[14] \oplus K_7^{\{8,10,12,14\}}$ and denote this value as t_6^m.
   (d) For any value of the 2^{24} values of $Y_6[8]\|Y_6[10]\|Y_6[12]\|Y_6[14]$ satisfying t_6^m, we deduce $X_6^m[8], X_6^m[10], X_6^m[12], X_6^m[14]$ from the corresponding S-boxes. Then, calculate the parity of:

$$0x0E \bullet (X_6^m[8] \oplus X_6^m[10] \oplus X_6^m[12] \oplus X_6^m[14])$$

   If the parity is 0 (resp. 1), increment (resp. decrement) the corresponding counter U_l by the value of U_m.
5. For l such that the value of U_l is maximal, output the value of the corresponding $K_9^{\{8,10,12,14\}}\| K_8^{\{8,10,12,14\}}\|K_7^{\{8,10,12,14\}}$ as the correct key information.

Attack complexity. The number of known plaintext/ciphertext pairs N required to perform the attack is estimated by the following formula, which is adopted from Corollary 1 in [22]:

$$N = \left(\frac{\Phi^{-1}(P_s) + \Phi^{-1}(1 - 2^{-a-1})}{2} \right)^2 \times \frac{4}{CO^2}, \tag{1}$$

where P_s is the probability of success, CO is the correlation of the linear hull, Φ^{-1} is the inverse cumulative function of the standard normal distribution, and a is the advantage of the adversary over the exhaustive search and equals $k - log_2 d$ if the correct key was ranked among the top d candidates out of the 2^k possible candidates of an k-bit key.

In our attack, we guess 120 bits, set the advantage a to 120, i.e., the correct key information is the first one of the list of candidates and set the probability of success to 0.95. Then, the number of plaintext/ciphertext pairs N equals $2^{5.68} \times \frac{4}{2^{-114.93}} = 2^{122.61}$. The time complexity of the attack is dominated by steps 2 and 4.(d). Therefore, the time complexity of the attack equals $2^{122.61} \times \frac{4}{16 \times 8} + 2^{24} \times 2^{32} \times 2^{24} \times 2^{24} \times 2^{24} \times \frac{4}{16 \times 8} \approx 2^{123.03}$ 8-round ARIA encryptions. The memory complexity of the attack is attributed to storing the counters U_l, where the size of each counter is set to 123 bits. Hence, the memory complexity of the attack is $2^{120} \times \frac{123}{128} \approx 2^{119.94}$ 128-bit blocks.

ARIA-192/256. The attack on 8-round ARIA-128 can be extended to 9-round ARIA-192 (resp. 11-round ARIA-256) with the post whitening key by utilizing the same 5-round linear hull and having 4 (resp. 6) analysis rounds. The attack procedure is similar to the attack on ARIA-128, except that in step 3, we initialize 2^{160} (resp. 2^{240}) counters U_l, where in this case, $0 \leq l \leq 2^{160} - 1$ (resp. $0 \leq l \leq 2^{240} - 1$) and represents the possible value of the 20 (resp. 30) bytes of $K_{r+1}^{\{8,10,12,14\}} \| Y_r[8] \| Y_r[10] \| Y_r[12] \| Y_r[14]$, where $6 \leq r \leq 9$ (resp. $6 \leq r \leq 11$) and we add 1 (resp. 3) more sub-step(s) in step 4 to accommodate the additional round(s).

In this case, for an advantage a of 160 (resp. 240) and P_s of 0.95, the number of known plaintext/ciphertext pairs N is $2^{6.06} \times \frac{4}{2^{-114.93}} = 2^{122.99}$ for ARIA-192 and is $2^{6.6} \times \frac{4}{2^{-114.93}} = 2^{123.53}$ for ARIA-256. The time complexity of the attack is $2^{122.99} \times \frac{4}{16 \times 9} + 2^{32} \times 2^{32} \times 2^{24} \times 2^{24} \times 2^{24} \times 2^{24} \times \frac{4}{16 \times 9} \approx 2^{154.83}$ 9-round ARIA encryptions for ARIA-192 and is $2^{123.53} \times \frac{4}{16 \times 11} + 2^{48} \times 2^{32} \times 2^{24} \times 2^{24} \times 2^{24} \times 2^{24} \times 2^{24} \times 2^{24} \times \frac{4}{16 \times 11} \approx 2^{218.54}$ 11-round ARIA encryptions for ARIA-256. The size of each counter of U_l is set to 123 (resp. 124) bits, therefore, the memory complexity of the attack is $2^{160} \times \frac{123}{128} \approx 2^{159.94}$ 128-bit blocks for ARIA-192 and $2^{240} \times \frac{124}{128} \approx 2^{239.95}$ 128-bit blocks for ARIA-256.

4.2   Recovering the Master Key

In this subsection, we show how the recovered bytes of information from the round keys can be used to recover the master key in all versions of ARIA.

ARIA-128. In the attack on 8-round ARIA-128, we recover 3 bytes of information from K_9, K_8, and K_7. Recall that in ARIA-128, KR is all zeros and KL is the 128-bit master key and at the same time it is $W0$. In order to recover the master key, we do the following:

- First, we guess 15 bytes of $W0$, i.e., all the bytes except $W0[7]$. These bytes enable us to compute $W1[0], W1[2], W1[4]$, with 6 other bytes, which gives us the first 5 bits of bytes $8, 10$, and 12 of $(W1 \lll 61)$.
- From the key schedule, we know that $K_9 = W0 \oplus (W1 \lll 61)$. As we recover $K_9^{\{8,10,12,14\}}$, i.e., $K_9[8] \oplus K_9[10] \oplus K_9[12] \oplus K_9[14]$, this means that we recover $W0[8] \oplus W0[10] \oplus W0[12] \oplus W0[14] \oplus (W1 \lll 61)[8] \oplus (W1 \lll 61)[10] \oplus (W1 \lll 61)[12] \oplus (W1 \lll 61)[14]$.
- As we guessed $W0[8], W0[10], W0[12]$ and $W0[14]$, recovered $K_9^{\{8,10,12,14\}}$ and computed the first 5 bits of bytes $8, 10$, and 12 of $(W1 \lll 61)$, we can deduce the first 5 bits of byte 14 of $(W1 \lll 61)$ which in turn enables us to deduce the last 5 bits of $SB(W0[7])$.
- Afterwards, we guess the 3 first bits of $SB(W0[7])$ which means that we have 2^{123} candidates for $W0$ or rather the master key.
- Then, we run the key schedule and use the remaining 3 bits of $K_9^{\{8,10,12,14\}}$ and the two bytes of $K_8^{\{8,10,12,14\}}$ and $K_7^{\{8,10,12,14\}}$ to discard the wrong guesses and so we end up with 2^{104} candidates for the master key which we can test using 2 plaintext/ciphertext pairs.

The time complexity of the master key recovery phase is dominated by the last step and equals $2^{123} \times \frac{3}{8} + 2 \times 2^{104} \approx 2^{121.59}$ 8-round ARIA encryptions as we need to compute 3 rounds of ARIA for the 2^{123} candidates to deduce $W2$ and $W3$ and then test the remaining 2^{104} candidates using 2 plaintext/ciphertext pairs. Therefore the total time complexity of the attack is $2^{123.03} + 2^{121.59} \approx 2^{123.48}$.

ARIA-192. In the attack on 9-round ARIA-192, we recover 4 bytes of information from K_{10}, K_9, K_8, and K_7. In order to recover the master key, we do the following:

- First, we guess the 16 bytes of $W0$ and calculate $F_o(W0, CK1)$. Then, to be able to compute bytes $8, 10, 12$ and 14 of $(W1 \lll 61)$, we guess 29 bits of KR as the 8 right bytes of KR are zeros.
- We use the recovered $K_9^{\{8,10,12,14\}}$ to discard the wrong guesses of $W0$ and the 29 bits guessed from KR and so we have 2^{149} for $W0$ along with the 29 bits of KR.
- Next, we guess the remaining 35 bits of the master key, i.e., the remaining 35 bits of KR so we have 2^{184} candidates for the master key.
- Then, we run the key schedule to compute $W1, W2$, and $W3$ and use the 3 bytes of $K_{10}^{\{8,10,12,14\}}$, $K_8^{\{8,10,12,14\}}$ and $K_7^{\{8,10,12,14\}}$ to discard the wrong guesses and we end up with 2^{160} candidates which we test using 2 plaintext/ciphertext pairs.

The time complexity of the master key recovery phase equals $2^{184} \times \frac{3}{9} + 2 \times 2^{160} \approx 2^{182.42}$ 9-round ARIA encryptions, hence the total time complexity of the attack is $2^{154.83} + 2^{182.42} \approx 2^{182.42}$.

ARIA-256. In the attack on 11-round ARIA-256, we recover 6 bytes of information from $K_{12}, K_{11}, K_{10}, K_9, K_8$, and K_7. In order to recover the master key, we do the following:

– First, we guess the 16 bytes of $W2$ and 14 bytes of $W3$ and use the recovered $K_7^{\{8,10,12,14\}}$ and $K_{11}^{\{8,10,12,14\}}$, both of them are deduced from $W2$ and $W3$, to calculate the remaining two bytes of $W3$ which means that we have 2^{240} candidates for both $W2$ and $W3$.

– Next, starting from $W2$ and $W3$, we run the key schedule to compute $W0$ and $W1$ and use the other 4 bytes of $K_{12}^{\{8,10,12,14\}}$, $K_{10}^{\{8,10,12,14\}}$, $K_9^{\{8,10,12,14\}}$ and $K_8^{\{8,10,12,14\}}$ to discard the wrong guesses and we end up with 2^{208} candidates for the master key which we test using 2 plaintext/ciphertext pairs.

The time complexity of the master key recovery phase is $2^{240} \times \frac{3}{11} + 2 \times 2^{208} \approx 2^{238.13}$ 11-round ARIA encryptions, therefore the total time complexity of the attack is $2^{218.54} + 2^{238.13} \approx 2^{238.13}$.

5   Conclusion

In this paper, we have revisited the security of round-reduced ARIA against linear cryptanalysis and presented the first 8-round attack on ARIA-128 and improved the previous 9 and 11-round attacks on ARIA-192/256 by including the post whitening key. We have achieved these results by constructing a 5-round linear hull on ARIA using the correlation matrix approach and exploiting the binary linear transformation layer in the analysis rounds. For all our attacks, we showed how the recovered bytes of information from the round keys can be used to recover the master key. This paper shows some weaknesses of reduced versions of ARIA, but the full round ARIA remains still secure.

References

1. Abdelraheem, M.A., Alizadeh, J., Alkhzaimi, H.A., Aref, M.R., Bagheri, N., Gauravaram, P.: Improved linear cryptanalysis of reduced-round SIMON-32 and SIMON-48. In: Biryukov, A., Goyal, V. (eds.) Progress in Cryptology - INDOCRYPT 2015. LNCS, vol. 9462, pp. 153–179. Springer, Cham (2015). http://dx.doi.org/10.1007/978-3-319-26617-6_9

2. Biryukov, A., Goyal, V. (eds.): Progress in Cryptology – INDOCRYPT 2015. LNCS, vol. 9462. Springer, Cham (2015). http://dx.doi.org/10.1007/978-3-319-26617-6_11

3. Bai, D., Yu, H.: Improved meet-in-the-middle attacks on round-reduced ARIA. In: Desmedt, Y. (ed.) ISC 2013. LNCS, vol. 7807, pp. 155–168. Springer, Heidelberg (2015). http://dx.doi.org/10.1007/978-3-319-27659-5_11

4. Biryukov, A., De Canniere, C., Lano, J., Ors, S.B., Preneel, B.: Security and performance analysis of ARIA, version 1.2. Technical report, Katholieke Universiteit Leuven, Belgium (2004).http://www.cosic.esat.kuleuven.be/publications/article-500.pdf

5. Daemen, J., Govaerts, R., Vandewalle, J.: Fast Software Encryption. LNCS, vol. 1008. Springer, Heidelberg (1995). http://dx.doi.org/10.1007/3-540-60590-8_21

6. Daemen, J., Rijmen, V.: The Design of Rijndael. Springer, New York (2002)

7. Du, C., Chen, J.: Impossible differential cryptanalysis of ARIA reduced to 7 rounds. In: Heng, S.-H., Wright, R.N., Goi, B.-M. (eds.) CANS 2010. LNCS, vol. 6467, pp. 20–30. Springer, Heidelberg (2010). http://dx.doi.org/10.1007/978-3-642-17619-7_2

8. Fleischmann, E., Forler, C., Gorski, M., Lucks, S.: New boomerang attacks on ARIA. In: Gong, G., Gupta, K.C. (eds.) INDOCRYPT 2010. LNCS, vol. 6498, pp. 163–175. Springer, Heidelberg (2010). http://dx.doi.org/10.1007/978-3-642-17401-8_13

9. Korean Agency for Technology and Standards (KATS): 128-bit Block Encryption Algorithm ARIA KS X 1213-1: December 2014 (in Korean)

10. Kim, W., Lee, J., Park, J., Kwon, D.: Addition of the ARIA cipher suites to Transport Layer Security (TLS). RFC 6209, RFC Editor, April 2011. http://www.rfc-editor.org/rfc/rfc6209.txt, http://www.rfc-editor.org/rfc/rfc6209.txt

11. Daesung, K., et al.: Information Security and Cryptology - ICISC 2003. LNCS, vol. 2971. Springer, Heidelberg (2004). http://dx.doi.org/10.1007/978-3-540-24691-6_32

12. Lee, J., Lee, J., Kim, J., Kwon, D., Kim, C.: A Description of the ARIA Encryption Algorithm. RFC 5794, RFC Editor, March 2010

13. Leurent, G.: Improved differential-linear cryptanalysis of 7-round chaskey with partitioning. Cryptology ePrint Archive, Report 2015/968 (2015). http://eprint.iacr.org/

14. Li, P., Sun, B., Li, C.: Integral cryptanalysis of ARIA. In: Bao, F., Yung, M., Lin, D., Jing, J. (eds.) Inscrypt 2009. LNCS, vol. 6151, pp. 1–14. Springer, Heidelberg (2010). http://dx.doi.org/10.1007/978-3-642-16342-5_1

15. Li, S., Song, C.: Improved impossible differential cryptanalysis of ARIA. In: A Description of the ARIA Encryption Algorithm. RFC 5794, RFC Editor International Conference on Information Security and Assurance, ISA 2008, pp. 129–132, April 2008

16. Li, Y., Wu, W., Zhang, L.: Integral attacks on reduced-round ARIA block cipher. In: Kwak, J., Deng, R.H., Won, Y., Wang, G. (eds.) ISPEC 2010. LNCS, vol. 6047, pp. 19–29. Springer, Heidelberg (2010). http://dx.doi.org/10.1007/978-3-642-12827-1_2

17. Liu, Z., Gu, D., Liu, Y., Li, J., Li, W.: Linear cryptanalysis of ARIA block cipher. In: Qing, S., Susilo, W., Wang, G., Liu, D. (eds.) ICICS 2011. LNCS, vol. 7043, pp. 242–254. Springer, Heidelberg (2011). http://dx.doi.org/10.1007/978-3-642-25243-3_20

18. Matsui, M.: Linear cryptanalysis method for DES cipher. In: Helleseth, T. (ed.) EUROCRYPT 1993. LNCS, vol. 765, pp. 386–397. Springer, Heidelberg (1994). http://dx.doi.org/10.1007/3-540-48285-7_33

19. Matsui, M., Yamagishi, A.: A new method for known plaintext attack of FEAL cipher. In: Rueppel, R.A. (ed.) EUROCRYPT 1992. LNCS, vol. 658, pp. 81–91. Springer, Heidelberg (1993). http://dx.doi.org/10.1007/3-540-47555-9_7

20. Nyberg, K.: Linear approximation of block ciphers. In: De Santis, A. (ed.) EUROCRYPT 1994. LNCS, vol. 950, pp. 439–444. Springer, Heidelberg (1995). http://dx.doi.org/10.1007/BFb0053460

21. Li, R., Bing Sun, P.Z., Li, C.: New Impossible Differential Cryptanalysis of ARIA. Cryptology ePrint Archive, Report 2008/227 (2008). http://eprint.iacr.org/2008/227.pdf

22. Selçuk, A.A.: On probability of success in linear and differential cryptanalysis. J. Cryptology $21(1)$, 131–147 (2007). http://dx.doi.org/10.1007/s00145-007-9013-7

23. Tang, X., Sun, B., Li, R., Li, C., Yin, J.: A meet-in-the-middle attack on reduced-round ARIA. J. Syst. Softw. **84**(10), 1685–1692 (2011). http://www.sciencedirect.com/science/article/pii/S016412121100104X
24. Wu, W.L., Zhang, W.T., Feng, D.G.: Impossible differential cryptanalysis of reduced-round ARIA and Camellia. J. Comput. Sci. Technol. **22**(3), 449–456 (2007). http://dx.doi.org/10.1007/s11390-007-9056-0
25. zhen Chen Tian-min Xu, S.: Biclique Attack of the Full ARIA-256. Cryptology ePrint Archive, Report 2012/011 (2012). http://eprint.iacr.org/2012/011.pdf

Partial Key Exposure Attacks on CRT-RSA: General Improvement for the Exposed Least Significant Bits

Atsushi Takayasu$^{(\boxtimes)}$ and Noboru Kunihiro

The University of Tokyo, Tokyo, Japan
a-takayasu@it.k.u-tokyo.ac.jp

Abstract. Blömer and May (Crypto 2003) used Coppersmith's lattice based method to study partial key exposure attacks on CRT-RSA, i.e., an attack on RSA with the least significant bits of a CRT exponent. The attack works for an extremely small public exponent e, however, improved attacks were proposed by Lu, Zhang, and Lin (ACNS 2014), Takayasu and Kunihiro (ACNS 2015). These attack works for $e < N^{0.375}$. For a smaller (resp. larger) e, an attack of Lu et al. (resp. Takayasu-Kunihiro's attack) requires less partial information to attack RSA.

In this paper, we propose a further improved attack. Indeed, our attack completely improves previous attacks in the sense that our attack requires less partial information than previous attacks for all $e < N^{0.375}$. We solve the same modular equation as Takayasu-Kunihiro, however, our attack can find larger roots. From the technical point of view, although the Takayasu-Kunihiro lattice follows the Jochemsz-May strategy (Asiacrypt 2006), we carefully analyze the algebraic structure of the underlying polynomial and propose better lattice constructions.

Keywords: CRT-RSA · Cryptanalysis · Partial key exposure · Coppersmith's method · Lattices

1 Introduction

1.1 Background

Let $N = pq$ be a public RSA modulus where prime factors p and q are the same bit-size [31]. A public exponent e and a secret exponent d satisfy $ed = 1$ mod $(p-1)(q-1)$. For encryption/verifying (resp. decryption/signing), the heavy modular exponentiation should be computed. To achieve a faster computation, a simple solution is to use a smaller public (resp. secret) exponent. However, Wiener [40] showed that a public RSA modulus can be factorized in polynomial time when a secret exponent is too small. Boneh and Durfee [3] used Coppersmith's lattice based method [6] and showed that the attack works for $d < N^{0.284}$. Moreover, in the same work, they proposed a better attack which works when $d < N^{0.292}$.

© Springer International Publishing Switzerland 2016
M. Bishop and A.C.A. Nascimento (Eds.): ISC 2016, LNCS 9866, pp. 35–47, 2016.
DOI: 10.1007/978-3-319-45871-7_3

To thwart the small secret exponent attack and achieve a faster decryption/signing simultaneously, the Chinese Remainder Theorem (CRT) is often used as described by Quisquater and Couvreur [29]. Instead of the original secret exponent d, CRT exponents d_p and d_q are used where

$$ed_p = 1 \mod (p-1) \quad \text{and} \quad ed_q = 1 \mod (q-1).$$

Analogous to a small secret exponent RSA [3,40], too small CRT exponents disclose the factorization of a public modulus [1,14,16,19,24]. The state-of-the-art Jochemsz-May attack [19] factorizes a public RSA modulus N in polynomial time for $d_p, d_q < N^{0.073}$ with a full size public exponent e. The attack showed the CRT-RSA becomes more vulnerable for a small public exponent. To thwart the attack for a practical setting such that e is small, e.g., $e = 2^{16} + 1$, CRT exponents should be $\approx N^{0.5}$.

Since RSA is one of the most famous cryptosystems and widely used, the hardness of the RSA/factorization problem with some *hints* have been studied in numerous papers, e.g., partial knowledge of prime factors [5,30,36,39], implicit hints of prime factors [13,22,27,33], and partial knowledge of a secret exponent [2,4,12,20,34,37]. In this paper, we focus on the problem with partial knowledge of CRT exponents and consider an attack scenario when the least significant bits of d_p are exposed to attackers although there exist several papers [2,23,32,38] whose attack scenarios are slightly different. More concretely, we study the following problem: given $\tilde{d}_p > N^{0.5-\delta}$, which is the least significant bits of d_p, then the goal of the problem is to factorize the public RSA modulus N for $e \approx N^{\alpha}$, $d_p \approx N^{0.5}$. Blömer and May (Crypto 2003) [2] studied the problem, however, the proposed attack works only for an extremely small public exponent, i.e., $e = poly(\log N)$. Lu, Zhang, and Lin (ACNS 2014) [23], Takayasu and Kunihiro (ACNS 2015) proposed improved attacks for larger α. The former attack works for $\alpha < 0.25$ and $\delta < 0.25$ for $\alpha \approx 0$ whereas the latter attack works for $\alpha < 0.375$ and $\delta < 0.207$ for $\alpha \approx 0$. Hence, although the latter attack works for larger e, the former attack works with less partial information for a small e[1].

1.2   Our Contributions

Our Results. In this paper, we propose an improved attack as follows:

Theorem 1. *Let $N = pq$ be a public RSA modulus where the prime factors p and q are the same bit-size. Let $e \approx N^{\alpha}$ denote a public exponent and $d_p \approx N^{0.5}$ denote a CRT exponent such that $ed_p = 1 \mod (p-1)$. Given the public elements (N, e) as well as $\tilde{d}_p > N^{0.5-\delta}$ which is the least significant bits of a CRT exponent. If*

$$-\ \delta < \tfrac{5-2\sqrt{1+14\alpha}}{14} \ \text{for} \ \tfrac{1}{18} < \alpha \le \tfrac{3}{8}, \ \text{or}$$

---

[1] Lu et al. [23] also proposed an attack for $\alpha < 0.375$, however, the attack is always weaker than Takayasu-Kunihiro's attack for all α.

$$- \eta \left(\alpha(1 - 2(\delta - \alpha)) - \delta\left(1 - 4(\delta - \alpha)\right)^2 \right) + \alpha(\delta - \alpha)(1 + 2\alpha - 4\delta) < 0 \; where$$

$$\eta = \frac{2\delta(1 - 4(\delta - \alpha)) + 2\sqrt{\delta(\delta - \alpha)(1 + 2\alpha - 8\delta(1 - 2\delta + 2\alpha))}}{1 - 2(\delta - \alpha)} \; for \; 0 < \alpha \leq \frac{1}{18},$$

then the public modulus N can be factorized in polynomial time.

The attack works for $\alpha < 0.375$ as the previous attacks, however, our attack completely improves previous two attacks proposed by Lu et. al. [23] and Takayasu-Kunihiro [38]. That means our attack requires less partial information than the previous attacks for all $\alpha < 0.375$.

Technical Overview. As the previous attacks[2] [23,38], our attack is based on Coppersmith's method to solve modular equations which have small solutions [6]. To solve a modular equation, the method constructs a lattice whose short vectors disclose the solutions. Appropriate lattice constructions are usually an important point of the method. The most famous lattice construction strategy was proposed by Jochemsz and May (Asiacrypt 2006) [18]. The strategy is simple, however, does not always offer the best bound. For example, in the context of small secret exponent attacks on RSA, the Boneh-Durfee weaker bound $d < N^{0.284}$ can be obtained by the Jochemsz-May strategy, however, the stronger bound $d < N^{0.292}$ cannot. To obtain better bounds, which cannot be obtained by the Jochemsz-May strategy, is technical and a challenging problem. In this paper, we also focus on the formulation of the equation. More concretely, how to formulate an attack scenario is crucial for a resulting attack condition since sizes of recoverable root bounds depend on the algebraic structure of the formulation.

Let d'_p and $\tilde{d}_p$ be the most/least significant bits of d_p, respectively. As we defined above, $\tilde{d}_p > N^{0.5-\delta}$ and $d'_p < N^\delta$. Then the CRT exponent can be rewritten as $d_p = d'_p M + \tilde{d}_p$ where $M \approx N^{0.5-\delta}$. To thwart the Jochemsz-May attack [19], we only consider the case $d_p \approx N^{0.5}$ in this paper and omit the analysis of the other case since the generalization is trivial. The CRT-RSA key generation can be rewritten as

$$e\left(d'_p M + \tilde{d}_p\right) = 1 + \ell(p - 1)$$

with some integer ℓ. Lu et al. [23] formulated the following equation:

$$1 - e\tilde{d}_p - eMx - y = 0 \quad \mod p$$

whose solution is $(x, y) = (d'_p, \ell)$. There are two algorithms known to solve the equation proposed by Herrmann and May [15], Takayasu and Kunihiro [35], respectively. Herrmann and May's algorithm is based on the Jochemsz-May strategy whereas Takayasu and Kunihiro's algorithm is not. Takayasu-Kunihiro's algorithm works for larger δ than Herrmann-May's algorithm for small α; when

---

[2] In [38], the attack is based on the other method; Coppersmith's method to solve integer equations [5,9,10]. However, we will show that their attack with the least significant bits of d_p or d_q can also be obtained from the modular method.

$\alpha \approx 0$, the latter algorithm works for $\delta < 1/4 = 0.25$ and the former algorithm works for $\delta < (\sqrt{2} - 1)/2 = 0.20710 \cdots$. The fact shows that when we can construct a better attack which cannot be obtained by the Jochemsz-May strategy, it works with less partial information for the same α.

As opposed to Lu et al., Takayasu and Kunihiro [38] formulated the following equation[3]:

$$1 - e\tilde{d}_p + x(y - 1) = 0 \quad \mod eM$$

whose solution is $(x, y) = (\ell, p)$. They solved the equation where the lattice construction is based on the Jochemsz-May strategy as the Herrmann-May. However, the formulation affects the resulting attack condition. Takayasu-Kunihiro's attack works for large α than an attack of Lu et al. with Herrmann-May's algorithm; when $\delta \approx 0$, the latter algorithm works for $\alpha < 3/8 = 0.375$ and the former attack works for $\alpha < (\sqrt{2} - 1)/2 = 0.20710 \cdots$. The fact shows that the latter formulation, i.e., mod eM equation, yields the attacks which work for larger α than the former equation, i.e., mod p equation.

Our improved attack in this paper is constructed by solving the mod eM equation and the lattice does not follow the Jochemsz-May strategy. The improvement is reasonable from the above discussion. Although the same mod eM equation is solved, Takayasu-Kunihiro's attack [38] is based on the Jochemsz-May strategy. Our proposed attack works for larger δ than Takayasu-Kunihiro's attack. Our proposed attack works for larger α than the attack of Lu et al. Therefore, our attack is better than the previous best attacks for all α.

1.3   Organization

In Sect. 2, we introduce the overview of Coppersmith's method; lattices, the LLL algorithm, and Howgrave-Graham's lemma. In Sect. 3, we briefly review a lattice construction of the Takayasu-Kunihiro attack [38]. Since we solve the same modular equation, i.e., the mod eM equation, and improve the attack, understanding of the previous lattice construction enables readers to understand the point of our improvement. To obtain possibly better attacks, our lattice constructions divide in two cases, i.e., the first and the second conditions of Theorem 1. In Sect. 4, due to the page limitation, we propose improved attacks for only large α, i.e., $1/18 < \alpha \leq 3/8$.

2   Preliminaries

In this section, we introduce Coppersmith's lattice based method to solve modular equations [6]. For simplicity, we explain the simpler reformulation due to Howgrave-Graham [17]. The method has been revealed numerous vulnerabilities of RSA. See [7,8,25,26] for more information.

---

[3] To be precise, the equation was formulated by Lu et al. in [23] and they solved the equation, however, Takayasu and Kunihiro corrected some mistakes of the paper and proposed an improved algorithm.

Let $b_1, \ldots, b_n \in \mathbb{Z}^d$ be linearly independent d-dimensional vectors. All vectors are row representations. The lattice $L(b_1, \ldots, b_n)$ spanned by the basis vectors $b_1, \ldots, b_n$ is defined as

$$L(b_1, \ldots, b_n) = \left\{ \sum_{j=1}^{n} c_j b_j : c_j \in \mathbb{Z} \right\}.$$

We also use the matrix representation for lattice bases. A basis matrix B is defined as the $n \times d$ matrix which has basis vectors $b_1, \ldots, b_n$ in each row. In this representation, a lattice spanned by the basis matrix B is defined as $L(B) = \{cB : c \in \mathbb{Z}^n\}$. We call n a rank of the lattice and d a dimension of the lattice. We call the lattice full-rank when $n = d$. In this paper, we only use full-rank lattices. A parallelepiped of a lattice is defined as $\mathcal{P}(B) = \{cB : 0 \leq c_j < 1\}$. We define a determinant of a lattice $\det(L(B))$ as an n-dimensional volume of the parallelepiped. In general, a determinant of a lattice can be computed as $\det(L(B)) = \sqrt{\det(BB^{\mathsf{T}})}$ where B^{T} is a transpose of B. A determinant of a full-rank lattice can be computed as $\det(L(B)) = |\det(B)|$.

In the context of cryptographic research, especially cryptanalysis, to find short lattice vectors is an important operation. See [28] for more information. Although to find the exact shortest vector is a computationally hard problem, LLL algorithm [21] proposed by Lenstra, Lenstra, and Lovász finds the approximate shortest lattice vectors in polynomial time.

Proposition 1 (LLL algorithm [25]). *Given a lattice basis $B \in \mathbb{Z}^{n \times n}$, the LLL algorithm finds short vectors v_1 and v_2 in $L(B)$ which satisfy*

$$\|v_1\| \leq 2^{(n-1)/4}(\det(L(B)))^{1/n} \quad and \quad \|v_2\| \leq 2^{n/4}(\det(L(B)))^{1/(n-1)}.$$

These norms are Euclidean norms. The running time of the LLL algorithm is polynomial time in n and input length.

Although output vectors by the LLL algorithm are not the exact shortest lattice vectors in the worst case, the quality is sufficient for Coppersmith's method [6,17].

In the rest of this section, we explain the overview of Coppersmith's method and how to use the LLL algorithm. For a k-variate polynomial $h(x_1, \ldots, x_k) = \sum h_{i_1,\ldots,i_k} x_1^{i_1} \cdots x_k^{i_k}$, let $\|h(x_1, \ldots, x_k)\| = \sqrt{\sum h_{i_1,\ldots,i_k}^2}$ and $\|h(x_1, \ldots, x_k)\|_\infty = \max_{i_1,\ldots,i_k} |h_{i_1,\ldots,i_k}|$ denote the norms of polynomials. To solve a modular equation $h(x_1, \ldots, x_k) = 0 \mod R$, it suffices to find k polynomials which have the same roots over the integers as the original equation. To derive such polynomials from the modular equation, we introduce the following Howgrave-Graham's lemma [17].

Lemma 1 (Howgrave-Graham's Lemma [17]). *Let $h(x_1, \ldots, x_k) \in \mathbb{Z}[x_1, \ldots, x_k]$ be a polynomial over the integers which consists of at most n monomials. Let $X_1, \ldots, X_k, R,$ and m be positive integers. If the polynomial $h(x_1, \ldots, x_k)$ satisfies*

- $h(\tilde{x}_1, \ldots, \tilde{x}_k) = 0 \mod R^m$, where $|\tilde{x}_1| < X_1, \ldots, |\tilde{x}_k| < X_k$,
- $\|h(x_1 X_1, \ldots, x_k X_k)\| < R^m/\sqrt{n}$.

Then $h(\tilde{x}_1, \ldots, \tilde{x}_k) = 0$ holds over the integers.

Hence, to find k polynomials which have the same roots as the original modular polynomial, it is sufficient to find algebraically independent k polynomials which have the same roots modulo R^m and whose norms are small enough to satisfy Howgrave-Graham's lemma. Then Gröbner bases or the resultant of the polynomials reveal the solution of the modular equation. For the purpose, we generate n polynomials $h_1(x_1, \ldots, x_k), \ldots, h_n(x_1, \ldots, x_k)$ and construct a lattice whose basis consists of coefficients of $h_1(x_1 X_1, \ldots, x_k X_k), \ldots, h_n(x_1 X_1, \ldots, x_k X_k)$. Then the LLL algorithm finds lattice vectors and the corresponding polynomials whose norms are small. Therefore, the original modular equation can be solved and all operations can be terminated in polynomial time.

To find larger roots, what we should do is to construct lattices which contain shorter lattice vectors. It is a challenging problem in this research area and the main focus of this paper. For the purpose, we introduce the lattice construction strategy which makes use of the notion of *helpful polynomials*. To solve a mod R equation, helpful polynomials, which was introduced by May [26], have smaller diagonals than the modulus R^m in a triangular basis matrix. Since the norms of short vectors output by the LLL algorithm are roughly $\det(L(B))^{1/n}$ and the determinant is computed by a product of all diagonals, helpful polynomials can reduce the norms. To summarize, Takayasu and Kunihiro [35] suggested that as many helpful polynomials as possible and as few unhelpful polynomials as possible should be selected in lattice bases as long as the basis matrix is triangular. Indeed, they used the strategy and improved the Herrmann-May algorithm [15].

We should note that the method requires heuristic argument. There are no assurance if new polynomials obtained by vectors output by the LLL algorithm are algebraically independent. In this paper, we assume that these polynomials are always algebraically independent and resultants of polynomials will not vanish as previous works [2,23,38]. Moreover, there have been few negative reports which contradict the assumption.

3   Lattice Construction of the Takayasu-Kunihiro

In this section, we summarize the Takayasu-Kunihiro lattice construction to solve a mod eM equation. To understand the spirit of the lattice construction helps readers to understand the point of our improvement easily in Sect. 4.

As discussed in Sect. 1, Takayasu and Kunihiro [38] found the small roots of the modular polynomial

$$f(x, y) := 1 - e\tilde{d}_p + x(y - 1) \mod eM$$

whose root is $(x, y) = (\ell, p)$. To obtain a better result, they also used an additional variable $z = q$ to make use of the algebraic relation $yz = N$ as [1,11].

The absolute values of the solutions are bounded above by $X := N^\alpha, Y := N^{1/2}, Z := N^{1/2}$ within constant factors. If two polynomials which have the roots $(x, y) = (\ell, p)$ over the integers are obtained, then the roots can be recovered. Notice that although there are three variables x, y, and z, two polynomials suffice to disclose the root since we already know the relation $yz = N$.

To solve the above modular equation $f(x, y) = 0$, they constructed lattices where the basis consists of coefficients of the following shift-polynomials:

$$g_{[i,j]}(x, y, z) := x^j z^s f(x, y)^i (eM)^{m-i}, \quad \text{and}$$

$$g'_{[i,j]}(x, y, z) := y^j z^s f(x, y)^i (eM)^{m-i},$$

with some positive integers m and $s = \eta m$ where $0 \le \eta \le 1$. These polynomials modulo $(eM)^m$ have the same root as the original modular polynomial, i.e., $g_{[i,j]}(\ell, p, q) = 0 \mod (eM)^m$ and $g'_{[i,j]}(\ell, p, q) = 0 \mod (eM)^m$. They collected shift-polynomials

$$g_{[i,j]}(x, y, z) \quad \text{for } i = 0, 1, \ldots, m; j = 0, 1, \ldots, m - i, \quad \text{and}$$

$$g'_{[i,j]}(x, y, z) \quad \text{for } i = 0, 1, \ldots, m; j = 1, 2, \ldots, t,$$

in a lattice basis with some positive integer $t = \tau m$ where $0 \le \tau \le \eta \le 1$. To reduce a determinant of the lattice, they multiply the inverse of N modulo $(eM)^m$. This operation eliminates the powers of N in diagonals. Properly ordered, the collection of polynomials generates a triangular basis matrix with diagonals

- $X^{i+j} Y^{i-s} (eM)^{m-i}$ for $g_{[i,j]}(x, y, z)$ and $i \ge s$,
- $X^{i+j} Z^{s-i} (eM)^{m-i}$ for $g_{[i,j]}(x, y, z)$ and $i < s$,
- $X^i Y^{i+j-s} (eM)^{m-i}$ for $g'_{[i,j]}(x, y, z)$ and $i + j \ge s$,
- $X^i Z^{s-i-j} (eM)^{m-i}$ for $g'_{[i,j]}(x, y, z)$ and $i + j < s$.

Then a dimension n and a determinant of the lattice $\det(L(B)) = X^{s_X} Y^{s_Y} Z^{s_Z} \cdot (eM)^{s_{eM}}$ are computed by

$$n = \sum_{i=0}^{m} \sum_{j=0}^{m-i} 1 + \sum_{i=0}^{m} \sum_{j=1}^{t} 1 = \left(\frac{1}{2} + \tau\right) m^2 + o(m^2),$$

$$s_X = \sum_{i=0}^{m} \sum_{j=0}^{m-i} (i + j) + \sum_{i=0}^{m} \sum_{j=1}^{t} i = \left(\frac{1}{3} + \frac{\tau}{2}\right) m^3 + o(m^3),$$

$$s_Y = \sum_{i=s}^{m} \sum_{j=0}^{m-i} (i - s) + \sum_{i=s-t}^{m} \sum_{j=s-t-i}^{t} (i + j - s) = \frac{(1 + \tau - \eta)^3}{6} m^3 + o(m^3),$$

$$s_Z = \sum_{i=0}^{s-1} \sum_{j=0}^{m-i} (s - i) + \sum_{i=0}^{s-1} \sum_{j=1}^{\min\{t, s-i\}} (s - i - j)$$

$$= \left(\frac{\eta^2}{2} - \frac{(\eta - \tau)^2}{6}\right) m^3 + o(m^3),$$

$$s_{eM} = \sum_{i=0}^{m} \sum_{j=0}^{m-i} (m - i) + \sum_{i=0}^{m} \sum_{j=1}^{t} (m - i) = \left(\frac{1}{3} + \frac{\tau}{2}\right) m^3 + o(m^3).$$

LLL outputs short lattice vectors and the corresponding polynomials satisfy Howgrave-Graham's lemma when $X^{s_X} Y^{s_Y} Z^{s_Z} (eM)^{s_{eM}} < (eM)^{mn}$. Ignoring low order terms of m, the condition becomes

$$\alpha \left(\frac{1}{3} + \frac{\tau}{2} \right) + \frac{1}{2} \left(\frac{(1 + \tau - \eta)^3}{6} + \frac{\eta^2}{2} - \frac{(\eta - \tau)^2}{6} \right)$$
$$< \left(\alpha + \frac{1}{2} - \delta \right) \left(\frac{1}{2} + \tau - \frac{1}{3} - \frac{\tau}{2} \right).$$

To maximize the right hand side of the inequality, optimizing the parameters

$$\eta = \frac{1 - 2\delta}{2} \quad \text{and} \quad \tau = \frac{\sqrt{1 - 4\delta} - 2\delta}{2},$$

then the above condition results in

$$-1 + 8\alpha + 8\delta - 12\delta^2 - 2(1 - 4\delta)\sqrt{1 - 4\delta} < 0.$$

4   Our Proposed Attack

In this section, we propose an improved attack for $1/18 < \delta \le 3/8$, i.e., the first condition of Theorem 1. We revisit the previous lattice construction in Sect. 4.2 and find the drawback. Then we solve the same modular equation as the Takayasu-Kunihiro [38] and find larger roots.

4.1   An Observation of the Previous Lattice

As we explained in Sect. 3, Takayasu and Kunihiro [38] constructed a lattice whose basis consists of polynomials which have the same roots as the original polynomial modulo $(eM)^m$. We want to analyze the validity of the lattice construction by using the helpful polynomials strategy [26, 35]. Based on the strategy, as many helpful polynomials (which have diagonals whose sizes are smaller than $(eM)^m$) as possible should be selected and as few unhelpful polynomials (which have diagonals whose sizes are larger than $(eM)^m$) as possible should be eliminated as long as a basis matrix to be triangular.

Then we observe the Takayasu-Kunihiro lattice after the parameter optimization. There are polynomials with diagonals

- $X^{i+j} Y^{i-s} (eM)^{m-i}$ for $i = s, s+1, \ldots, m; j = 0, 1, \ldots, m - i$,
- $X^{i+j} Z^{s-i} (eM)^{m-i}$ for $i = 0, 1, \ldots, s - 1; j = 0, 1, \ldots, m - i$,
- $X^i Y^{i+j-s} (eM)^{m-i}$ for $i = s-t, s-t+1, \ldots, m; j = s-t-i, s-t-i+1, \ldots, t$,
- $X^i Z^{s-i-j} (eM)^{m-i}$ for $i = 0, 1, \ldots, s - 1; j = 1, 2, \ldots, \min\{t, s - i\}$.

We focus on the bottom two families of polynomials, i.e., $g'_{[i,j]}(x, y, z)$. The lattice basis does not contain as many helpful polynomials as possible since when polynomials

$$g'_{[i,j]}(x, y, z) \quad \text{for } i = 1, 2, \ldots, s - 1; j = s - i$$

are added in the basis, the corresponding diagonals become

- $X^i(eM)^{m-i}$ for $i = 1, 2, \ldots, s-1$

and

$$X^i(eM)^{m-i} = N^{\left(\alpha+\frac{1}{2}-\delta\right)m-\left(\frac{1}{2}-\delta\right)i} < N^{\left(\alpha+\frac{1}{2}-\delta\right)m} = (eM)^m.$$

Similarly, the lattice basis contains some unhelpful polynomials which do not contribute for the basis matrix to be triangular. Indeed, the basis matrix is still triangular without polynomials

$$g'_{[i,j]}(x, y, z) \quad \text{for } i = \left\lceil \frac{1 - \sqrt{1-4\delta}}{4\delta}m \right\rceil, \left\lceil \frac{1 - \sqrt{1-4\delta}}{4\delta}m \right\rceil + 1 \ldots, m; j = t$$

whose corresponding diagonals are

- $X^i Y^{i+t-s}(eM)^{m-i}$ for $i = \left\lceil \frac{1-\sqrt{1-4\delta}}{4\delta}m \right\rceil, \left\lceil \frac{1-\sqrt{1-4\delta}}{4\delta}m \right\rceil + 1, \ldots, m$

and the following inequality holds:

$$X^i Y^{i+t-s}(eM)^{m-i} = N^{\left(\alpha+\frac{1}{2}-\delta\right)m+\delta i-\frac{1}{2}(s-t)} > N^{\left(\alpha+\frac{1}{2}-\delta\right)m} = (eM)^m.$$

Notice that

$$\delta i - \frac{1}{2}(s-t) = \delta i - \frac{1}{2}\left(\frac{1-2\delta}{2} - \frac{\sqrt{1-4\delta}-2\delta}{2}\right)m$$

$$= \delta\left(i - \frac{1 - \sqrt{1-4\delta}}{4\delta}m\right) > 0$$

for all $i = \left\lceil \frac{1-\sqrt{1-4\delta}}{4\delta}m \right\rceil, \left\lceil \frac{1-\sqrt{1-4\delta}}{4\delta}m \right\rceil + 1, \ldots, m$.

The above examples are not all the helpful polynomials which are not selected and all the unhelpful polynomials which are selected. Hence, if we can construct more appropriate lattices, the resulting attack condition should be improved.

4.2   Our Lattice Construction

Based on the observation in Sect. 4.1, we construct more appropriate lattices than the Takayasu-Kunihiro [38]. More concretely, we select all helpful $g'_{[i,j]}(x, y, z)$ for $i+j \geq s$ and do not select any unhelpful $g'_{[i,j]}(x, y, z)$ for $i+j \geq s$.

At first, we analyze which $g'_{[i,j]}(x, y, z)$ for $i+j \geq s$ are helpful or not. As we explained in Sect. 3, the corresponding diagonals are $X^i Y^{i+j-s}(eM)^{m-i}$. Then the polynomials are helpful when

$$X^i Y^{i+j-s}(eM)^{m-i} < (eM)^m \Leftrightarrow \alpha i + \frac{1}{2}(i+j-s) < \left(\alpha + \frac{1}{2} - \delta\right)i$$

$$\Leftrightarrow j < s - 2\delta i.$$

Therefore, we collect the following shift-polynomials:

$$g_{[i,j]}(x, y, z) \quad \text{for } i = 0, 1, \ldots, m; j = 0, 1, \ldots, m - i \text{ and}$$
$$g'_{[i,j]}(x, y, z) \quad \text{for } i = 0, 1, \ldots, m; j = 1, 2, \ldots, \lfloor s - 2\delta i \rfloor$$

in a lattice basis. Here, we do not take into account if polynomials $g_{[i,j]}(x, y, z)$ and $g'_{[i,j]}(x, y, z)$ for $i + j < s$ are helpful or not, however, these polynomials contribute the basis matrix to be triangular. Hence, we use the above collection of shift-polynomials only when $\eta > 2\delta$. Otherwise, polynomials $g_{[i,j]}(x, y, z)$ for $i + j > \frac{\eta}{2\delta}m$ do not contribute the basis matrix to be triangular. We will analyze the other case in Sect. 5.

We compute the resulting attack condition. A dimension n and a determinant of the lattice $\det(L(\boldsymbol{B})) = X^{s_X} Y^{s_Y} Z^{s_Z} (eM)^{s_{eM}}$ are computed by

$$n = \sum_{i=0}^{m} \sum_{j=0}^{m-i} 1 + \sum_{i=0}^{m} \sum_{j=1}^{\lfloor s-2\delta i \rfloor} 1 = \left(\frac{1}{2} - \delta + \eta \right) m^2 + o(m^2),$$

$$s_X = \sum_{i=0}^{m} \sum_{j=0}^{m-i} (i+j) + \sum_{i=0}^{m} \sum_{j=1}^{\lfloor s-2\delta i \rfloor} i = \left(\frac{1 - 2\delta}{3} + \frac{\eta}{2} \right) m^3 + o(m^3),$$

$$s_Y = \sum_{i=s}^{m} \sum_{j=0}^{m-i} (i-s) + \sum_{i=0}^{m} \sum_{j=\max\{s-i+1,0\}}^{\lfloor s-2\delta i \rfloor} (i+j-s) = \frac{(1-2\delta)^2}{6} m^3 + o(m^3),$$

$$s_Z = \sum_{i=0}^{s-1} \sum_{j=0}^{m-i} (s-i) + \sum_{i=0}^{s-1} \sum_{j=0}^{s-i} (s-i-j) = \frac{\eta^2}{2} m^3 + o(m^3),$$

$$s_{eM} = \sum_{i=0}^{m} \sum_{j=0}^{m-i} (m-i) + \sum_{i=0}^{m} \sum_{j=1}^{\lfloor s-2\delta i \rfloor} (m-i) = \left(\frac{1-\delta}{3} + \frac{\eta}{2} \right) m^3 + o(m^3).$$

LLL outputs short lattice vectors and the corresponding polynomials satisfies Howgrave-Graham's lemma when $X^{s_X} Y^{s_Y} Z^{s_Z} (eM)^{s_{eM}} < (eM)^{mn}$. Ignoring low order terms of m, the condition becomes

$$\alpha \left(\frac{1 - 2\delta}{3} + \frac{\eta}{2} \right) + \frac{1}{2} \left(\frac{(1 - 2\delta)^2}{6} + \frac{\eta^2}{2} \right) < \left(\alpha + \frac{1}{2} - \delta \right) \left(\frac{1}{6} - \frac{2\delta}{3} + \frac{\eta}{2} \right).$$

To maximize the right hand side of the inequality, we set the parameter η to be a solution of

$$\alpha \frac{1}{2} + \frac{1}{2}\eta = \left(\alpha + \frac{1}{2} - \delta \right) \frac{1}{2},$$

that is,

$$\eta = \frac{1 - 2\delta}{2}.$$

By substituting the parameter, the above attack condition becomes

$$7(1 - 2\delta)^2 - 4(1 - 2\delta) - 8(\alpha + 1/2) + 4 > 0.$$

Therefore, the attack works when

$$\delta < \frac{5 - 2\sqrt{1 + 14\alpha}}{14}$$

as required.

Notice that the attack works only when $\frac{1-2\delta}{2} > 2\delta$, that leads to $\delta < \frac{1}{6}$ and equivalent to

$$\alpha > \frac{1}{18}.$$

5 Concluding Remarks

In this paper, we study the partial key exposure attacks on CRT-RSA when the least significant bits of CRT exponent are exposed to attackers. We solve the same modular equation as Takayasu and Kunihiro [38], however, make use of the property of helpful polynomials and provide better lattice constructions. Our attack is better than previous attacks for all $0 < e \leq N^{0.375}$; the attack of Lu et al. [23] as well as the Takayasu-Kunihiro [38].

There are similar works for partial key exposure attacks on CRT-RSA which are not the scope of this paper; attacks when the most significant bits of CRT exponent are exposed to attackers. There are attacks with the exposed most significant bits by Blömer and May [2] and the Takayasu-Kunihiro [38] which are analogous to the attack with the exposed least significant bits by Lu et al. [23] and the Takayasu-Kunihiro [38], respectively. An interesting open problem is whether there exists a partial key exposure attack on CRT-RSA with the exposed most significant bits which is analogous to our attack with the exposed least significant bits.

Acknowledgement. The first author is supported by a JSPS Fellowship for Young Scientists. This research was supported by JSPS Grant-in-Aid for JSPS Fellows 14J08237, CREST, JST, and KAKENHI Grant Number 25280001.

References

1. Bleichenbacher, D., May, A.: New attacks on RSA with small secret CRT-exponents. In: Yung, M., Dodis, Y., Kiayias, A., Malkin, T. (eds.) PKC 2006. LNCS, vol. 3958, pp. 1–13. Springer, Heidelberg (2006)
2. Blömer, J., May, A.: New partial key exposure attacks on RSA. In: Boneh, D. (ed.) CRYPTO 2003. LNCS, vol. 2729, pp. 27–43. Springer, Heidelberg (2003)
3. Boneh, D., Durfee, G.: Cryptanalysis of RSA with private key d less than $N^{0.292}$. IEEE Trans. Inf. Theor. **46**(4), 1339–1349 (2000)
4. Boneh, D., Durfee, G., Frankel, Y.: An attack on RSA given a small fraction of the private key bits. In: Ohta, K., Pei, D. (eds.) ASIACRYPT 1998. LNCS, vol. 1514, pp. 25–34. Springer, Heidelberg (1998)
5. Coppersmith, D.: Finding a small root of a bivariate integer equation; factoring with high bits known. In: Maurer, U.M. (ed.) EUROCRYPT 1996. LNCS, vol. 1070, pp. 178–189. Springer, Heidelberg (1996)

6. Coppersmith, D.: Finding a small root of a univariate modular equation. In: Maurer, U.M. (ed.) EUROCRYPT 1996. LNCS, vol. 1070, pp. 155–165. Springer, Heidelberg (1996)
7. Coppersmith, D.: Small solutions to polynomial equations, and low exponent RSA vulnerabilities. J. Cryptol. 10(4), 233–260 (1997)
8. Coppersmith, D.: Finding small solutions to small degree polynomials. In: Silverman, J.H. (ed.) CaLC 2001. LNCS, vol. 2146, pp. 20–31. Springer, Heidelberg (2001)
9. Coron, J.-S.: Finding small roots of bivariate integer polynomial equations revisited. In: Cachin, C., Camenisch, J.L. (eds.) EUROCRYPT 2004. LNCS, vol. 3027, pp. 492–505. Springer, Heidelberg (2004)
10. Coron, J.-S.: Finding small roots of bivariate integer polynomial equations: a direct approach. In: Menezes, A. (ed.) CRYPTO 2007. LNCS, vol. 4622, pp. 379–394. Springer, Heidelberg (2007)
11. Durfee, G., Nguyên, P.Q.: Cryptanalysis of the RSA schemes with short secret exponent from Asiacrypt 1999. In: Okamoto, T. (ed.) ASIACRYPT 2000. LNCS, vol. 1976, pp. 14–29. Springer, Heidelberg (2000)
12. Ernst, M., Jochemsz, E., May, A., de Weger, B.: Partial key exposure attacks on RSA up to full size exponents. In: Cramer, R. (ed.) EUROCRYPT 2005. LNCS, vol. 3494, pp. 371–386. Springer, Heidelberg (2005)
13. Faugère, J.-C., Marinier, R., Renault, G.: Implicit factoring with shared most significant and middle bits. In: Nguyen, P.Q., Pointcheval, D. (eds.) PKC 2010. LNCS, vol. 6056, pp. 70–87. Springer, Heidelberg (2010)
14. Galbraith, S.D., Heneghan, C., McKee, J.F.: Tunable balancing of RSA. In: Boyd, C., González Nieto, J.M. (eds.) ACISP 2005. LNCS, vol. 3574, pp. 280–292. Springer, Heidelberg (2005)
15. Herrmann, M., May, A.: Solving linear equations modulo divisors: on factoring given any bits. In: Pieprzyk, J. (ed.) ASIACRYPT 2008. LNCS, vol. 5350, pp. 406–424. Springer, Heidelberg (2008)
16. Herrmann, M., May, A.: Maximizing small root bounds by linearization and applications to small secret exponent RSA. In: Nguyen, P.Q., Pointcheval, D. (eds.) PKC 2010. LNCS, vol. 6056, pp. 53–69. Springer, Heidelberg (2010)
17. Howgrave-Graham, N.: Finding small roots of univariate modular equations revisited. In: Darnell, M.J. (ed.) Cryptography and Coding 1997. LNCS, vol. 1355. Springer, Heidelberg (1997)
18. Jochemsz, E., May, A.: A strategy for finding roots of multivariate polynomials with new applications in attacking RSA variants. In: Lai, X., Chen, K. (eds.) ASIACRYPT 2006. LNCS, vol. 4284, pp. 267–282. Springer, Heidelberg (2006)
19. Jochemsz, E., May, A.: A polynomial time attack on RSA with private CRT-exponents smaller than $N^{0.073}$. In: Menezes, A. (ed.) CRYPTO 2007. LNCS, vol. 4622, pp. 395–411. Springer, Heidelberg (2007)
20. Joye, M., Lepoint, T.: Partial key exposure on RSA with private exponents larger than N. In: Ryan, M.D., Smyth, B., Wang, G. (eds.) ISPEC 2012. LNCS, vol. 7232, pp. 369–380. Springer, Heidelberg (2012)
21. Lenstra, A., Lenstra, H., Lovász, L.: Factoring polynomials with rational coefficients. Math. Ann. 261, 515–534 (1982)
22. Lu, Y., Peng, L., Zhang, R., Hu, L., Lin, D.: Towards optimal bounds for implicit factorization problem. In: Dunkelman, O., Keliher, L. (eds.) SAC 2015. LNCS, vol. 9566, pp. 462–476. Springer, Heidelberg (2016). doi:10.1007/978-3-319-31301-6_26

23. Lu, Y., Zhang, R., Lin, D.: New partial key exposure attacks on CRT-RSA with large public exponents. In: Boureanu, I., Owesarski, P., Vaudenay, S. (eds.) ACNS 2014. LNCS, vol. 8479, pp. 151–162. Springer, Heidelberg (2014)
24. May, A.: Cryptanalysis of unbalanced RSA with small CRT-exponent. In: Yung, M. (ed.) CRYPTO 2002. LNCS, vol. 2442, pp. 242–256. Springer, Heidelberg (2002)
25. May, A.: New RSA vulnerabilities using lattice reduction methods. Ph.D. thesis, University of Paderborn (2003)
26. May, A.: Using LLL-reduction for solving RSA and factorization problems. In: Nguyen, P.Q., Vallée, B. (eds.) The LLL Algorithm - Survey and Applications. Information Security and Cryptography, pp. 315–348. Springer, Heidelberg (2010)
27. May, A., Ritzenhofen, M.: Implicit factoring: on polynomial time factoring given only an implicit hint. In: Jarecki, S., Tsudik, G. (eds.) PKC 2009. LNCS, vol. 5443, pp. 1–14. Springer, Heidelberg (2009)
28. Nguyên, P.Q., Stern, J.: The two faces of lattices in cryptology. In: Silverman, J.H. (ed.) CaLC 2001. LNCS, vol. 2146, pp. 146–180. Springer, Heidelberg (2001)
29. Quisquater, J.J.: Fast decipherment algorithm for RSA public-key cryptosystem. Electron. Lett. **18**(2), 905–907 (1982)
30. Rivest, R.L., Shamir, A.: Efficient factoring based on partial information. In: Pichler, F. (ed.) EUROCRYPT 1985. LNCS, vol. 219, pp. 31–34. Springer, Heidelberg (1986)
31. Rivest, R.L., Shamir, A., Adleman, L.M.: A method for obtaining digital signatures and public-key cryptosystems. Commun. ACM **21**(2), 120–126 (1978)
32. Sarkar, S., Maitra, S.: Partial key exposure attack on CRT-RSA. In: Abdalla, M., Pointcheval, D., Fouque, P.-A., Vergnaud, D. (eds.) ACNS 2009. LNCS, vol. 5536, pp. 473–484. Springer, Heidelberg (2009)
33. Sarkar, S., Maitra, S.: Approximate integer common divisor problem relates to implicit factorization. IEEE Trans. Inf. Theor. **57**(6), 4002–4013 (2011)
34. Sarkar, S., Sen Gupta, S., Maitra, S.: Partial key exposure attack on RSA – improvements for limited lattice dimensions. In: Gong, G., Gupta, K.C. (eds.) INDOCRYPT 2010. LNCS, vol. 6498, pp. 2–16. Springer, Heidelberg (2010)
35. Takayasu, A., Kunihiro, N.: Better lattice constructions for solving multivariate linear equations modulo unknown divisors. IEICE Trans. **97**(A(6)), 1259–1272 (2014)
36. Takayasu, A., Kunihiro, N.: General bounds for small inverse problems and its applications to multi-prime RSA. In: Lee, J., Kim, J. (eds.) Information Security and Cryptology - ICISC 2014. LNCS, vol. 8949, pp. 3–17. Springer, Heidelberg (2014)
37. Takayasu, A., Kunihiro, N.: Partial key exposure attacks on RSA: achieving the boneh-durfee bound. In: Joux, A., Youssef, A. (eds.) SAC 2014. LNCS, vol. 8781, pp. 345–362. Springer, Heidelberg (2014)
38. Takayasu, A., Kunihiro, N.: Partial key exposure attacks on CRT-RSA: better cryptanalysis to full size encryption exponents. In: Malkin, T., Kolesnikov, V., Lewko, A., Polychronakis, M. (eds.) ACNS 2015. LNCS, vol. 9092, pp. 518–537. Springer, Heidelberg (2015). doi:10.1007/978-3-319-28166-7_25
39. de Weger, B.: Cryptanalysis of RSA with small prime difference. Appl. Algebra Eng. Commun. Comput. **13**(1), 17–28 (2002)
40. Wiener, M.J.: Cryptanalysis of short RSA secret exponents. IEEE Trans. Inf. Theor. **36**(3), 553–558 (1990)

Cryptanalysis and Improved Construction of a Group Key Agreement for Secure Group Communication

Jun Xu[1,2], Lei Hu[1,2]([✉]), Xiaona Zhang[1,2], Liqiang Peng[1,2], and Zhangjie Huang[1,2]

[1] State Key Laboratory of Information Security,
Institute of Information Engineering,
Chinese Academy of Sciences, Beijing 100093, China
{xujun,hulei,zhangxiaona,pengliqiang,huangzhangjie}@iie.ac.cn
[2] Data Assurance and Communications Security Research Center,
Chinese Academy of Sciences, Beijing 100093, China

Abstract. In this paper, we give a ciphertext-only attack on a NTRU-based group key agreement. Our attack can recover the plaintext without having access to the secret decryption key of any group member even when there are only two group members. In order to overcome this drawback, we propose an improved group key agreement and make the corresponding cryptanalysis, which shows that it is secure and resilient to this ciphertext-only attack as well as other attacks under some constraints.

Keywords: Group key agreement · Secure group communication · NTRU cryptosystem · Ciphertext-only attack

1 Introduction

An symmetric group key agreement enables two or more members to create a common secret key, which is used by group members for encryption and decryption. Diffie and Hellman [5] constructed the first symmetric group key agreement in 1976, which is a one-round two-party protocol. Joux [11] proposed a tripartite version of the Diffie-Hellman protocol in 2004. Upon to now, a mass of schemes have been put forward in order to generalize the Diffie-Hellman protocol to the multi-party situations, such as [2,4,10,12,13,15].

In Eurocrypt 2009, Wu et al. [20] proposed the concept of asymmetric group agreement, which is a new class of group key agreement. In this sense, all group members negotiate and publish a common group encryption key but withhold respective secret decryption keys. Each member can correctly decrypt any ciphertext encrypted with the common encryption key by using his own secret key. Wu et al. also gave the generic construction and a concrete one-round asymmetric group agreement based on bilinear pairings. However, their protocol is unauthenticated and only fits for static groups.

© Springer International Publishing Switzerland 2016
M. Bishop and A.C.A. Nascimento (Eds.): ISC 2016, LNCS 9866, pp. 48–58, 2016.
DOI: 10.1007/978-3-319-45871-7_4

A different generic construction about the asymmetric group key agreement was designed in [16,17]. The common group encryption key is generated by combing public keys of all members using the idea of Chinese remainder theorem. The designers in [16] presented an instantiation based on the ElGamal scheme with both confidentiality and nonrepudiation for mobile ad-hoc networks. The authors in [17] proposed another concrete NTRU-based asymmetric group key agreement for dynamic peer systems, which is an one-round, distributed and self-organizing protocol. The designers pointed out that the security of this protocol depends on NTRU cryptosystem.

NTRU is a lattice-based post-quantum encryption algorithm with its security depends on the intractability of solving certain lattice problems. The first version of NTRU was proposed by Hoffstein, Pipher and Silverman in 1998 [7], that is denoted by NTRU-1998. Subsequently, two main variants with different parameter sets were presented in 2001 and 2005, which are respectively NTRU-2001 [8] and NTRU-2005 [9]. NTRU has been adopted as the IEEE 1363.1 Standard [19] and the X9.98 Standard [1] for use in the financial services industry.

In this paper, we demonstrate that the NTRU-based group key agreement proposed in [17] is not secure. Any attacker can recover the plaintext when only getting access to the corresponding ciphertext. Furthermore, we put forward an improved design to resist this ciphertext-only attack, and reduce its security to NTRU cryptosystem.

The rest of this paper is organized as follows. In Sect. 2, we introduce some terminology and NTRU cryptosystem. In Sect. 3, we recall a NTRU-based group key agreement. The ciphertext-only attack and experimental results are presented in Sect. 4. In Sect. 5, we give an improved construction and security analysis. Section 6 is the conclusion.

2   Preliminary

For the rest of the paper, $\mathbb{Z}$, $\mathbb{Z}_m$ and $\mathbb{Q}$ are respectively denoted as integer ring, the residue class ring and the rational field, where m is a positive integer. Here, $\mathbb{Z}_m$ of m elements is regarded as the set of integers from $(-\frac{m}{2}, \frac{m}{2}]$. For positive integer N, $\mathcal{R}$, $\mathcal{R}_m$ and $\mathcal{R}'$ are respectively presented $\mathbb{Z}[x]/(x^N - 1)$, $\mathbb{Z}_m[x]/(x^N - 1)$ and $\mathbb{Q}[x]/(x^N - 1)$, which are the set of polynomials with degree less than N and coefficients in $\mathbb{Z}$, $\mathbb{Z}_m$ and $\mathbb{Q}$ respectively. Let $*$ be cyclic multiplication. For $a(x) = \sum_{i=0}^{N-1} a_i x^i, b(x) = \sum_{j=0}^{N-1} b_j x^j \in \mathcal{R}'$, the product $c(x) = \sum_{k=0}^{N-1} c_k x^k$ of $a(x)$ and $b(x)$ is given by

$$c(x) = a(x) * b(x) \text{ with } c_k = \sum_{i+j=k \pmod{N}} a_i b_j.$$

For any polynomial $a(x) = \sum_{i=0}^{N-1} a_i x^i \in \mathcal{R}'$, its vector and matrix forms are respectively written as $\mathbf{a} = (a_0, a_1 \cdots, a_{N-1})^T$ and

$$M_a = \begin{bmatrix} a_0 & a_{N-1} & \cdots & a_1 \\ a_1 & a_0 & \cdots & a_2 \\ & & \cdots & \\ a_{N-1} & a_{N-2} & \cdots & a_0 \end{bmatrix}.$$

Thus, the corresponding vector and matrix forms of $a(x) * b(x)$ are $M_a \cdot \mathbf{b}$ and $M_a \cdot M_b$.

2.1 NTRU Public Key Cryptosystem

We give a description of NTRU cryptosystem. Please refer to [7–9] for detailed information.

Parameters: The tuple $(N, p, q, L_f, L_g, L_r, L_m)$ is a set of parameters. N is an odd prime number. p is an integer or a polynomial. q is a positive integer. L_f, L_g, L_r, L_m are subsets of $\mathcal{R}$. L_f, L_g are related to private key space. L_r, L_m are respectively random polynomial space and plaintext space.

Key generation: Randomly choose a polynomial $f(x) \in L_f$ such that there exist polynomials $f_p(x), f_q(x) \in \mathcal{R}$ satisfying

$$f(x) * f_p(x) = 1 \ (\mathrm{mod}\ p) \text{ and } f(x) * f_q(x) = 1 \ (\mathrm{mod}\ q).$$

Randomly choose a polynomial $g(x) \in L_g$ and compute polynomial $h(x) \in \mathcal{R}_q$ by $h(x) = p * f_q(x) * g(x) \mathrm{mod}\ q$.

Public and private key: Polynomial $h(x)$ is public key and tuple $(f(x), f_p(x))$ is private key.

Encryption: For a plaintext $m(x) \in L_m$, randomly choose a polynomial $r(x) \in L_r$ and compute ciphertext $c(x) \in \mathcal{R}_q$ as

$$c(x) = h(x) * r(x) + m(x) \ \mathrm{mod}\ q. \tag{1}$$

Decryption: The receiver first computes polynomial $a(x) \in \mathcal{R}_q$ by $a(x) = f(x) * c(x) \ \mathrm{mod}\ q$, which is equal to $a(x) = p * g(x) * r(x) + f(x) * m(x) \ (\mathrm{mod}\ q)$. Note that almost all coefficients of $p * g(x) * r(x) + f(x) * m(x)$ are in $(-q/2, q/2]$, one has that $a(x) = p * g(x) * r(x) + f(x) * m(x)$ with a very high possibility. Finally, the receiver recovers plaintext $m(x)$ due to $m(x) = a(x) * f_p(x) \ \mathrm{mod}\ p$.

Mol and Yung summarized the parameters of three main instantiations of NTRU in [18] and we list their comparison in Table 1.

Here, d_f, d_g, d_r are known positive integers. $B, B(d), T$ are subsets of $\mathcal{R}$. The coefficients of all polynomials in B belong to $\{0, 1\}$. $B(d)$ is a subset of B. For each polynomial in $B(d)$, there are exactly d 1's in its coefficients. The coefficients of all polynomials in T are in $\{-1, 0, 1\}$. For integers d_1 and d_2, $L(d_1, d_2)$ is a subset of T. For each polynomial in $L(d_1, d_2)$, there are exactly d_1 1's and d_2 -1's in its coefficients.

Table 1. Parameters of NTRU variants

Variants	q	p	L_f	L_g	L_r	L_m
NTRU-1998	$2^k \in [\frac{N}{2}, N]$	3	$L(d_f, d_f - 1)$	$L(d_g, d_g)$	$L(d_r, d_r)$	T
NTRU-2001	$2^k \in [\frac{N}{2}, N]$	$x + 2$	$1 + p * B(d_f)$	$B(d_g)$	$B(d_r)$	B
NTRU-2005	prime	2	$1 + p * B(d_f)$	$B(d_g)$	$B(d_r)$	B

3 Description of an NTRU-Based Group Key Agreement

In this section, we review an NTRU-based group key agreement proposed in [17]. Please refer to [17] for further information. Here we denote the set of group members as $\{u_1, u_2, \ldots, u_n\}$.

3.1 An NTRU-Based Group Key Agreement

1: **Every member's public and private key.** Every member u_i $(i = 1, \ldots, n)$ randomly generates a public key $h_i(x)$ and the corresponding private key $(f_i(x), f_{ip}(x))$ of the NTRU cryptosystem. Every member broadcasts the corresponding public key to other group members.

2: **Public parameters.** A sponsor member picks n considerably larger positive integers $m_1, m_2, \ldots, m_n$ compared with q in the NTRU cryptosystem. These integers are relatively prime in pairs. The sponsor member broadcasts his public key along with $m_1, m_2, \ldots, m_n$.

3: **Group encryption key.** Let $M = \prod\limits_{i=1}^{n} m_i$, $M_i = M/m_i$ and $y_i = M_i^{-1} \bmod m_i$. Every member u_i $(i = 1, \ldots, n)$ picks the corresponding m_i and locally computes the common group encryption key $h(x) \in \mathcal{R}_M$ by

$$h(x) = \sum_{i=1}^{n} M_i y_i h_i(x) \bmod M. \tag{2}$$

4: **Encryption.** Given a plaintext $m(x) \in L_m$, any group member can randomly choose a polynomial $r(x)$ in L_r and compute ciphertext $c(x) \in \mathcal{R}_M$ as

$$c(x) = \left(h(x) * r(x) + \left(\sum_{i=1}^{n} M_i y_i \right) * m(x) \right) \bmod M. \tag{3}$$

5: **Decryption.** For a given ciphertext $c(x)$, every group member u_i $(i = 1, \ldots, n)$ compute $c_i(x) = c(x) \bmod m_i$, where the coefficients of $c_i(x)$ lie in $(-m_i/2, m_i/2]$. This equation can be reduced to

$$c_i(x) = h_i(x) * r(x) + m(x) \pmod{m_i}.$$

According to the m_i are sufficiently larger than q, there is

$$c_i(x) = h_i(x) * r(x) + m(x). \tag{4}$$

Then, he can recover the plaintext $m(x)$ by using the corresponding private key $\left(f_i(x), f_{ip}(x)\right)$ like the decryption procedure in the NTRU cryptosystem.

Remark 1. In practice, the designers of [17] pointed out that each member u_i should broadcast $h_i'(x) = h_i(x) * \varphi_i(x)$ instead of $h_i(x)$ for $i = 1, \ldots, n$, where the $\varphi_i(x)$ are unknown random polynomials in L_r. In fact, this $\varphi_i(x)$ can be also regarded as the ephemeral secret keys to prevent possible attacks due to the exposure of public key $h_i(x)$. In this situation, the common group encryption key $h(x)$ is generated by

$$h(x) = \sum_{i=1}^{n} M_i y_i h_i'(x) \bmod M. \tag{5}$$

It is obvious that (5) degrades into (2) if $\varphi_i(x) = 1$ for $i = 1, \ldots, n$.

4   Attack on NTRU-Based Group Key Agreement

In this section, we give an efficient ciphertext-only attack on the NTRU-based group key agreement.

4.1   Ciphertext-Only Attack

Noting that public keys $h_1(x), \ldots, h_n(x)$ and positive integers $m_1, \ldots, m_n$ are publicly broadcasted, they can easily be obtained by any attacker, which means that the attacker can get Eq. (4), namely,

$$c_i(x) = h_i(x) * r(x) + m(x) \text{ for } i = 1, \ \ldots, n.$$

Under the usual sense of polynomial multiplication, the above equation can be transformed into

$$c_i(x) = h_i(x)r(x) + m(x) \ (\bmod \ x^N - 1) \text{ for } i = 1, \ldots, n. \tag{6}$$

It is clear that the attacker can recover the plaintext $m(x)$ once the random polynomial $r(x)$ is revealed. According to this idea, the attacker first eliminates $m(x)$ in (6) and gets the following equations

$$c_i(x) - c_j(x) = \left(h_i(x) - h_j(x)\right) r(x) \ (\bmod \ x^N - 1), where \ 1 \leq i < j \leq n.$$

Since N is a prime and $x^N - 1 = (x^{N-1} + \cdots + x + 1)(x - 1)$, we know that $x^{N-1} + \cdots + x + 1$ and $x - 1$ are coprime and irreducible over $\mathbb{Q}$ according to linear algebra. Hence, there exist different i and j such that polynomials $h_i(x) - h_j(x)$ and $x^N - 1$ are coprime with the overwhelming possibility. In other words, $h_i(x) - h_j(x)$ is likely to be invertible in $\mathcal{R}'$. Therefore, the attacker can compute out

$$r(x) = \left(h_i(x) - h_j(x)\right)^{-1} \left(c_i(x) - c_j(x)\right) \ (\bmod \ x^N - 1).$$

Furthermore, the attacker recovers plaintext $m(x)$ by computing

$$m(x) = \left(c_i(x) - h_i(x) \left(h_i(x) - h_j(x) \right)^{-1} \left(c_i(x) - c_j(x) \right) \right) \pmod{x^N - 1}. \quad (7)$$

In fact, let $\mathbf{c}_i = (c_{i,0}, \ldots, c_{i,N-1})^T$, $\mathbf{r} = (r_0, \ldots, r_{N-1})^T$ and $\mathbf{m} = (m_0, \ldots, m_{N-1})^T$ be the corresponding vectors of polynomials $c_i(x) = \sum_{k=0}^{N-1} c_{i,k} x^k$, $r(x) = \sum_{k=0}^{N-1} r_k x^k$ and $m(x) = \sum_{k=0}^{N-1} m_k x^k$. Let the corresponding matrices of polynomials $h_i(x) = \sum_{k=0}^{N-1} h_{i,k} x^k$ be

$$M_{h_i} = \begin{bmatrix} h_{i,0} & h_{i,N-1} & \cdots & h_{i,1} \\ h_{i,1} & h_{i,0} & \cdots & h_{i,2} \\ & & \cdots & \\ h_{i,N-1} & h_{i,N-2} & \cdots & h_{i,0} \end{bmatrix} \quad \text{with } i = 1, \ldots, n.$$

The polynomial equation (7) can also be rewritten as the following linear form

$$\mathbf{m} = \mathbf{c}_i - M_{h_i} \cdot \left(M_{h_i} - M_{h_j} \right)^{-1} \cdot (\mathbf{c}_i - \mathbf{c}_j), \quad (8)$$

where $1 \leq i < j \leq n$.

4.2 Further Analysis

For the group encryption key $h(x)$ in Remark 1, the above ciphertext-only attack still works well. Similarly, the attacker can obtain the following polynomial equations

$$c_i(x) = h'_i(x) * r(x) + m(x), i = 1, \ldots, n.$$

The form of these equations is the same as (4) except that the public keys $h_i(x)$ of the member u_i is replaced with known $h'_i(x)$. By performing the method in Sect. 4.1, plaintext $m(x)$ can be recovered by computing

$$m(x) = \left(c_i(x) - h'_i(x) \left(h'_i(x) - h'_j(x) \right)^{-1} \left(c_i(x) - c_j(x) \right) \right) \pmod{x^N - 1},$$

or

$$\mathbf{m} = \mathbf{c}_i - M_{h'_i} \cdot \left(M_{h'_i} - M_{h'_j} \right)^{-1} \cdot (\mathbf{c}_i - \mathbf{c}_j),$$

where $1 \leq i < j \leq n$.

4.3 Experiment Results

Our experiments were done by using the Magma computer algebra system [3] on a PC with Intel(R) Core(TM) Quad CPU (2.83 GHz, 3.25 GB RAM, Windows XP). Each experiment for a random instance were finished within 1 s. We performed the above ciphertext-only attack against group key agreement based on three main instantiations of NTRU. We found that polynomial $h_1(x) - h_2(x)$ and matrix $M_{h_1} - M_{h_2}$ are always invertible even when $n = 2$. For the case that $h(x)$ is generated by (5), we can also always obtain invertible polynomial $h'_1(x) - h'_2(x)$ and invertible matrix $M_{h'_1} - M_{h'_2}$ even when $n = 2$. Some results are listed in Table 2:

Table 2. Experiment results of attacking schemes based on different variants of NTRU

Variants	N	q	p	d_f	d_g	d_r	$M_{h_1} - M_{h_2}$	$M_{h'_1} - M_{h'_2}$	Results
NTRU-1998	167	128	3	60	20	18	Invertible	Invertible	Success
NTRU-2001	167	128	$x + 2$	60	19	18	Invertible	Invertible	Success
NTRU-2005	107	97	2	25	24	25	Invertible	Invertible	Success

5   An Improved NTRU-Based Group Key Agreement

In this section, we propose an improved NTRU-based group key agreement to prevent the above ciphertext-only attack and give the corresponding security analysis.

5.1   The Improved Group Key Agreement

1: **Every member's public and private key.** Every member u_i ($i = 1, \ldots, n$) randomly choose a polynomial $f_i(x) \in L_f$ such that there exist polynomials $f_{ip}(x), f_{iq}(x) \in \mathcal{R}$ satisfying

$$f_i(x) * f_{ip}(x) = 1 \pmod{p} \text{ and } f_i(x) * f_{iq}(x) = 1 \pmod{q}.$$

Randomly choose a polynomial $g_i(x) \in L_g$ and compute invertible polynomial $h_i(x) \in \mathcal{R}_q$ by $h_i(x) = p * f_{iq}(x) * g_i(x) \bmod q$. Every member broadcasts the corresponding public key to other group members.

2: **Public parameters.** A sponsor member picks n different primes m_1, m_2, $\ldots$, m_n satisfying $min > (N + 2)q + 2$, where $min = \min\{m_1, \cdots, m_n\}$. The sponsor member broadcasts his public key along with $m_1, m_2, \ldots, m_n$. Every member u_i picks up the corresponding m_i for $i = 1, \ldots, n$.

3: **Group encryption key.** Let $M = \prod\limits_{i=1}^{n} m_i$, $M_i = M/m_i$ and $y_i = M_i^{-1} \bmod m_i$. All members locally compute the common group encryption key $h(x) \in \mathcal{R}_M$ according to

$$h(x) = \sum_{i=1}^{n} M_i y_i h_i(x) \bmod M.$$

4: **Encryption.** Given a plaintext $m(x) \in L_m$, any group member chooses n random polynomials $r_1(x), \ldots, r_n(x) \in L_r$ and n random polynomials $s_1(x), \ldots, s_n(x)$ with all coefficients are from the interval $(-\lfloor \frac{min-Nq-2}{2q} \rfloor, \lfloor \frac{min-Nq-2}{2q} \rfloor]$. Then, the member computes ciphertext $c(x) \in \mathcal{R}_M$ as

$$c(x) = \left(h(x) * m(x) + \sum_{i=1}^{n} M_i y_i \left(r_i(x) + q * s_i(x) \right) \right) \bmod M. \qquad (9)$$

Remark 2. Our encryption procedure is different from [17]. This new design can prevent the above ciphertext-only attack. Detailed analysis is presented in next subsection.

5: **Decryption.** For a given ciphertext $c(x)$, every group member u_i ($i = 1, \ldots, n$) can recover plaintext $m(x)$ with the corresponding private key $(f_i(x), f_{ip}(x))$.

Next, we present the detailed decryption process and explain why it works.

Step 1. Every group member u_i firstly computes the polynomial $c_i(x) \in \mathcal{R}_M$ as $c_i(x) = c(x) \bmod m_i$. According to (9), there is

$$c_i(x) = h_i(x) * m(x) + r_i(x) + q * s_i(x) (\bmod m_i).$$

Notice that all coefficients of $h_i(x)$ and $s_i(x)$ are respectively from $(-q/2, q/2]$ and $(-\lfloor \frac{min-Nq-2}{2q} \rfloor, \lfloor \frac{min-Nq-2}{2q} \rfloor]$, $m(x) \in L_m$ and $r_i(x) \in L_r$, according to the triangular inequality, all coefficients of polynomial $h_i(x) * m(x) + r_i(x) + q * s_i(x)$ are bounded by

$$\frac{q}{2}N + 1 + q\frac{min - Nq - 2}{2q} = \frac{min}{2} \leq \frac{m_i}{2}.$$

Since all coefficients of $c_i(x)$ are from $(-m_i/2, m_i/2]$, there is

$$c_i(x) = h_i(x) * m(x) + r_i(x) + q * s_i(x).$$

Furthermore, the member u_i can get the following equation

$$c_i(x) = h_i(x) * m(x) + r_i(x) \ (\bmod q). \qquad (10)$$

Step 2. The member u_i computes polynomial $a(x) = f_i(x) * c_i(x) \ (\bmod q)$, where $a(x) \in \mathcal{R}_q$. From (10), the member u_i knows that

$$a(x) = p * g_i(x) * m(x) + f(x) * r_i(x) \bmod q.$$

Due to almost all coefficient of polynomial $p * g_i(x) * m(x) + f(x) * r_i(x)$ lie in $(-q/2, q/2]$ with a very high possibility. Therefore, $a(x) = p*g_i(x)*m(x)+f(x)* r_i(x)$. Furthermore, u_i can acquire $r_i(x)$ according to $r_i(x) = a(x)*f_{ip}(x) \bmod p$.
Step 3. The member u_i recovers plaintext $m(x)$ by computing

$$m(x) = h_i^{-1}(x) * (c_i(x) - r_i(x)) \bmod q.$$

5.2 Security Analysis

In this subsection, we give an analysis on the security of this improved group key agreement based on NTRU cryptosystem.

First, we explain why this improved group key agreement is secure against the ciphertext-only attack in Sect. 4. Notice that the key equations which can be acquired by the attacker are (4) and (10) respectively. As for the former, there are n polynomial equations with two unknown polynomials, thus, the plaintext can be obtained only by using the eliminating technique. But for the latter, there are n polynomial equations with $n + 1$ unknown polynomials. Hence, the attacker can not find the plaintext with the same technique.

Next, we give the following theorem to show that the security of our improved group key agreement relies on the NTRU public key cryptosystem.

Theorem 1. *The improved group key agreement in Sect. 5.1 is secure if the underlying NTRU public key vcryptosystem is secure.*

Proof. On one hand, public key $h_i(x)$ and private key $\big(f_i(x), f_{i_p}(x)\big)$ of every group member u_i $(1 \leq i \leq n)$ for the improved group key agreement are randomly generated from the underlying NTRU public key cryptosystem. Thus, if NTRU public key cryptosystem can prevent the attacks of recovering private keys, the improved group key agreement in Sect. 5.1 is also secure under the same situation.

On the other hand, according to the encryption equation (9) in the improved group key agreement, the attacker can directly get the system of Eq. (10), i.e.,

$$c_i(x) = h_i(x) * m(x) + r_i(x) \pmod{q} \text{ for } i = 1, \dots, n.$$

Since all public keys $h_i(x)$ are invertible in $\mathcal{R}_q$, the attacker can get the following equations

$$h_i^{-1}(x) * c_i(x) = h_i^{-1}(x) * r_i(x) + m(x) \pmod{q} \text{ with } i = 1, \dots, n.$$

Note that the form of the above equations is identical with the encryption equation (1) in NTRU public key cryptosystem. Hence, the improved group key agreement in Sect. 5.1 is as secure as the NTRU public key cryptosystem on the attacks about recovering the plaintext. □

Actually, there is constraints on the number n of the group members in our improved group key agreement. When n is sufficiently large, the broadcast attacks on the NTRU cryptosystem in [6,14] may affect the security of this scheme. From the key equation (10), each $c_i(x)$ can be regarded as the ciphertext of the same plaintext $m(x)$ by different public keys $h_i(x)$. Once the ciphertext $c(x)$ is exposed, the $c_i(x)$ can be easily obtained. In this circumstance, the attack wants to recover the plaintext $m(x)$. This is the so called broadcast attacks.

The authors in [6,14] gave lower bounds on n in order to make the broadcast attacks work well, we present their results in Table 3.

Table 3. The least number of ciphertexts for success of broadcast attacks

Number of ciphertexts	NTRU-1998	NTRU-2001	NTRU-2005
n [6]	$\frac{N^2+11}{6}$	$\frac{N+1}{2}$	$\frac{N+1}{2}$
n [14]	$\frac{3N-3}{2}$	$\frac{3N-5}{2}$	$\frac{3N-5}{2}$

In order to prevent this attack, we give the constraints on the number n of group members about this improved group key agreement as follows:

$$n < \begin{cases} \frac{3N-3}{2}, & \text{NTRU-1998,} \\ \frac{N+1}{2}, & \text{NTRU-2001,} \\ \frac{N+1}{2}, & \text{NTRU-2005,} \end{cases}$$

where prime N is a parameter of NTRU public key cryptosystem.

6 Conclusion

In this paper, we proposed an efficient ciphertext-only attack on a NTRU-based group agreement. Our attack can recover the plaintext even when there are only two group members. In order to overcome this disadvantage, we put forward an improved construction. And finally, we presented detailed analyses on the security of this improved scheme.

Acknowledgements. The authors would like to thank anonymous reviewers for their helpful comments and suggestions. The work of this paper was supported by the National Key Basic Research Program of China (Grants 2013CB834203), the National Natural Science Foundation of China (Grants 61472417, 61472415 and 61502488), the Strategic Priority Research Program of Chinese Academy of Sciences under Grant XDA06010702, and the State Key Laboratory of Information Security, Chinese Academy of Sciences.

References

1. Security innovations NTRUEncrypt adopted as X9 standard for data protection. Businesswire.com. Accessed 7 Dec 2014
2. Askoxylakis, I., Sauveron, D., Markantonakis, K., Tryfonas, T., Traganitis, A.: A body-centered cubic method for key agreement in dynamic mobile ad hoc networks. In: Second International Conference on Emerging Security Information, Systems and Technologies, SECURWARE 2008, pp. 193–202, August 2008
3. Bosma, W., Cannon, J., Playoust, C.: The magma algebra system I: the user language. J. Symbolic Comput. **24**(3–4), 235–265 (1997)
4. Burmester, M., Desmedt, Y.G.: A secure and efficient conference key distribution system. In: De Santis, A. (ed.) EUROCRYPT 1994. LNCS, vol. 950, pp. 275–286. Springer, Heidelberg (1995)

5. Diffie, W., Hellman, M.: New directions in cryptography. IEEE Trans. Inf. Theory **22**(6), 644–654 (1976)
6. Ding, J., Pan, Y., Deng, Y.: An algebraic broadcast attack against NTRU. In: Susilo, W., Mu, Y., Seberry, J. (eds.) ACISP 2012. LNCS, vol. 7372, pp. 124–137. Springer, Heidelberg (2012)
7. Hoffstein, J., Pipher, J., Silverman, J.H.: NTRU: a ring-based public key cryptosystem. In: Buhler, J.P. (ed.) ANTS 1998. LNCS, vol. 1423, pp. 267–288. Springer, Heidelberg (1998)
8. Hoffstein, J., Silverman, J.: Optimizations for NTRU. Technical report, NTRU Cryptosystems (2001)
9. Howgrave-Graham, N., Silverman, J.H., Whyte, W.: Choosing parameter sets for NTRUEncrypt with NAEP and SVES-3. In: Menezes, A. (ed.) CT-RSA 2005. LNCS, vol. 3376, pp. 118–135. Springer, Heidelberg (2005)
10. Ingemarsson, I., Tang, D., Wong, C.: A conference key distribution system. IEEE Trans. Inf. Theory **28**(5), 714–720 (1982)
11. Joux, A.: A one round protocol for tripartite Diffie-Hellman. J. Cryptology **17**(4), 263–276 (2004)
12. Kim, Y., Perrig, A., Tsudik, G.: Communication-efficient group key agreement. In: Proceedings of the 16th International Conference on Information Security: Trusted Information: The New Decade Challenge, SEC 2001, Norwell, MA, USA, pp. 229–244. Kluwer Academic Publishers (2001)
13. Kim, Y., Perrig, A., Tsudik, G.: Tree-based group key agreement. ACM Trans. Inf. Syst. Secur. **7**(1), 60–96 (2004)
14. Li, J., Pan, Y., Liu, M., Zhu, G.: An efficient broadcast attack against NTRU. In: Proceedings of the 7th ACM Symposium on Information, Computer and Communications Security, ASIACCS 2012, pp. 22–23. ACM, New York (2012)
15. Li-ping, Z., Guo-Hua, C., Zhi-Gang, Y.: An efficient group key agreement protocol for ad hoc networks. In: 4th International Conference on Wireless Communications, Networking and Mobile Computing, WiCOM 2008, pp. 1–5, October 2008
16. Lv, X., Li, H.: Secure group communication with both confidentiality and non-repudiation for mobile ad-hoc networks. IET Inf. Secur. **7**(2), 61–66 (2013)
17. Lv, X., Li, H., Wang, B.: Group key agreement for secure group communication in dynamic peer systems. J. Parallel Distrib. Comput. **72**(10), 1195–1200 (2012)
18. Mol, P., Yung, M.: Recovering NTRU secret key from inversion oracles. In: Cramer, R. (ed.) PKC 2008. LNCS, vol. 4939, pp. 18–36. Springer, Heidelberg (2008)
19. Whyte, W., Howgrave-Graham, N., Hoffstein, J., Pipher, J., Silverman, J.H., Hirschhorn, P.S.: IEEE p. 1363.1 draft 10: Draft standard for public key cryptographic techniques based on hard problems over lattices. IACR Cryptology ePrint Archive 2008/361 (2008)
20. Wu, Q., Mu, Y., Susilo, W., Qin, B., Domingo-Ferrer, J.: Asymmetric group key agreement. In: Joux, A. (ed.) EUROCRYPT 2009. LNCS, vol. 5479, pp. 153–170. Springer, Heidelberg (2009)

Enhanced Correlation Power Analysis by Biasing Power Traces

Changhai Ou[1,2]($\boxtimes$), Zhu Wang[1]($\boxtimes$), Degang Sun[1], Xinping Zhou[1,2], Juan Ai[1,2], and Na Pang[1,2]

[1] Institute of Information Engineering, Chinese Academy of Sciences, Beijing, China
{ouchanghai,wangzhu,sundegang,zhouxinping,aijuan,pangna}@iie.ac.cn
[2] University of Chinese Academy of Sciences, Beijing, China

Abstract. Biasing power traces with high Signal to Noise Ratio (SNR) proposed by K. Yongdae et al. can significantly improve the efficiency of the CPA. But it is still a problem to be solved that how to efficiently select power traces with high SNR. Through the analysis of the statistical characteristics of power traces, we propose three methods to better solve this problem in this paper. We bias power traces by using the Minkowski distance (i.e. Euclidean distance or Manhattan distance) between each power trace and mean power trace. Biasing power traces can also be carried out by using probability density function values of power consumption of interesting points, or even directly using power consumption of interesting points. Our schemes can blindly select power traces with high SNR in a high probability. The efficiency of the CPA by using the three of our methods is significantly improved. Thus, our schemes are more effective compared to the one proposed by K. Yongdae et al.

Keywords: Side channel · CPA · Power trace · Biasing power trace · Success rate

1 Introduction

There exist side channel leakages when cryptographic devices are in operation, such as power consumption [12], electromagnetic [1], acoustic [8], et al. The attacker can successfully recover the key used in the encryption algorithm by using these information.

In 1996, Kocher et al. made full use of timing information leaked from cryptographic devices and successfully recovered the key of RSA [13]. They proposed Simple Power Analysis (SPA) and Differential Power Analysis (DPA) in 1999 [12]. Side channel attacks, such as Template Attack [7], Correlation Power Analysis (CPA) [6], Mutual Information Analysis (MIA) [9], Collision Attack [16] and side channel based watermarks [3], developed quickly then. Side channel attacks, pose serious threats to the security of cryptographic devices.

© Springer International Publishing Switzerland 2016
M. Bishop and A.C.A. Nascimento (Eds.): ISC 2016, LNCS 9866, pp. 59–72, 2016.
DOI: 10.1007/978-3-319-45871-7_5

1.1 Related Works

The idea of biasing power traces was firstly proposed by K. Tiri et al. [19]. The power consumption of the interesting point most relevant to the outputs of S-box can be approximated by a normal distribution [11]. K. Yongdae et al. made full use of this characteristic [11]. By calculating the probability density function values of power consumption of this interesting point, they obtained the distribution. The smaller the probability density function value is, the higher the SNR of the power trace will be. However, the result of the calculation is the only reference to bias power traces. The power consumption of an interesting point on a power trace may not have a significant linear correlation with its Hamming weight of S-box output since the presence of noise. Suppose that an encryption algorithm has k bit-length S-box output. Hamming weights close to 0 or k may correspond to power traces with greater probability density function values of power consumption of interesting points. Accuracy of biasing power traces is greatly reduced if noise on the power traces is large.

B. Noura et al. used a new power model to bias power traces and improved the efficiency of the CPA [4]. They did not make improvements to the methods of choosing power traces, but instead of improving the power consumption model. The scheme is not perfect.

H. Wenjing et al. proposed an adaptive chosen plaintext correlation power analysis [10]. They solved the problem of discarding too many power traces in the scheme proposed by K. Yongade et al. [11]. However, this scheme was very complex.

1.2 Our Contributions

In this paper, we analyze the statistical characteristics of power traces and propose three methods to bias power traces which we will introduce in Sect. 3:

1. Minkowski distance. We analyze the correlation between Hamming weights of the outputs of S-box and power consumption of interesting points, and we get a conclusion that Hamming weights 0 and 8 correspond to power traces with highest or lowest power consumption. We can calculate the Minkowski distance between each power trace and mean power trace to estimate the corresponding Hamming weight of a power trace.

2. The sum of probability density function values of power consumption of interesting points. The power consumption of an interesting point can be approximated by a normal distribution. Power traces with high SNR are located in two ends of the distribution. We use probability density function values of power consumption of interesting points to bias power traces.

3. Power consumption of interesting points. The method of biasing power traces by directly adding up power consumption of interesting points on each power trace and then sorting the sums also works very well according to the approximated normal distribution.

Not only the feasibility but also the high efficiency of our schemes are proved and verified in this paper.

1.3   Organization

This paper is organized as follows. The statistical characteristics of power traces are given in Sect. 2. In Sect. 3, we introduce the three of our methods to bias power traces. Then, in Sect. 4, experiments are carried out to compare our schemes with the scheme proposed by K. Yongdae et al. Finally, we conclude this paper in Sect. 5.

2   The Statistical Characteristics of Power Traces

2.1   Composition of Power Consumption

The power consumption of each interesting point of a power trace can be modeled as the sum of exploitable power consumption P_{exp}, the noise component P_{noise}, and the constant component P_{const} [14].

$$P_{total} = P_{exp} + P_{noise} + P_{const} \tag{1}$$

The component P_{exp} corresponds to the sum of the operation-dependent component P_{op} and the data-dependent component P_{data}:

$$P_{exp} = P_{op} + P_{data}. \tag{2}$$

The component P_{noise} corresponds to the sum of the switching noise $P_{sw.noise}$ and the electronic noise $P_{el.noise}$:

$$P_{noise} = P_{el.noise} + P_{sw.noise}. \tag{3}$$

The switching noise $P_{sw.noise}$ is not exploitable noise and it is not electronic noise. If we consider all bits of the outputs of a S-box, the switching noise will not exist. Referring to [10,14], the SNR of an interesting point of a power trace is

$$SNR = \frac{var(P_{exp})}{var(P_{noise})}. \tag{4}$$

S. Mangard et al. proposed that the $var(P_{exp})$ and $var(P_{noise})$ quantified how much an interesting point of power traces varied because of the exploitable signal and noise respectively [14]. To continue the work that S. Mangard et al. have not finished in their paper, we derive the formula (4). Because P_{op}, P_{data}, $P_{sw.noise}$ and $P_{el.noise}$ are independent of each other, the variance

$$var(P_{exp}) = var(P_{op} + P_{data}) = var(P_{data}) + var(P_{op}), \tag{5}$$

and the variance

$$var(P_{noise}) = var(P_{sw.noise} + P_{el.noise}) = var(P_{sw.noise}) + var(P_{el.noise}). \tag{6}$$

Suppose that we use all bits of the output of a S-box of an encryption algorithm, then the variance of switching noise $var(P_{sw.noise}) = 0$. If we consider the same interesting point on different power traces, the same operation (i.e. looking-up table S-box) is executed. So, the variance $var(P_{op}) = 0$. Then,

$$SNR = \frac{var(P_{op} + P_{data})}{var(P_{sw.noise} + P_{el.noise})}, \tag{7}$$

then

$$SNR = \frac{var(P_{data})}{var(P_{el.noise})}. \tag{8}$$

The electronic noise follows normal distribution $N(0, \sigma_{el.noise})$. The greater (or smaller) the Hamming weight of the output of S-box is, the larger the variance $var(P_{data})$ will be, and so is the SNR.

2.2 Correlation Between SNR and Power Consumption of an Interesting Point

Suppose that the attacker encrypts m random known plaintexts and acquires m power traces using an oscilloscope when a cryptographic device is in operation. We also assume that there are n interesting points on each power trace. Then, the attacker obtains an $m \times n$ power consumption matrix. Each point consists of the sum of exploitable power consumption P_{exp}, the noise component P_{noise}, and the constant component P_{const}.

We only consider serial processor in this paper. Let P_{total}^{j} donate the power consumption of the jth interesting point of power traces. Then, the correlation coefficient between P_{total}^{j} and P_{data}^{j} is

$$\rho(P_{data}^{j}, P_{total}^{j}) = \frac{cov(P_{data}^{j}, P_{total}^{j})}{\sqrt{var(P_{data}^{j})var(P_{total}^{j})}}. \tag{9}$$

Let $var(P_{data}^{j})$ denote the variance of P_{data}^{j} and $var(P_{total}^{j})$ denote the variance of P_{total}^{j}. Then,

$$
\begin{aligned}
cov(P_{data}^{j}, P_{total}^{j}) &= cov(P_{data}^{j}, P_{data}^{j} + P_{op}^{j} + P_{el.noise}^{j} + P_{const}^{j}) \\
&= E\Big[\big(P_{data}^{j} + P_{op}^{j} + P_{el.noise}^{j} + P_{const}^{j} - E(P_{data}^{j} + P_{op}^{j} \\
&\quad + P_{el.noise}^{j} + P_{const}^{j})\big)\big(P_{data}^{j} - E\left(P_{data}^{j}\right)\big)\Big] \\
&= E\Big[\big(P_{op}^{j} - E\left(P_{op}^{j}\right) + P_{data}^{j} - E\left(P_{data}^{j}\right) + P_{el.noise}^{j} \\
&\quad - E\left(P_{el.noise}^{j}\right) + P_{const}^{j} - E\left(P_{const}^{j}\right)\big)\big(P_{data}^{j} - E\left(P_{data}^{j}\right)\big)\Big].
\end{aligned}
\tag{10}
$$

For each interesting point, the operation is exactly the same. So, $P_{op}^{j} - E\left(P_{op}^{j}\right) = 0$. The constant component of this column is equal to $E\left(P_{const}^{j}\right)$. So, $P_{const}^{j} - E\left(P_{const}^{j}\right) = 0$. Suppose that the noise component follows normal distribution $N(0, \sigma_{el.noise})$, and data-dependent component P_{data}^j follows binomial distribution which can be approximated by a normal distribution. Then,

$$
\begin{aligned}
cov\left(P_{data}^j, P_{total}^j\right) &= E\left[\left(P_{data}^j - E\left(P_{data}^j\right) + P_{el.noise}^j - E\left(P_{el.noise}^j\right)\right)\right.\\
&\qquad\left.\left(P_{data}^j - E\left(P_{data}^j\right)\right)\right]\\
&= E\left[\left(P_{data}^j - E\left(P_{data}^j\right)\right)\left(P_{data}^j - E\left(P_{data}^j\right)\right)\right.\\
&\qquad\left. + \left(P_{el.noise}^j - E\left(P_{el.noise}^j\right)\right)\left(P_{data}^j - E\left(P_{data}^j\right)\right)\right]\\
&= \sigma^2 + E\left[\left(P_{el.noise}^j - 0\right)\left(P_{data}^j - E\left(P_{data}^j\right)\right)\right]\\
&= \sigma^2 + E\left[\left(P_{el.noise}^j - 0\right)\right] E\left[\left(P_{data}^j - E\left(P_{data}^j\right)\right)\right]\\
&= \sigma^2.
\end{aligned}
\tag{11}
$$

The variance

$$
var\left(P_{total}^j\right) = var\left(P_{op}^j + P_{data}^j + P_{el.noise}^j + P_{const}^j\right).
\tag{12}
$$

Because P_{data}^j, P_{op}^j, $P_{el.noise}^j$ and P_{const}^j are independent of each other, the variance

$$
var\left(P_{total}^j\right) = var\left(P_{op}^j\right) + var\left(P_{data}^j\right) + var\left(P_{el.noise}^j\right) + var\left(P_{const}^j\right).
\tag{13}
$$

Because of the variance $var\left(P_{op}^j\right) = 0$ and $var\left(P_{const}^j\right) = 0$. Then,

$$
var\left(P_{total}^j\right) = var\left(P_{data}^j\right) + var\left(P_{el.noise}^j\right) = \sigma^2 + \sigma_{el.noise}^2.
\tag{14}
$$

Then,

$$
\rho(P_{data}^j, P_{total}^j) = \frac{\sigma}{\sqrt{\sigma^2 + \sigma_{el.noise}^2}}.
\tag{15}
$$

Then, the correlation coefficient between P_{data}^j and P_{total}^j will be

$$
\rho(P_{data}^j, P_{total}^j) = \frac{1}{\sqrt{1 + \frac{\sigma_{el.noise}^2}{\sigma^2}}} = \frac{1}{\sqrt{1 + \frac{1}{SNR}}}.
\tag{16}
$$

From the above formula (16), we know that, the higher the SNR, the higher the correlation coefficient between P_{data} and P_{total}. If we consider a column of power consumption matrix, the correlation coefficient $\rho(P_{data}, P_{total})$ between P_{data} and P_{total} will be higher after biasing power traces. So, selecting power traces with high SNR is conducive to improve the efficiency of the CPA.

3   Our Methods to Bias Power Traces

If a cryptographic algorithm has k bit-length S-box output, then the number of output Hamming weights follow binomial distribution (as shown in Table 1). Let $T = C_k^N$ denote the number of S-box outputs with a Hamming weight of N. The power component P_{data} of an interesting point of S-box outputs follows binomial distribution.

Table 1. The binomial distribution of Hamming weights of S-box outputs

HW	0	1	2	...	k-2	k-1	k
T	$C_k^{N=0}$	$C_k^{N=1}$	$C_k^{N=2}$	...	$C_k^{N=k-2}$	$C_k^{N=k-1}$	$C_k^{N=k}$

In the next three sub-sections, we will introduce 4 metric methods to bias power traces.

3.1   Biasing Power Traces by Using Minkowski Distance

Suppose that the power consumption of a cryptographic device can be modeled using Hamming weight model. The Hamming weights of S-box outputs being close to k or 0 will correspond to power traces with higher SNR than those being close to $k/2$.

Suppose that the Hamming weights of S-box outputs corresponding to power traces T_a and T_b are HW_a and HW_b respectively. Suppose that there are k interesting points on the power traces. Let $P^{(a,j)}$ and $P^{(b,j)}$ donate the data-dependent power consumption component of the jth interesting points of power trace a and b respectively. The absolute value of HW_a (or HW_b) minus the mean of Hamming weights HW_{mean} is $|HW_a - HW_{mean}|$ (or $|HW_b - HW_{mean}|$).

Condition 1: If HW_a, HW_b and HW_{mean} meet $HW_a \geq HW_b \geq HW_{mean}$ or $HW_a \leq HW_b \leq HW_{mean}$, then $|HW_a - HW_{mean}| \geq |HW_b - HW_{mean}|$, then $\left| P^{(a,j)} - P_{mean}^j \right| \geq \left| P^{(b,j)} - P_{mean}^j \right|$, then $\sum_{j=1}^k \left| P^{(a,j)} - P_{mean}^j \right| \geq \sum_{j=1}^k \left| P^{(b,j)} - P_{mean}^j \right|$.

Condition 2: $HW_a \leq HW_b$ and $HW_b \geq HW_{mean}$ (or $HW_a \geq HW_b$ and $HW_b \leq HW_{mean}$). If $|HW_a - HW_{mean}| \geq |HW_b - HW_{mean}|$, then we get $\left| P^{(a,j)} - P_{mean}^j \right| \geq \left| P^{(b,j)} - P_{mean}^j \right|$. Then $\sum_{j=1}^k \left| P^{(a,j)} - P_{mean}^j \right| \geq \sum_{j=1}^k \left| P^{(b,j)} - P_{mean}^j \right|$ still be supported.

Biasing power traces can be done in the above two conditions. Power traces corresponding to Hamming weights being close to 0 or k will be preferentially selected.

Suppose that we use t interesting points to profile templates. Let $P^{(m,j)}$ denote data-dependent power consumption component of the jth interesting point on the mth power trace, and $\mu_{(j)}$ denote the mean power consumption of the jth interesting point. Then, we can measure the distance between the mth power trace and the mean power trace by using Manhattan distance

$$Distance_{(Manh,m)} = \sum_{j=1}^{n} \left| P^{(m,j)} - \mu_{(j)} \right|. \tag{17}$$

Since $\left| P^{(a,j)} - P^j_{mean} \right| \geq \left| P^{(b,j)} - P^j_{mean} \right|$. So, $\left(P^{(a,j)} - P^j_{mean} \right)^2 \geq \left(P^{(b,j)} - P^j_{mean} \right)^2$, then we can get a conclusion that $\sum_{j=1}^{t} \left(P^{(a,j)} - P^j_{mean} \right)^2 \geq \sum_{j=1}^{t} \left(P^{(b,j)} - P^j_{mean} \right)^2$. Absolutely, we can also measure the distance between the mth power trace and the mean power trace by using Euclidean distances

$$Distance_{(Euc,m)} = \sqrt{\sum_{j=1}^{t} \left(P^{(m,j)} - \mu_{(j)} \right)^2}. \tag{18}$$

Actually, Manhattan distance and Euclidean distance are the first and second order of Minkowski distance respectively. Higher order Minkowski distance

$$Distance_{(Mink,m)} = \sqrt[q]{\sum_{j=1}^{t} \left(P^{(m,j)} - \mu_{(j)} \right)^q} \tag{19}$$

can also be used to measure the distance between the mth power trace and the mean power trace.

3.2 Biasing Power Traces by Using the Sum of Probability Density Function Values of Power Consumption of Interesting Points

Let $\sigma^2_{(j)}$ denote the variance of the jth interesting point. If HW_a, HW_b and HW_{mean} satisfy **Condition 1** or **Condition 2** in the Sect. 3.1, then $\left(P^{(a,j)} - P^j_{mean} \right)^2 \geq \left(P^{(b,j)} - P^j_{mean} \right)^2$. Let

$$f(a,j) = \frac{1}{\sqrt{2\pi\sigma^2_{(j)}}} \left\{ \frac{-\left(P^{(a,j)} - P^j_{mean} \right)^2}{2\sigma^2_{(j)}} \right\} \tag{20}$$

and

$$f(b,j) = \frac{1}{\sqrt{2\pi\sigma^2_{(j)}}} \left\{ \frac{-\left(P^{(b,j)} - P^j_{mean} \right)^2}{2\sigma^2_{(j)}} \right\} \tag{21}$$

denote the probability density function value of data-dependent power consumption component of jth interesting points of power trace a and power trace b respectively. Then,

$$f(a,j) \leq f(b,j) \tag{22}$$

If there are n interesting points on the power traces, then

$$\sum_{j=1}^{n} f(a,j) \leq \sum_{j=1}^{n} f(b,j) \tag{23}$$

still be supported. So, power traces with high SNR can still be filtered out.

We can measure the distance between the mth power trace and the mean power consumption trace by using the sum of probability density function (PDF) values

$$Distance_{(PDF,m)} = \sum_{j=1}^{n} \frac{1}{\sqrt{2\pi\sigma_{(j)}^2}} exp\left\{\frac{-\left(P^{(m,j)} - \mu_{(j)}\right)^2}{2\sigma_{(j)}^2}\right\}. \tag{24}$$

It is worth noting that, the greater the sum of probability density function (PDF) values is, the closer to $\mu_{(j)}$ the power consumption $P^{(m,j)}$ is. In order to bias power traces with high SNR, we calculate a $Distance_{(PDF,m)}$ for each power trace, then sort the sums in ascending order. Smaller sums correspond to power traces with higher SNR.

3.3   Biasing Power Traces by Directly Using Power Consumption of Interesting Points

Referring to [11], the power consumption of an interesting point can be approximated by a normal distribution. That is to say, the Hamming weight 0 and k correspond to power traces with the highest or lowest power consumption. We can directly sort the power consumption of an interesting point on the power traces, then extract a small number of power traces from two ends of the distribution. SNR of power traces on both these two ends is generally high. We add up all data-dependent power consumption component $P_{(m,j)}$ $(1 \leq j \leq n)$ of interesting points for each power trace and then sort the sums. Thus,

$$Distance_{(Power,m)} = \sum_{j=1}^{n} P_{(m,j)}. \tag{25}$$

Compared to the previous three methods, this scheme is very simple.

3.4   How to Bias Power Traces Using the Above 4 Schemes

We only consider the data-dependent power consumption component of power traces when we introduce the four of our schemes to bias power traces. If we repeatedly encrypt a plaintext and acquire a number of power traces. We average these power traces to remove the noise. Through the above two kinds of measurements using Euclidean distance and Manhattan distance respectively, we can measure the distance between the mth power trace and the mean power trace. Greater distances correspond to power traces with higher SNR. Biasing power traces can also be done by using the sum of probability density function values, greater sums correspond to power traces with higher SNR, too.

However, if we want to bias power traces using the sum of power consumption of interesting points, we should select power traces from two ends of the approximated normal distribution. Hamming weights 0 and 8 correspond to power traces with the greatest or smallest power consumption on their interesting points. Biasing power traces by using the above 4 metric methods can improve correlation coefficients, so as to improve the efficiency of the CPA.

If we encrypt each random known plaintext only once and acquire only one power trace, this power trace may contain a lot of noise. The power consumption of interesting points is no longer in linear relationship with their corresponding S-box output Hamming weights. This will result in relatively low attack efficiency. However, our schemes can still be utilized.

4   Experimental Results

We carry out our experiments on AT89S52 micro-controller. The clock frequency of this chip is 12 MHz. The minimum instructions takes 12 clock cycles for execution. We utilize Tektronix DPO 7254 oscilloscope, and the sampling rate is set to 1 GS/s. The outputs of the first S-box in the first round of AES encryption is chosen as the attack point. We test the instruction '$MOVCA, @A + DPTR$', which treats the value of register A as the offset value of S-box and treats the S-box of AES algorithm as the base address, then looks up table S-box and writes the result back to register A.

4.1   Interesting Points Extraction

So far, there exist various kinds of methods to extract interesting points. For example, C. Rechberger et al. extracted interesting points by adding up all deviations of every two power traces [15]. D. Agraval et al. used DPA to extract interesting points [2]. L. Bohy et al. used PCA to extract interesting points [5]. F. X. Standaert et al. extracted interesting points by using FLDA [17]. We randomly encrypt 100 known plaintexts and acquire 1000 power traces. We calculate the Pearson correlation coefficients between these power traces and their corresponding S-box outputs' Hamming weights, which is equivalent to $\rho(P_{data}, P_{total})$ (as shown in Fig. 1. Then we sort the correlation coefficients by descending order and choose the maximum 100 interesting points in the front of the sorted sequence.

In Sects. 4.2 and 4.3, we will compare our schemes with the scheme proposed by K. Yongdae et al. from two aspects: CRS which we will introduce in Sect. 4.2 and success rates (SR) proposed by F. X. Standaert et al. [18] which we will introduce in Sect. 4.3. Specific comparison will be given in Sects. 4.2 and 4.3.

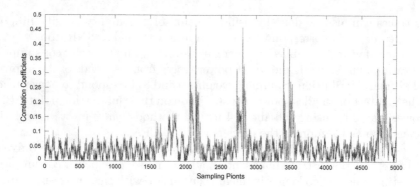

Fig. 1. Correlation coefficients between S-box output Hamming weights and each column of power consumption matrix.

4.2 Comparison of Screening Accuracy

Suppose that we encrypt n random known plaintexts. The total number of S-box outputs' Hamming weights 0, 1, 7 and 8 of the first S-box is k. We sort the sums by using a scheme (i.e. Euclidean distance) and select k power traces which satisfy the selection conditions of this scheme. Suppose that we find out t power traces corresponding to S-box outputs' Hamming weights 0, 1, 7, or 8 from these k power traces. Then, we define correct rates of selection as

$$CRS = \frac{t}{k} \tag{26}$$

If we consider 30 interesting points and choose the 30 points most relevant to S-box outputs. We encrypt 700 random known plaintexts. The correct rates of selection of all 16 S-boxes are shown in Fig. 2. Both the scheme proposed by K. Yongdae et al. [11] and our schemes are used.

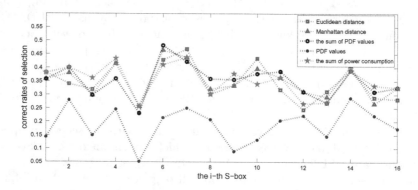

Fig. 2. The CRS of all 16 S-boxes outputs' Hamming weights 0, 1, 7 and 8.

As shown in Fig. 2, by using 30 interesting points, the four schemes proposed by us have higher correct rates of selection of all 16 S-boxes than the scheme proposed by Kongdae et al. [11]. We also compare our schemes with the scheme proposed by K. Yongdae et al. by using different number of power traces. The experimental results by using the same 30 interesting points are shown in Fig. 3. The correct rates of selection by using Euclidean distance, Manhattan Distance, the sum of probability density function (PDF) values and the sum of power consumption under 30 interesting points are about 15 % higher than that of considering the probability density function value of only one interesting point corresponding to the scheme of K. Yongdae et al.

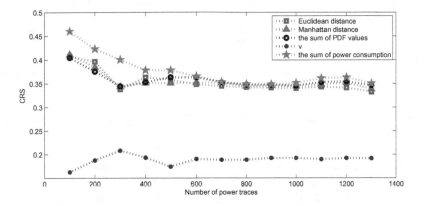

Fig. 3. CRS under different number of power traces by using the four schemes of ours and the scheme proposed by K. Yongdae et al.

We also compare the CRS in our schemes with the scheme proposed by K. Yongdae et al. [11] by using different number of interesting points. We choose the 100 points most relevant to S-box outputs and sort the correlation coefficients of these points in descending order. We then extract 1, 11, 21,..., 91 interesting points respectively. By calculating the average CRS of 16 S-boxes under 700 power traces, we get the experimental results as shown in Fig. 4.

As shown in Fig. 4, if we extract more interesting points, we will get higher CRS when the number of interesting points is less than or equal to 21. When the number of interesting points is greater than 21, the correct rates will be very close to each other. Compared with the scheme proposed by K. Yongdae et al. [11] corresponding to the sum of probability density function values where the number of interesting points is equal to 1 in Fig. 4, the CRS in the four schemes proposed by us have been significantly improved.

4.3 Comparison of Success Rates

We also utilize success rate (SR) proposed by F. X. Standaert et al. [18] to evaluate the attack efficiency in our experiments. Bying using the traditional

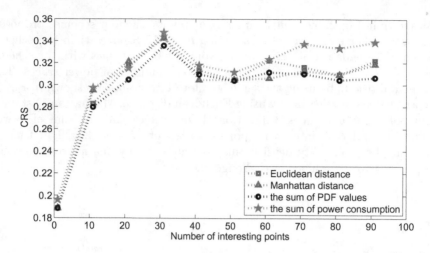

Fig. 4. The average CRS of 16 S-boxes by using different number of interesting points

CPA, we can recover the key used in AT89S52 with a probability of about 1.00 when the number of power traces reaches 136. We encrypt 20000 random known plaintexts and acquire 20000 power traces. We randomly choose 136 power traces to perform CPA attacks by using the four of our schemes and scheme proposed by K. Yongdae et al. [11]. The output of the first S-box is chosen as the attack point. By repeating 100 times using the 30 interesting points most relevant to S-box outputs, the experimental results are shown in Fig. 5.

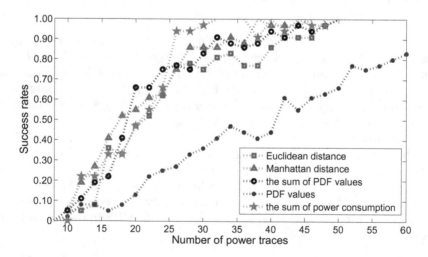

Fig. 5. Success rates by using the four schemes proposed by us and the scheme proposed by Kongdae et al.

As shown in Fig. 5, we get higher success rates by using Euclidean distance, Manhattan distance and the sum of probability density function values proposed by us than the scheme proposed by K. Yongdae et al. [11]. Our method of directly using the sum of power consumption of interesting points also gets very good attack efficiency.

5 Conclusion

In this paper, we analyze the statistical characteristics of power traces, and the relationship between SNR and S-box outputs' Hamming weight in serial processors. We use Minkowski distance, the sum of probability density function values of power consumption of interesting points, power consumption of interesting points to bias power traces. Compared with the scheme proposed by K. Yongdae et al., the three of our schemes can blindly screen power traces with high SNR with a higher CRS of 15 %. Thus, significantly improve the efficiency of CPA. Signal preprocessing can improve the accuracy of screening, which is the content of our future research. Better and more effective methods to bias power traces are still our research goal in the future.

Acknowledgment. This research is supported by the Nation Natural Science Foundation of China (No. 61372062). The authors would like to thank the anonymous referees of ISC 2016 for the suggestions to improve this paper.

References

1. Agrawal, D., Archambeault, B., Rao, J.R., Rohatgi, P.: The EM side—channel(s). In: Kaliski Jr., B.S., Koç, Ç.K., Paar, C. (eds.) CHES 2002. LNCS, vol. 2523. Springer, Heidelberg (2003)
2. Agrawal, D., Rao, J.R., Rohatgi, P., Schramm, K.: Templates as master keys. In: Rao, J.R., Sunar, B. (eds.) CHES 2005. LNCS, vol. 3659, pp. 15–29. Springer, Heidelberg (2005)
3. Becker, G.T., Kasper, M., Moradi, A., Paar, C.: Side-channel based watermarks for integrated circuits. In: IEEE International Symposium on Hardware-Oriented Security and Trust, pp. 30–35 (2010)
4. Benhadjyoussef, N., Machhout, M., Tourki, R.: Optimized power trace numbers in CPA attacks. In: 2011 8th International Multi-Conference on Systems, Signals and Devices (SSD), pp. 1–5 (2011)
5. Bohy, L., Neve, M., Samyde, D., Quisquater, J.J.: Principal and independent component analysis for crypto-systems with hardware unmasked units. In: Proceedings of e-Smart 2003 (2003)
6. Brier, E., Clavier, C., Olivier, F.: Correlation power analysis with a leakage model. In: Joye, M., Quisquater, J.-J. (eds.) CHES 2004. LNCS, vol. 3156, pp. 16–29. Springer, Heidelberg (2004)
7. Chari, S., Rao, J.R., Rohatgi, P.: Template attacks. In: Kaliski Jr., B.S., Koç, Ç.K., Paar, C. (eds.) CHES 2002. LNCS, vol. 2523. Springer, Heidelberg (2003)
8. Genkin, D., Shamir, A., Tromer, E.: Acoustic cryptanalysis. J. Cryptol. 1–52 (2016). doi:10.1007/s00145-015-9224-2

9. Gierlichs, B., Batina, L., Tuyls, P., Preneel, B.: Mutual information analysis. In: Oswald, E., Rohatgi, P. (eds.) CHES 2008. LNCS, vol. 5154, pp. 426–442. Springer, Heidelberg (2008)

10. Hu, W., Wu, L., Wang, A., Xie, X., Zhu, Z., Luo, S.: Adaptive chosen-plaintext correlation power analysis. In: Tenth International Conference on Computational Intelligence and Security, pp. 494–498 (2014)

11. Kim, Y., Sugawara, T., Homma, N., Aoki, T., Satoh, A.: Biasing power traces to improve correlation in power analysis attacks. ESRC Centre Population Change **2**(3), 10–16 (2014)

12. Kocher, P., Jaffe, J., Jun, B.: Differential power analysis. Int. Cryptol. Conf. Adv. Cryptol. **1666**, 388–397 (1999)

13. Kocher, P.C.: Timing attacks on implementations of Diffie-Hellman, RSA, DSS, and other systems. Int. Cryptol. Conf. Adv. Cryptol. **1109**, 104–113 (2010)

14. Mangard, S., Oswald, E., Popp, T.: Power Analysis Attacks: Revealing the Secrets of Smart Cards. Springer, New York (2007)

15. Rechberger, C., Oswald, E.: Practical template attacks. In: Lim, C.H., Yung, M. (eds.) WISA 2004. LNCS, vol. 3325, pp. 440–456. Springer, Heidelberg (2005)

16. Schramm, K., Wollinger, T., Paar, C.: A new class of collision attacks and its application to DES. Fast Softw. Encryp. FSE **2887**(6), 206–222 (2003)

17. Standaert, F.-X., Archambeau, C.: Using subspace-based template attacks to compare and combine power and electromagnetic information leakages. In: Oswald, E., Rohatgi, P. (eds.) CHES 2008. LNCS, vol. 5154, pp. 411–425. Springer, Heidelberg (2008)

18. Standaert, F.-X., Malkin, T.G., Yung, M.: A unified framework for the analysis of side-channel key recovery attacks. In: Joux, A. (ed.) EUROCRYPT 2009. LNCS, vol. 5479, pp. 443–461. Springer, Heidelberg (2009)

19. Tiri, K., Schaumont, P.: Changing the odds against masked logic. In: Biham, E., Youssef, A.M. (eds.) SAC 2006. LNCS, vol. 4356, pp. 134–146. Springer, Heidelberg (2007)

Damaging, Simplifying, and Salvaging p-OMD

Tomer Ashur[1,2(✉)] and Bart Mennink[1,2]

[1] Department of Electrical Engineering, ESAT/COSIC, KU Leuven, Leuven, Belgium
{tomer.ashur,bart.mennink}@esat.kuleuven.be
[2] iMinds, Ghent, Belgium

Abstract. One of the submissions to the CAESAR competition for the design of a new authenticated encryption scheme is Offset Merkle-Damgård (OMD). At FSE 2015, Reyhanitabar et al. introduced p-OMD, an improvement of OMD that processes the associated data almost for free. As an extra benefit, p-OMD was claimed to offer integrity against nonce-misusing adversaries, a property that OMD does not have. In this work we show how a nonce-misusing adversary can forge a message for the original p-OMD using only 3 queries (including the forgery). As a second contribution, we generalize and simplify p-OMD. This is done via the introduction of the authenticated encryption scheme Spoed. The most important difference is the usage of a generalized padding function GPAD, which neatly eliminates the need for a case distinction in the design specification and therewith allows for a significantly shorter description of the scheme and a better security bound. Finally, we introduce the authenticated encryption scheme Spoednic, a variant of Spoed providing authenticity against a nonce-misusing adversary at a modest price.

Keywords: Authenticated encryption · CAESAR · p-OMD · Nonce-misuse · Forgery · Simplification

1 Introduction

The principle of authenticated encryption, where both the confidentiality as well as the integrity of data is guaranteed has gained renewed attention in the last couple of years. Emerged from this is the CAESAR competition for the design of new authenticated encryption schemes [6]. CAESAR has received 57 submissions, 30 of which have recently advanced to the second round. Many of these designs have already received further attention via attacks, supporting security proofs, or generalizations.

One of the second round candidates of the CAESAR competition is Offset Merkle-Damgård (OMD) by Cogliani et al. [9,10]. It is characterized by the usage of a full-fledged compression function, and in fact the CAESAR submission takes the SHA256 compression function. OMD is proven to achieve birthday-bound security on the state against adversaries that are not allowed to re-use the nonce. At ProvSec 2014, Reyhanitabar et al. [18] showed how to generalize the

© Springer International Publishing Switzerland 2016
M. Bishop and A.C.A. Nascimento (Eds.): ISC 2016, LNCS 9866, pp. 73–92, 2016.
DOI: 10.1007/978-3-319-45871-7_6

scheme to achieve security against nonce-misusing adversaries. On the downside, these schemes are not online and are less efficient than OMD. At FSE 2015 Reyhanitabar et al. [19] presented p-OMD (pure OMD). p-OMD improves over classical OMD in that the associated data is processed almost for free. This is achieved by processing the message blocks as normal message inputs to the compression function and by XORing the associated data into the state. The authors prove that p-OMD inherits all security features of OMD, particularly birthday-bound security against nonce-respecting adversaries. In [20], an early version of [19], it was suggested that p-OMD also offers integrity against nonce-misusing adversaries.

1.1 Nonce-Misuse Forgery on p-OMD (Damaging)

As first contribution of this work, we point out that this claim is incorrect. In more detail, we present a nonce-misusing adversary that can forge a message for p-OMD in only 3 queries, including the forgery itself. At a high level, the attack relies on the observation that if an evaluation for p-OMD is made for a certain nonce, the adversary learns (most of) the corresponding state values. If the adversary is allowed to misuse the nonces, this means that it can effectively influence the state values, and henceforth generate a forgery. We also point out where the mistake occurs in the proof. We stress that this attack *does not* invalidate the security of p-OMD (nor OMD) in the nonce-respecting setting: that proof seems sound and the scheme achieves confidentiality and integrity.

1.2 Speed (Simplifying)

One may argue that the flaw slipped into [20] in part due to the complex character of p-OMD. Indeed, the specification of p-OMD consists of 6 cases (or in fact 13, if you consider the scheme in full detail), depending on the number of associated data and message blocks. The forking of one scheme into a plurality of cases entails difficulties both on the theory side, leading to longer and more cumbersome proofs (which are incidentally harder to verify, as in the case above), and on the practical side, forcing less efficient and error-prone implementations. Additionally, it does not particularly facilitate an easy understanding and adoption of the scheme.

Driven by these conclusions and the potential that p-OMD offers for certain scenarios, we next explore the possibilities to generalize and simplify p-OMD. In more detail, we introduce Speed,[1] a variant of p-OMD that aims to provide a higher level of simplicity at the same efficiency as p-OMD. In more detail, Speed is an authenticated encryption mode that can use any keyed compression function $F_K : \{0,1\}^{2n} \rightarrow \{0,1\}^n$ for $n \geq 1$ as its underlying primitive. Speed differs from p-OMD in the following aspects:

- Most importantly, Speed uses a generalized padding scheme GPAD. It takes as input the associated data A and the message M, and injectively maps those

---

[1] The name is an acronym for "Simplified Pure OMD Encryption and Decryption".

to generalized message blocks of size $2n$ bits. As GPAD includes the length encodings of A and M as one of the generalized message blocks, it allows to give a unified description of the scheme: *one* scheme for all variants;
– p-OMD relies on Gray codes for case separation. Due to the usage of the generalized padding scheme, we can resort to the simpler-to-grasp powering-up approach [21] or word-based LFSR approach by Granger et al. [12]. Note that the usage of these approaches for state sizes larger than 128 bits has only been validated recently [12].

We prove that, assuming F_K is a sufficiently secure keyed compression function, Spoed redeems the security results of p-OMD in the nonce-respecting setting. To instantiate F_K, one can use the SHA256 or SHA512 compression functions. These functions have been the target of extensive cryptanalysis, and their security is well understood. They are also in wide use and efficient implementations of them can be found for practically any platform.

Spoed makes exactly the same number of compression function calls as p-OMD, except in the rare case where $a > m$ and $a + m$ is odd (in which case Spoed makes one extra compression function call due to the length encoding of A and M). We see this as a modest price to pay for achieving a scheme that (i) has a shorter and simpler description, making it easier to implement, (ii) requires less precomputational overhead, and (iii) has a proof of about $1/4$th the size of the proof of p-OMD, making it easier to verify. The fact that Spoed has a slightly improved security bound can be seen as a bonus.

1.3 Spoednic (Salvaging)

For the cases where nonce-misuse resistance is needed, we introduce Spoednic.[2] Spoednic is a variant of Spoed preserving integrity, up to the birthday bound, in the nonce-misuse scenario at the cost of one additional finite field multiplication per primitive call. Intuitively, the finite field multiplication is used to obfuscate the value XORed into the state, thus preventing an adversary from choosing a "convenient" value.

We prove that Spoednic inherits all security traits of Spoed, and has the added benefit of preserving the integrity against a nonce-reusing adversary. Surprisingly, the proof for this case leads to a better security bound than the flawed one claimed for p-OMD [20].

We stress that the reader should not take Spoednic as a recommendation for allowing the nonce to repeat. The question about who is responsible for dealing with the uniqueness of the nonce is debated in the cryptographic community. One side to this discussion believes that making sure the nonce is unique is an implementation matter while the other side believes that it should be dealt with by the algorithms designers. Both sides agree that a repeating nonce is an unwanted scenario, and the contribution of Spoednic is to allow for a graceful fail rather than a disastrous one in this unwanted event.

---

[2] The name is an acronym for "Simplified Pure OMD Encryption and Decryption with Nonce-misuse Integrity Conserved".

2   Security Model

Throughout, $n \geq 1$ denotes the state size. By $\oplus$ we denote the exclusive-or (XOR) operation, and by $\otimes$ or $\cdot$ finite field multiplication over 2^n. Concatenation is denoted using $\|$. Denote by $\{0,1\}^*$ the set of binary strings of arbitrary length and by $\{0,1\}^n$ the set of blocks of size n. Denote by $(\{0,1\}^n)^+$ the set of strings of length a *positive* multiple of n. For an arbitrary string X, $|X|$ denotes its length, and $\langle |X| \rangle_n$ denotes its encoding in $n \geq 1$ bits. By $\text{left}_n(X)$ (resp. $\text{right}_n(X)$) we denote its n leftmost (resp. rightmost) bits. We use little-endian notation, which means the three notations "bit position 0", "the rightmost bit", and "the least significant bit" all refer to the same bit.

In Sect. 2.1, we describe our model for the security of authenticated encryption. Then, we present some theoretical background on keyed compression functions in Sect. 2.2.

2.1   Authenticated Encryption

Let $\Pi = (\mathcal{E}, \mathcal{D})$ be an authenticated encryption scheme, where

$$\mathcal{E} : (K, N, A, M) \mapsto (C, T) \text{ and}$$
$$\mathcal{D} : (K, N, A, C, T) \mapsto M/\perp$$

are the encryption and decryption functions of Π. Let \$ be a random function that returns $(C, T) \xleftarrow{\$} \{0,1\}^{|M|} \times \{0,1\}^\tau$ on every new tuple (N, A, M). In other words, $\mathcal{E}$ and \$ have the same interface, but the latter outputs a uniformly randomly drawn ciphertext and tag for every new input.

An adversary $\mathcal{A}$ is a probabilistic algorithm that has access to one or more oracles $\mathcal{O}$, denoted $\mathcal{A}^{\mathcal{O}}$. By $\mathcal{A}^{\mathcal{O}} = 1$ we denote the event that $\mathcal{A}$, after interacting with $\mathcal{O}$, outputs 1. In below games, the adversaries have oracle access to $\mathcal{E}_K$ or its counterpart \$, and possibly $\mathcal{D}_K$. The key K is randomly drawn from $\{0,1\}^k$ at the beginning of the security experiment. We say that $\mathcal{A}$ is *nonce-respecting* (nr) if it never queries its encryption oracle under the same nonce twice, and *nonce-misusing* (nm) if it is allowed to make multiple encryption queries with the same nonce. The security definitions below follow [1,5,11,13,14].

We define the advantage of $\mathcal{A}$ in breaking the confidentiality of Π as follows:

$$\mathbf{Adv}_{\Pi}^{\text{conf}}(\mathcal{A}) = \left| \mathbf{Pr}\left(K \xleftarrow{\$} \{0,1\}^k, \mathcal{A}^{\mathcal{E}_K} = 1 \right) - \mathbf{Pr}\left(\mathcal{A}^{\$} = 1 \right) \right|.$$

For $n \in \{nr, nm\}$, we denote by $\mathbf{Adv}_{\Pi}^{\text{conf}}(n, q, \ell, \sigma, t)$ the maximum advantage over all n-adversaries that make at most q queries, each of length at most ℓ generalized message blocks and together of length at most σ generalized message blocks, and that run in time t.

For integrity, we consider an adversary that tries to forge a ciphertext, which means that $\mathcal{D}_K$ ever returns a valid message (other than $\perp$) on input (N, A, C, T)

and no previous encryption query $\mathcal{E}_K(N, A, M)$ returned (C, T) for any M. Formally:

$$\mathbf{Adv}_{\Pi}^{\mathrm{int}}(\mathcal{A}) = \mathbf{Pr}\left(K \xleftarrow{\$} \{0, 1\}^k, \mathcal{A}^{\mathcal{E}_K, \mathcal{D}_K} \text{ forges}\right).$$

For $\mathsf{n} \in \{\mathsf{nr}, \mathsf{nm}\}$, we denote by $\mathbf{Adv}_{\Pi}^{\mathrm{int}}(\mathsf{n}, q_{\mathcal{E}}, q_{\mathcal{D}}, \ell, \sigma, t)$ the maximum advantage over all n-adversaries that make at most $q_{\mathcal{E}}$ encryption and $q_{\mathcal{D}}$ decryption queries, each of length at most ℓ generalized message blocks and together of length at most σ generalized message blocks, and that run in time t. We remark that the nonce-respecting condition *only* applies to encryption queries: the adversary is always allowed to make decryption queries for "old" nonces, and to make an encryption query using a nonce which is already used in a decryption query before.

2.2 (Tweakable) Keyed Compression Function

Let $F : \{0, 1\}^k \times \{0, 1\}^{n+m} \to \{0, 1\}^n$ be a keyed compression function. Denote by $\mathrm{Func}(\{0, 1\}^{n+m}, \{0, 1\}^n)$ the set of all compression functions from $n + m$ to n bits. We define the PRF security of F as

$$\mathbf{Adv}_F^{\mathrm{prf}}(\mathcal{A}) = \left| \begin{array}{l} \mathbf{Pr}\left(K \xleftarrow{\$} \{0, 1\}^k, \mathcal{A}^{F_K} = 1\right) \\ \quad - \mathbf{Pr}\left(R \xleftarrow{\$} \mathrm{Func}(\{0, 1\}^{n+m}, \{0, 1\}^n), \mathcal{A}^R = 1\right) \end{array} \right|.$$

We denote by $\mathbf{Adv}_F^{\mathrm{prf}}(q, t)$ the maximum advantage over all adversaries that make at most q queries and that run in time t.

A tweakable keyed compression function $\widetilde{F} : \{0, 1\}^k \times \mathcal{T} \times \{0, 1\}^{n+m} \to \{0, 1\}^n$ takes as additional input a tweak $t \in \mathcal{T}$. Denote by $\widetilde{\mathrm{Func}}(\mathcal{T}, \{0, 1\}^{n+m}, \{0, 1\}^n)$ the set of all tweakable compression functions from $n + m$ to n bits, where the tweak inputs come from $\mathcal{T}$. Formally, a tweakable keyed compression function is equivalent to a keyed compression function with a larger input, but for our analysis it is more convenient to adopt dedicated notation. We define the tweakable PRF ($\widetilde{\mathrm{PRF}}$) security of $\widetilde{F}$ as

$$\mathbf{Adv}_{\widetilde{F}}^{\widetilde{\mathrm{prf}}}(\mathcal{A}) = \left| \begin{array}{l} \mathbf{Pr}\left(K \xleftarrow{\$} \{0, 1\}^k, \mathcal{A}^{\widetilde{F}_K} = 1\right) \\ \quad - \mathbf{Pr}\left(\widetilde{R} \xleftarrow{\$} \widetilde{\mathrm{Func}}(\mathcal{T}, \{0, 1\}^{n+m}, \{0, 1\}^n), \mathcal{A}^{\widetilde{R}} = 1\right) \end{array} \right|.$$

We denote by $\mathbf{Adv}_{\widetilde{F}}^{\widetilde{\mathrm{prf}}}(q, t)$ the maximum advantage over all adversaries that make at most q queries and that run in time t.

3 p-OMD

Let $k, m, n, \tau \in \mathbb{N}$ such that $m \leq n$. Let $F : \{0, 1\}^k \times \{0, 1\}^{n+m} \to \{0, 1\}^n$ be a keyed compression function. p-OMD is a mapping that takes as input a

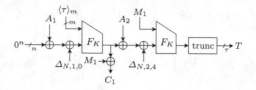

Fig. 1. p-OMD for the specific case of $|A| = 2n$ and $|M| = m$.

key $K \in \{0,1\}^k$, a nonce $N \in \{0,1\}^{\leq n-1}$, an arbitrarily sized associated data $A \in \{0,1\}^*$, and an arbitrarily sized message $M \in \{0,1\}^*$, and it returns a ciphertext $C \in \{0,1\}^{|M|}$ and tag $T \in \{0,1\}^\tau$.

For our attack it suffices to describe p-OMD for the specific case where $|A| = 2n$ and $|M| = m$ (or in other words, the associated data consists of two integral blocks and the message of one integral block). It is depicted in Fig. 1 (and corresponds to Case A of [20]). Here,

$$\Delta_{N,1,0} = F_K(N\|10^{n-1-|N|}, 0^m) \oplus 16 F_K(0^n, 0^m),$$
$$\Delta_{N,2,4} = F_K(N\|10^{n-1-|N|}, 0^m) \oplus (32 \oplus 16 \oplus 4) F_K(0^n, 0^m),$$

but our attack will not effectively use these masking values.

3.1 Preliminary Security Claims of p-OMD

In [20], Reyhanitabar et al. proved the following security levels for p-OMD:

Theorem 1. *We have*

$$\mathbf{Adv}^{\mathrm{conf}}_{\text{p-OMD}}(\mathsf{nr}, q, \ell, \sigma, t) \leq \frac{3\sigma^2}{2^n} + \mathbf{Adv}^{\mathrm{prf}}_F(2\sigma, t'),$$

$$\mathbf{Adv}^{\mathrm{int}}_{\text{p-OMD}}(\mathsf{nr}, q_{\mathcal{E}}, q_{\mathcal{D}}, \ell, \sigma, t) \leq \frac{3\sigma^2}{2^n} + \frac{\ell q_{\mathcal{D}}}{2^n} + \frac{q_{\mathcal{D}}}{2^\tau} + \mathbf{Adv}^{\mathrm{prf}}_F(2\sigma, t'),$$

$$\mathbf{Adv}^{\mathrm{int}}_{\text{p-OMD}}(\mathsf{nm}, q_{\mathcal{E}}, q_{\mathcal{D}}, \ell, \sigma, t) \leq \frac{3\sigma^2}{2^n} + \frac{\ell(q_{\mathcal{E}}^2 + q_{\mathcal{E}})q_{\mathcal{D}}}{2^n} + \frac{q_{\mathcal{D}}}{2^\tau} + \mathbf{Adv}^{\mathrm{prf}}_F(2\sigma, t'),$$

where $t' \approx t$.

In the updated version [19], the authors removed the last claim of the three as a result of this attack also presented in [2]. In the remainder of the section, we demonstrate why the bound does not hold.

3.2 Nonce-Misusing Attack on p-OMD

We consider a nonce-misusing adversary that operates as follows:

(i) Fix $N = \varepsilon$ and choose arbitrary $M \in \{0,1\}^m$ and $A_1, A_2, A'_1 \in \{0,1\}^n$ such that $A_1 \neq A'_1$;

(ii) Query p-OMD$_K(N, A_1A_2, M) \to (C, T)$;
(iii) Query p-OMD$_K(N, A_1'A_2, M) \to (C', T')$;
(iv) Set $A_2' = C \oplus C' \oplus A_2$;
(v) Query forgery p-OMD$_K^{-1}(N, A_1'A_2', C', T)$.

For the first and second evaluation of p-OMD, it holds that the state difference right *before* the second F-evaluation equals $C \oplus C'$. The forgery is formed simply by adding this value to A_2. Consequently, it holds that the first call to p-OMD and the forgery attempt have the exact same input to the second F-evaluation, and thus the same tag. Therefore, the forgery attempt succeeds as

$$\text{p-OMD}_K^{-1}(N, A_1'A_2', C', T) = M$$

by construction. In other words, for some negligibly small t,

$$\mathbf{Adv}_{\text{p-OMD}}^{\text{int}}(\text{nm}, 2, 1, 2, 6, t) = 1.$$

The issue appears in the proof of [20] in Lemma 4 case 4, and more specifically the analysis of probability $\mathbf{Pr}(\text{intcol} \mid \mathsf{E}_4)$. The authors claim that an adversary can, indeed, find an internal collision, but that any such collision happens with a birthday bound only. This reasoning, however, assumes that the input to every F-call is random, which is not the case given that the adversary can re-use the nonce and thus observe and modify the state using encryption queries.

4   Spoed

We introduce the authenticated encryption scheme Spoed with the motivation of generalizing and simplifying p-OMD. As a bonus, the simplification allows for a better bound and a significantly shorter proof, making the scheme less susceptible to mistakes hiding in one of the lemmas.

4.1   Syntax

Let $k, n, \tau \in \mathbb{N}$ such that $\tau \leq n$. Here and throughout, we assume Spoed to process blocks of $m = n$ bits. However, the results easily generalize to arbitrary (but fixed) block sizes. Let $F : \{0,1\}^k \times \{0,1\}^{2n} \to \{0,1\}^n$ be a keyed compression function. Spoed consists of an encryption function $\mathcal{E}$ and a decryption function $\mathcal{D}$.

- The encryption function $\mathcal{E}$ takes as input a key $K \in \{0,1\}^k$, a nonce $N \in \{0,1\}^n$, an arbitrarily sized associated data $A \in \{0,1\}^*$, and an arbitrarily sized message $M \in \{0,1\}^*$. It returns a ciphertext $C \in \{0,1\}^{|M|}$ and a tag $T \in \{0,1\}^\tau$;
- The decryption function $\mathcal{D}$ takes as input a key $K \in \{0,1\}^k$, a nonce $N \in \{0,1\}^n$, an arbitrarily sized associated data $A \in \{0,1\}^*$, an arbitrarily sized ciphertext $C \in \{0,1\}^*$, and a tag $T \in \{0,1\}^\tau$. It returns either a message $M \in \{0,1\}^{|C|}$ such that M satisfies $\mathcal{E}(K, N, A, M) = (C, T)$ or a dedicated failure sign $\perp$.

The encryption and decryption function are required to satisfy

$$\mathcal{D}(K, N, A, \mathcal{E}(K, N, A, M)) = M$$

for any K, N, A, M.

4.2 Generalized Padding

Spoed uses a generalized padding function

$$\text{GPAD}_{n,\tau} : \{0,1\}^* \times \{0,1\}^* \to \left(\{0,1\}^{2n}\right)^+.$$

It is indexed by state sizes n, τ, and it maps the associated data and message to generalized message blocks. Formally, it is defined as follows: First, A (associated data) and X (message or ciphertext) are padded into n-bit message blocks $A_1 \| \cdots \| A_a = A \| 0^{n-|A| \bmod n}$ and $X_1 \| \cdots \| X_m = X \| 0^{n-|X| \bmod n}$, respectively. Denote $\ell = \max\left\{m, \lceil \frac{a+m}{2} \rceil\right\} + 1$, and define $\text{len}(A, X) = \langle |A| \rangle_{n/2} \| \langle |X| \rangle_{n/2}$.[3] The function $\text{GPAD}_{n,\tau}(A, X)$ outputs $Z_1, \ldots, Z_\ell$ as follows:

if $a \leq m$:		if $a > m$, $a + m$ even:		if $a > m$, $a + m$ odd:	
Z_1	$= \langle \tau \rangle_n \| A_1$	Z_1	$= \langle \tau \rangle_n \| A_1$	Z_1	$= \langle \tau \rangle_n \| A_1$
Z_2	$= X_1 \| A_2$	Z_2	$= X_1 \| A_2$	Z_2	$= X_1 \| A_2$
$\ldots$		$\ldots$		$\ldots$	
Z_a	$= X_{a-1} \| A_a$	Z_{m+1}	$= X_m \| A_{m+1}$	Z_{m+1}	$= X_m \| A_{m+1}$
Z_{a+1}	$= X_a \| 0^n$	Z_{m+2}	$= A_{m+2} \| A_{m+3}$	Z_{m+2}	$= A_{m+2} \| A_{m+3}$
$\ldots$		$\ldots$		$\ldots$	
$Z_{\ell-1}$	$= X_{m-1} \| 0^n$	$Z_{\ell-1}$	$= A_{a-2} \| A_{a-1}$	$Z_{\ell-1}$	$= A_{a-1} \| A_a$
Z_ℓ	$= X_m \| \text{len}(A, X)$	Z_ℓ	$= A_a \| \text{len}(A, X)$	Z_ℓ	$= 0^n \| \text{len}(A, X)$

The encoding of the message length is included in order to circumvent the need for a case distinction in the description of Spoed. Note that, in fact, almost any injective padding rule would do the job; however, for our purposes the described $\text{GPAD}_{n,\tau}$ is the most suitable. We generically write $Z_i = Z_i^0 \| Z_i^1$, and denote $Z^\beta = Z_1^\beta \| \cdots \| Z_\ell^\beta$ for $\beta \in \{0, 1\}$.

4.3 Data Processing

Spoed is designed with the SHA256 and SHA512 compression functions in mind as its underlying primitive. SHA256 is a compression function

$$\text{SHA256} : \{0,1\}^{256} \times \{0,1\}^{512} \to \{0,1\}^{256}.$$

---

[3] As we show in Sect. 6, Spoed achieves birthday bound security, and the limitation of the length of X and A to $2^{n/2} - 1$ does not pose any issues.

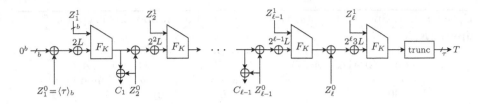

Fig. 2. Spoed encryption, which outputs $C = \text{left}_{|M|}(C_1\| \cdots \|C_{\ell-1})$ and T. Here, $L = F_K(N\|0)$

Similarly, SHA512 is a compression function SHA512 : $\{0,1\}^{512} \times \{0,1\}^{1024} \to \{0,1\}^{512}$. In the sequel, we will define Spoed using SHA256, or in other words used keyed compression function

$$F_K(Z) = \text{SHA256}(K, Z),$$

where K is injected through the chaining value interface, and the block is injected through the message interface. Note that this implicitly means that we take $k = n = 256$. We nevertheless continue to use k and n for clarity. Note that Spoed can be equivalently designed using the SHA512 compression function, but a proper change in the sizes of Z_i^0 and Z_i^1 should be introduced.

We now informally describe how to use Spoed, and refer the reader to Algorithms 1 and 2 for a formal specification. Define $L = F_K(N\|0)$. First, the associated data and message are padded into

$$(Z_1, \dots, Z_\ell) = \text{GPAD}_{n,\tau}(A, M),$$

where each Z_i is a $(2n = 512)$-bit block consisting of two blocks $Z_i = Z_i^0 \| Z_i^1$ of size 256 bits each. Spoed reads all blocks but the last one sequentially and processes them by

$$t_i = F_K(t_{i-1} \oplus 2^i L \oplus Z_i^0 \| Z_i^1),$$

where $t_0 = 0^n$. The ciphertext block C_i is generated as $C_i = t_i \oplus M_i$, chopped to the appropriate length if M_i does not contain a full amount of n message bits. The last block Z_ℓ contains the lengths of the message and the associated data and is processed through

$$t_\ell = F_K(t_{\ell-1} \oplus 2^\ell 3L \oplus Z_\ell^0 \| Z_\ell^1).$$

The tag T is generated by removing the leftmost 256-τ bits of t_ℓ. Spoed is depicted in Fig. 2.

Decryption goes fairly the same way: a t_i and a C_i value are used to recover M_i, and the state is set to C_i. This eventually leads to a verification tag T', and the message M is released if $T = T'$.

Algorithm 1. Spoed encryption $\mathcal{E}$

Input: (K, N, A, M)
Output: (C, T)
1: $(Z_1, \ldots, Z_\ell) = \text{GPAD}_{n,\tau}(A, M)$
2: $m = \lceil |M|/n \rceil$
3: $L = F_K(N \| 0)$
4: $t_0 = 0^n$
5: **for** $i = 1, \ldots, \ell - 1$ **do**
6:     $t_i = F_K(t_{i-1} \oplus 2^i L \oplus Z_i^0 \| Z_i^1)$
7:     $C_i = t_i \oplus Z_{i+1}^0$
8: $t_\ell = F_K(t_{\ell-1} \oplus 2^\ell 3L \oplus Z_\ell^0 \| Z_\ell^1)$
9: $C = \text{left}_{|M|}(C_1 \| \cdots \| C_{\ell-1})$
10: $T = \text{left}_\tau(t_\ell)$
11: **return** (C, T)

Algorithm 2. Spoed decryption $\mathcal{D}$

Input: (K, N, A, C, T)
Output: M or $\perp$
1: $(Z_1, \ldots, Z_\ell) = \text{GPAD}_{n,\tau}(A, C)$
2: $m = \lceil |C|/n \rceil$, $\rho = |C| \bmod n$
3: $L = F_K(N \| 0)$
4: $t_0 = 0^n$, $M_0 = Z_1^0$
5: **for** $i = 1, \ldots, \ell - 1$ **do**
6:     $t_i = F_K(t_{i-1} \oplus 2^i L \oplus M_{i-1} \| Z_i^1)$
7:     **if** $i < m$ **then** $M_i = t_i \oplus Z_{i+1}^0$
8:     **if** $i = m$ **then** $M_i = \text{left}_\rho(t_i) \oplus Z_{i+1}^0$
9:     **if** $i > m$ **then** $M_i = Z_{i+1}^0$
10: $t_\ell = F_K(t_{\ell-1} \oplus 2^\ell 3L \oplus Z_\ell^0 \| Z_\ell^1)$
11: $M = \text{left}_{|C|}(M_1 \| \cdots \| M_{\ell-1})$
12: $T' = \text{left}_\tau(t_\ell)$
13: **return** $T = T'$? $M : \perp$

4.4 Security of Spoed

Spoed achieves confidentiality and integrity against nonce-respecting adversaries. Note that we do not claim security against nonce-misusing adversaries.

Theorem 2. *We have*

$$\mathbf{Adv}^{\text{conf}}_{\text{Spoed}}(\text{nr}, q, \ell, \sigma, t) \leq \frac{1.5\sigma^2}{2^n} + \mathbf{Adv}^{\text{prf}}_F(2\sigma, t'),$$

$$\mathbf{Adv}^{\text{int}}_{\text{Spoed}}(\text{nr}, q_\mathcal{E}, q_\mathcal{D}, \ell, \sigma, t) \leq \frac{1.5\sigma^2}{2^n} + \frac{\ell q_\mathcal{D}}{2^n} + \frac{q_\mathcal{D}}{2^\tau} + \mathbf{Adv}^{\text{prf}}_F(2\sigma, t'),$$

where $t' \approx t$.

These bounds, surprisingly, improve over the ones of p-OMD (see Theorem 1), but with a much shorter proof. The proof is given in Sect. 6.

5 Spoednic

Spoed is simpler and more efficient than p-OMD, but it also falls victim to nonce-misuse attacks. In this section, we introduce Spoednic, a strengthened version of Spoed that retains some level of security if the nonce gets reused. As a matter of fact, Spoednic differs from Spoed in (and only in) the fact that it uses an additional subkey $L' = F_K(N \| 1)$ and that the input values Z_i^0 are blinded by L'. More formally, Spoednic inherits the syntax and generalized padding from Spoed (see Sects. 4.1 and 4.2). The data processing is fairly similar to that of Spoed (Sect. 4.3); we only present the depiction in Fig. 3 and the formal description in Algorithms 3 and 4. Both algorithms differ from Algorithms 1 and 2 *only in* lines 3 and 6. Spoednic boils down to Spoed (Fig. 2) if one would use $L' = 1$ instead of $L' = F_K(N \| 1)$.

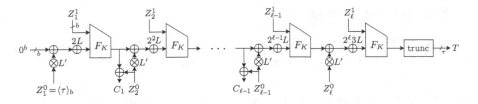

Fig. 3. Spoednic encryption, which outputs $C = \text{left}_{|M|}(C_1\|\cdots\|C_{\ell-1})$ and T. Here, $L = F_K(N\|0)$ and $L' = F_K(N\|1)$

Algorithm 3. Spoednic encryption $\mathcal{E}$	**Algorithm 4.** Spoednic decryption $\mathcal{D}$
Input: (K, N, A, M)	**Input:** (K, N, A, C, T)
Output: (C, T)	**Output:** M or $\perp$
1: $(Z_1, \ldots, Z_\ell) = \text{GPAD}_{n,\tau}(A, M)$	1: $(Z_1, \ldots, Z_\ell) = \text{GPAD}_{n,\tau}(A, C)$
2: $m = \lceil |M|/n \rceil$	2: $m = \lceil |C|/n \rceil$, $\rho = |C| \bmod n$
3: $L = F_K(N\|0)$, $L' = F_K(N\|1)$	3: $L = F_K(N\|0)$, $L' = F_K(N\|1)$
4: $t_0 = 0^n$	4: $t_0 = 0^n$, $M_0 = Z_1^0$
5: **for** $i = 1, \ldots, \ell - 1$ **do**	5: **for** $i = 1, \ldots, \ell - 1$ **do**
6: $\quad t_i = F_K(t_{i-1} \oplus 2^i L \oplus (Z_i^0 \cdot L') \| Z_i^1)$	6: $\quad t_i = F_K(t_{i-1} \oplus 2^i L \oplus (M_{i-1} \cdot L') \| Z_i^1)$
7: $\quad C_i = t_i \oplus Z_{i+1}^0$	7: $\quad$ **if** $i < m$ **then** $M_i = t_i \oplus Z_{i+1}^0$
8: $t_\ell = F_K(t_{\ell-1} \oplus 2^\ell 3L \oplus Z_\ell^0 \| Z_\ell^1)$	8: $\quad$ **if** $i = m$ **then** $M_i = \text{left}_\rho(t_i) \oplus Z_{i+1}^0$
9: $C = \text{left}_{|M|}(C_1 \| \cdots \| C_{\ell-1})$	9: $\quad$ **if** $i > m$ **then** $M_i = Z_{i+1}^0$
10: $T = \text{left}_\tau(t_\ell)$	10: $t_\ell = F_K(t_{\ell-1} \oplus 2^\ell 3L \oplus Z_\ell^0 \| Z_\ell^1)$
11: **return** (C, T)	11: $M = \text{left}_{|C|}(M_1 \| \cdots \| M_{\ell-1})$
	12: $T' = \text{left}_\tau(t_\ell)$
	13: **return** $T = T'$? $M : \perp$

5.1 Security of Spoednic

We prove that Spoednic achieves confidentiality against nonce-respecting adversaries and integrity against both nonce-respecting and nonce-misusing adversaries. Note that we do not claim confidentiality against nonce-misusing adversaries.

Theorem 3. *We have*

$$\mathbf{Adv}_{\text{Spoednic}}^{\text{conf}}(\text{nr}, q, \ell, \sigma, t) \leq \frac{1.5\sigma^2}{2^n} + \mathbf{Adv}_F^{\text{prf}}(3\sigma, t'),$$

$$\mathbf{Adv}_{\text{Spoednic}}^{\text{int}}(\text{nr}, q_\mathcal{E}, q_\mathcal{D}, \ell, \sigma, t) \leq \frac{1.5\sigma^2}{2^n} + \frac{\ell q_\mathcal{D}}{2^n} + \frac{q_\mathcal{D}}{2^\tau} + \mathbf{Adv}_F^{\text{prf}}(3\sigma, t'),$$

$$\mathbf{Adv}_{\text{Spoednic}}^{\text{int}}(\text{nm}, q_\mathcal{E}, q_\mathcal{D}, \ell, \sigma, t) \leq \frac{1.5\sigma^2}{2^n} + \frac{\ell q_\mathcal{E}^2/2}{2^n} + \frac{\ell q_\mathcal{E} q_\mathcal{D}}{2^n} + \frac{q_\mathcal{D}}{2^\tau} + \mathbf{Adv}_F^{\text{prf}}(3\sigma, t'),$$

where $t' \approx t$.

The proof is given in Sect. 7.

6   Security of Spoed (Theorem 2)

The proof of Theorem 2 is given in Sect. 6.2. It relies on a preliminary result on a tweakable keyed compression function, which will be given in Sect. 6.1.

6.1   Security of Tweakable Keyed Compression Function

In the proof of Spoed we will use the following result. It is in fact an abstraction of the XE tweakable blockcipher [21] to compression functions, and it has also been used for OMD [10] and p-OMD [19], albeit with a worse bound.

Lemma 1. *Let $F : \{0,1\}^k \times \{0,1\}^{2n} \rightarrow \{0,1\}^n$ be a keyed compression function. Let $\mathcal{T} = [1, 2^{n/2}] \times [0,1] \times \{0,1\}^n$, and define $\widetilde{F} : \{0,1\}^k \times \mathcal{T} \times \{0,1\}^{2n} \rightarrow \{0,1\}^n$ as*

$$\widetilde{F}(K, (\alpha, \beta, N), S) = F(K, (2^\alpha 3^\beta \cdot F_K(N\|0) \| 0^n) \oplus S). \tag{1}$$

Then, we have

$$\mathbf{Adv}_{\widetilde{F}}^{\widetilde{\mathrm{prf}}}(q, t) \leq \frac{1.5q^2}{2^n} + \mathbf{Adv}_F^{\mathrm{prf}}(2q, t'),$$

where $t' \approx t$.

Proof. The proof is performed using the H-coefficient technique [8,17]. It closely follows the proof of [12, Theorem 2]; the only significant differences appear in the fact that the underlying primitive is a one-way function instead of a permutation, and hence various bad events have become redundant. To wit, in the terminology of [12, Theorem 2], the events $\mathrm{bad}_{1,2}$ and $\mathrm{bad}_{2,K}$ are inapplicable (as the adversary has no access to the underlying primitive), and for the events $\mathrm{bad}_{1,1}$, $\mathrm{bad}_{1,K}$, and $\mathrm{bad}_{K,K}$, we only have to consider input collisions to the primitives. Checking the corresponding bounds reveals a term $1.5q^2/2^n$.

As a first step, we replace the evaluations of F_K for $K \xleftarrow{\$} \{0,1\}^k$ by a random function $R : \{0,1\}^{2n} \rightarrow \{0,1\}^n$. As every evaluation of $\widetilde{F}$ renders at most 2 evaluations of F_K, this step costs us $\mathbf{Adv}_F^{\mathrm{prf}}(2q, t')$, where $t' \approx t$, and allows us to consider

$$\widetilde{F} : ((\alpha, \beta, N), S) \mapsto R((2^\alpha 3^\beta \cdot R(N\|0) \| 0^n) \oplus S), \tag{2}$$

based on $R \xleftarrow{\$} \mathrm{Func}(\{0,1\}^{2n}, \{0,1\}^n)$. As we have replaced the underlying function F by a secret random primitive, we can focus on adversaries with unbounded computational power, and consider them to be information theoretic. Without loss of generality, any such adversary is deterministic. For the remainder of the analysis, consider any fixed deterministic adversary $\mathcal{A}$. Without loss of generality, we assume that $\mathcal{A}$ does not repeat any queries.

Let $R \xleftarrow{\$} \mathrm{Func}(\{0,1\}^{2n}, \{0,1\}^n)$ and $\widetilde{R} \xleftarrow{\$} \widetilde{\mathrm{Func}}(\mathcal{T}, \{0,1\}^{2n}, \{0,1\}^n)$. Consider any fixed deterministic adversary $\mathcal{A}$. In the real world, it has access to $\widetilde{F}$

of (2), while in the ideal world it has access to $\widetilde{R}$, and its goal is to distinguish both worlds. It makes q queries to the oracle, which are summarized in a view

$$\nu_F = \{(\alpha_1, \beta_1, N_1, S_1, T_1), \ldots, (\alpha_q, \beta_q, N_q, S_q, T_q)\}.$$

Note that, as $\mathcal{A}$ is deterministic, this view ν_F properly summarizes the interaction with the oracle. To suit the analysis, we will provide $\mathcal{A}$ with additional information *after* its interaction with its oracle. In more detail, it is given a *subkey transcript* ν_L that includes the computations of $R(N\|0)$ for all $N \in \{N_1, \ldots, N_q\}$. As the latter set may include duplicates, i.e., it may be that $N_i = N_j$, the formalism of ν_L requires some notation. Let $\{M_1, \ldots, M_r\}$ be a minimal set that includes $N_1, \ldots, N_q$. Then, after the interaction of $\mathcal{A}$ with its oracle, we reveal

$$\nu_L = \{(M_1, L_1), \ldots, (M_r, L_r)\},$$

In the real world the values $L_1, \ldots, L_r$ are defined as $L_i = R(M_i\|0)$, while in the ideal world, these values are randomly generated dummy subkeys $L_i \xleftarrow{\$} \{0,1\}^n$. Clearly, the disclosure of ν_L is without loss of generality as it only increases the adversary's chances. The complete view is defined as $\nu = (\nu_F, \nu_L)$. It is important to note that, as $\mathcal{A}$ never repeats queries, ν_F does not contain any duplicate elements. Neither does ν_L, by minimality of the set $\{M_1, \ldots, M_r\}$.

H-Coefficient Technique. For brevity, denote $\mathcal{A}$'s distinguishing advantage by $\Delta_{\mathcal{A}}(\widetilde{F}; \widetilde{R})$. Denote by $X_{\widetilde{F}}$ the probability distribution of views when $\mathcal{A}$ is interacting with $\widetilde{F}$ and by $X_{\widetilde{R}}$ the probability distribution of views when $\mathcal{A}$ is interacting with $\widetilde{R}$. Let $\mathcal{V}$ be the set of all attainable views, being the views that can be generated from $\widetilde{R}$ with non-zero probability. Let $\mathcal{V} = \mathcal{V}_{\text{good}} \cup \mathcal{V}_{\text{bad}}$ be a partition of the set of attainable views. The H-coefficient technique states the following. Let $0 \leq \varepsilon \leq 1$ be such that for all $\nu \in \mathcal{V}_{\text{good}}$ we have

$$\frac{\mathbf{Pr}\left(X_{\widetilde{F}} = \nu\right)}{\mathbf{Pr}\left(X_{\widetilde{R}} = \nu\right)} \geq 1 - \varepsilon.$$

Then, the distinguishing advantage of $\mathcal{A}$ satisfies

$$\Delta_{\mathcal{A}}(\widetilde{F}; \widetilde{R}) \leq \varepsilon + \mathbf{Pr}\left(X_{\widetilde{R}} \in \mathcal{V}_{\text{bad}}\right). \tag{3}$$

We refer to [7] for a proof.

Bad Transcripts. Note that every tuple in ν_F uniquely fixes a subkey in ν_L and therewith uniquely fixes one evaluation $R(s) = t$. On the other hand, the evaluations in ν_L represent evaluations of R themselves. Informally, we will consider a transcript as *bad* if there exist two different tuples that have the same input to R. Formally, we say that a view ν is *bad* if it satisfies one of the following conditions:

Bad1. There exist $(\alpha, \beta, N, S, T) \in \nu_F$ and $(N, L), (M^*, L^*) \in \nu_L$ such that:

$$(2^\alpha 3^\beta \cdot L \| 0^n) \oplus S = M^* \| 0^n;$$

Bad2. There exist distinct $(\alpha, \beta, N, S, T), (\alpha^*, \beta^*, N^*, S^*, T^*) \in \nu_F$ and (not necessarily distinct) $(N, L), (N^*, L^*) \in \nu_L$ such that:

$$(2^\alpha 3^\beta \cdot L \parallel 0^n) \oplus S = (2^{\alpha^*} 3^{\beta^*} \cdot L^* \parallel 0^n) \oplus S^*.$$

Probability of Bad Transcripts. Consider a view ν in the ideal world $\widetilde{R}$. We will consider both bad events separately.

Bad1. Consider any query $(\alpha, \beta, N, S, T) \in \nu_F$ with corresponding subkey $(N, L) \in \nu_L$, and let $(M^*, L^*) \in \nu_L$ (q^2 choices in total). The queries render a bad view if

$$2^\alpha 3^\beta \cdot L = S^0 \oplus M^*.$$

As in the ideal world $L \xleftarrow{\$} \{0, 1\}^n$, this equation is satisfied with probability $1/2^n$. Summing over all possible choices of queries, Bad1 is satisfied with probability at most $q^2/2^n$;

Bad2. Consider any distinct $(\alpha, \beta, N, S, T), (\alpha^*, \beta^*, N^*, S^*, T^*) \in \nu_F$ with corresponding $(N, L), (N^*, L^*) \in \nu_L$ ($\binom{q}{2}$ choices in total). The queries render a bad view if

$$2^\alpha 3^\beta \cdot L \oplus S^0 = 2^{\alpha^*} 3^{\beta^*} \cdot L^* \oplus S^{*0} \wedge S^1 = S^{*1}.$$

Clearly, if $N \neq N^*$, then $L \xleftarrow{\$} \{0, 1\}^n$ is generated independently of the remaining values, and the first part of the condition holds with probability $1/2^n$. Similar for the case where $N = N^*$ but $2^\alpha 3^\beta \neq 2^{\alpha^*} 3^{\beta^*}$. On the other hand, if $N = N^*$ and $2^\alpha 3^\beta =^{\alpha^*} 3^{\beta^*}$, we necessarily have $(N, \alpha, \beta) = (N^*, \alpha^*, \beta^*)$ (due to the non-colliding property of $2^\alpha 3^\beta$). As the two queries in ν_F are distinct, we have $S \neq S^*$, making above condition false. Concluding, Bad2 is satisfied with probability at most $\binom{q}{2}/2^n$.

We thus obtained that $\mathbf{Pr}\left(X_{\widetilde{R}} \in \mathcal{V}_{\text{bad}}\right) \leq 1.5 q^2/2^n$.

Good Transcripts. Consider a good view ν. Denote by $\Omega_{\widetilde{F}}$ the set of all possible oracles in the real world and by $\text{comp}_{\widetilde{F}}(\nu) \subseteq \Omega_{\widetilde{F}}$ the set of oracles compatible with view ν. Define $\Omega_{\widetilde{R}}$ and $\text{comp}_{\widetilde{R}}(\nu)$ similarly. The probabilities $\mathbf{Pr}\left(X_{\widetilde{F}} = \nu\right)$ and $\mathbf{Pr}\left(X_{\widetilde{R}} = \nu\right)$ can be computed as follows:

$$\mathbf{Pr}\left(X_{\widetilde{F}} = \nu\right) = \frac{|\text{comp}_{\widetilde{F}}(\nu)|}{|\Omega_{\widetilde{F}}|} \text{ and } \mathbf{Pr}\left(X_{\widetilde{R}} = \nu\right) = \frac{|\text{comp}_{\widetilde{R}}(\nu)|}{|\Omega_{\widetilde{R}}|}.$$

Note that $|\Omega_{\widetilde{F}}| = (2^n)^{2^{2n}}$ and $|\Omega_{\widetilde{R}}| = (2^n)^{|T|+2^{2n}} \cdot (2^n)^r$ (taking into account that in the ideal world ν contains r dummy subkeys). The computation of the number of compatible oracles is a bit more technical. Starting with $\text{comp}_{\widetilde{F}}(\nu)$, as ν is a good view, every tuple in ν represents *exactly one* evaluation of R, $q+r$ in total, and hence the number of functions R compatible with ν is $|\text{comp}_{\widetilde{F}}(\nu)| = (2^n)^{2^{2n}-(q+r)}$. Next, for $\text{comp}_{\widetilde{R}}(\nu)$, the tuples in ν_F all define *exactly one* evaluation of $\widetilde{R}$, q in total, and ν_L fixes all dummy keys. Therefore, the number of compatible oracles in the ideal world is $|\mathrm{comp}_{\widetilde{R}}(\nu)| = (2^n)^{|\mathcal{T}|+2^{2n}-q}$. We consequently obtain

$$\frac{\mathbf{Pr}\left(X_{\widetilde{F}} = \nu\right)}{\mathbf{Pr}\left(X_{\widetilde{R}} = \nu\right)} = \frac{|\mathrm{comp}_{\widetilde{F}}(\nu)| \cdot |\Omega_{\widetilde{R}}|}{|\Omega_{\widetilde{F}}| \cdot |\mathrm{comp}_{\widetilde{R}}(\nu)|} = \frac{(2^n)^{2^{2n}-(q+r)} \cdot (2^n)^{|\mathcal{T}|+2^{2n}} \cdot (2^n)^r}{(2^n)^{2^{2n}} \cdot (2^n)^{|\mathcal{T}|+2^{2n}-q}} = 1,$$

putting $\varepsilon = 0$.

Conclusion. The proof is concluded via (3) and above computations. □

Note that p-OMD uses tweaks of the form 2^α, while we use $2^\alpha 3^\beta$. This is not a problem as long as the offsets are unique [21] (i.e., there is no $(\alpha, \beta) \neq (\alpha', \beta')$ such that $2^\alpha 3^\beta = 2^{\alpha'} 3^{\beta'}$). For the case of $n = 128$, Rogaway [21] proved—via the computation of discrete logarithms—that the tweak domain $[1, 2^{n/2}] \times [0, 1]$ works properly, but this result is inadequate for our purposes as we use a compression function with $n \in \{256, 512\}$. Granger et al. [12] recently computed discrete logarithms for $n \leq 1024$, therewith confirming properness of the tweak set domain. Note that the tweak sets computed in [12,21] commonly exclude the all-zero tweak $(\alpha, \beta) = (0, 0)$ because it is a representative of 1 and hence problematic for XEX: see also [21, Sect. 6] and [16, Sect. 4]. Because F is a one-way function, its security analysis follows the one of XE, and this issue does not apply.

Also from an efficiency point of view, there is a difference between the masking of $\widetilde{F}$ in p-OMD and in Spoed. In more detail, p-OMD uses the Gray code masking (also used in OCB1 and OCB3) while for Spoed we have opted to describe it with powering-up (used in OCB2 and in various CAESAR candidates). Krovetz and Rogaway demonstrated that Gray codes are more efficient than powering-up [15], but on the downside they require more precomputation. Granger et al. [12] revisited the principle of masking of tweakable blockciphers, and presented a masking technique based on word-based linear feedback shift registers that improves over both Gray codes and powering-up in terms of efficiency and simplicity. The new masking technique can be implemented with Spoed with no sacrifice in security (and the result of Lemma 1 still applies).

6.2 Proof of Theorem 2

Let $K \in \{0, 1\}^k$. Note that all evaluations of F_K are done in a tweakable manner, namely via (1). We replace these tweakable evaluations of F_K by a random tweakable compression function $\widetilde{R} \xleftarrow{\$} \widetilde{\mathrm{Func}}([1, 2^{n/2}] \times [0, 1] \times \{0, 1\}^n, \{0, 1\}^{2n}, \{0, 1\}^n)$. Note that for both confidentiality and integrity, the underlying $\widetilde{F}_K$ is invoked at most σ times. In other words, this step costs (cf. Lemma 1)

$$\mathbf{Adv}_{\widetilde{F}}^{\widetilde{\mathrm{prf}}}(\sigma, t) \leq \frac{1.5\sigma^2}{2^n} + \mathbf{Adv}_F^{\mathrm{prf}}(2\sigma, t'),$$

where $t' \approx t$. This step has led us to an idealized version of Spoed, called IdSpoed. IdSpoed is depicted in Fig. 4. Concretely, we have obtained that

$$\mathbf{Adv}_{\text{Spoed}}^{\text{conf}}(\text{nr}, q, \ell, \sigma, t) \le \mathbf{Adv}_{\text{IdSpoed}}^{\text{conf}}(\text{nr}, q, \ell, \sigma) + \frac{1.5\sigma^2}{2^n} + \mathbf{Adv}_F^{\text{prf}}(2\sigma, t'),$$

$$\mathbf{Adv}_{\text{Spoed}}^{\text{int}}(\text{nr}, q_{\mathcal{E}}, q_{\mathcal{D}}, \ell, \sigma, t) \le \mathbf{Adv}_{\text{IdSpoed}}^{\text{int}}(\text{nr}, q_{\mathcal{E}}, q_{\mathcal{D}}, \ell, \sigma) + \frac{1.5\sigma^2}{2^n} + \mathbf{Adv}_F^{\text{prf}}(2\sigma, t'),$$

where t dropped out of the advantage function for IdSpoed because it has become irrelevant (formally, we proceed by considering an adversary that is unbounded in time). We prove in Lemma 2 that its confidentiality security satisfies $\mathbf{Adv}_{\text{IdSpoed}}^{\text{conf}}(\text{nr}, q, \ell, \sigma) = 0$, and in Lemma 3 that it provides integrity up to bound $\mathbf{Adv}_{\text{IdSpoed}}^{\text{int}}(\text{nr}, q_{\mathcal{E}}, q_{\mathcal{D}}, \ell, \sigma) \le \frac{\ell q_{\mathcal{D}}}{2^n} + \frac{q_{\mathcal{D}}}{2^\tau}$.

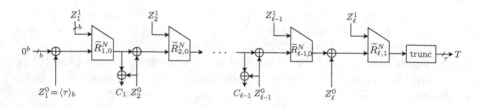

Fig. 4. IdSpoed encryption, which outputs $C = \text{left}_{|M|}(C_1 \| \cdots \| C_{\ell-1})$ and T

Lemma 2. *The advantage of any nonce-respecting adversary trying to break the confidentiality of* IdSpoed *is bounded as:*

$$\mathbf{Adv}_{\text{IdSpoed}}^{\text{conf}}(\text{nr}, q, \ell, \sigma) = 0.$$

Proof. The functions $\widetilde{R}_{i,j}^N$ for $i = 1, \ldots, \ell-1$, $j = 0, 1$, and $N \in \{0,1\}^n$ are independently and randomly distributed compression functions. As the adversary is assumed to be nonce-respecting, every nonce is used at most once. Every nonce is used in at most ℓ calls to R, but these calls are by design all for different tweaks $(i, j) \in [1, 2^{n/2}] \times [0, 1]$. Therefore, all responses are randomly generated from $\{0,1\}^n$, and all ciphertext blocks and tag values are perfectly random. $\square$

Lemma 3. *The advantage of any nonce-respecting adversary trying to break the integrity of* IdSpoed *is bounded as:*

$$\mathbf{Adv}_{\text{IdSpoed}}^{\text{int}}(\text{nr}, q_{\mathcal{E}}, q_{\mathcal{D}}, \ell, \sigma) \le \frac{\ell q_{\mathcal{D}}}{2^n} + \frac{q_{\mathcal{D}}}{2^\tau}.$$

Proof. Assume that $\mathcal{A}$ has made encryption queries (N^j, A^j, M^j) for $j = 1, \ldots, q_{\mathcal{E}}$, and denote the ciphertexts and tags by (C^j, T^j). Write $(Z_1^j, \ldots, Z_{\ell^j}^j) = \text{GPAD}_{n,\tau}(A^j, M^j)$ and denote the in- and outputs of the random functions by (s_i^j, t_i^j) for $i = 1, \ldots, \ell^j$.

Consider any forgery attempt (N, A, C, T), and denote its length by ℓ. Denote the message computed upon decryption by M. Refer to the state values as (s_i, t_i)

for $i = 1, \ldots, \ell$, and write $(Z_1, \ldots, Z_\ell) = \mathrm{GPAD}_{n,\tau}(A, M)$. The forgery is successful if $T = \mathrm{left}_\tau(t_\ell)$.

Denote by col the event that there exists an encryption query j with $N^j = N$, $\ell^j = \ell$, and an index $i \in \{1, \ldots, \ell\}$, such that

$$t_{i-1}^j \oplus Z_i^{0j} \,\|\, Z_i^{1j} \neq t_{i-1} \oplus Z_i^0 \,\|\, Z_i^1 \ \wedge\ t_i^j = t_i.$$

Note that, as the adversary is nonce-respecting, there is at most one query j with $N^j = N$. We have, using shorthand notation $[i = \ell]$ for 0 if $i \neq \ell$ and 1 if $i = \ell$,

$$\mathbf{Pr}\,(\mathsf{col}) \le \sum_{i=1}^{\ell} \mathbf{Pr}\left(s_i^j \neq s_i \wedge \widetilde{R}_{i,[i=\ell]}^N(s_i^j) = \widetilde{R}_{i,[i=\ell]}^N(s_i) \right) \le \frac{\ell}{2^n}. \tag{4}$$

We make the following, fairly simple, case distinction:

(i) $N \notin \{N^1, \ldots, N^{q_\varepsilon}\}$. This particularly means that $\widetilde{R}$ has never been queried for tweak $(\ell, 1, N)$, and thus that $\widetilde{R}_{\ell,1}^N$ responds with $t_\ell \xleftarrow{\$} \{0,1\}^n$. The forgery is successful with probability $1/2^\tau$;

(ii) $N = N^j$ for some (unique) j. As the different evaluations of IdSpoed for different tweaks are independent, it suffices to focus on these two construction queries (the jth encryption query and the forgery). We proceed with a further case distinction:

  – $\ell \neq \ell^j$. This, again, means that $\widetilde{R}$ has never been queried for tweak $(\ell, 1, N)$. The forgery is successful with probability $1/2^\tau$;

  – $\ell = \ell^j$. We proceed with a further case distinction:

    • $s_\ell \neq s_{\ell^j}^j$. In this case, $\widetilde{R}$ has been queried before for tweak $(\ell, 1, N)$, but only once (as the adversary must be nonce-respecting) and never on input s_ℓ. Consequently, the response t_ℓ is uniformly randomly drawn from $\{0,1\}^n$ and the forgery is successful with probability $1/2^\tau$;

    • $s_\ell = s_{\ell^j}^j$. As the forgery must be different from the encryption queries, and as $\mathrm{GPAD}_{n,\tau}$ is an injective mapping, this case implies the existence of a non-trivial state collision. Hence, the forgery is successful with probability at most $\mathbf{Pr}\,(\mathsf{col})$.

Concluding, the forgery is successful with probability at most $\mathbf{Pr}\,(\mathsf{col}) + 1/2^\tau$, where $\mathbf{Pr}\,(\mathsf{col})$ is bounded in (4). A summation over all $q_\mathcal{D}$ forgery attempts (cf. [4]) gives our final bound. $\qquad\square$

7   Security of Spoednic (Theorem 3)

The proof of Theorem 3 is given in Sect. 7.2. It relies on a preliminary result on a tweakable keyed compression function, which will be given in Sect. 7.1.

7.1 Security of Tweakable Keyed Compression Function

We will use a slightly more complex version of the tweakable keyed compression function of Sect. 6.1, where the masking using $Z_i^0 \cdot L'$ is included within the function. The proof is a fairly straightforward extension of the one of Lemma 1.

Lemma 4. Let $F : \{0,1\}^k \times \{0,1\}^{2n} \to \{0,1\}^n$ be a keyed compression function. Let $\mathcal{T} = [1, 2^{n/2}] \times [0,1] \times \{0,1\}^n \times \{0,1\}^n$, and define $\widetilde{F} : \{0,1\}^k \times \mathcal{T} \times \{0,1\}^{2n} \to \{0,1\}^n$ as

$$\widetilde{F}(K, (\alpha, \beta, A, N), S) = F(K, (2^\alpha 3^\beta \cdot F_K(N\|0) \oplus A \cdot F_K(N\|1) \parallel 0^n) \oplus S). \quad (5)$$

Then, we have

$$\mathbf{Adv}_{\widetilde{F}}^{\mathrm{prf}}(q, t) \leq \frac{1.5q^2}{2^n} + \mathbf{Adv}_F^{\mathrm{prf}}(3q, t'),$$

where $t' \approx t$.

Proof. The proof is a slight extension of the one of Lemma 1, where now we have twice as many subkeys. Consequently, this means that the transcript contains twice as many subkeys, but because the different subkey generations never collide (due to domain separation $0/1$), and for collisions between construction queries and subkey evaluations of F we focus on the leftmost n bits (i.e., N) in the first place, this does not affect the security bound. The proof is included in the full version of the paper [3]. □

7.2 Proof of Theorem 3

Let $K \in \{0,1\}^k$. Note that all evaluations of F_K are done in a tweakable manner, namely via (5). We replace these tweakable evaluations of F_K by a random tweakable compression function $\widetilde{R} \xleftarrow{\$} \widetilde{\mathrm{Func}}([1, 2^{n/2}] \times [0,1] \times \{0,1\}^n \times \{0,1\}^n, \{0,1\}^{2n}, \{0,1\}^n)$. Note that for both confidentiality and integrity, the underlying $\widetilde{F}_K$ is invoked at most σ times. In other words, this step costs (cf. Lemma 4)

$$\mathbf{Adv}_{\widetilde{F}}^{\mathrm{prf}}(\sigma, t) \leq \frac{1.5\sigma^2}{2^n} + \mathbf{Adv}_F^{\mathrm{prf}}(3\sigma, t'),$$

where $t' \approx t$. This step has led us to an idealized version of Spoednic, called IdSpoednic. IdSpoednic is depicted in Fig. 5. Concretely, we have obtained that

$$\mathbf{Adv}_{\mathrm{Spoednic}}^{\mathrm{conf}}(\mathrm{nr}, q, \ell, \sigma, t) \leq \mathbf{Adv}_{\mathrm{IdSpoednic}}^{\mathrm{conf}}(\mathrm{nr}, q, \ell, \sigma) + \frac{1.5\sigma^2}{2^n} + \mathbf{Adv}_F^{\mathrm{prf}}(3\sigma, t'),$$

$$\mathbf{Adv}_{\mathrm{Spoednic}}^{\mathrm{int}}(\mathrm{n}, q_{\mathcal{E}}, q_{\mathcal{D}}, \ell, \sigma, t) \leq \mathbf{Adv}_{\mathrm{IdSpoednic}}^{\mathrm{int}}(\mathrm{n}, q_{\mathcal{E}}, q_{\mathcal{D}}, \ell, \sigma) + \frac{1.5\sigma^2}{2^n} + \mathbf{Adv}_F^{\mathrm{prf}}(3\sigma, t'),$$

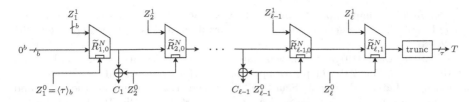

Fig. 5. IdSpoednic encryption, which outputs $C = \text{left}_{|M|}(C_1 \| \cdots \| C_{\ell-1})$ and T. The boxes in $\widetilde{R}$ indicate that Z_i^0 also functions as a tweak.

where $\mathsf{n} \in \{\mathsf{nr}, \mathsf{nm}\}$, and where t dropped out of the advantage function for IdSpoednic because it has become irrelevant. The remainder of the proof centers around this scheme. For the nonce-respecting setting, the bounds of Lemmas 2 and 3 carry over almost verbatim, with the same security bound. We consider integrity in the nonce-misuse setting in Lemma 5 and prove that $\mathbf{Adv}^{\text{int}}_{\text{IdSpoednic}}(\mathsf{nm}, q_{\mathcal{E}}, q_{\mathcal{D}}, \ell, \sigma) \leq \frac{\ell q_{\mathcal{E}}^2/2}{2^n} + \frac{\ell q_{\mathcal{E}} q_{\mathcal{D}}}{2^n} + \frac{q_{\mathcal{D}}}{2^\tau}$.

Lemma 5. *The advantage of any nonce-misusing adversary trying to break the integrity of* IdSpoednic *is bounded as:*

$$\mathbf{Adv}^{\text{int}}_{\text{IdSpoed}}(\mathsf{nm}, q_{\mathcal{E}}, q_{\mathcal{D}}, \ell, \sigma) \leq \frac{\ell q_{\mathcal{E}}^2/2}{2^n} + \frac{\ell q_{\mathcal{E}} q_{\mathcal{D}}}{2^n} + \frac{q_{\mathcal{D}}}{2^\tau}.$$

Proof. At a high level, the proof follows the one of Lemma 3, with the difference that now, potentially, nonces may be the same leading to a slightly different bound. A formal proof of Lemma 5 is included in the full version of the paper [3]. $\qquad\square$

Acknowledgments. This work was supported in part by the Research Council KU Leuven: GOA TENSE (GOA/11/007). In addition, this work was partially supported by the Research Fund KU Leuven, OT/13/071, and by European Unions Horizon 2020 research and innovation programme under No. H2020-MSCA-ITN-2014-643161 ECRYPT-NET. Bart Mennink is a Postdoctoral Fellow of the Research Foundation – Flanders (FWO).

References

1. Andreeva, E., Bogdanov, A., Luykx, A., Mennink, B., Tischhauser, E., Yasuda, K.: Parallelizable and authenticated online ciphers. In: Sako, K., Sarkar, P. (eds.) ASIACRYPT 2013, Part I. LNCS, vol. 8269, pp. 424–443. Springer, Heidelberg (2013)
2. Ashur, T., Mennink, B.: Trivial nonce-misusing attack on pure OMD. Cryptology ePrint Archive, Report 2015/175 (2015)
3. Ashur, T., Mennink, B.: Damaging, simplifying, and salvaging p-OMD. Cryptology ePrint Archive, Report 2016/534 (2016). http://eprint.iacr.org/2016/534
4. Bellare, M., Goldreich, O., Mityagin, A.: The power of verification queries in message authentication and authenticated encryption. Cryptology ePrint Archive, Report 2004/309 (2004)

5. Bellare, M., Namprempre, C.: Authenticated encryption: relations among notions and analysis of the generic composition paradigm. J. Cryptology **21**(4), 469–491 (2008)
6. CAESAR: Competition for Authenticated Encryption: Security, Applicability, and Robustness, May 2014. http://competitions.cr.yp.to/caesar.html
7. Chen, S., Lampe, R., Lee, J., Seurin, Y., Steinberger, J.: Minimizing the two-round even-mansour cipher. In: Garay, J.A., Gennaro, R. (eds.) CRYPTO 2014, Part I. LNCS, vol. 8616, pp. 39–56. Springer, Heidelberg (2014)
8. Chen, S., Steinberger, J.: Tight security bounds for key-alternating ciphers. In: Nguyen, P.Q., Oswald, E. (eds.) EUROCRYPT 2014. LNCS, vol. 8441, pp. 327–350. Springer, Heidelberg (2014)
9. Cogliani, S., Maimut, D., Naccache, D., do Canto, R.P., Reyhanitabar, R., Vaudenay, S., Vizár, D.: Offset Merkle-Damgrd (OMD) version 1.0, submission to CAESAR competition (2014)
10. Cogliani, S., Maimuţ, D.-S., Naccache, D., do Canto, R.P., Reyhanitabar, R., Vaudenay, S., Vizár, D.: OMD: a compression function mode of operation for authenticated encryption. In: Joux, A., Youssef, A. (eds.) SAC 2014. LNCS, vol. 8781, pp. 112–128. Springer, Heidelberg (2014)
11. Fleischmann, E., Forler, C., Lucks, S.: McOE: a family of almost foolproof on-line authenticated encryption schemes. In: Canteaut, A. (ed.) FSE 2012. LNCS, vol. 7549, pp. 196–215. Springer, Heidelberg (2012)
12. Granger, R., Jovanovic, P., Mennink, B., Neves, S.: Improved masking for tweakable blockciphers with applications to authenticated encryption. In: Fischlin, M., Coron, J.-S. (eds.) EUROCRYPT 2016. LNCS, vol. 9665, pp. 263–293. Springer, Heidelberg (2016). doi:10.1007/978-3-662-49890-3_11
13. Iwata, T., Ohashi, K., Minematsu, K.: Breaking and repairing GCM security proofs. In: Safavi-Naini, R., Canetti, R. (eds.) CRYPTO 2012. LNCS, vol. 7417, pp. 31–49. Springer, Heidelberg (2012)
14. Jovanovic, P., Luykx, A., Mennink, B.: Beyond $2^{c/2}$ security in sponge-based authenticated encryption modes. In: Sarkar, P., Iwata, T. (eds.) ASIACRYPT 2014. LNCS, vol. 8873, pp. 85–104. Springer, Heidelberg (2014)
15. Krovetz, T., Rogaway, P.: The software performance of authenticated-encryption modes. In: Joux, A. (ed.) FSE 2011. LNCS, vol. 6733, pp. 306–327. Springer, Heidelberg (2011)
16. Minematsu, K.: Improved security analysis of XEX and LRW modes. In: Biham, E., Youssef, A.M. (eds.) SAC 2006. LNCS, vol. 4356, pp. 96–113. Springer, Heidelberg (2007)
17. Patarin, J.: The "coefficients H" technique. In: Avanzi, R.M., Keliher, L., Sica, F. (eds.) SAC 2008. LNCS, vol. 5381, pp. 328–345. Springer, Heidelberg (2009)
18. Reyhanitabar, R., Vaudenay, S., Vizár, D.: Misuse-resistant variants of the OMD authenticated encryption mode. In: Chow, S.S.M., Liu, J.K., Hui, L.C.K., Yiu, S.M. (eds.) ProvSec 2014. LNCS, vol. 8782, pp. 55–70. Springer, Heidelberg (2014)
19. Reyhanitabar, R., Vaudenay, S., Vizár, D.: Boosting OMD for almost free authentication of associated data. In: Leander, G. (ed.) FSE 2015. LNCS, vol. 9054, pp. 411–427. Springer, Heidelberg (2015)
20. Reyhanitabar, R., Vaudenay, S., Vizár, D.: Boosting OMD for almost free authentication of associated data. In: FSE 2015 preprint version (2015)
21. Rogaway, P.: Efficient instantiations of tweakable blockciphers and refinements to modes OCB and PMAC. In: Lee, P.J. (ed.) ASIACRYPT 2004. LNCS, vol. 3329, pp. 16–31. Springer, Heidelberg (2004)

Cryptographic Protocols

Blind Password Registration for Two-Server Password Authenticated Key Exchange and Secret Sharing Protocols

Franziskus Kiefer[1]([✉]) and Mark Manulis[2]

[1] Mozilla, Berlin, Germany
mail@franziskuskiefer.de
[2] Department of Computer Science, Surrey Center for Cyber Security,
University of Surrey, Guildford, UK
mark@manulis.eu

Abstract. Many organisations enforce policies on the length and formation of passwords to encourage selection of strong passwords and protect their multi-user systems. For Two-Server Password Authenticated Key Exchange (2PAKE) and Two-Server Password Authenticated Secret Sharing (2PASS) protocols, where the password chosen by the client is secretly shared between the two servers, the initial remote registration of policy-compliant passwords represents a major problem because none of the servers is supposed to know the password in clear.

We solve this problem by introducing *Two-Server Blind Password Registration (2BPR)* protocols that can be executed between a client and the two servers as part of the remote registration procedure.

2BPR protocols guarantee that secret shares sent to the servers belong to a password that matches their combined password policy and that the plain password remains hidden from any attacker that is in control of at most one server. We propose a security model for 2BPR protocols capturing the requirements of *policy compliance* for client passwords and their *blindness* against the servers. Our model extends the adversarial setting of 2PAKE/2PASS protocols to the registration phase and hence closes the gap in the formal treatment of such protocols. We construct an efficient 2BPR protocol for ASCII-based password policies, prove its security in the standard model, give a proof of concept implementation, and discuss its performance.

1 Introduction

Password policies set by organisations aim to rule out potentially "weak" passwords and by this contribute to the protection of multi-user systems. In traditional web-based password authentication mechanisms a password policy chosen by the server is typically enforced during the password registration phase — the corresponding compliance check is performed either by the client or on the server side, depending on the available trust assumptions. If the client is not trusted with the selection of a policy-compliant password, then the compliance check

© Springer International Publishing Switzerland 2016
M. Bishop and A.C.A. Nascimento (Eds.): ISC 2016, LNCS 9866, pp. 95–114, 2016.
DOI: 10.1007/978-3-319-45871-7_7

must be performed by the server. The most common approach in this case is to transmit chosen passwords over a secure channel, e.g., TLS channel, to the server that performs the check on the received (plain) password. The drawback of this approach, however, is that the client's password is disclosed to the server. Although this approach represents a common practice nowadays, its main drawback is the necessity to trust the server to process and store the received password in a protected way, e.g., by hashing it. This trust assumption often does not hold in practice as evident from the frequent server-compromise attacks based on which plain password databases have been disclosed [8,26,31,33].

Considering that "password-cracking tools" such as Hashcat [15] and John the Ripper [28] are very efficient, it is safe to assume that leaked password hashes are not safer than un-hashed ones when compromised by an attacker [3,9,11,24]. The notion of threshold and two-server password authenticated key-exchange [12, 25] has been proposed where the password is not stored on a single server but split between a number of servers such that leakage of a password database on a non-qualified subset does not reveal the password. The two-server setting is regarded as more practical (in comparison to a more general threshold setting) given that if one server is compromised a notification to change the password can be sent out to the clients. Two-server password authenticated key-exchange protocols (2PAKE) [4,20,32] split the client's password pw into two shares s_1 and s_2 such that each share is stored on a distinct server. During the authentication phase both servers collaborate in order to authenticate the client. Yet, no server alone is supposed to learn the plain password. A second, more recent development in two-server (and threshold) password protocols is password authenticated secret sharing (PASS) [2,6,17] where a client stores shares of a (high-entropy) secret key on a number of servers and uses a (low-entropy) password to authenticate the retrieval process.

Registering password shares for 2PAKE/2PASS protocols however makes it impossible for the servers to verify their password policies upon registration unless the password is transferred to each of them in plain. This however, would imply that the client trusts both servers to securely handle its password, which contradicts the purpose and trust relationships of multi-server protocols. The use of two-server password protocols in a remote authentication setting, therefore, requires a suitable password registration procedure in which none of the servers would receive information enabling it (or an attacker in control of the server) to deliberately or inadvertently recover the client's password. This registration procedure must further allow for policy compliance checks to be performed by the servers since secret sharing per se does not protect against "weak" passwords. A trivial approach of sending s_1 and s_2 to the corresponding servers over secure channels is not helpful here since it is not clear how the two servers can perform the required compliance check. To alleviate a similar problem in the verifier-based PAKE setting, Kiefer and Manulis [22] introduced the concept of zero-knowledge password policy checks (ZKPPC), where upon registration the client can prove to the server the compliance of its chosen password with respect to the server's policy without disclosing the actual password. In this work, we propose the concept of blind password registration for two-server password protocols and thus show how to realise secure registration of password shares in a way that protects against at most one malicious server (if both servers are malicious, the attacker obviously gets the password), yet allows both servers to check password compliance against their mutual password policy. It bases on techniques introduced in the framework for ZKPPC from [22] but uses a security model for the entire blind setup process and is based in the two-server setting which brings additional challenges. Two-server Blind Password Registration (2BPR) is not vulnerable to offline dictionary attacks as long as one server remains honest. This is in contrast to the single-server setting where an attacker is always able to perform offline dictionary attacks on password verifiers after compromising a server. Our main contribution is the 2BPR security model and the corresponding protocol for secure registration of 2PAKE/2PASS passwords. We show how secure distribution of password shares can be combined with an appropriate policy-compliance proof for the chosen password in a way that does not reveal the password and can still be verified by both servers. Our 2BPR protocol can be used to enforce policies over the alphabet of all 94 printable ASCII characters[1], including typical requirements on password length and character types.

2   Preliminaries

In this section we recall the underlying primitives and concepts that are used in the construction of our two-server blind password registration protocol.

2.1   Commitments

Let $C = (\mathtt{CSetup}, \mathtt{Com})$ denote a commitment scheme and $C \leftarrow \mathtt{Com}(x; r)$ a commitment on x using randomness r, with $\mathtt{CSetup}$ generating parameters for C. A commitment scheme $C = (\mathtt{CSetup}, \mathtt{Com})$ is *efficient* if $\mathtt{CSetup}(\lambda)$ and $(C, d) \leftarrow \mathtt{Com}(x; r)$ are computable in polynomial time, *complete* if $\mathtt{Com}(d) = (C, d)$ for $(C, d) \leftarrow \mathtt{Com}(x; r)$, and secure if it is

- Binding: For all PPT adversaries $\mathcal{A}$ there exists a negligible function $\varepsilon_{\mathsf{bi}}(\cdot)$ such that for all $(x, x', r, r', C) \leftarrow A$: $\Pr[x \neq x' \wedge (C, d) = \mathtt{Com}(x; r) \wedge (C, d') = \mathtt{Com}(x'; r')] \leq \varepsilon_{\mathsf{bi}}(\lambda)$,
- Hiding: For all PPT adversaries $\mathcal{A}$ there exists a negligible function $\varepsilon_{\mathsf{hi}}(\cdot)$ such that for all x_0, x_1 with $|x_0| = |x_1|$ and $b \in_R \{0, 1\}, (C, d) \leftarrow \mathtt{Com}(x_b; r)$ and $b' \leftarrow A(C, x_1, x_2)$: $\Pr[b = b'] \leq 1/2 + \varepsilon_{\mathsf{hi}}(\lambda)$.

Pedersen Commitments [29]. We use perfectly hiding, computationally binding, homomorphic Pedersen commitments [29] defined as follows. Let $C_P = (\mathtt{CSetup}, \mathtt{Com})$ with $(g, h, q, \lambda) \leftarrow \mathtt{CSetup}(\lambda)$ and $C \leftarrow \mathtt{Com} = (x; r) = g^x h^r$

---

[1] Note that using other encodings such as UTF-8 is possible but might influence performance due to a different size of possible characters.

denote the Pedersen commitment scheme where g and h are generators of a cyclic group G of prime order q with bit-length in the security parameter λ and the discrete logarithm of h with respect to base g is not known. Pedersen commitments are *additively homomorph*, i.e. for all $(C_i, d_i) \leftarrow \text{Com}(x_i; r_i)$ for $i \in 0, \ldots, m$ it holds that $\prod_{i=0}^{m} C_i = \text{Com}(\sum_{i=0}^{m} x_i; \sum_{i=0}^{m} r_i)$.

Trapdoor Commitments. In order to build zero-knowledge proofs of knowledge with malicious verifiers we require a trapdoor commitment scheme, which allows a party knowing the correct trapdoor to open a commitment to any value. Fortunately, Pedersen commitments are trapdoor commitments as they can be opened to any element using the discrete logarithm $\log_g(h)$ as trapdoor.

2.2   Zero Knowledge Proofs

A zero-knowledge proof is executed between a prover and a verifier, proving that a word x is in a language L, using a witness w proving so. An interactive protocol ZKP for a language L between prover P and verifier V is a zero knowledge proof if the following holds:

– Completeness: If $x \in L$, V accepts if P holds a witness proving so.
– Soundness: For every malicious prover $P^*(x)$ with $x \in L$ that the probability of making V accept is negligible.
– Zero-Knowledge: If $x \in L$, then there exists an efficient simulator Sim that on input of x is able to generate a view, indistinguishable from the view of a malicious verifier V^*.

A zero-knowledge proof of knowledge ZKPoK is a zero-knowledge proof with the following special soundness definition:

– Soundness: There exists an efficient knowledge extractor Ext that can extract a witness from any malicious prover $P^*(x)$ with $x \in L$ that has non-negligible probability of making V accept.

We use the following committed Σ-protocol to ensure extractability (ZKPoK) and simulatability when interacting with a malicious verifier [7,18]. Let $P_1(x, w, r)$ and $P_2(x, w, r, c)$ denote the two prover steps of a Σ-protocol and $H : \{0, 1\}^* \mapsto \mathbb{Z}_q$ a collision-resistant hash function. A committed Σ-protocol based on Pedersen commitments is then given by the following steps:

– The prover computes $m_1 \leftarrow P_1(x, w, r)$, $\text{Co} \leftarrow \text{Com}(H(x, m_1); r_1) = g^{H(x, m_1)} h^{r_1}$, and sends Co to the verifier.
– The verifier picks random challenge $\text{Ch} = c$ and returns it to the prover.
– The prover computes $m_2 \leftarrow P_2(x, w, r, c)$, $\text{Rs}_1 \leftarrow \text{Com}(H(m_2); r_2) = g^{H(m_2)} h^{r_2}$, and sends Rs to the verifier.
– Further, the prover opens the commitments Co and Rs_1 by sending $\text{Rs}_2 = (x, m_1, m_2, r_1, r_2)$ to the verifier.

– The verifier accepts if both commitments are valid and if the verification of the Σ-protocol (x, m_1, c, m_2) is successful.

We note that in the malicious verifier setting, this type of protocol is a concurrent zero-knowledge proof since its security proof does not require rewinding [7,18]. We observe that all zero-knowledge protocols used in this work are committed Σ-protocols those security relies on the hardness of the discrete logarithm problem in G and the collision resistance property of H.

Passwords. We adopt the reversible, structure-preserving encoding scheme from [22] that (uniquely) maps strings of printable ASCII characters to integers. We use pw for the ASCII password string, $c_i = \text{pw}[i]$ for the i-th ASCII character in pw, and integer π for the encoded password string. The encoding proceeds as follows: $\pi \leftarrow \text{PWDtoINT}(\text{pw}) = \sum_{i=0}^{n-1} b^i(\text{ASCII}(c_i) - 32)$ for the password string pw and $\pi_i \leftarrow \text{CHRtoINT}(c_i) = \text{ASCII}(c_i) - 32$ for the i-th *unshifted* ASCII character in pw. Note that n denotes the length of pw and $b \in \mathbb{N}$ is used as shift base. (We refer to [22] for a discussion on the shift base b. Note, however, that shift base related attacks on the password verifier from [22] are not possible in our two-server setting.) The ASCII function returns the decimal ASCII code of a character.

Remark 1. While password distribution is important for the security of password registration protocols for Verifier-based PAKE [22], the role of password distribution in the two-server setting is different. Since each server stores only a random-looking password share, offline dictionary attacks from an attacker who compromises at most one of the two servers become infeasible. Security of 2BPR protocols defined in this work is therefore independent of client passwords. Note however that the password strength still continues to play an important role for the security of 2PAKE/2PASS protocols, where it influences the probability of successful online dictionary attacks.

Password Sharing. We focus on the additive password sharing of client passwords, i.e. $\pi = \mathfrak{s}_0 + \mathfrak{s}_1 \bmod q$ over a prime-order group G_q. Such sharing has been used in various 2PAKE protocols, including [19–21,34]. To be used in combination with 2PASS protocols such as [6] one can define the password as g^π and thus adopts the multiplicative sharing $g^\pi = g^{\mathfrak{s}_0} g^{\mathfrak{s}_1}$. Password shares are created as $\mathfrak{s}_0 \in_R \mathbb{Z}_q$ and $\mathfrak{s}_1 = \pi - \mathfrak{s}_0 \bmod q$. We remark that other sharing options such as XOR have been used in literature [4,32] but are not supported by our 2BPR protocol.

Password Policies. We represent password policies as in [22], i.e. a password policy $f = (R, n_{\min})$ consists of a simplified regular expression R that defines ASCII subsets that must be present in the chosen password string and the minimum length $n_{\min}$ of the password string. The expression R is defined over the *four* ASCII subsets $\Sigma = \{d, u, l, s\}$ with digits d, upper case letters u, lower case letters l and symbols s, and gives the minimum frequency of a character from the subset that is necessary to fulfil the policy; for instance, $R = ulld$ means that policy-conform password strings must contain at least one upper case letter, two lower case letters and one digit. In the two-server setting, if each of the servers has its own password policy, i.e. f_0 and f_1, then registered passwords would need to comply with the *mutual password policy* defined as $f = f_0 \cap f_1 = (\max(R_0, R_1), \max(n_{\min 0}, n_{\min 1}))$, where $\max(R_0, R_1)$ is the regular expression with the maximum number of characters from each of the subsets u, l, d, s from R_0 and R_1. A mutual policy is fulfilled, i.e. $f(\text{pw}) = \text{true}$, iff $f_0(\text{pw}) = \text{true}$ and $f_1(\text{pw}) = \text{true}$, and not fulfilled, i.e. $f(\text{pw}) = \text{false}$, iff $f_0(\text{pw}) = \text{false}$ or $f_1(\text{pw}) = \text{false}$. We mainly operate on the integer representation π of a password string pw throughout this paper and sometimes write $f(\pi)$, which means $f(\text{pw})$ for $\pi \leftarrow \text{PWDtoINT}(\text{pw})$. Further note that a character $c_i \in \text{pw}$ is called *significant* if this character is necessary to fulfil R and we denote the corresponding set $R_j \in R$ as a *significant set* for the policy.

Password Dictionaries. A password dictionary $\mathcal{D}_f$, if not specified otherwise, is a set of password strings adhering to a given policy $f = (R, n_{\min})$, i.e. their length is limited by $n_{\min} \leq |\text{pw}|$ and the required types of characters are identified by R. We denote the size of a dictionary $\mathcal{D}$ by $|\mathcal{D}|$. We omit index f if the policy is clear from the context. We further define dictionary $\mathcal{D}_{f,n}$ holding policy-conform passwords according to f of length n and will use it throughout the paper. In order to be able to use the optimal dictionary $\mathcal{D}_f$, the client would either have to prove correctness of password characters that are not necessary for R without revealing their number (which seems impossible with the approach used in this paper), or use a fixed password length to hide n in it (which is inefficient). (Note that we only consider reasonable dictionaries sizes, i.e. $|\mathcal{D}_{f,n}| > 1$).

3 Two-Server Blind Password Registration

Two-server Blind Password Registration (2BPR) allows a client to register password shares with two servers for later use in 2PAKE/2PASS protocols and prove that the shares can be combined to a password that complies with the mutual password policy of both servers, without disclosing the password. A 2BPR protocol is executed between client $\mathcal{C}$ and two servers S_0 with password policy f_0 and S_1 with password policy f_1. $\mathcal{C}$ interacts with S_0 and S_1 in order to distribute shares of a freshly chosen password string pw and prove its compliance with the mutual policy, i.e. $f_0(\text{pw}) = \text{true}$ and $f_1(\text{pw}) = \text{true}$. A 2BPR protocol between an honest client $\mathcal{C}$ and two honest servers S_0 and S_1 is correct if S_0 and S_1 accept their password shares if and only if the client is able to prove the following statement for $f = f_0 \cap f_1$:

$$(\text{pw}, \mathfrak{s}_0, \mathfrak{s}_1) : \text{PWDtoINT}(\text{pw}) = \mathfrak{s}_0 + \mathfrak{s}_1 \ \wedge \ f(\text{pw}) = \text{true}. \tag{1}$$

Note that the 2BPR protocol can be used to register new clients or to register new passwords for existing clients. The following definition formally captures the functionality of 2BPR protocols.

Definition 1 (Two-Server Blind Password Registration). *A 2BPR protocol is executed between a client C and two servers S_0 and S_1, holding a password policy f_b each, such that the servers, when honest, eventually accept password shares $\mathfrak{s}_b$ of a policy compliant, client chosen password* pw *iff $f(\text{pw}) = \text{true}$ for $f = f_0 \cap f_1$, PWDtoINT(pw) $= \mathfrak{s}_b + \mathfrak{s}_{1-b}$ and $b \in \{0, 1\}$.*

Definition 1 requires that password shares $\mathfrak{s}_0$ and $\mathfrak{s}_1$ can be combined to the policy-compliant integer password π. The corresponding verification must therefore be part of the 2BPR protocol. Otherwise, the client could register password shares $\mathfrak{s}_0$ and $\mathfrak{s}_1$ that can both be combined to a policy compliant password in the respective proofs with the servers, but combining $\mathfrak{s}_0$ and $\mathfrak{s}_1$ might result in a password that is *not* policy compliant, i.e. $f(\mathfrak{s}_0 + \mathfrak{s}') = \text{true}$ and $f(\mathfrak{s}_1 + \mathfrak{s}'') = \text{true}$ but $f(\pi) \neq \text{true}$. This further ensures that servers hold valid password shares, which is crucial for the security of 2PAKE/2PASS protocols that should be executed later with these password shares. We assume that the protocol is initiated by servers (possibly after the client expresses his interest to register). This allows each server to send its password policy to the client. We further assume that both servers can communicate with each other over an authenticated and confidential channel. This communication can either be done directly between the servers or indirectly using the client to transmit messages.

3.1 Security Model for 2BPR Protocols

2BPR protocols must guarantee that the client knows the sum PWDtoINT(pw) of the password shares $\mathfrak{s}_0$ and $\mathfrak{s}_1$, and that pw fulfils both password policies f_0 and f_1 if both servers accept the registration procedure. We translate Eq. (1) into a game-based security model that captures 2BPR security in form of two security requirements. The first requirement is called *Policy Compliance (PC)* of the registered password. In particular, if both servers are honest while accepting their password shares in the 2BPR protocol, the combination π of the shares represents a password compliant with their mutual policy $f = f_0 \cap f_1$, i.e. $f(\mathfrak{s}_b + \mathfrak{s}_{1-b}) = \text{true}$. The second requirement relates to the fact that servers should not learn anything about the registered password and is therefore called *Password Blindness (PB)*, i.e. a malicious server S_b may only learn whether a registered password is compliant with the mutual policy and nothing else. We observe that the blindness property must hold for all possible password policies and all compliant passwords. PB also implies impossibility of mounting an offline dictionary attack after observing 2BPR executions or through gaining access to and controlling at most one of the servers.

Setup and Participants. Protocol participants C, S_0, S_1 with C from the universe of clients and S_0, S_1 from the universe of servers have common inputs, necessary for the execution of the protocol. Instances of protocol participants C or S are denoted C_i, $S_{0,i}$ or $S_{1,i}$. Protocol participants without specified role are denoted by P, and S_b and S_{1-b} for unspecified servers. A client can register one password with any pair of servers from the universe. We use $\mathcal{C}$ and S_b as unique identifiers for the client and servers (e.g. $\mathcal{C}$ can be seen as a *username* that will be stored by servers alongside with password shares). We say a client $\mathcal{C}$ registers a password share for $(\mathcal{C}, S_{1-b})$ at server S_b and a password share for $(\mathcal{C}, S_b)$ at server S_{1-b}. There can be only at most one (most recent) password share registered at S_b resp. S_{1-b} for $(\mathcal{C}, S_{1-b})$ resp. $(\mathcal{C}, S_b)$ at any given time. A tuple $(\mathcal{C}, S_{1-b}, \mathfrak{s}_b)$ is stored on server S_b and tuple $(\mathcal{C}, S_b, \mathfrak{s}_{1-b})$ on server S_{1-b} only if the 2BPR protocol is viewed as successful by the servers.

Oracles. A PPT adversary $\mathcal{A}$ has access to Setup, Send, Execute and Corrupt oracles for interaction with the protocol participants.

- Setup$(\mathcal{C}, S_0, S_1, \mathrm{pw}')$ creates new instances of all participants and stores identifiers of the other parties to each participant. To this end the client receives the server policies $f_0 \cap f_1 = f$ and either chooses a new policy compliant password $\mathrm{pw} \in \mathcal{D}_f$ if $\mathrm{pw}' = \bot$ or uses $\mathrm{pw} = \mathrm{pw}'$.
- Execute$(\mathcal{C}, S_0, S_1)$ models a passive attack and executes a 2BPR protocol between new instances of $\mathcal{C}$, S_0 and S_1. It returns the protocol transcript and the internal state of all corrupted parties.
- Send$_{\mathcal{C}}(\mathcal{C}_i, S_{b,j}, m)$ sends message m, allegedly from client instance $\mathcal{C}_i$, to server instance $S_{b,j}$ for $b \in \{0,1\}$. If $\mathcal{C}_i$ or $S_{b,j}$ does not exist, the oracle aborts. Note that any instance $\mathcal{C}_i$ and $S_{b,j}$ was thus set up with Setup and therefore has an according partner instance $S_{1-b,j}$. If all participants exist, the oracle returns the server's answer m' if there exists any. Necessary inter server communication is performed in Send$_{\mathcal{C}}$ queries. If $m = \bot$, server $S_{b,j}$ returns its first protocol message if it starts the protocol.
- Send$_S(S_{b,i}, \mathcal{C}_j, m)$ sends message m, allegedly from server instance $S_{b,i}$ for $b \in \{0,1\}$, to client instance $\mathcal{C}_j$. If $S_{b,i}$ or $\mathcal{C}_j$ does not exist, the oracle aborts. Note that any instance $S_{b,i}$ and $\mathcal{C}_j$ was thus set up with Setup and therefore has an according partner instance $S_{1-b,i}$. If all participants exist, the oracle returns the client's answer m' if there exists any. If $m = \bot$, server $S_{b,i}$ returns its first message if he starts the protocol.
- Send$_{SS}(S_{b,i}, S_{1-b,j}, m)$ sends message m, from server instance $S_{b,i}$ for $b \in \{0,1\}$, to server instance $S_{1-b,j}$. If $S_{b,i}$ or $S_{1-b,j}$ does not exist, the oracle aborts. Note that any instance $S_{b,i}$ and $S_{1-b,j}$ was thus set up with Setup. If all participants exist, the oracle returns the server's answer m' if there exists any.
- Corrupt(S_b) allows the adversary to corrupt a server S_b and retrieve its internal state, i.e. stored messages and randomness, and the list of stored password shares $(\mathcal{C}, S_{1-b}, \mathfrak{s}_b)$. S_b is marked *corrupted*.

Note that we allow the adversary to register passwords with servers without requiring existence of a client instance $\mathcal{C}_i$ in a successful registration session. This is because we do not assume authenticated clients, i.e. client identifiers $\mathcal{C}$ are unique but not secret and can therefore be used by the adversary.

Policy Compliance. This is a natural security property of 2BPR protocols, requiring that registered client passwords comply with the mutual policy $f(\text{pw}) = \texttt{true}$. The attacker here plays the role of the client trying to register a password pw that is *not* policy compliant at two honest servers.

Definition 2 (Policy Compliance). *Policy compliance of a 2BPR protocol holds if for every PPT adversary $\mathcal{A}$ with access to* Setup *and* Send$_C$ *oracles the probability that two server instances $S_{b,i}$ and $S_{1-b,j}$ exist after $\mathcal{A}$ stopped that accepted $(\mathcal{C}, S_{1-b}, \mathfrak{s}_b)$, $(\mathcal{C}, S_b, \mathfrak{s}_{1-b})$ respectively, with $f(\mathfrak{s}_b + \mathfrak{s}_{1-b}) = \texttt{false}$ is negligible.*

Password Blindness. This property requires that every password, chosen and set-up by an honest client must remain hidden from an adversary who may corrupt at most one of the two servers, thus obtaining the internal state and taking full control over the corrupted server. We model password blindness through a distinguishing experiment where the attacker, after interacting with the oracles, outputs a challenge comprising of two passwords (pw_0 and pw_1), two clients ($\mathcal{C}_0$ and $\mathcal{C}_1$), and a pair of servers (S_0 and S_1). After a random assignment of passwords to the two clients, the adversary interacts with the oracles again and has to decide which client used which password in the 2BPR protocol execution. This is formalised in the following definition.

Definition 3 (Password Blindness). *The password blindness property of a 2BPR protocol Π holds if for every PPT adversary $\mathcal{A}$ there exists a negligible function $\varepsilon(\cdot)$ such that*

$$\mathsf{Adv}^{\text{PB}}_{\Pi,\mathcal{A}} = \left| \Pr[\mathsf{Exp}^{\text{PB}}_{\Pi,\mathcal{A}} = 1] - \frac{1}{2} \right| \leq \varepsilon(\lambda).$$

$\mathsf{Exp}^{\text{PB}}_{\Pi,\mathcal{A}}$:

$\quad (\mathcal{C}_0, \mathcal{C}_1, S_0, S_1, \text{pw}_0, \text{pw}_1) \leftarrow \mathcal{A}_1^{\mathsf{Setup},\mathsf{Send}_S,\mathsf{Send}_{SS},\mathsf{Execute},\mathsf{Corrupt}}$

$\quad check\ \text{pw}_0, \text{pw}_1 \in \mathcal{D}_{f_0 \cap f_1}, |\text{pw}_0| = |\text{pw}_1|, \mathcal{C}_0, \mathcal{C}_1 \in \{\mathcal{C}\}\ and\ S_0, S_1 \in \{S\}$

$\quad b' \leftarrow \mathcal{A}^{\mathsf{Setup}',\mathsf{Send}_S,\mathsf{Send}_{SS},\mathsf{Execute},\mathsf{Corrupt}}(\lambda, \mathcal{D}, \{\mathcal{C}\}, \{S\})$

$\quad if\ S_0\ or\ S_1\ is\ uncorrupted,\ return\ b = b';\ otherwise\ return\ 0$

where the modified oracle Setup' *(in contrast to* Setup*) picks a random bit $b \in_R \{0,1\}$ and uses pw_b as a password for client $\mathcal{C}_0$ and pw_{1-b} for $\mathcal{C}_1$.*

4 An Efficient Two-Server BPR Protocol

Before diving into technical details, we give a high-level description of our 2BPR protocol. We assume that client $\mathcal{C}$ selected two servers S_0 and S_1 to register with. We also assume the existence of server-authenticated and confidential channels (e.g. TLS channels [10, 16, 23]) between $\mathcal{C}$ and each S_b, $b \in \{0, 1\}$ as well as between S_0 and S_1. These channels prevent active impersonation of any server S_b,

$b \in \{0, 1\}$ and hide the contents of exchanged messages unless the corresponding server is corrupted.

Our 2BPR protocol further assumes a common reference string $\mathtt{crs} = (g, h, q)$ containing two generators g and h of a cyclic group of prime order q where $\log_g(h)$ is not known.

At the beginning of the registration phase the client C commits to the integer representation π of the chosen password string pw and sends this commitment together with a password share $\mathfrak{s}_b$ to the corresponding server S_b, $b \in \{0, 1\}$, along with auxiliary information that is needed to perform the policy compliance proof. For the latter, the client needs to prove the knowledge of π in the commitment such that $\pi = \mathfrak{s}_0 + \mathfrak{s}_1$ and that it fulfills both policies f_1 and f_2. Thus, servers S_0 and S_1 eventually register the new client, accept and store the client's password share, iff each S_b holds $\mathfrak{s}_b$ such that $\mathfrak{s}_0 + \mathfrak{s}_1 = \pi$ for $\pi \leftarrow \mathtt{PWDtoINT}(\mathtt{pw})$ and $f(\mathtt{pw}) = \mathtt{true}$ for $f = f_0 \cap f_1$.

4.1 Protocol Overview

In Fig. 1 we give an overview of the 2BPR protocol involving a client C and two servers S_b, $b \in \{0, 1\}$. The protocol proceeds in three phases. In the first phase *(client preparation)* the client chooses $\mathtt{pw} \in_R \mathcal{D}_f$, encodes it to π, computes shares $\mathfrak{s}_0$ and $\mathfrak{s}_1$, and computes commitments $\mathfrak{C}_0, \mathfrak{C}_1, \mathfrak{D}_0, \mathfrak{D}_1$ to the shares and the password. In the second phase *(password registration)* C interacts with each server S_b, $b \in \{0, 1\}$ over a server-authenticated and confidential channel. C computes a commitment C_i for each encoded character $\pi_i \leftarrow \mathtt{CHRtoINT}(c_i)$, $c_i \in \mathtt{pw}$, and a second commitment C_i' as a re-randomised version of C_i. The set $\boldsymbol{C'}$ containing the re-randomised commitments C_i', is then shuffled and used to prove through the Proof of Membership (**PoM**) protocol that each character committed to in $C_i' \in \boldsymbol{C'}$ is a member of some character set $\omega_{\phi(i)}$, chosen according to policy f. Note that **PoM** must be performed over the *shuffled* set $\boldsymbol{C'}$ of commitments as the server would otherwise learn the type (i.e. lower/upper case, digit, or symbol) of each password character. To further prove that transmitted commitments $\boldsymbol{C}, \mathfrak{C}_b$, and $\mathfrak{D}_b$ are correct, namely that the product of commitments in $\boldsymbol{C}$ commits to the password pw, $\mathfrak{C}_b$ contains the correct share $\mathfrak{s}_b$, and $\mathfrak{D}_b$ contains pw, client and server execute the Proof of Correctness (**PoC**) protocol. Finally, the client proves to each server that set $\boldsymbol{C'}$ is a shuffle of set $\boldsymbol{C}$ by executing the Proof of Shuffle (**PoS**) protocol. This proof is necessary to finally convince both servers that (1) the characters committed to in $\boldsymbol{C'}$ are the same as the characters in the commitments in $\boldsymbol{C}$, which can be combined to password pw (as follows from the **PoC** protocol) and (2) each commitment C_i is for a character $c_i \in \mathtt{pw}$ from some set ω_i, chosen according to policy f (as follows from the **PoM** protocol). For all three committed Σ-protocols (**PoM**, **PoC**, **PoS**) we use variables as defined in Sect. 2. If each server S_b, $b \in \{0, 1\}$ successfully verifies all three committed Σ-protocols and the length of the committed password pw is policy-conform, then both servers proceed with the last phase. In the third phase *(share verification)* the two servers S_0 and S_1 interact with each other over a mutually-authenticated and confidential channel. Each

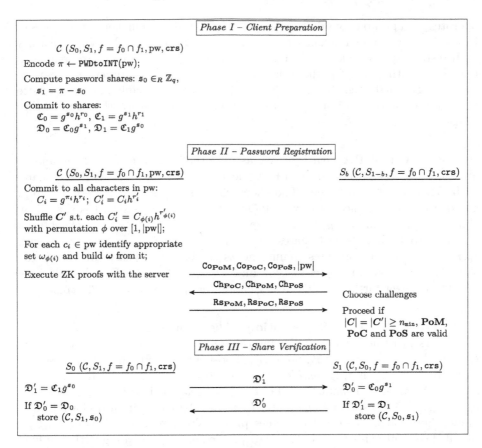

Fig. 1. Two-Server BPR Protocol — A high-level overview ω contains character sets of $c_{\phi(i)}$ ordered according to permutation ϕ used in **PoM**

S_b computes its verification value $\mathfrak{D}'_{1-b}$ and sends it to S_{1-b}. Upon receiving $\mathfrak{D}'_b$, S_b checks it against $\mathfrak{D}_b$ to verify that the client used the same password with both servers in the second phase, i.e. that $\mathfrak{s}_b + \mathfrak{s}_{1-b} = \pi$. If this verification is successful, S_b stores the client's password share $(\mathcal{C}, S_{1-b}, \mathfrak{s}_b)$ and considers $\mathcal{C}$ as being registered.

4.2 Two-Server BPR Specification

In the following we give a detailed description of the 2BPR protocol. To this end we describe the three proofs **PoC**, **PoM** and **PoS** detailing on their computations. We describe the interaction between client $\mathcal{C}$ and server S_b and therefore only consider one policy f_b. Note that $\mathcal{C}$ and each server S_b perform the same protocol. If both servers accept, the password fulfils the policy $f = f_b \cap f_{1-b}$.

We first describe the client's pre-computations such as password encoding and sharing before giving a detailed description of the proofs. The protocol operates on a group G of prime-order q with generator g. Further, let $h, f_i \in_R G$ for $i \in [-4, m]$ denote random group elements such that their discrete logarithm with respect to g is unknown. Public parameters of the protocol are defined as $(q, g, h, \boldsymbol{f})$ with $\boldsymbol{f} = \{f_i\}$ where m is at least $n = |pw|$. In practice m can be chosen big enough, e.g., 100, in order to process all reasonable passwords. Note that we use the range $i \in [0, n-1]$ for characters $pw[i]$, but $[1, x]$ for most other ranges.

Phase I – Client Preparation. We assume that password policies f_0 and f_1 are known by the client. This can be achieved by distributing them beforehand with other set-up parameters. The client chooses a password $pw \in_R \mathcal{D}_f$ from the dictionary and encodes it $\pi \leftarrow \texttt{PWDtoINT}(pw)$. The password is shared by choosing a random $\mathfrak{s}_b \in_R \mathbb{Z}_q$ and computing $\mathfrak{s}_{1-b} = \pi - \mathfrak{s}_b$. The client then commits to both password shares $\mathfrak{C}_b = g^{\mathfrak{s}_b} h^{r_b}$ and $\mathfrak{C}_{1-b} = g^{\mathfrak{s}_{1-b}} h^{r_{1-b}}$ with $r_b, r_{1-b} \in_R \mathbb{Z}_q$ and computes commitments to the entire password π with the same randomness, i.e. $\mathfrak{D}_b = \mathfrak{C}_b g^{\mathfrak{s}_{1-b}}$ and $\mathfrak{D}_{1-b} = \mathfrak{C}_{1-b} g^{\mathfrak{s}_b}$. For the following proofs the client further encodes every character $c_i \in pw$ as $\pi_i \leftarrow \texttt{CHRtoINT}(c_i)$.

Phase II – Password Registration. The client iterates over all encoded characters π_i to perform the following operations: commit to π_i by computing $C_i = g^{\pi_i} h^{r_i}, C_i' = C_i h^{r_i'}$ for $r_i, r_i' \in_R \mathbb{Z}_q^*$; choose a random permutation $\phi(i)$ over $[1, n]$ to shuffle C_i'; if π_i is *significant* for any $R_j \in R$, set $\omega_{\phi(i)} \leftarrow R_j$, otherwise $\omega_{\phi(i)} \leftarrow \Sigma$ (all ASCII characters). Let $l_i \in \mathbb{N}$ denote the index in $\omega_{\phi(i)}$ such that $c_i = \omega_{\phi(i)}[l_i]$. Values $(C_i, C_i', \omega_{\phi(i)}, \phi(i), l_i, \pi_i, r_i, r_i')$ are used in the following zero-knowledge proofs. The client combines previously computed values $C = \{C_i\}$. Shuffled commitments $C_{\phi(i)}'$ and sets $\omega_{\phi(i)}$ are combined according to the shuffled index $\phi(i)$, i.e. $C' = \{C_{\phi(i)}'\}$ and $\omega = \{\omega_{\phi(i)}\}$. Once these computations are finished $\mathcal{C}$ and S_b proceed with the protocol. In the following we describe the three proofs **PoM**, **PoC** and **PoS** and define their messages.

Proof of Correctness (PoC). This proof links the password shares, sent to each server, to the proof of policy compliance and shows knowledge of the other password share. We define the proof of correctness for an encoded password π, which proves that share $\mathfrak{s}_b$ can be combined with a second share $\mathfrak{s}_{1-b}$ such that $\pi = \mathfrak{s}_b + \mathfrak{s}_{1-b}$ and that the received commitments to password characters c_i can be combined to a commitment to that same password π. **PoC** is defined as a committed zero-knowledge proof between $\mathcal{C}$ and S_b for the statement

$$\mathsf{ZKP}\{(\pi, r_{1-b}, r_b, r_{Cb}) : \mathfrak{C}_{1-b} g^{\mathfrak{s}_b} = g^\pi h^{r_{1-b}} \wedge \prod_{i=0}^{n-1} C_i^{b^i} = g^\pi h^{r_{Cb}} \wedge \mathfrak{D}_b = g^\pi h^{r_b}\}.$$

$C_i = g^{\pi_i} h^{r_i}$ are character commitments from the set-up stage and $r_{Cb} = \sum_{i=0}^{n-1} b^i \cdot r_i$ is the combined randomness from the character commitments C_i. $\mathfrak{C}_{1-b} = g^{\mathfrak{s}_{1-b}} h^{r_{1-b}}$, $\mathfrak{D}_b = \mathfrak{C}_b g^{\mathfrak{s}_{1-b}}$, and $\mathfrak{C}_b = g^{\mathfrak{s}_b} h^{r_b}$ are the share and password commitments from the client preparation phase. This incorporates the link of the password commitment to the product of the commitments to the single characters with the proof of knowledge of the combined password $\pi = \mathfrak{s}_b + \mathfrak{s}_{1-b}$. The messages for **PoC** are computed as follows:

1. The client chooses random k_π, $k_{\rho b}$, $k_{\rho(1-b)}, k_{\rho C} \in_R \mathbb{Z}_q$, computes $t_{C(1-b)} = g^{k_\pi} h^{k_{\rho(1-b)}}$, $t_C = g^{k_\pi} h^{k_{\rho C}}$ and $t_{Db} = g^{k_\pi} h^{k_{\rho b}}$. The first message with $r_{\text{ComPoC}} \in_R \mathbb{Z}_q$ is then given by commitment

$$\text{Com}_{\textbf{PoC}} = g^{H(\mathfrak{C}_{1-b} g^{\mathfrak{s}b}, \{C_i\}, \mathfrak{D}_b, t_{C(1-b)}, t_C, t_{Db})} h^{r_{\text{ComPoC}}}.$$

2. After receiving $\text{Com}_{\textbf{PoC}}$ from the client the server chooses a random challenge $\textbf{Ch}_{\textbf{PoC},b} \in_R \mathbb{Z}_q$ and sends it back to the client.
3. After receiving the challenge $\textbf{Ch}_{\textbf{PoC},b}$, the client computes $s_\pi = k_\pi + \textbf{Ch}_{\textbf{PoC},b}\pi$, $s_{\rho(1-b)} = k_{\rho(1-b)} + \textbf{Ch}_{\textbf{PoC},b} r_{1-b}$, $s_{\rho C} = k_{\rho C} + \textbf{Ch}_{\textbf{PoC},b} \sum_{i=0}^{n-1} b^i r_i$ and $s_{\rho b} = k_{\rho b} + \textbf{Ch}_{\textbf{PoC},b} r_b$ before computing the next message with $r_{\text{RsPoC}} \in_R \mathbb{Z}_q$

$$\text{Rs}_{\textbf{PoC}}1 = g^{H(s_\pi, s_{\rho(1-b)}, s_{\rho C}, s_{\rho b})} h^{r_{\text{RsPoC}}}.$$

4. Eventually the client sets the decommitment message

$$\text{Rs}_{\textbf{PoC}}2 = (\mathfrak{s}_b, \mathfrak{C}_{1-b}, \{C_i\}, \mathfrak{D}_b, t_{C(1-b)}, t_C, t_{Db}, s_\pi, s_{\rho(1-b)}, s_{\rho C}, s_{\rho b}, r_{\text{ComPoC}}, r_{\text{RsPoC}}).$$

$\text{Rs}_{\textbf{PoC}}1$ and $\text{Rs}_{\textbf{PoC}}2$ form together client's response message $\text{Rs}_{\textbf{PoC}}$. The server verifies the proof by checking the following:

- $\text{Com}_{\textbf{PoC}} \overset{?}{=} g^{H(\mathfrak{C}_{1-b} g^{\mathfrak{s}b}, \{C_i\}, \mathfrak{D}_b, t_{C(1-b)}, t_C, t_{Db})} h^{r_{\text{ComPoC}}}$

- $\text{Rs}_{\textbf{PoC}}1 \overset{?}{=} g^{H(s_\pi, s_{\rho(1-b)}, s_{\rho C}, s_{\rho b})} h^{r_{\text{RsPoC}}}$; $\quad g^{s_\pi} h^{s_{\rho(1-b)}} \overset{?}{=} t_{C(1-b)} (\mathfrak{C}_{1-b} g^{\mathfrak{s}b})^{\textbf{Ch}_{\textbf{PoC},b}}$

- $g^{s_\pi} h^{s_{\rho C}} \overset{?}{=} t_C (\prod_{i=0}^{n-1} C_i^{b^i})^{\textbf{Ch}_{\textbf{PoC},b}}$; $\quad g^{s_\pi} h^{s_{\rho b}} \overset{?}{=} t_{Db} \mathfrak{D}_b^{\textbf{Ch}_{\textbf{PoC},b}}$

Proof of Membership (PoM). The proof of membership **PoM** proves for every password character $c_{\phi(i)} \in \text{pw}$ that its integer value $\pi_{\phi(i)} \in \omega_{\phi(i)}$ using the shuffled commitments $C'_{\phi(i)}$, i.e.

$$ZKP\{\{\pi_i, r_i\}_{i \in [0, n-1]} : \quad C'_{\phi(i)} = g^{\pi_i} h^{r_i} \wedge \pi_{\phi(i)} \in \omega_{\phi(i)}\}.$$

This proof consists of the following steps:

1. To prove that every $C'_{\phi(i)}$ commits to a value in the according set $\omega_{\phi(i)}$ the client computes the following values for the first move of the proof:

   - $\forall \pi_j \in \omega_{\phi(i)} \wedge \pi_j \neq \pi_{\phi(i)} :$ $\quad s_j \in_R \mathbb{Z}_q^*, c_j \in_R \mathbb{Z}_q^*$ and $t_j = g^{\pi_j} h^{s_j} (C'_{\phi(i)}/g^{\pi_j})^{c_j}$
   - $k_{\rho_i} \in_R \mathbb{Z}_q^*;$ $\quad t_{l_{\phi(i)}} = g^{\pi_i} h^{k_{\rho_i}}$

   Values $(\boldsymbol{t}_{\phi(i)}, \boldsymbol{s}_{\phi(i)}, \boldsymbol{c}_{\phi(i)}, k_{\rho_i})$, with $\boldsymbol{t}_{\phi(i)} = \{t_j, t_{l_{\phi(i)}}\}$, $\boldsymbol{s}_{\phi(i)} = \{s_j\}$, and $\boldsymbol{c}_{\phi(i)} = \{c_j\}$ are stored for future use. Note that $t_{l_{\phi(i)}}$ has to be added at the correct position $l_{\phi(i)}$ in $\boldsymbol{t}_{\phi(i)}$. A commitment $\text{Co}_{\textbf{PoM}} = g^{H(\omega, C', \boldsymbol{t}_{\phi(i)})} h^{r_{\text{CoPoM}}}$ with $r_{\text{CoPoM}} \in_R \mathbb{Z}_q$ is computed as output with $\omega = \{\omega_{\phi(i)}\}$.

2. The server stores received values, checks them for group membership, and chooses a random challenge $Ch_{PoM} = c \in_R \mathbb{Z}_q^*$.
3. After receiving the challenge c from the server, the client computes the following verification values for all commitments $C'_{\phi(i)}$ (note that s_j and c_j for all $j \neq l_{\phi(i)}$ are chosen already):

$$cl_{\phi(i)} = c \oplus \bigoplus_{j=1, j \neq l_{\phi(i)}}^{|\omega_{\phi(i)}|} c_j; \quad sl_{\phi(i)} = k_{\rho_{\phi(i)}} - cl_{\phi(i)}(r_i + r'_{\phi(i)}),$$

where i is the index of $C'_{\phi(i)}$ before shuffling. The client then combines $s = \{s_{\phi(i)} \cup \{sl_{\phi(i)}\}\}$ and $c = \{c_{\phi(i)} \cup \{cl_{\phi(i)}\}\}$. Note again that the set union has to consider the position of $l_{\phi(i)}$ to add the values at the correct position. A commitment $Rs_{PoM1} = g^{H(s,c)} h^{r_{RsPoM}}$ with $r_{RsPoM} \in_R \mathbb{Z}_q$ is computed as output.
4. Eventually the client sets the decommitment message with $t = \{t_{\phi(i)}\}$, $\omega = \{\omega_{\phi(i)}\}$, $r_{CoPoM} = \{r_{CoPoMi}\}$, $r_{RsPoM} = \{r_{RsPoMi}\}$, and $C' = \{C'_{\phi(i)}\}$ to

$$Rs_{PoM2} = (\omega, C', t, s, c, r_{CoPoM}, r_{RsPoM}).$$

Rs_{PoM1} and Rs_{PoM2} form together Rs_{PoM}. To verify the proof, i.e. to verify that every commitment $C'_{\phi(i)}$ in C' commits to a character c_i from either a subset of Σ if significant or Σ if not, the server verifies the following for every set $\omega_{\phi(i)} \in \omega$ with $i \in [1, n]$ and $x = \phi(i)$:

- Let $c_j \in c_i$ for $c_i \in c$ and verify $c \stackrel{?}{=} \bigoplus_{j=1}^{|\omega_i|} c_j$
- Let $\pi_j \in \omega_{\phi(i)}$, $s_i \in s$, $t_i \in t$, and $c_i \in c$, and verify $t_i[j] \stackrel{?}{=} g^{\pi_j} h^{s_i[j]} (C'_i / g^{\pi_j})^{c_i[j]}$ for all $j \in [1, |\omega_{\phi(i)}|]$

The server further verifies commitments $Co_{PoM} \stackrel{?}{=} g^{H(\omega, C', t)} h^{r_{CoPoM}}$ and $Rs_{PoM} \stackrel{?}{=} g^{H(s,c)} h^{r_{RsPoM}}$. The verification of the proof is successful iff all equations above are true *and* ω contains all significant characters for f_b.

Proof of Shuffle (PoS). The proof of correct shuffling **PoS** is based on the proofs from [13, 14]. In the following we specify the proof to work with Pedersen commitments instead of ElGamal ciphertexts. Note that indices for commitments C and C' run from 1 to n and index ranges in the following change frequently.

1. In the first move, the client (prover) builds a permutation matrix and commits to it. First he chooses random $A'_j \in_R \mathbb{Z}_q^*$ for $j \in [-4, n]$. Let A_{ij} denote a matrix with $i \in [-4, n]$ and $j \in [0, n]$, i.e. of size $(n+5) \times (n+1)$, such that a $n \times n$ sub-matrix of A_{ij} is the permutation matrix (built from permutation ϕ). Further, let ϕ^{-1} be the inverse shuffling function. This allows us to write the shuffle as $C'_i = \prod_{j=0}^n C_j^{A_{ji}} = C_{\kappa_i} h^{r'_{\kappa_i}}$ with $C_0 = h$ and $\kappa_i = \phi^{-1}(i)$

for $i \in [1, n]$. The matrix A_{ij} is defined with $A_{w0} \in_R \mathbb{Z}_q^*$, $A_{-1v} \in_R \mathbb{Z}_q^*$ and $A_{0v} = r'_{\phi(v)}$ for $w \in [-4, n]$ and $v \in [1, n]$. The remaining values in A_{ij} are computed as follows for $v \in [1, n]$:

$$- A_{-2v} = \sum_{j=1}^{n} 3A_{j0}^2 A_{jv}; \quad A_{-3v} = \sum_{j=1}^{n} 3A_{j0} A_{jv}; \quad A_{-4v} = \sum_{j=1}^{n} 2A_{j0} A_{jv}$$

After generating A_{ij} the client commits to it in $(C_0', \tilde{f}, \boldsymbol{f}', w, \tilde{w})$ for $\boldsymbol{f}' = \{f_v'\}$ with $v \in [0, n]$:

$$- f_v' = \prod_{j=-4}^{n} f_j^{A_{jv}}; \quad \tilde{f} = \prod_{j=-4}^{n} f_j^{A_j'}; \quad \tilde{w} = \sum_{j=1}^{n} A_{j0}^2 - A_{-40}$$

$$- C_0' = g^{\sum_{j=1}^n \pi_j A_{j0}} h^{A_{00} + \sum_{j=1}^n r_j A_{j0}}; \quad w = \sum_{j=1}^{n} A_{j0}^3 - A_{-20} - A_{-3}' \quad (2)$$

Note that $C_0' = \prod_{j=0}^{n} C_j^{A_{j0}} = h^{A_{00}} \prod_{j=1}^{n} C_j^{A_{j0}}$, but Eq. 2 saves $n - 1$ exponentiations. The output is then created as $\mathrm{Co_{PoS}} = g^{H(\{C_i\}, \{C_{\phi(i)}'\}, C_0', \tilde{f}, \boldsymbol{f}', w, \tilde{w})} h^{r_{\mathrm{Co_{PoS}}}}$ with $r_{\mathrm{Co_{PoS}}} \in_R \mathbb{Z}_q$.

2. When receiving $\mathrm{Co_{PoS}}$ the server chooses $\boldsymbol{c} = \{c_v\}$ with $c_v \in_R \mathbb{Z}_q^*$ for $v \in [1, n]$ and sets $\mathrm{Ch_{PoS}} = \boldsymbol{c}$.

3. After receiving challenges $\boldsymbol{c}$ from the server, the client computes the following verification values $(\boldsymbol{s}, \boldsymbol{s}')$ for $\boldsymbol{s} = \{s_v\}$ and $\boldsymbol{s}' = \{s_v'\}$ with $v \in [-4, n]$ and $c_0 = 1$:

$$s_v = \sum_{j=0}^{n} A_{vj} c_j; \qquad\qquad s_v' = A_v' + \sum_{j=1}^{n} A_{vj} c_j^2$$

The client sets $\mathrm{Rs_{PoS1}} = g^{H(\boldsymbol{s}, \boldsymbol{s}')} h^{r_{\mathrm{Rs_{PoS}}}}$ with $r_{\mathrm{Rs_{PoS}}} \in_R \mathbb{Z}_q$.

4. Eventually, the client sends the decommitment message to the server

$$\mathrm{Rs_{PoS2}} = (C_0', \tilde{f}, \boldsymbol{f}', w, \tilde{w}, \boldsymbol{s}, \boldsymbol{s}', r_{\mathrm{Co_{PoS}}}, r_{\mathrm{Rs_{PoS}}}).$$

Note that $\{C_i\}$ and $\{C_{\phi(i)}'\}$ are omitted here as they are part of $\mathrm{Rs_{PoC2}}$, $\mathrm{Rs_{PoM2}}$ respectively, already. If this proof is used stand-alone, those values have to be added to $\mathrm{Rs_{PoS2}}$.

$\mathrm{Rs_{PoS1}}$ and $\mathrm{Rs_{PoS2}}$ form together $\mathrm{Rs_{PoS}}$. The server verifies now that the correctness of the commitments $\mathrm{Co_{PoS}} \overset{?}{=} g^{H(\{C_i\}, \{C_{\phi(i)}'\}, C_0', \tilde{f}, \boldsymbol{f}', w, \tilde{w})} h^{r_{\mathrm{Co_{PoS}}}}$ and $\mathrm{Rs_{PoS1}} \overset{?}{=} g^{H(\boldsymbol{s}, \boldsymbol{s}')} h^{r_{\mathrm{Rs_{PoS}}}}$, and that the following equations hold for a randomly chosen $\alpha \in_R \mathbb{Z}_q^*$ and $C_0 = h$:

$$- \prod_{v=-4}^{n} f_v^{s_v + \alpha s_v'} \overset{?}{=} f_0' \tilde{f}^\alpha \prod_{j=1}^{n} f_j^{c_j + \alpha c_j^2}; \quad \prod_{v=0}^{n} C_v^{s_v} \overset{?}{=} \prod_{j=0}^{n} C_j^{c_j};$$

$$- \sum_{j=1}^{n} (s_j^3 - c_j^3) \overset{?}{=} s_{-2} + s_{-3}' + w; \quad \sum_{j=1}^{n} (s_j^2 - c_j^2) \overset{?}{=} s_{-4} + \tilde{w}$$

The server accepts the proof if and only if all those verifications succeed. This concludes the proof of correct shuffling.

Phase III – Share Verification. To verify that the client used the same password pw and shares $\mathfrak{s}_0, \mathfrak{s}_1$ with both servers S_0 and S_1, the servers compute the commitment $\mathfrak{D}'_b$ from the share commitment $\mathfrak{C}_b$ and their share $\mathfrak{s}_{1-b}$, and exchange it. Comparing $\mathfrak{D}'_b$ with the value $\mathfrak{D}_b$ received from the client, the server verifies share correctness. This concludes the 2BPR protocol and each server S_b stores $(\mathcal{C}, S_{1-b}, \mathfrak{s}_b)$ if all checks were successful.

4.3 Security Analysis

We show that our 2BPR protocol is secure in the model from Sect. 3.1 and thus offers policy compliance and password blindness. For space limitations we include only the proofs of Theorems 1 and 2 note that **PoM** and **PoC** protocols are standard concurrent ZK proofs and **PoS** is a slightly modified concurrent ZK proof from [13,14].

Lemma 1. *The* **PoC** *protocol from Sect. 4.2 is a concurrent zero-knowledge proof if the discrete logarithm problem in the used group G is hard and $H : \{0,1\}^* \mapsto \mathbb{Z}_q$ is a collision resistant hash function.*

Lemma 2. *The* **PoM** *protocol from Sect. 4.2 is a concurrent zero-knowledge proof if the discrete logarithm problem in the used group G is hard and $H : \{0,1\}^* \mapsto \mathbb{Z}_q$ is a collision resistant hash function.*

Lemma 3 ([13,14]). *The* **PoS** *protocol from Sect. 4.2 is a concurrent zero-knowledge proof of knowledge of shuffling ϕ if the discrete logarithm problem in the used group G is hard and $H : \{0,1\}^* \mapsto \mathbb{Z}_q$ is a collision resistant hash function.*

Theorem 1. *If G is a DL-hard group of prime-order q with generators g and h, and H a collision resistant hash function, the construction in Fig. 1 provides policy compliance according to Definition 2.*

Proof. We show how to build a successful attacker on the soundness of **PoC**, **PoM** and **PoS** using a successful attacker against policy compliance who has access to Setup and $\mathsf{Send}_\mathcal{C}$ oracles.

Game_0 : This game corresponds to the correct execution of the protocol.
Game_1 : In this game we change how $\mathsf{Send}_\mathcal{C}(\mathcal{C}_i, S_{b,j}, m)$ queries are answered. If m is parsed as $(\mathsf{Co}_{\mathbf{PoM}}, \mathsf{Co}_{\mathbf{PoC}}, \mathsf{Co}_{\mathbf{PoS}})$ the $\mathsf{Co}_{\mathbf{PoM}}$ is used by the challenger as output to the **PoM** verifier who returns challenge $\mathsf{Ch}_{\mathbf{PoM}}$ which is then returned in response to the $\mathsf{Send}_\mathcal{C}$ query (other challenges are generated at random). If m is parsed as $(\mathsf{Rs}_{\mathbf{PoM}1}, \mathsf{Rs}_{\mathbf{PoC}1}, \mathsf{Rs}_{\mathbf{PoS}1})$ or $(\mathsf{Rs}_{\mathbf{PoM}2}, \mathsf{Rs}_{\mathbf{PoC}2}, \mathsf{Rs}_{\mathbf{PoS}2})$ and the first $\mathsf{Send}_\mathcal{C}$ query from that session was forwarded to the verifier then $\mathsf{Rs}_{\mathbf{PoM}1}, \mathsf{Rs}_{\mathbf{PoM}2}$ respectively, is used as output to the **PoM** verifier. It is easy to see that the

challenger breaks soundness of **PoM** if the adversary uses a password pw $\notin \mathcal{D}_f$ and **PoM** verifies successfully. We can therefore assume for the remaining games that pw $\in \mathcal{D}_f$.

Game$_2$: In this game we introduce another change to the processing of $\mathsf{Send}_{\mathcal{C}}(\mathcal{C}_i, S_{b,j}, m)$ queries. If m is parsed as $(\mathrm{Co}_{\mathbf{PoM}}, \mathrm{Co}_{\mathbf{PoC}}, \mathrm{Co}_{\mathbf{PoS}})$ then $\mathrm{Co}_{\mathbf{PoC}}$ is used by the challenger as output to the **PoC** verifier who returns challenge $\mathrm{Ch}_{\mathbf{PoC}}$ that is then used as response to the $\mathsf{Send}_{\mathcal{C}}$ query (other challenges are generated at random). If m is parsed as $(\mathrm{Rs}_{\mathbf{PoM}1}, \mathrm{Rs}_{\mathbf{PoC}1}, \mathrm{Rs}_{\mathbf{PoS}1})$ or $(\mathrm{Rs}_{\mathbf{PoM}2}, \mathrm{Rs}_{\mathbf{PoC}2}, \mathrm{Rs}_{\mathbf{PoS}2})$ and the first $\mathsf{Send}_{\mathcal{C}}$ query from that session was forwarded to the verifier then $\mathrm{Rs}_{\mathbf{PoC}1}, \mathrm{Rs}_{\mathbf{PoC}2}$ respectively, is used as output to the **PoC** verifier. It is easy to see that the challenger breaks soundness of **PoC** if $\mathfrak{s}_0 + \mathfrak{s}_1 \neq \pi$, i.e. the password share $\mathfrak{s}_b$ can not be used with a second share $\mathfrak{s}_{1-b}$ to rebuild the password π committed to in C, i.e. $\sum_i b^i \pi_i \neq \pi$. Observe further that the second share $\mathfrak{s}_{1-b}$ has to be stored on server S_{1-b}, i.e. the attacker has not performed the set-up with S_b and S_{1-b} with shares that do not combine to the same encoded password π. Otherwise we can break the binding property of Pedersen commitments. In particular, the attacker has to generate commitments $\mathfrak{C}_0, \mathfrak{C}_1, \mathfrak{D}_0$ and $\mathfrak{D}_1$ such that $\mathfrak{C}_0 g^{\mathfrak{s}_1} = \mathfrak{D}_0$ or $\mathfrak{C}_1 g^{\mathfrak{s}_0} = \mathfrak{D}_1$. We can therefore for the remaining games that the password share $\mathfrak{s}_b$ received by server S_b can be combined with the second share $\mathfrak{s}_{1-b}$ of server S_{1-b} to an encoded password π with according character commitments C_i.

Game$_3$: In this game we change once more how $\mathsf{Send}_{\mathcal{C}}(\mathcal{C}_i, S_{b,j}, m)$ queries are answered. If m is parsed as $(\mathrm{Co}_{\mathbf{PoM}}, \mathrm{Co}_{\mathbf{PoC}}, \mathrm{Co}_{\mathbf{PoS}})$ then $\mathrm{Co}_{\mathbf{PoS}}$ is used by the challenger as output to the **PoS** verifier who returns challenge $\mathrm{Ch}_{\mathbf{PoS}}$ that is then out in response to the $\mathsf{Send}_{\mathcal{C}}$ query (other challenges are generated at random). If m from the adversary is parsed as $(\mathrm{Rs}_{\mathbf{PoM}1}, \mathrm{Rs}_{\mathbf{PoC}1}, \mathrm{Rs}_{\mathbf{PoS}1})$ or $(\mathrm{Rs}_{\mathbf{PoM}2}, \mathrm{Rs}_{\mathbf{PoC}2}, \mathrm{Rs}_{\mathbf{PoS}2})$ and the first $\mathsf{Send}_{\mathcal{C}}$ query from that session was forwarded to the verifier then $\mathrm{Rs}_{\mathbf{PoS}1}, \mathrm{Rs}_{\mathbf{PoS}2}$ respectively, is used as the output to the **PoS** verifier. In this case if the attacker is rewindable, the challenger can act as a knowledge extractor for **PoS**. In particular, we can extract shuffling function ϕ and re-randomiser $\{r_i'\}$ to break soundness of **PoS**. This implies that C' is a correct shuffle of C. We conclude the proof by observing that the password shares stored on both servers can be combined to a policy compliant password.

Theorem 2. *If G is a DL-hard group of prime-order q with generators g and h, and H a collision resistant hash function, the construction in Fig. 1 provides password blindness according to Definition 3.*

Proof. We prove this theorem through a sequence of games. In the last game simulated interactions between servers and clients are simulated and password independent, thus requiring the attacker to perform a random guess of the bit b.

Game$_0$: This is the correct execution of the protocol.
Game$_1$: The challenger computes crs with the knowledge of the trapdoor $\tau = \log_g(h)$.
Game$_2$: The challenger simulates the proofs **PoC**, **PoM** and **PoS** and messages exchanged between the servers as part of the Execute oracle but stores two correct

shares on the servers to allow consistency if servers become corrupted. Since at least one server must remain uncorrupted the probability difference between both games is negligible due to the zero-knowledge property of the proofs.

Game₃ : This game modifies Send_S and Send_{SS} responses if the second participating server is uncorrupted by simulating zero-knowledge proofs and answering Send_{SS} queries using $\mathfrak{D}'_b = \mathfrak{D}_b$. To guarantee consistency in case of corruptions the challenger still stores appropriate shares. The probability difference between both games is negligible due to the zero-knowledge property of the proofs. Since all proofs are password-independent and Pedersen commitments offer unconditional hiding the attacker can only win by guessing b.

5   Performance and Use with 2PAKE/2PASS Protocols

An unoptimised prototype of the 2BPR protocol from Sect. 4 was implemented over the NIST P-192 elliptic curve [27] in Python using the Charm framework [1] to estimate the performance. The tests (completed on a laptop with an Intel Core Duo P8600 at 2.40 GHz for both client and server) underline the claim that the protocol is practical. For instance, for a password of length 10 and policies $(dl, 5)$ and $(ds, 7)$ computations take 1.4 s on the client and 0.68 s on each server. The overall computing time for a password of length 10 was 2.76 s and increased to 6.34 s for a password of length 20. Also note that the execution can be parallelised if the client performs the proofs with S_0 and S_1 at the same time. The source code is available from https://goo.gl/XfIZtn.

Application to Existing 2PAKE/2PASS Protocols. Our 2BPR protocol can be used to register passwords for 2PAKE and 2PASS protocols that adopt additive password sharing in $\mathbb{Z}_q$ or multiplicative sharing in G. This includes 2PAKE protocols from [20,21] for which no password registration procedures were addressed. Integration of 2BPR into 2PASS protocols is more involved since password registration is considered to be part of the 2PASS protocol during the secret sharing phase. 2PASS protocols in general can be divided in two stages: password and secret registration/sharing and secret reconstruction. While the approach from [2] and subsequent works [5,17,30] do not actually share the password and could therefore use other means to verify policy compliance of a passwords used, the UC-secure 2PASS protocol from [6] uses multiplicative password sharing in G. In order to use our 2BPR protocol withing the setup procedure of [6] we can redefine the encoded password to g^π with $\pi \leftarrow \text{PWDtoINT}(pw)$ such that shares are computed as $g^\pi = g^{s_0} g^{s_1}$. The first message (step 1) from the setup protocol in [6] can piggyback the first 2BPR protocol message. The subsequent three messages between the client and each server are performed between step 1 and step 2, while the inter-server communication can be piggybacked on step 2 and step 3. In addition to checking correctness of shares in the setup of [6] the servers can now verify the 2BPR proofs to check policy compliance. This would adds three flows to the setup protocol of [6].

6   Conclusion

In this work we introduced the notion of two-server blind password registration (2BPR), which is a solution for secure registration of policy-compliant, user-selected passwords for 2PAKE/2PASS protocols where each server is supposed to learn only its own share of the password and whether the combined password is conform with his password policy. Our efficient 2BPR protocol can be used to register 2PAKE/2PASS passwords satisfying server-chosen policies over the alphabet of all 94 printable ASCII characters.

References

1. Akinyele, J.A., Garman, C., Miers, I., Pagano, M.W., Rushanan, M., Green, M., Rubin, A.D.: Charm: a framework for rapidly prototyping cryptosystems. J. Crypt. Eng. **3**(2), 111–128 (2013)
2. Bagherzandi, A., Jarecki, S., Saxena, N., Lu, Y.: Password-protected secret sharing. In: CCS 2011, pp. 433–444. ACM (2011)
3. Bonneau, J.: The science of guessing: analyzing an anonymized corpus of 70 million passwords. In: IEEE S&P, pp. 538–552. IEEE Computer Society (2012)
4. Brainard, J.G., Juels, A., Kaliski, B., Szydlo, M.: A new two-server approach for authentication with short secrets. In: USENIX Security Symposium, USENIX Association (2003)
5. Camenisch, J., Lehmann, A., Lysyanskaya, A., Neven, G.: Memento: how to reconstruct your secrets from a single password in a hostile environment. In: Garay, J.A., Gennaro, R. (eds.) CRYPTO 2014, Part II. LNCS, vol. 8617, pp. 256–275. Springer, Heidelberg (2014)
6. Camenisch, J., Lysyanskaya, A., Neven,G.: Practical yet universally composable two-server password-authenticated secret sharing. In: CCS 2012, pp. 525–536. ACM (2012)
7. Damgård, I.B.: Efficient concurrent zero-knowledge in the auxiliary string model. In: Preneel, B. (ed.) EUROCRYPT 2000. LNCS, vol. 1807, pp. 418–430. Springer, Heidelberg (2000)
8. Goodin, D., Hack of cupid media dating website exposes 42 million plaintext passwords. http://goo.gl/ImLE1C. Accessed 01 Apr 2015
9. Dell'Amico, M., Michiardi, P., Roudier, Y.: Password strength: an empirical analysis. In: INFOCOM, pp. 983–991. IEEE (2010)
10. Dierks, T., Rescorla, E.: The Transport Layer Security (TLS) protocol version 1.2. RFC 5246 (proposed standard), updated by RFCs 5746, 5878, 6176, 7465, August 2008
11. Dürmuth, M., Kranz, T.: On password guessing with GPUs and FPGAs. In: PASSWORDS 2014, pp. 19–38 (2014)
12. Ford, W., Kaliski, Jr. B.S.: Server-assisted generation of a strong secret from a password. In: WETICE, pp. 176–180. IEEE (2000)
13. Furukawa, J.: Efficient and verifiable shuffling and shuffle-decryption. IEICE Trans. **88−A**(1), 172–188 (2005)
14. Furukawa, J., Sako, K.: An efficient scheme for proving a shuffle. In: Kilian, J. (ed.) CRYPTO 2001. LNCS, vol. 2139, pp. 368–387. Springer, Heidelberg (2001)
15. hashcat: hashcat - advanced password recovery. http://hashcat.net/. Accessed 01 Apr 2015

16. Jager, T., Kohlar, F., Schäge, S., Schwenk, J.: On the security of TLS-DHE in the standard model. In: Safavi-Naini, R., Canetti, R. (eds.) CRYPTO 2012. LNCS, vol. 7417, pp. 273–293. Springer, Heidelberg (2012)
17. Jarecki, S., Kiayias, A., Krawczyk, H.: Round-optimal password-protected secret sharing and T-PAKE in the password-only model. In: Sarkar, P., Iwata, T. (eds.) ASIACRYPT 2014, Part II. LNCS, vol. 8874, pp. 233–253. Springer, Heidelberg (2014)
18. Jarecki, S., Lysyanskaya, A.: Adaptively secure threshold cryptography: introducing concurrency, removing erasures (extended abstract). In: Preneel, B. (ed.) EUROCRYPT 2000. LNCS, vol. 1807, p. 221. Springer, Heidelberg (2000)
19. Jin, H., Wong, D.S., Xu, Y.: An efficient password-only two-server authenticated key exchange system. In: Qing, S., Imai, H., Wang, G. (eds.) ICICS 2007. LNCS, vol. 4861, pp. 44–56. Springer, Heidelberg (2007)
20. Katz, J., MacKenzie, P.D., Taban, G., Gligor, V.D.: Two-server password-only authenticated key exchange. In: Ioannidis, J., Keromytis, A.D., Yung, M. (eds.) ACNS 2005. LNCS, vol. 3531, pp. 1–16. Springer, Heidelberg (2005)
21. Kiefer, F., Manulis, M.: Distributed smooth projective hashing and its application to two-server password authenticated key exchange. In: Boureanu, I., Owesarski, P., Vaudenay, S. (eds.) ACNS 2014. LNCS, vol. 8479, pp. 199–216. Springer, Heidelberg (2014)
22. Kiefer, F., Manulis, M.: Zero-knowledge password policy checks and verifier-based PAKE. In: Kutyłowski, M., Vaidya, J. (eds.) ICAIS 2014, Part II. LNCS, vol. 8713, pp. 295–312. Springer, Heidelberg (2014)
23. Krawczyk, H., Paterson, K.G., Wee, H.: On the security of the TLS protocol: a systematic analysis. In: Canetti, R., Garay, J.A. (eds.) CRYPTO 2013, Part I. LNCS, vol. 8042, pp. 429–448. Springer, Heidelberg (2013)
24. Ma, J., Yang, W., Luo, M., Li, N.: A study of probabilistic password models. In: IEEE S&P, pp. 689–704 (2014)
25. MacKenzie, P.D., Shrimpton, T., Jakobsson, M.: Threshold password-authenticated key exchange. In: Yung, M. (ed.) CRYPTO 2002. LNCS, vol. 2442, pp. 385–400. Springer, Heidelberg (2002)
26. Cubrilovic, N., Hack, R.: From bad to worse (2014). http://goo.gl/AF5ZDM. Accessed 01 Apr 2015
27. NIST: National Institute of Standards and Technology. Recommended elliptic curves for federal government use (1999). http://goo.gl/M1q10h
28. Openwall: John the Ripper password cracker. http://www.openwall.com/john/. Accessed 01 Apr 2015
29. Pedersen, T.P.: Non-interactive and information-theoretic secure verifiable secret sharing. In: Feigenbaum, J. (ed.) CRYPTO 1991. LNCS, vol. 576, pp. 129–140. Springer, Heidelberg (1992)
30. Pryvalov, I., Kate, A.: Introducing fault tolerance into threshold password-authenticated key exchange. Cryptology ePrint Archive, report 2014/247 (2014)
31. Reuters: Trove of Adobe user data found on web after breach: security firm (2014). http://goo.gl/IC4lu8. Accessed 01 Apr 2015
32. Szydlo, M., Kaliski, B.: Proofs for two-server password authentication. In: Menezes, A. (ed.) CT-RSA 2005. LNCS, vol. 3376, pp. 227–244. Springer, Heidelberg (2005)
33. Reuters, T.: Microsoft India store down after hackers take user data. http://goo.gl/T7puD1. Accessed 01 Apr 2015
34. Yang, Y., Deng, R.H., Bao, F.: A practical password-based two-server authentication and key exchange system. IEEE Trans. Dependable Sec. Comput. 3(2), 105–114 (2006)

Chip Authentication for E-Passports: PACE with Chip Authentication Mapping v2

Lucjan Hanzlik[✉] and Mirosław Kutyłowski

Wrocław University of Technology,
Wybrzeże Wyspiańskiego 27, 50-370 Wrocław, Poland
{lucjan.hanzlik,miroslaw.kutylowski}@pwr.edu.pl

Abstract. According to the European Commission Decision C(2006) 2909, EU Member States must implement Supplemental Access Control (SAC) on biometric passports. The SAC standard describes two versions of a password based authenticated key exchange protocol called PACE-GM and PACE-IM. Moreover, it defines an extension called PACE-CAM. Apart from password authentication and establishing a session key, the PACE-CAM protocol executes an active authentication of the ePassport with just one extra modular multiplication. However, it uses PACE-GM as a building block and does not work with the more efficient protocol PACE-IM. In this paper we propose an active authentication extension, which can be used with both PACE-GM and PACE-IM. Moreover, the protocol's overhead on the side of the ePassport, remains the same despite more universality.

Keywords: ePassport · Supplemental Access Control · PACE · Active Authentication · Chip Authentication Mapping · ICAO

1 Introduction

Electronic passport (ePassport) is a combination of a traditional paper passport and an electronic layer. This layer includes a microprocessor (also called chip or tag) and an RFID antenna used for power supply and communication. A chip is not merely a data storage, but also a secure suite for performing cryptographic algorithms.

Since 2004, standards regarding *machine-readable travel documents* (MRTD) are issued by the International Civil Aviation Organization (ICAO). These guidelines define cryptographic algorithms used during the inspection procedure as well as the format of the document. In particular, it defines a special machine-readable zone (MRZ), which can be read by an optical reader and is usually placed on the bottom of the first page of the ePassport. The ICAO standard has been implemented by most countries and is the *de-facto* standard for ePassports.

In 2008 the German Federal Office for Information Security proposed changes to the ePassport standard [BSI15]. The new standard defines a password-based authentication key exchange protocol called *Password Authenticated Connection Establishment* (PACE). Unlike the still used Basic Access Control protocol

© Springer International Publishing Switzerland 2016
M. Bishop and A.C.A. Nascimento (Eds.): ISC 2016, LNCS 9866, pp. 115–129, 2016.
DOI: 10.1007/978-3-319-45871-7_8

(BAC) it is secure against offline dictionary attacks. The new standard was successfully implemented on new German personal identity documents (nPA - *neuer Personalausweis*). On the downside, due to the interoperability requirement, all ePassports must implement the old ICAO standard. Fortunately, in 2011 ICAO proposed a new standard called Supplemental Access Control (SAC) [ISO11], which proposes PACE as a *recommended* replacement for BAC. The standard defines two protocol versions: PACE with Generic Mapping (PACE-GM) and PACE with Integrated Mapping (PACE-IM). It also defines additional means to authenticate the chip. Active Authentication and Chip Authentication (from the standard [BSI15]) are recommended. However, those protocol require at least one exponentiation. In 2014 ICAO updated the SAC standard [ISO14]. This update introduced PACE with Chip Authentication Mapping (PACE-CAM). This protocol verifies the authenticity of the ePassport chip at a cost of one extra modular multiplication and is integrated with PACE-GM. Note that this is more efficient than using Active Authentication or Chip Authentication as a separate protocol. However, this version is limited to PACE-GM and cannot be used with the more efficient PACE-IM. Moreover, BSI patented PACE-CAM [BK12]. In this paper we will focus on the PACE protocol and in particular on the PACE-CAM version.

Related Work. Password-authenticated key exchange protocols were first introduced by Bellovin and Merritt [BM92]. Their solution, called Encrypted Key Exchange (EKE), uses a combination of symmetric and public-key cryptography and provides resistance against offline-dictionary attacks. In 1996, David Jablon proposed SPEKE (Simple Password Encrypted Key Exchange) [Jab96] as an extension of EKE. Apart from security against dictionary attacks, the SPEKE protocol achieves perfect forward secrecy, i.e. disclosure of a password does not compromise the remaining session keys.

The PACE protocol introduced by BSI is somewhat similar to the SPEKE protocol. Both parties use a shared password and a mapping function to derive a secret group generator. This generator is then used in a DH key exchange protocol with some additional steps based on message authentication codes.

PACE-GM was proven to be a secure authenticated key exchange protocol under a new assumption called *Password-Based Chosen-Element DH* [BFK09]. The authors also proposed a general version of this problem and used it to prove security of PACE with an arbitrary mapping function. Coron et al. proved security of the PACE-IM protocol [CGIP11] using a newly introduced cryptographic assumption called *Gap Chosen-Base Diffie-Hellman*. This problem implies that the protocol is secure even for groups with a DDH oracle (in real life it can be instantiated with groups with admissible bilinear maps).

The first extension of the PACE protocol that additionally authenticates the chip was proposed by Bender et al. in [BDFK12] and called PACE|AA protocol. Their solution was based on a combination of the PACE-GM protocol with ECDSA signature scheme. The chip reuses some random coins from the PACE-GM protocol to produce the signature, minimizing the cost of its generation.

The authors showed that PACE|AA inherits the AKE security from PACE-GM and that it is secure against impersonation attacks.

Independently, Bender et al. [BFK13] and Hanzlik et al. [HKK13] published an extension that instead of using signatures, uses the random coins from PACE-GM to bind them to the public key. In other words, the user gives a "proof" that, if it knows the random coins used (which is implicitly verified by PACE-GM), it also knows the secret key. The authors of [BFK13] propose two roughly equivalent versions: with a multiplicative binding (same as the one proposed by [HKK13]) and an additive binding. The multiplicative version was later on added to the ICAO SAC standard and renamed as Chip Authentication Mapping version (PACE-CAM). The security against impersonation of PACE-CAM, described in [BFK13] is based on non-standard assumptions. On the other hand, the authors of [HKK13] propose only security arguments without a reduction based proof and security model. However, they discuss privacy of the solution and show in particular that a reader cannot convince a third party that it has interacted with an ePassport. Moreover, they propose a leakage-resilient version of the protocol, i.e. the secret key is secure even in the case of leakage of all ephemeral value (this is not the case for the standard version and the one proposed in [BDFK12]).

Motivation. Today, the scale of usage of travel documents is rapidly growing. At the same time, the share of ID documents equipped with electronic layer is growing as well. Speed-up and accuracy of document inspection comes with a price: we have to ensure that the personal and biometric data are secure and that their privacy is well protected. For this reason, there have been significant efforts of the ICAO organization to improve security and interoperability of travel documents. This work has a big impact, since most state authorities issuing ePassports are using the ICAO standard as a guidance.

Unfortunately, as shown by many authors, this work is not always perfect. Formal security proofs of the introduced cryptographic protocols are frequently published after appearance of the standard and after deployment in the ePassports. Since a passport is usually issued with validity period of 10 years, lack of a formal security proof or its incompleteness is a major risk. Indeed, in case of a security flaw, there is no way to install a security patch in the ePassports and there is no way to exchange in a reasonable time the flawed passports, unless their number is small. In this situation any mistake or unnoticed weakness might have profound consequences in the future.

Last not least, due to the application scale, there is also a strong motivation for deep efficiency optimization of the protocols. Indeed, the saving for the production cost of a single chip on ePassport might be a few cents, but we have to take into account billions of travel documents issued. Similar arguments apply for the time necessary to inspect a travel document in an automatic border control booth – ideally the travellers should simply go through these booths and the inspection should take a small fraction of a second.

Paper Contribution. In this paper we consider the Password Authenticated Connection Establishment (PACE) protocol. In particular, we focus on the Chip Authentication Mapping introduced in the latest revision of the Supplemental Access Control document [ISO14].

Our contribution is protocol called PACE-CAM v2. Thanks to bilinear maps we can apply the same trick as in PACE-CAM, but independent of the mapping function used in PACE. Moreover, the efficiency on side of the ePassport remains the same. Note that the standard PACE-CAM is limited to PACE with the generic mapping function (based on Diffie-Hellman key exchange protocol). Moreover, our results indicate that currently PACE-CAM v2 has better security guarantees than PACE-CAM. In particular, we prove that under the 2-Strong Diffie-Hellman assumption PACE-CAM v2 (with Integrated Mapping or Generic Mapping) is secure against impersonation.

2   Preliminaries

In this chapter we recall basic definitions, number-theoretic assumptions and cryptographic primitives necessary to formally define and analyze the cryptographic protocol presented in this paper.

2.1   Bilinear Maps

Let us consider cyclic groups $(\mathbb{G}_1, +)$, $(\mathbb{G}_2, +)$, $(\mathbb{G}_T, \cdot)$ of a prime order q. Let P, Q be generators of respectively $\mathbb{G}_1$ and $\mathbb{G}_2$. We say that $(\mathbb{G}_1, \mathbb{G}_2, \mathbb{G}_T)$ are *bilinear map groups* if there exists a bilinear map $e : \mathbb{G}_1 \times \mathbb{G}_2 \to \mathbb{G}_T$ satisfying the following properties:

Bilinearity: for all $(S, T) \in \mathbb{G}_1 \times \mathbb{G}_2$ and $a, b \in \mathbb{Z}_q$, we have
$e(aS, bT) = e(S, T)^{a \cdot b}$,
Non-degeneracy: $e(P, Q) \neq 1$ is a generator of group $\mathbb{G}_T$,
Computability: $e(S, T)$ is efficiently computable for all $(S, T) \in \mathbb{G}_1 \times \mathbb{G}_2$.

Moreover, we say that the map e is an *admissible bilinear map* or *admissible pairing function*. Depending on the choice of groups we say that map e is of:

Type 1: if $\mathbb{G}_1 = \mathbb{G}_2$,
Type 2: if $\mathbb{G}_1$ and $\mathbb{G}_2$ are distinct groups and there exists an efficiently computable isomorphism $\psi : \mathbb{G}_2 \to \mathbb{G}_1$,
Type 3: if $\mathbb{G}_1$ and $\mathbb{G}_2$ are distinct groups and no efficiently computable isomorphism $\psi : \mathbb{G}_2 \to \mathbb{G}_1$ is known.

Bilinear map groups are known to be instantiable with ordinary elliptic curves introduced by Barreto and Naehrig [BN05]. Thus, Hereafter, by the group description $\mathbb{G}$ we also mean all the parameters that are required to perform computations in $\mathbb{G}$. Moreover, in the course of this paper we will only use the multiplicative notation. This means that, while using elliptic curves, we denote scalar multiplication as exponentiation and point addition as multiplication.

2.2   Assumptions and Cryptographic Primitives

First, we recall the Strong Diffie-Hellman assumption formulated by Boneh and Boyen in [BB08]:

Definition 1 (ℓ-Strong Diffie-Hellman ($\ell - \mathsf{SDH}$)). *Given $(\ell + 3)$ elements $(g_1, g_1^x, g_1^{x^2} \ldots, g_1^{x^\ell}, g_2, g_2^x) \in \mathbb{G}_1^{\ell+1} \times \in \mathbb{G}_2^2$, output $(c, g_1^{1/(x+c)}) \in \mathbb{Z}_q \times \mathbb{G}_1$. We say that an algorithm $\mathcal{A}$ has advantage ϵ in solving the $\ell - \mathsf{SDH}_1$ in $\mathbb{G}_1$, $\mathbb{G}_2$ of prime order q if:*

$$\Pr[(c, g_1^{1/(x+c)}) \leftarrow \mathcal{A}(g_1, g_1^x, g_1^{x^2} \ldots, g_1^{x^\ell}, g_2, g_2^x)] \geq \epsilon,$$

where the probability is taken over the random choice of the generators $g_1 \in \mathbb{G}_1$, $g_2 \in \mathbb{G}_2$, the random choice of $x \in \mathbb{Z}_q$, and the random bits of $\mathcal{A}$. By $\mathbf{Adv}^{\ell-\mathsf{SDH}}(t)$ we denote the maximal advantage for any adversary running in time t in solving the ℓ-Strong Diffie-Hellman problem.

Definition 2 (Certification Scheme). *A certification scheme consists of three PPT algorithms ($\mathsf{KeyGen}_{\mathsf{cert}}$, $\mathsf{Certify}$, CVer). $\mathsf{KeyGen}_{\mathsf{cert}}$ takes as input security parameter 1^λ and outputs a key pair ($sk_{\mathcal{CA}}$, $pk_{\mathcal{CA}}$) of the CA. Algorithm $\mathsf{Certify}$ takes as input secret key $sk_{\mathcal{CA}}$, user's public key $pk_{\mathcal{U}}$ and other information info$_{\mathcal{U}}$ and outputs a certificate cert$_{\mathcal{U}}$. Deterministic algorithm CVer is used to verify certificate cert$_{\mathcal{U}}$. We denote by $\mathbf{Adv}_{CA}^{forge}(t, q_{CA})$ a bound on the value ϵ for which no attacker in time t can forge a certificate (while making at most q_{CA} certifying queries).*

Definition 3 (Message Authentication Codes). *A message authentication codes consists of three algorithms ($\mathsf{KeyGen}_{\mathsf{Mac}}$, Mac, MVer) defined as follows:*

$\mathsf{KeyGen}_{\mathsf{Mac}}(1^\lambda)$: *on input of the security parameter 1^λ, the probabilistic algorithm $\mathsf{KeyGen}_{\mathsf{Mac}}$ outputs a secret key sk.*
$\mathsf{Mac}(sk, m)$: *on input of the secret key sk and message m, algorithm Mac outputs a tag T.*
$\mathsf{MVer}(sk, m, T)$: *on input of the secret key sk, message m and tag T, the deterministic algorithm MVer outputs either 1 (when T is valid) or 0 (when T is invalid).*

We say that a message authentication code is (t, q_m, q_v, ϵ)-unforgeable against adaptively-chosen-message attacks (UNF-CMA) if no algorithm $\mathcal{A}$, running in time t, making at most q_m queries to a tagging oracle and q_v queries to a verifying oracle, has advantage (denoted by $\mathbf{Adv}_{\mathsf{Mac}}^{forge}(t, q_m, q_v)$) at most ϵ in outputting a valid tag for a not queried message.

Definition 4 (Symmetric Encryption Scheme). *A symmetric encryption scheme consists of three polynomial time algorithms ($\mathsf{KeyGen}_{\mathsf{Enc}}$, Enc, Dec) defined as follows:*

$\mathsf{KeyGen}_{\mathsf{Enc}}(1^\lambda)$: *on input of the security parameter 1^λ, the probabilistic algorithm $\mathsf{KeyGen}_{\mathsf{Enc}}$ outputs a secret key sk.*

Enc(sk, m) : *on input of the secret key sk and a message m, algorithm* Enc
outputs a ciphertext c of m.

Dec(sk, c) : *on input of the secret key sk and ciphertext c of m the deterministic
algorithm* Dec *outputs the plaintext m.*

We say the advantage $\mathbf{Adv}_{\mathcal{E}}^{ind-cpa}(\mathcal{A})$ of the adversary $\mathcal{A}$ in attacking this
encryption scheme is the probability that he wins the IND-CPA game. We
assume that encryption scheme can be used to encrypt blocks of messages from
$\{0, 1\}^{\ell}$.

Definition 5 (Key Derivation Function). *Key derivation functions can be
used to derive keys in a way that is indistinguishable from probing with the
uniform distribution in the key space. In practice, such functions are implemented using hash functions or message authentication codes. The ICAO stan-
dard [ISO14] defines the following key derivation function:* KDF(K, c) = H($K\|c$),
where K is the shared secret, c is a counter and H *is a hash function. We
will use the notations* KDF$_{\mathsf{Enc}}(K)$ = KDF($K, 1$), KDF$_{\mathsf{Mac}}(K)$ = KDF($K, 2$) *and*
KDF$_{\pi}(\pi)$ = KDF($f(\pi), 3$), *where f is some encoding of password π defined in
[ISO14].*

2.3 Security Model

As already mentioned, PACE is a password-based authenticated key exchange
protocol. We follow the authors of [BDFK12,BFK13], who provide a security
model built on top of the BPR model [BPR00], which at the same time can be
used to capture session key confidentiality and *impersonation resistance*.

To prove security in this model, we have to show that there exists no adversary $\mathcal{A}$ that wins the real-or-random game with a non-negligible probability. The
game is played in a system with a set of users $U \in \mathcal{U}$. The set $\mathcal{U}$ is divided into
the set of clients $\mathcal{C}$ and the set of servers $\mathcal{S}$. Each user $C \in \mathcal{C}$ is given a key pair
(sk_C, pk_C). Moreover, each pair of a client and a server shares a secret password
π[1]. We assume that this password is randomly chosen from a dictionary with
N elements. The adversary is given all public keys of *honest* users $C \in \mathcal{C}$. Note
that the adversary can register new users which we call *adversarially controlled*.

During the game the adversary may create several instances of a user. The
i-th instance of a user U is denoted by U^i. After successful termination, the
instance U^i outputs a session key K, a session ID sid, and a user ID pid identifying the intended partner. The session identifier sid contains the entire transcript
of the communication.

The goal of the adversary $\mathcal{A}$ is to distinguish between real session keys,
derived by honest parties, from random keys. This is modeled using a test oracle,
which verifies the capability of the adversary to distinguish the session keys
from the random keys. To achieve his goals, the adversary controls the entire
communication in the system. Formally, this is modeled by the following oracles:

---

[1] For the protocols concerned, in fact we may assume there is one *server* with many
instances.

Execute(C^i, S^j) - this query allows to model passive adversaries. The output consists of all messages exchanged during a protocol execution between client C^i and server S^j.

Send(U^i, m) - this query allows to model an active adversary. The output of this query is the message that U^i would generate upon receipt of the message m.

The adversary is also given the power to gain control over honest users and reveal session keys. This is modeled by the following oracles:

Reveal(U, i) - reveals the session key computed by U^i in an accepting state.

Corrupt.pw(U) - returns the secret password π of the user U.

Corrupt.key(U) - returns the secret key sk_C^* of user $U \in C$.

Register(C, pk) - allows to register a public key pk in the name of a new, adversarially controlled, client (identity) $C \in C$,

Test(U, i) - at the beginning of the real-or-random game, this oracle is initialized with a bit b. Assume that at some point of the game, the adversary makes a test query about (U, i). In addition, let U^i terminate in accepting state, holding a session key K (otherwise this oracle returns $\perp$). Then this oracle returns K if $b = 0$ or a random key K' if $b = 1$. Without loss of generality we assume that the adversary never queries twice for the same user instance.

Two instances A^i and B^j are called *partnered* if they both have terminated in an accepting state with the same output. An instance A^i is called *fresh* if: (a) there has been no Reveal(A, i) query, (b) there has been no Reveal(B, j) query, for the partner B^j of A^i, (c) A^i and the partner of A^i are not adversarially controlled (i.e. there were no Corrupt.pw queries for A or B). Otherwise, it is called *unfresh*. Informally, an instance is called fresh if the session key (computed by this instance) has not been leaked and both A^i and its partner B^j are not controlled by the adversary.

AKE Security. Finally, the adversary outputs a bit b'. We say that the adversary wins the real-or-random game if $b = b'$ (b is the bit chosen internally by the Test oracle) and instance (U, i) queried to the Test oracle is *fresh*. We measure the resources Q of the adversary by the maximum number initiated executions q_e, hash oracle queries q_h and cipher oracle queries q_c.

The advantage of an AKE adversary $\mathcal{A}$ (i.e. advantage in winning the real-or-random game by $\mathcal{A}$) for a protocol Π is defined by:

$$\mathbf{Adv}_\Pi^{ake}(\mathcal{A}) = 2 \cdot \Pr[\mathcal{A} \text{ wins the ROR Game}] - 1$$
$$\mathbf{Adv}_\Pi^{ake}(t, Q) = \max\{\mathbf{Adv}_\Pi^{ake}(\mathcal{A}) \mid \mathcal{A} \text{ is } (t, Q) - \text{bounded}\}.$$

Impersonation Resistance (IKE Security). Informally, an adversary *successfully impersonates*, if he succeeds to impersonate an honest client, make the server accept a fake certificate (without knowing the corresponding secret key) or perform any kind of man-in-the-middle attacks.

Formally, the adversary impersonates if an honest reader accepts with partner id pid and session id sid such that: (a) the intended partner C in pid is not adversarially controlled or the public key pk_C has not been registered, (b) there have been no Corrupt.key query for C, before the reader accepted, (c) the session id sid has not appeared in any other accepting session.

The advantage of IKE adversaries against the protocol Π is defined as follows:

$$\mathbf{Adv}_{\Pi}^{ike}(\mathcal{A}) = \Pr[\mathcal{A} \text{ successfully impersonates}]$$
$$\mathbf{Adv}_{\Pi}^{ike}(t,Q) = \max\{\mathbf{Adv}_{\Pi}^{ike}(\mathcal{A}) \mid \mathcal{A} \text{ is } (t,Q) - \text{bounded}\}.$$

3   Generic Version of the Chip Authentication Mapping

The major disadvantage of PACE-CAM is that it heavily relies on the computation performed during the PACE-GM instantiation of the function **Map**. This means that we cannot use the more efficient instantiation based on hash functions used in PACE-IM. Moreover, the same trick cannot be simply applied to the value y'_C as the public key pk_C and the value Y'_C are computed using different group generators (g and $\hat{g}$, respectively). However, it can be done using bilinear maps as we will now show.

Protocol Description. In the table below we present the consequtive steps and communicates exchanged between the ePassport and the Reader during the PACE-CAM v2 protocol. Both parties share a password π and the pairing friendly elliptic curve parameters $(\mathbb{G}_1, \mathbb{G}_2, \mathbb{G}_T, e)$. It is worth noting that the ePassport must only store the definition of group $\mathbb{G}_1$. Moreover, in case of integrated mapping both parties share a hash function H_{EC} that maps arbitrary strings to elements of group $\mathbb{G}_1$. The ePassport also receives a private/public key pair which is certified by the Document Issuer. Similar to Active Authentication and Chip Authentication, the public key is stored in data groups of the ePassport and signed by the Document Issuer. We follow the approach of [BDFK12, BFK13, HKK13] and use a certificate on the public key to model this in an easy-to-follow way. We will denote the public key of the Document Issuer as pk_{CA}.

Our protocol makes minimal changes in the PACE protocol – which itself is one of important design objectives from the practical point of view (e.g. it allows to reuse a major part of the work done for formal certification of PACE products).

ePassport:	Reader:
password π	password π
$\text{spar} = (\mathbb{G}_1, \mathbb{G}_2, \mathbb{G}_T, e, pk_{CA})$	$\text{spar} = (\mathbb{G}_1, \mathbb{G}_2, \mathbb{G}_T, e, pk_{CA})$
$sk_C^* = sk_C$	
$pk_C^* = (pk_C, cert_C)$	
$K_\pi = \mathsf{KDF}_\pi(\pi)$	$K_\pi = \mathsf{KDF}_\pi(\pi)$
choose $s \leftarrow_R \{0,1\}^\ell \subseteq \mathbb{Z}_q$	
$z = \mathsf{Enc}(K_\pi, s)$	

$$\xrightarrow{\quad z \quad}$$

	$s = \mathsf{Dec}(K_\pi, z)$

.... Mapping Function (**Map$_{\mathsf{GM}}$** or **Map$_{\mathsf{IM}}$**) ....

create parameters $\hat{\mathbb{G}}_1$ as	create parameters $\hat{\mathbb{G}}_1$ as
$(a_1, b_1, p_1, q_1, \hat{g}_1, k_1) = \mathbf{Map}(\mathbb{G}_1, s)$	$(a_1, b_1, p_1, q_1, \hat{g}_1, k_1) = \mathbf{Map}(\mathbb{G}_1, s)$

.... Generic Mapping (**Map$_{\mathsf{GM}}$**) ....

choose $y_C \leftarrow_R \mathbb{Z}_q^*$	choose $y_R \leftarrow_R \mathbb{Z}_q^*$
$Y_C = g_1^{y_C}$	$Y_R = g_1^{y_R}$

$$\xleftarrow{\quad Y_R \quad}$$
$$\xrightarrow{\quad Y_C \quad}$$

abort if $Y_R \notin \langle g_1 \rangle \backslash \{1\}$	abort if $Y_C \notin \langle g_1 \rangle \backslash \{1\}$
$h = Y_R^{y_C}$	$h = Y_C^{y_R}$
$\hat{g}_1 = h \cdot g_1^s$	$\hat{g}_1 = h \cdot g_1^s$

.... Integrated Mapping (**Map$_{\mathsf{IM}}$**) ....

	choose $r \leftarrow_R \mathbb{Z}_q^*$

$$\xleftarrow{\quad r \quad}$$

$\hat{g}_1 = \mathsf{H}_{EC}(s, r)$	$\hat{g}_1 = \mathsf{H}_{EC}(s, r)$

....

choose $y_C' \leftarrow_R \mathbb{Z}_q^*$	choose $y_R' \leftarrow_R \mathbb{Z}_q^*$
$Y_C' = \hat{g}_1^{y_C'}$	$Y_R' = \hat{g}_1^{y_R'}$

$$\xleftarrow{\quad Y_R' \quad}$$
$$\xrightarrow{\quad Y_C' \quad}$$

abort if $Y_R' \notin \langle g_1 \rangle \backslash \{1_{\mathbb{G}_1}\}$	abort if $Y_C' \notin \langle g_1 \rangle \backslash \{1_{\mathbb{G}_1}\}$
$K = Y_R'^{y_C'}$	$K = Y_C'^{y_R'}$
$K_{\mathsf{Enc}} = \mathsf{KDF}_{\mathsf{Enc}}(K)$	$K_{\mathsf{Enc}} = \mathsf{KDF}_{\mathsf{Enc}}(K)$
$K_{\mathsf{Mac}} = \mathsf{KDF}_{\mathsf{Mac}}(K)$	$K_{\mathsf{Mac}} = \mathsf{KDF}_{\mathsf{Mac}}(K)$
$K_{\mathsf{Enc}}' = \mathsf{KDF}(K, 2)$	$K_{\mathsf{Enc}}' = \mathsf{KDF}(K, 2)$
$K_{\mathsf{Mac}}' = \mathsf{KDF}(K, 4)$	$K_{\mathsf{Mac}}' = \mathsf{KDF}(K, 4)$
$T_C = \mathsf{Mac}(K_{\mathsf{Mac}}', (Y_R', \mathbb{G}_1))$	$T_R = \mathsf{Mac}(K_{\mathsf{Mac}}', (Y_C', \mathbb{G}_1))$

$$\xleftarrow{\quad T_R \quad}$$
$$\xrightarrow{\quad T_C \quad}$$

if $\mathsf{MVer}(K_{\mathsf{Mac}}', (Y_C', \mathbb{G}), T_R) = 0$,	if $\mathsf{MVer}(K_{\mathsf{Mac}}', (Y_R', \mathbb{G}), T_C) = 0$,
$\qquad$ then abort	$\qquad$ then abort
$w = y_C' / sk_C$	
$c = \mathsf{Enc}(K_{\mathsf{Enc}}', (w, cert_C))$	

$$\xrightarrow{\quad c \quad}$$

	$(w, cert_C) = \mathsf{Dec}(K_{\mathsf{Enc}}', c)$
	abort if $\mathsf{CVer}(pk_{CA}, cert_C) = 0$
	extract pk_C from $cert_C$
	abort if $e(Y_C', g_2) \neq e(\hat{g}_1, pk_C)^w$

....

key $= (K_{\mathsf{Enc}}, K_{\mathsf{Mac}})$	key $= (K_{\mathsf{Enc}}, K_{\mathsf{Mac}})$
sid $= (Y_C', Y_R', \hat{\mathbb{G}}_1, \mathbb{G}_1)$	sid $= (Y_C', Y_R', \hat{\mathbb{G}}_1, \mathbb{G}_1)$
pid $= pk_C^*$	pid $= pk_C^*$

Remark 1. It is worth noting that the ePassport only uses computations in group $\mathbb{G}_1$, which is a ordinary elliptic curve if pairing friendly BN-curves are used [BN05]. Thus, the efficiency on side of the ePassport is similar to the original scheme PACE-CAM.

Security Analysis. Due to space reasons, we just argue that the AKE security relies on the security of the PACE protocol and the same reasoning as in [BFK13] can be applied. Note that PACE-CAM v2 extends PACE only by one encryption c. We now focus on the main issue, namely IKE security of PACE-CAM v2. We present a combined proof for both cases, i.e. when the function **Map** is instantiated as in Integrated Mapping (**Map$_{\text{IM}}$**) or as in Generic Mapping (**Map$_{\text{GM}}$**) and use markings to identify the instantiation for which the part of the proof applies.

Theorem 1. *In the random oracle model, the PACE-CAM v2 (with **Map$_{\text{IM}}$** or **Map$_{\text{GM}}$**) satisfies:*

$$\mathbf{Adv}^{ike}_{\text{PACE-CAMv2}}(t,Q) \leq q_e^2 \cdot q_h \cdot \mathbf{Adv}^{2-\text{SDH}}(t^*) + \frac{q_h}{2^\lambda} + \frac{q_e^2}{q}$$
$$+ 2q_e \cdot \mathbf{Adv}^{forge}_{\text{Mac},}(t^*, 2q_e, 2q_e) + \mathbf{Adv}^{forge}_{CA}(t^*, q_e)$$

where $t^* \sim t$ and $Q = (q_e, q_h, q_c)$.

Proof. First we use the game based techniques [Sho04] to evade simple attacks. The idea is that at the end the adversary cannot perform any replay attacks, the adversary can only impersonate honest ePassports and any malicious reader or ePassport must query the random oracle for the key K'_{Mac} in order to compute a valid message authentication code. We sketch a proof that these constraints only lower the adversary's advantage by a negligible fraction. Finally, for the last game we will show a reduction to the $2 - \text{SDH}$ problem.

Game 0: The original impersonation game as defined in Subsect. 2.3.
Game 1: Similar to the above but we abort the game in case any honest reader accepts an unregistered public key.

Note that the public key of the ePassport is contained in the certificate, which are issued by the Document Issuer. Thus, we lower the adversary's advantage by the advantage of a successful forgery of a certificate.
Game 2: Abort if an honest ePassport computes the same key K that any honest ePassport has computed before.

The key K depends on the values Y'_C and Y'_R. Note that the honest ePassport sends the random element Y'_C after it receives Y'_R from an honest or malicious reader. Hence, the key K is a uniformly random key and the probability that it matches any of the previous i keys, is at most i/q. It follows that the adversary's advantage can drop by a fraction of $\frac{1}{2}q_e^2/q$, where q_e is the maximal number of executions.
Game 3: Abort if there are collisions among the values Y'_R computed by the honest readers.

By the birthday bound this again lowers the adversary's advantage by a fraction of $\frac{1}{2}q_e^2/q$.

Game 4: Abort if a malicious reader submits a valid T_R to the honest ePassport, and such that the adversary has not made a hash query to the key K derived by the honest ePassport in the execution before and T_R was not computed by an honest reader.

We will sketch a proof that an adversary submitting such T_R can be used to break the UNF-CMA security of the message authentication code. Note that by **Game 2** honest ePassports use unique Y_C' and the adversary has to compute a tag for a new message $(Y_C', \mathbb{G})$. Thus, we can simulate the whole protocol and with probability $1/q_e$ choose the execution in which the adversary submits this T_R. In this particular execution we use the UNF-CMA oracles to compute T_C. Thus, we conclude that the adversary's loss can be at most q_e times the advantage of forging a Mac.

Game 5: Abort if a malicious ePassport submits a valid T_C to the honest reader, and such that the adversary has not made a hash query to the key K derived by the honest reader in the execution before and T_C was not computed by an honest ePassport.

Similar reasoning as above can be applied. Thus, we conclude that the adversary's loss can be again at most q_e times the advantage of forging a Mac.

Let us assume that there exists an algorithm $\mathcal{A} = \mathcal{A}^{ike}_{\text{PACE−CAMv2}}$ running in time t and making at most $Q = (q_e, q_h, q_c)$ queries has advantage ϵ in breaking the IKE security in **Game 5**. We show how to use $\mathcal{A}$ to create reduction $\mathcal{R}$, running in time t^*, that has advantage ϵ in solving the $2 - \text{SDH}$ problem on input $(h_1, h_1^x, h_1^{x^2}, h_2, h_2^x)$. First, $\mathcal{R}$ prepares the following impersonation game:

- it sets the system parameters $g_1 = h_1^x$ and $g_2 = h_2^x$ using the problem's instance,
- it takes the public parameters of CA,
- for a given number of ePassports n, $\mathcal{R}$ creates and certifies their key pairs,
- for each pair of users, it chooses passwords from a, possibly small, set N,
- $\mathcal{R}$ chooses the ePassport to be impersonated and replaces his public key with $(h_2)^r$ where $r \leftarrow_R \mathbb{Z}_q$. Note that this implies that $r \cdot x^{-1}$ is the private key of the impersonated user. We will use (sk_C^*, pk_C^*) to denote this particular private/public keypair.

Random Oracle queries. The hash function used for key derivation is similar to the one from the previous proof.

Map$_{\text{IM}}$ Specific Steps:
....
The hash function H_{EC} used by **Map** is programmed as follows. For the i-th unique oracle query of $\mathcal{A}$, the reduction chooses a random number $r_i \leftarrow_R \mathbb{Z}_q^*$ and it outputs $g_1^{r_i}$. For repeated queries, the reduction uses the query table based approach.
....

Standard executions. All oracle queries for regular ePassports (all but the impersonated one) can be simulated by $\mathcal{R}$. Note that it knows all the necessary data, i.e. the private keys and the certificates.

Executions for the impersonated ePassport. Since $\mathcal{R}$ does not know the exponent x^{-1}, the way $\mathcal{R}$ handles oracle queries of $\mathcal{A}$ must be shown. Again, two cases can be distinguished, i.e. $\mathcal{A}$ uses the Execute oracle to create transcripts of communication and $\mathcal{A}$ uses the Send oracle to interact with the ePassport. In both cases we will use the following idea. $\mathcal{R}$ does not know the private key α, it follows that it cannot compute the value $w = y_C'/sk_C^*$. However, if $\mathcal{R}$ uses $Y_C' = \hat{g}_1^{sk_C^* \cdot y_C'}$, then it can compute w as y_C'. Note that this only works if we are able to compute $\hat{g}_1^{sk_C^*}$.

The transcripts received from the Execute oracle for the ePassport and a reader can be easily computed by $\mathcal{R}$. First, it computes $\hat{g}_1$ according to the protocol. Let us denote by γ the discrete logarithm of $\hat{g}_1$ to the base g_1, i.e. $\gamma = \log_{g_1} \hat{g}_1$. Note that in case of $\mathbf{Map_{IM}}$ γ is one of exponents r_i used in definition of H_{EC}. On the other hand, in case of $\mathbf{Map_{GM}}$ $\gamma = y_C \cdot y_R + s$ and y_C, y_R and s are chosen by $\mathcal{R}$. It follows that in both cases $\mathcal{R}$ knows γ.

$\mathcal{R}$ then uses $Y_C' = (h_1)^{r \cdot \gamma \cdot y_C'}$, for a random $y_C' \leftarrow_R \mathbb{Z}_q^*$. Then, it can compute the key K using the formula of the reader's side $(Y_C')^{y_R}$. Finally, $\mathcal{R}$ sets $w = y_C'$. Note that we have $Y_C' = (g_1)^{sk_C^* \cdot \gamma \cdot y_C'} = \hat{g}_1^{sk_C^* \cdot y_C'}$ and $e(Y_C', g_2) = e(\hat{g}_1^{sk_C^* \cdot y_C'}, g_2) = e(\hat{g}_1, g_2^{sk_C^*})^{y_C'} = e(\hat{g}_1, pk_C^*)^w$.

The second case, i.e. $\mathcal{A}$ uses the Send oracle to interact with the ePassport, is a bit more tricky. This time γ is only known when using $\mathbf{Map_{IM}}$. Therefore depending on the instantiation of $\mathbf{Map}$ $\mathcal{R}$ will perform differently.

$\mathbf{Map_{GM}}$ Specific Steps:

. . . .

First, $\mathcal{R}$ computes the random value y_C according to protocol but intead of computing $Y_C = g_1^{y_C}$, it computes $Y_C = (h_1^{x^2})^{r^{-1} \cdot y_C} = g_1^{x \cdot r^{-1} \cdot y_C}$. It follows that $\hat{g}_1 = g_1^{(sk_C^*)^{-1} \cdot y_C \cdot y_R + s}$, where only y_C and s are known to $\mathcal{R}$. In order to compute $\hat{g}_1^{sk_C^*}$ $\mathcal{R}$ has to compute $g_1^{y_C \cdot y_R + sk_C^* \cdot s}$. Note that $\mathcal{R}$ can compute $(Y_R)^{y_C} = g_1^{y_C \cdot y_R}$. It remains to show that knowing s, $\mathcal{R}$ can use it and $h_1^r = g_1^{x^{-1} \cdot r} = g_1^{sk_C^*}$ to compute the remaining part.

. . . .

$\mathbf{Map_{IM}}$ Specific Steps:

. . . .

For $\mathbf{Map_{IM}}$ this part is easier. $\mathcal{R}$ computes $\hat{g} = g_1^{r_j}$ according to the protocol. Note that $\mathcal{R}$ knows the inputs to the mapping function and can find the right exponent r_j. Moreover, $\mathcal{R}$ can compute $\hat{g}_1^{sk_C^*}$ by computing $h_1^r = g_1^{x^{-1} \cdot r} = g_1^{sk_C^*}$ and $(g_1^{sk_C^*})^{r_j} = \hat{g}_1^{sk_C^*}$.

. . . .

Knowing $\hat{g}_1^{sk_C^*}$, $\mathcal{R}$ sets $Y_C' = (\hat{g}_1^{sk_C^*})^{y_C'}$. It follows that $\mathcal{R}$ cannot compute K, since in this case the exponent y_R' is unknown. The key computed by $\mathcal{A}$ is $K = ((\hat{g}_1^{sk_C^*})^{y_C'})^{y_R'}$ and $\mathcal{R}$ cannot compute it. However, the ePassport only uses this key, if the tag T_R sent by the reader is valid. It follows that $\mathcal{A}$ must query the random oracle for the key K (this is ensured by **Game 4**). Thus, $\mathcal{R}$ may search the query table with entries of the form $(query, output)$ for an $output$ such that the tag is valid. Then $\mathcal{R}$ can extract the key K from the $query$ corresponding to the found $output$ and derive all keys. Finally, $\mathcal{R}$ sets $w = y_C'$. Note that in both cases the pairing equation (checked by the reader) holds.

Impersonation execution. Since $\mathcal{A}$ wins the impersonation game, there must be a protocol execution i (among at most q_e executions) in which $\mathcal{A}$ tries to impersonate the ePassport and $\mathcal{R}$ plays the role of the reader. $\mathcal{R}$ guesses this execution and succeeds with probability at least $1/q_e$.

Unlike in the regular executions, $\mathcal{R}$ chooses $y_R' \leftarrow_R \mathbb{Z}_q^*$ and sets $Y_R' = h_1^{y_R'}$. It follows that $\mathcal{A}$ computes $K = (Y_R')^{y_C'} = (h_1^{y_R'})^{y_C'}$. Note that $\mathcal{R}$ cannot compute this K and derive key K_{Mac}' used to create the tag T_R. However, with probability at least $1/q_h$, $\mathcal{R}$ may guess the step at which $\mathcal{A}$ queries the random oracle for K,4 and returns a random and known value. In this way $\mathcal{R}$ is able to compute T_R that will be accepted by $\mathcal{A}$. The only way $\mathcal{A}$ may notice the difference in the random oracle is when he queries with the key K in another query. However, since the key K is always distinct (for the honest ePassports), it follows that the probability that $\mathcal{A}$ queries this key is at most $q_h/2^\lambda$. Then, since the adversary must compute a valid ciphertext c, $\mathcal{R}$ is able to find the key K in the oracle table, derive the key K_{Enc}' and decrypt the last message of $\mathcal{A}$ receiving $w = y_C'/sk_C^*$.

Finally, $\mathcal{R}$ has the following data: $K = (h_1^{y_R'})^{y_C'}$, $w = y_C'/sk_C^*$, y_R', so $\mathcal{R}$ can compute $K^{w^{-1} \cdot y_R'^{-1}} = h_1^{sk_C^*}$. Since $sk_C^* = x^{-1} \cdot r$, $\mathcal{R}$ can compute $(h_1^{sk_C^*})^{r^{-1}} = h_1^{x^{-1}}$. Thereby $\mathcal{R}$ solves the $2 - \mathsf{SDH}$ problem with probability $\frac{\epsilon}{q_h \cdot q_e \cdot \min\{n, q_e\}}$ ($\mathcal{R}$ must choose the correct execution, the correct ePassport and the right oracle query of $\mathcal{A}$) by returning $(0, h_1^{1/x})$.

4   Conclusion

In this paper we propose a new version of the PACE protocol with Chip Authentication Mapping based on bilinear maps. It is not only as efficient as the standard version but it can also be used with an arbitrary **Map** function, e.g. based on hash functions like in PACE with Integrated Mapping.

Although, the Chip Authentication Mapping v2 seems to be interesting from industrial (patent free) and academic (provable security under the 2-Strong Diffie-Hellman assumption) perspective, it might require some time and acknowledgment from involved parties in order for the protocol to become a part of the ICAO standard. Moreover, bilinear maps are not commonly used by the industry which may additionally slow down and even prevent the protocol from being implemented in a near future. On the bright side, bilinear maps are not required

on the ePassport but only on the readers, which have the computational power
to compute such maps in reasonable time.

Acknowledgment. The research was supported by the Polish National Science Centre based on the decision DEC-2013/08/M/ST6/00928. Initial work of the first author has been supported by Foundation for Polish Science project VENTURES/2012-9/4.

References

[BB08]   Boneh, D., Boyen, X.: Short signatures without random oracles and the
         SDH assumption in bilinear groups. J. Cryptology **21**(2), 149–177 (2008)
[BDFK12] Bender, J., Dagdelen, Ö., Fischlin, M., Kügler, D.: The PACE—AA protocol for machine readable travel document, and its security. In: Proceedings
         of the 16th International Conference on Financial Cryptography and Data
         Security (2012)
[BFK09] Bender, J., Fischlin, M., Kügler, D.: Security analysis of the PACE keyagreement protocol. In: Samarati, P., Yung, M., Martinelli, F., Ardagna,
         C.A. (eds.) ISC 2009. LNCS, vol. 5735, pp. 33–48. Springer, Heidelberg
         (2009)
[BFK13] Bender, J., Fischlin, M., Kügler, D.: The PACE|CA protocol for machine
         readable travel documents. In: Bloem, R., Lipp, P. (eds.) INTRUST 2013.
         LNCS, vol. 8292, pp. 17–35. Springer, Heidelberg (2013)
[BK12]   Bender, J., Kügler, D.: Verfahren zur Authentisierung, RF-chip-Dokument,
         RF-Chip-Lesegerät und Computerprogrammprodukte, 13 September 2012.
         WO Patent App. PCT/EP2012/001,076 (2012)
[BM92]   Bellovin, S.M., Merritt, M.: Encrypted key exchange: password-based protocols secure against dictionary attacks. In: IEEE Symposium on Research
         in Security and Privacy, pp. 72–84 (1992)
[BN05]   Barreto, P.S.L.M., Naehrig, M.: Pairing-friendly elliptic curves of prime
         order. In: Preneel, B., Tavares, S. (eds.) SAC 2005. LNCS, vol. 3897, pp.
         319–331. Springer, Heidelberg (2006)
[BPR00] Bellare, M., Pointcheval, D., Rogaway, P.: Authenticated key exchange
         secure against dictionary attacks. In: Preneel, B. (ed.) EUROCRYPT 2000.
         LNCS, vol. 1807, pp. 139–155. Springer, Heidelberg (2000)
[BSI15] BSI. Advanced Security Mechanisms for Machine Readable Travel Documents and eIDAS Token 2.20. Technical Guideline TR-03110-2 (2015)
[CGIP11] Coron, J.-S., Gouget, A., Icart, T., Paillier, P.: Supplemental Access Control
         (PACE v2): Security Analysis of PACE Integrated Mapping. Cryptology
         ePrint Archive, Report 2011/058 (2011)
[HKK13] Hanzlik, L., Krzywiecki, Ł., Kutyłowski, M.: Simplified PACE|AA protocol.
         In: Deng, R.H., Feng, T. (eds.) ISPEC 2013. LNCS, vol. 7863, pp. 218–232.
         Springer, Heidelberg (2013)
[ISO11] ISO/IEC JTC1 SC17 WG3/TF5 for the International Civil Aviation Organization. Supplemental access control for machine readable travel documents v1.01. Technical report, 08 March 2011
[ISO14] ISO/IEC JTC1 SC17 WG3/TF5 for the International Civil Aviation Organization. Supplemental access control for machine readable travel documents v1.1. Technical report, 15 April 2014

[Jab96] David, P.: Jablon: strong password-only authenticated key exchange. SIG-COMM Comput. Commun. Rev. **26**(5), 5–26 (1996)

[Sho04] Shoup, V.: Sequences of games: a tool for taming complexity in security proofs. Cryptology ePrint Archive, Report 2004/332 (2004). http://eprint.iacr.org/

AEP-M: Practical Anonymous E-Payment for Mobile Devices Using ARM TrustZone and Divisible E-Cash

Bo Yang[1], Kang Yang[1(✉)], Zhenfeng Zhang[1], Yu Qin[1], and Dengguo Feng[1,2]

[1] Trusted Computing and Information Assurance Laboratory, Institute of Software, Chinese Academy of Sciences, Beijing, China
{yangbo,yangkang,zfzhang,qin_yu,feng}@tca.iscas.ac.cn
[2] State Key Laboratory of Computer Science, Institute of Software, Chinese Academy of Sciences, Beijing, China

Abstract. Electronic payment (e-payment) has been widely applied to electronic commerce and has especially attracted a large number of mobile users. However, current solutions often focus on protecting users' money security without concerning the issue of users' privacy leakage. In this paper, we propose AEP-M, a practical anonymous e-payment scheme specifically designed for mobile devices using TrustZone. On account of the limited resources on mobile devices and time constraints of electronic transactions, we construct our scheme based on efficient divisible e-cash system. Precisely, AEP-M allows users to withdraw a large coin of value 2^n at once, and then spend it in several times by dividing it without revealing users' identities to others, including banks and merchants. Users' payments cannot be linked either. AEP-M utilizes bit-decomposition technique and pre-computation to further increase the flexibility and efficiency of spending phase for mobile users. As a consequence, the frequent online spending process just needs at most n exponentiations on elliptic curve on mobile devices. Moreover, we elaborately adapt AEP-M to TrustZone architecture for the sake of protecting users' money and critical data. The methods about key derivation and sensitive data management relying on a root of trust from SRAM Physical Unclonable Function (PUF) are presented. We implement a prototype system and evaluate AEP-M using Barreto-Naehrig (BN) curve with 128-bit security level. The security analysis and experimental results indicate that our scheme could meet the practical requirement of mobile users in respects of security and efficiency.

Keywords: E-Payment · Privacy · TrustZone · Divisible e-cash · PUF

1 Introduction

Depending on the development and achievements of wireless network as well as modern mobile devices, electronic commerce (e-commerce) is benefiting more and more people's daily lives. As e-commerce becomes a major component of

© Springer International Publishing Switzerland 2016
M. Bishop and A.C.A. Nascimento (Eds.): ISC 2016, LNCS 9866, pp. 130–146, 2016.
DOI: 10.1007/978-3-319-45871-7_9

business operations, e-payment, which builds up e-commerce, has become one of the most critical issues for successful business and financial services. Defined as the transfer of an electronic value of payment from a payer to a payee through the Internet, e-payment has been already realized in different ways and applied to mobile devices by intermediaries such as PayPal, Google Wallet, Apple Pay and Alipay [8]. Unfortunately, with the widespread use of e-payment, users are faced with the risk of privacy disclosure.

Although the intermediaries and online banks try the best to enhance the security of their e-payment solutions, the privacy-preserving scheme is often neglected or weakened in the implementation [9]. Generally, the spending procedure is associated with the authenticated identity to indicate who withdraws digital coins from banks, so that all the user's relevant consuming behaviors are identified and linked. In reality, the most of current deployed e-payment solutions unintentionally reveal user personal information, perhaps involving user real identity, billing and shopping records etc., to banks, intermediaries or payees [11]. Such sensitive information implies one's political view, location, religion or health condition. Statistically, mobile users account for a high proportion among all the e-payment users [10]. Thus, the issue of information leakage is threatening mobile e-payment users' personal privacy.

In theory, constructing anonymous e-payment scheme is able to effectively solve the above problem. Some anonymous protocols are the candidates here including direct anonymous attestation (DAA) and U-Prove. Based on DAA, Yang et al. [19] put forward LAMS for anonymous mobile shopping. However, these protocols hardly fulfill the anonymous e-payment from the perspectives of both anonymity and flexibility for payment. Acting as a targeted component for e-payment, electronic cash (e-cash), introduced by Chaum [5], allows users to withdraw digital coins from a bank and to spend them to merchants in an anonymous way, thus perfectly emulating conventional cash transactions. Derived from e-cash, divisible e-cash systems are proposed to address the issue of splitting coins of large values. Depending on it, users could withdraw a large coin of value 2^n at once and spend it in several times by dividing it. In practice, divisible e-cash makes the cash transactions more efficient and flexible. In regard to mobile devices, the limited resources along with the strong time constraints of electronic transactions indeed require the practical withdrawal and spending procedures. Therefore, it is advisable to build anonymous e-payment scheme upon efficient divisible e-cash for mobile devices.

It is commonly believed that good security and trust will ultimately increase the use of e-payment. Nevertheless, the direct application of anonymous e-payment scheme on mobile devices would bring potential security risks. Without the dedicated protection, the scheme's executing codes and sensitive data are easily either compromised or stolen by the malwares. In some cases, the attacks on mobile e-payment could cause user's great loss of property. The technique of Trusted Execution Environment (TEE) on mobile devices is able to lend us a helping hand. Isolated from a Rich Execution Environment (REE) where the Guest OS runs, TEE aims to protect sensitive codes execution and assets.

As a prevalent example of providing TEE for embedded devices, ARM TrustZone [1] has been used to execute security-critical services [17]. Actually, TrustZone enables a single physical processor to execute codes in one of two possible isolated operating worlds: the *normal world* (NW) for REE and the *secure world* (SW) for TEE. The two worlds have independent memory address spaces and different privileges. As a hardware-based security extension of ARM architecture, Trust-Zone is widely supported and applied by mobile devices. But there is a fly in the ointment that TrustZone does not definitely provide the root of trust with inside root key for sensitive data management. To the best of our knowledge, there is no anonymous e-payment scheme specially designed for mobile devices using TrustZone.

1.1   Our Contribution

Based on ARM TrustZone and the divisible e-cash scheme with the best efficiency by Canard et al. [4], we propose AEP-M, a practical anonymous e-payment scheme for mobile devices, which enables a user to spend his digital coins securely and efficiently while preserving his privacy. This is the first complete work to design an efficient anonymous e-payment scheme integrated with TrustZone. We substantially modify the original e-cash scheme for adapting it to the executing mode of TrustZone and guaranteeing its security on mobile devices.

For device-centered design, we make following steps towards practical and secure usage:

– the sensitive codes on the user side of AEP-M are isolated and executed in TEE provided by TrustZone for the possibility that the guest OS is compromised;
– AEP-M utilizes some secret keys, which are derived from a root key seed reproduced via an on-chip SRAM PUF [6], to protect users' coins and data;
– in AEP-M, online banks could authenticate a user who holds a mobile device with available TrustZone and a valid account-password pair.

AEP-M elaborately protects the security of the user's passwords and coins even if the NW of his mobile device is corrupted while the SW still keeps honest. The pre-computation stage is carefully added into our scheme such that the computation amounts of the frequent online spending phase for mobile users are decreased. Furthermore, our scheme supports that one spends a coin of value v for any $1 \leq v \leq 2^n$ at once by using the bit-decomposition technique, while the original scheme [4] cannot, where the maximum denomination of a coin is 2^n.

We implement a prototype of AEP-M and evaluate its efficiency using BN curve at the security level of 128-bit. The experimental results show that our scheme is efficient enough for practical usage, even from the perspective of mobile devices.

1.2   Related Work

E-Payment Scheme. Different from pre-paid cards, credit cards and electronic checks, e-cash system does a better job to construct anonymous e-payment. After

Chaum first introduced e-cash [5], Camenisch et al. [2] presented the compact e-cash system allowing users to withdraw wallets with 2^n coins at once. Unfortunately, its spending procedure should be done coin by coin. Afterwards, some truly anonymous divisible e-cash systems [3] were described, but quite inefficiency. Recently, Canard et al. [4] proposed the first really efficient divisible e-cash system by defining one global binary tree. Our scheme takes it as a reference and further increases its efficiency and security according to our architecture of trusted mobile device.

TrustZone Technology. ARM TrustZone technology for the mobile devices can guarantee codes integrity and data security. Relying on TrustZone, many practical mobile schemes are proposed. For instance, AdAttester [7] was presented specially for secure mobile advertisement on a TrustZone-enabled device. To date, TrustZone has been popularized and applied by many mainstream mobile manufacturers, such as Apple, Samsung and Huawei, to achieve secure applications [7].

2   Preliminaries

2.1   Notation

Throughout the paper, we use the notation shown in Table 1.

Let $\Lambda = (p, \mathbb{G}_1, \mathbb{G}_2, \mathbb{G}_T, e, g, \tilde{g})$ be a description of bilinear groups which consist of three (multiplicatively written) groups $\mathbb{G}_1$, $\mathbb{G}_2$ and $\mathbb{G}_T$ of prime order p equipped with a bilinear map $e : \mathbb{G}_1 \times \mathbb{G}_2 \to \mathbb{G}_T$, where g and $\tilde{g}$ is the generator of $\mathbb{G}_1$ and $\mathbb{G}_2$ respectively. In this paper, we only consider the Type-3 pairings [16].

2.2   ARM TrustZone

ARM TrustZone [1] is a hardware-based security extension technology incorporated into ARM processors. The whole system is separated into two worlds and each world has banked registers and memory to run the domain-dedicated OS and software. As a result, access permissions are strictly under the control of the *secure world* that the *normal world* components cannot access the *secure world* resources. As the processor only runs in one world at a time, to run in the other world requires context switch. A secure monitor mode exists in the *secure world* to control the switch and migration between the two worlds.

2.3   Physical Unclonable Functions

Physical Unclonable Functions (PUFs) [12] are functions where the relationship between input (or challenge) and output (or response) is decided by a physical system. Randomness and unclonability are two significant properties of PUFs. PUFs are able to implicitly "store" a piece of secret data. PUFs provide much higher physical security by extracting the secret data from complex physical systems rather than directly reading them from non-volatile memory.

Table 1. Notation used in this paper

Notation	Descriptions
λ	Security parameter
$x \xleftarrow{\$} \mathbb{S}$	x chosen uniformly at random from a set $\mathbb{S}$
$y := x$	y assigned as x
$x\|y$	Concatenation of x and y
$(y_1, ..., y_j) \leftarrow \mathsf{A}(x_1, ..., x_i)$	A (randomized) algorithm with input $(x_1, ..., x_i)$ and output $(y_1, ..., y_j)$
$1_\mathbb{G}$	The identity element of a group $\mathbb{G}$
$\mathbb{G}^*$	$\mathbb{G} \setminus \{1_\mathbb{G}\}$ for a group $\mathbb{G}$
$\Sigma_1 = (\mathsf{KeyGen}, \mathsf{Sign}, \mathsf{Verify})$	Digital signature algorithm
$\Sigma_2 = (\mathsf{MAC})$	Message authentication code
$\Sigma_3 = (\mathsf{Enc_{asym}}, \mathsf{Dec_{asym}})$	Asymmetric (public key) encryption and decryption algorithm
$\Sigma_4 = (\mathsf{Enc_{sym}}, \mathsf{Dec_{sym}})$	Symmetric encryption and decryption algorithm

Strictly speaking, only equipped with a root of trust, TrustZone becomes a real "trusted" execution environment (TEE) [22]. Because TrustZone almost does not internally install an available root key, it loses the capability to offer a root of trust. Employing a PUF can cover this shortage. In this paper, AEP-M takes the secret data extracted from the PUF as a root key seed to generate other keys. We adopt SRAM PUF [6] that leverages the relationship between an SRAM cell's address for the challenge and its power up value for the response.

3 System Model and Assumptions

3.1 System Model

The system model of AEP-M is composed of five kinds of entities: mobile device $\mathcal{D}$, merchant $\mathcal{M}$, trusted authority $\mathcal{T}$, central bank $\mathcal{B}$ and traditional commercial bank. In practice, there could be a number of mobile devices and merchants participating in our system. For the sake of brevity and clarity, we use $\mathcal{D}$ and $\mathcal{M}$ to represent an individual instantiation respectively. $\mathcal{D}$ is equipped with ARM processor having TrustZone extension technology. $\mathcal{B}$ is responsible for issuing digital coins to legitimate (or trusted) $\mathcal{D}$ through **Withdraw** phase. $\mathcal{B}$ could be a bank card organization supporting e-payment or an intermediary serving electronic transactions. In the background, several commercial banks, where users actually deposit money, are in cooperation with $\mathcal{B}$ for dealing with money transfers in the real world. Service or product providers play the role of $\mathcal{M}$ in this interactive model. They collect digital coins from $\mathcal{D}$ via **Spend** phase and redeem them from $\mathcal{B}$ via **Deposit** phase. Note that $\mathcal{M}$ verifies the digital coins of some

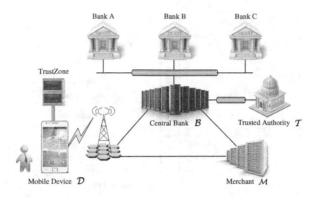

Fig. 1. System model of AEP-M.

user without revealing user's identity to any entities including $\mathcal{M}$ itself. Managed by the government or the industry administration, in **Identify** phase $\mathcal{T}$ performs revealing identity of the users who attempt to double-spend digital coins. Figure 1 illustrates the system model for our scheme.

3.2 Assumptions and Threat Model

To simplify our design in the system model, we assume that data communications between $\mathcal{B}$ and traditional bank, and between $\mathcal{B}$ and $\mathcal{T}$ build on secure transport protocols, such as TLS, which can provide confidentiality, authenticity and integrity protection for data transmission. Also, $\mathcal{M}$, $\mathcal{D}$ and $\mathcal{B}$ are able to acquire public parameters from $\mathcal{T}$ in the correct way. Public Key Infrastructure (PKI) is supposed to be already realized for authenticating $\mathcal{B}$ and $\mathcal{M}$. As a consequence, (1) $\mathcal{D}$ and $\mathcal{M}$ can accurately obtain the public key of $\mathcal{B}$ by verifying its certificate; (2) $\mathcal{D}$ and $\mathcal{B}$ can accurately obtain the public key of $\mathcal{M}$ similarly.

Based on the assumptions, AEP-M protects against the following adversary:

- The adversary can attack the scheme itself by attempting to pretend entities, manipulate data transmission between entities and forge data.
- The adversary can perform software-based attacks which compromise the mobile Rich OS or existing applications running in REE. AEP-M interfaces in REE are also available for the adversary.
- The adversary can physically access the mobile device. He can reboot the device and gain access to data residing on persistent storage.

However, we ignore the malicious behaviors of tampering with the TrustZone hardware or mounting side-channel attacks on PUF.

4 AEP-M Scheme for Mobile Devices

In this section, we provide the specific architecture of trusted mobile device, and then present the key derivation and sensitive data management. Depending on

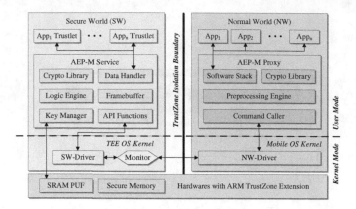

Fig. 2. Architecture of trusted mobile device for AEP-M.

these, the construction of AEP-M scheme is detailed next. Finally, the security properties of AEP-M is analyzed.

4.1   The Architecture of Trusted Mobile Device

Leveraging TrustZone and PUF technology, we design the architecture of trusted mobile device specifically for AEP-M based on our previous work [20]. The software-based implementation of AEP-M functionality on existing hardwares targets at economy, flexibility and extensibility. Meanwhile, our architecture is designed to be compatible with the conventional running model of secure applications using TrustZone. Figure 2 shows the detailed architecture with the way components interact with each other.

AEP-M functionality in the architecture contains two components: untrusted AEP-M Proxy in *normal world* (NW) and security-sensitive AEP-M Service in *secure world* (SW). In reality, SW instantiates TEE, while NW implements REE. Depending on the whitelist and integrity protection mechanism, only the trusted codes of programs in SW could be loaded and executed. Thus, AEP-M Service resides in a relatively secure environment isolated from other codes running in NW. The different components from [20] are formally described as follows.

AEP-M Proxy. This is the component visible for mobile (e-payment) applications in NW. Waiting for their AEP-M service requests, the proxy handles the parameters and preprocesses them. **Preprocessing Engine** executes precomputation for AEP-M after digital coins are withdrawn from central bank to the mobile device.

AEP-M Service. This is the core component to perform AEP-M critical computations and operations. The execution of the component codes is under the well protection of TrustZone isolation mechanism. **Framebuffer** stores the image of confirmation message (e.g., the identity of merchant to be paid) to be securely displayed for the user. Different from the general frame buffer in NW, Frame-buffer is devoted to the reliable graphical user interface (GUI) for SW.

Application and Application Trustlet. The corresponding application should be launched if the user wants to enjoy e-payment service. For upper-level interaction, the application released by $\mathcal{B}$ consists of two parts: App for NW and App Trustlet for SW. App provides the general GUI and basic functions, while App Trustlet is securely loaded by SW and trusted for processing security-sensitive user inputs and data operations.

Components in Hardwares. Protected by TrustZone mechanism, SRAM PUF component and Secure Memory component are only accessible for SW. Secure Memory contributes to temporally saving sensitive data.

4.2 Key Derivation and Sensitive Data Management

Prior to describing the concrete construction of our AEP-M scheme, we show how to derive various keys for different purposes using the root key seed extracted from SRAM PUF and how to utilize the derived keys to protect sensitive data.

Root Key Seed Extraction. We use the technique of SRAM PUF in [22] to extract the secret root key *seed*, which is a unique bit string picked randomly by the OEM who "stores" it in $\mathcal{D}$ through the physical features of one SRAM inside $\mathcal{D}$. From SRAM PUF component, *seed* is only reproduced and securely cached by Key Manager when $\mathcal{D}$ starts up every time in normal use.

Key Derivation. Key Manager has the deterministic key derivation function KDF: $\widetilde{\mathcal{S}} \times \{0,1\}^* \to \widetilde{\mathcal{K}}$, where $\widetilde{\mathcal{S}}$ is the key seed space, and $\widetilde{\mathcal{K}}$ is the derived key space. Using the KDF, the device key pair and the storage root key is derived as $(\mathsf{dsk}, \mathsf{dpk}) \leftarrow \mathsf{KDF}_{seed}(\texttt{"identity"})$ and $\mathsf{srk} \leftarrow \mathsf{KDF}_{seed}(\texttt{"storage\_root"})$ respectively. The unique device key pair is analogous to the endorsement key defined in trusted computing but supports encryption and decryption. The storage root key srk is used for generating specific storage keys to preserve sensitive data. The hierarchical structure of storage keys enhances the security for key usage. Note that all the derived keys are never stored permanently. Instead, they are regained via KDF with *seed* at the same way when needed.

Sensitive Data Management. We can utilize the storage keys derived from the storage root key srk to seal the AEP-M's public parameters *params*, $\mathcal{D}$'s digital coin σ, the secret key m, and other related variables CT and δ. What these variables represent will be explained in Sect. 4.3. The sealed results of these data are stored in the insecure positions of $\mathcal{D}$.

- Protect integrity for *params*: $\mathsf{mk}_{params} \leftarrow \mathsf{KDF}_{srk}(\texttt{"storage\_key"}||\texttt{"MAC"}||$ *params*), and $blob_{params} \leftarrow \mathsf{Data\_Seal}(\texttt{"MAC"}, \mathsf{mk}_{params}, params)$, where

$$blob_{params} := params || \mathsf{MAC}(\mathsf{mk}_{params}, params).$$

- Protect integrity for σ: $\mathsf{mk}_\sigma \leftarrow \mathsf{KDF}_{\mathsf{srk}}(\texttt{"storage\_key"}||\texttt{"MAC"}||\sigma)$, and $blob_\sigma \leftarrow \mathsf{Data\_Seal}(\texttt{"MAC"}, \mathsf{mk}_\sigma, \sigma)$, where

$$blob_\sigma := \sigma||\mathsf{MAC}(\mathsf{mk}_\sigma, \sigma).$$

- Protect both confidentiality and integrity for m, CT and δ with the aid of U: $(\mathsf{sk}_m, \mathsf{mk}_m) \leftarrow \mathsf{KDF}_{\mathsf{srk}}(\texttt{"storage\_key"}||\texttt{"Enc+MAC"}||U)$, and $blob_m \leftarrow \mathsf{Data\_Seal}(\texttt{"Enc+MAC"}, \mathsf{sk}_m, \mathsf{mk}_m, m||CT||\delta, U)$, where

$$blob_m := \mathsf{Enc}_{\mathsf{sym}}(\mathsf{sk}_m, m||CT||\delta)||U||\mathsf{MAC}(\mathsf{mk}_m, \mathsf{Enc}_{\mathsf{sym}}(\mathsf{sk}_m, m||CT||\delta)||U).$$

Data Handler can use $\mathsf{Data\_Unseal}()$ to recover and verify the sensitive data from blobs with the related keys regained by Key Manager.

4.3 The Details of AEP-M Scheme

Following the divisible e-cash scheme [4], a unique and public global tree of depth n is used for all coins of value $V = 2^n$ as illustrated in Fig. 3. So each leaf denotes the smallest unit of value to spend. We define $\mathcal{S}_n$ as the set of bit strings of size smaller than or equal to n and $\mathcal{F}_n$ as the set of bit strings of size exactly n. Thus, each node of the tree refers to an element $s \in \mathcal{S}_n$, the root to the empty string ϕ, and each leaf to an element $f \in \mathcal{F}_n$. For any node $s \in \mathcal{S}_n$, $\mathcal{F}_n(s)$ $= \{f \in \mathcal{F}_n | s \text{ is a prefix of } f\}$ contains all the leaves in the subtree below s.

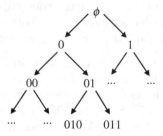

Fig. 3. Public global tree for all coins.

Assume, before leaving the factory, $\mathcal{D}$ is initialized by the OEM in SW to generate the unique device key $(\mathsf{dsk}, \mathsf{dpk})$ which could uniquely identify $\mathcal{D}$. Then, the OEM issues a certificate $\mathsf{cert}_\mathcal{D}$ w.r.t. the public key dpk to indicate the OEM's recognition for $\mathcal{D}$. The certificate $\mathsf{cert}_\mathcal{D}$ also contains some $\mathcal{D}$'s configuration information (e.g., whether TrustZone is available).

AEP-M scheme consists of six phases: **Setup**, **KeyGen**, **Withdraw**, **Spend**, **Deposit** and **Identify**. First of all, **Setup** is executed to create the public parameters by $\mathcal{T}$. After that, $\mathcal{B}$ and $\mathcal{M}$ can execute **KeyGen** to generate their public-private key pairs according to the public parameters. Then, other phases are enabled to be executed according to requirements. The phases of the scheme are presented in detail as follows.

Setup. In this phase, the trusted authority $\mathcal{T}$ creates the public parameters. Given a security parameter λ, $\mathcal{T}$ picks the suitable bilinear groups parameters $\Lambda := (p, \mathbb{G}_1, \mathbb{G}_2, \mathbb{G}_T, e, g, \tilde{g})$ described in Sect. 2.1 such that $|p| \geq 2\lambda$. And then, according to the global tree, $\mathcal{T}$ generates (1) $r_s \xleftarrow{\$} \mathbb{Z}_p$ and $g_s := g^{r_s}$ for each $s \in \mathcal{S}_n$, and (2) $l_f \xleftarrow{\$} \mathbb{Z}_p$ and $\tilde{g}_{s \mapsto f} := \tilde{g}^{l_f/r_s}$ for each $s \in \mathcal{S}_n$ and each $f \in \mathcal{F}_n(s)$. $\mathcal{T}$ keeps $\mathbf{sck} = \{r_s | s \in \mathcal{S}_n\}$ as its secret keys to be used in **Identify**

phase. Also, $\mathcal{T}$ determines a series of algorithms Ψ including the algorithms covering from Σ_1 to Σ_4 in Table 1, and four independent collision-resistant hash functions:

$$H_1 : \{0,1\}^* \to \mathbb{Z}_p, \ H_2 : \{0,1\}^* \to \mathbb{Z}_p, \ H_3 : \{0,1\}^* \to \{0,1\}^{2\lambda}, \ H_4 : \{0,1\}^* \to \{0,1\}^{2\lambda}.$$

Finally, $\mathcal{T}$ sets $(\Lambda, n, \Psi, \{r_s | s \in \mathcal{S}_n\}, \{\tilde{g}_{s \mapsto f} | s \in \mathcal{S}_n \wedge f \in \mathcal{F}_n(s)\})$ as the public parameters, where $\mathcal{D}$ and $\mathcal{M}$ only need to know $params := (\Lambda, n, \Psi, \{r_s | s \in \mathcal{S}_n\})$, while $\mathcal{B}$ requires $params' := (\Lambda, n, \Psi, \{r_s | s \in \mathcal{S}_n\}, \{\tilde{g}_{s \mapsto f} | s \in \mathcal{S}_n \wedge f \in \mathcal{F}_n(s)\})$. After obtaining $params$, $\mathcal{D}$ calls Data_Seal() to seal it and stores the output $blob_{params}$.

KeyGen. This phase initializes the public-private key pair for the central bank $\mathcal{B}$ and a merchant $\mathcal{M}$.

- *Key Generation for Central Bank.* First, given $params'$ as input, $\mathcal{B}$ picks $x, y \xleftarrow{\$} \mathbb{Z}_p^*$, and computes $X := \tilde{g}^x$ and $Y := \tilde{g}^y$. $\mathcal{B}$ sets (x,y) as the private key $\overline{sk}_{\mathcal{B}}$ and publishes (X,Y) as the public key $\overline{pk}_{\mathcal{B}}$. Second, $\mathcal{B}$ uses KeyGen() in Σ_1 to generate key pair for establishing sessions with $\mathcal{D}$: $(sk_{\mathcal{B}}, pk_{\mathcal{B}}) \leftarrow \text{KeyGen}(1^\lambda)$, where $sk_{\mathcal{B}}$ is the private key.
- *Key Generation for Merchant.* Similarly, $\mathcal{M}$ uses KeyGen() to generate key pair for establishing sessions with $\mathcal{D}$: $(sk_{\mathcal{M}}, pk_{\mathcal{M}}) \leftarrow \text{KeyGen}(1^\lambda)$.

Accordingly, $\mathcal{D}$ could get the correct $pk_{\mathcal{B}}$ and $pk_{\mathcal{M}}$ from $\mathcal{B}$ and $\mathcal{M}$ via verifying their certificates. And likewise, $\mathcal{M}$ and $\mathcal{B}$ could acquire the correct $\overline{pk}_{\mathcal{B}}$ and $pk_{\mathcal{M}}$ respectively as well as $\mathcal{T}$ obtains $\overline{pk}_{\mathcal{B}}$.

Withdraw. In this phase, a user with mobile device $\mathcal{D}$ could withdraw some digital coins from the central bank $\mathcal{B}$ as follows.

1. The user operates App in NW of $\mathcal{D}$ to prepare for withdrawing some digital coins. $\mathcal{D}$ switches into SW and chooses a nonce $n_{\mathcal{D}} \xleftarrow{\$} \{0,1\}^\lambda$. $n_{\mathcal{D}}$ is saved in Secure Memory and delivered to AEP-M Proxy that sends $n_{\mathcal{D}}$, $\mathcal{D}$'s dpk with its certificate $cert_{\mathcal{D}}$ to $\mathcal{B}$.

2. $\mathcal{B}$ checks whether dpk is valid with $cert_{\mathcal{D}}$ and checks the configuration information on $cert_{\mathcal{D}}$. If the check is passed, $\mathcal{B}$ chooses a nonce $n_{\mathcal{B}} \xleftarrow{\$} \{0,1\}^\lambda$, a key $k_{mac} \xleftarrow{\$} \{0,1\}^\lambda$ for MAC and a key $k_{enc} \xleftarrow{\$} \{0,1\}^\lambda$ for Enc_{sym} and Dec_{sym}. Then, $\mathcal{B}$ encrypts $n_{\mathcal{B}}$, k_{mac} and k_{enc} using dpk to get a cipher text $C_{\mathcal{B}} \leftarrow \text{Enc}_{asym}(\text{dpk}, n_{\mathcal{B}} || k_{mac} || k_{enc})$ and signs dpk, $n_{\mathcal{D}}$ and $C_{\mathcal{B}}$ using $sk_{\mathcal{B}}$ to output a signature $\alpha \leftarrow \text{Sign}(sk_{\mathcal{B}}, \text{dpk} || n_{\mathcal{D}} || C_{\mathcal{B}})$. Finally, $\mathcal{B}$ sends a commitment request $comm_{req} := (C_{\mathcal{B}}, \alpha)$ to $\mathcal{D}$.

3. AEP-M Proxy invokes AEP-M Service with input $comm_{req}$. In SW, App Trustlet waits for the user to input his bank account $account_{\mathcal{D}}$, the corresponding password pwd and the amount of digital coins to withdraw. For simplicity, we only describe how to withdraw one coin. The **Withdraw** phase could be easily extended to support withdrawing multiple coins at once. After the user finishes inputting, Logic Engine calls the API AEPM_SW_Withdraw() to generate a commitment response:

$$comm_{res} \leftarrow \text{AEPM\_SW\_Withdraw}(blob_{params}, n_{\mathcal{D}}, \text{pk}_{\mathcal{B}}, comm_{req}, account_{\mathcal{D}}, pwd),$$

where the API is executed as follows:

(1) Unseal the blob $blob_{params}$ to get $params$ by calling Data_Unseal().
(2) Verify α using $\mathsf{pk}_\mathcal{B}$: $res \leftarrow$ Verify($\mathsf{pk}_\mathcal{B}, \mathsf{dpk}||n_\mathcal{D}||C_\mathcal{B}, \alpha$). If $res = false$, $comm_{res} := \perp$ and return.
(3) Decrypt $C_\mathcal{B}$ using dsk: $(n'_\mathcal{B}, \mathsf{k}_{mac}, \mathsf{k}_{enc}) \leftarrow \mathsf{Dec}_{asym}(\mathsf{dsk}, C_\mathcal{B})$.
(4) Choose $m \overset{\$}{\leftarrow} \mathbb{Z}_p^*$ as the secret key for a coin, and compute the commitment $U := g^m$.
(5) Set $\delta := V$ where δ denotes the current balance of the coin.
(6) Set CT as a string of $2^{n+1} - 1$ bits where each bit is 1. CT denotes the current tree structure of the unspent coin.
(7) Call Data_Seal() to seal m, CT and δ, and generate $blob_m$ (see Sect. 4.2).
(8) Choose a random number $r_\mathcal{D} \overset{\$}{\leftarrow} \mathbb{Z}_p^*$ and compute $R_\mathcal{D} := g^{r_\mathcal{D}}$.
(9) Compute $c_\mathcal{D} := \mathsf{H}_1(g||U||R_\mathcal{D}||C_\mathcal{B}||\alpha||n'_\mathcal{B})$.
(10) Compute $s_\mathcal{D} := r_\mathcal{D} + c_\mathcal{D} \cdot m \pmod{p}$.
(11) Generate a cipher context $C_\mathcal{D} \leftarrow \mathsf{Enc}_{sym}(\mathsf{k}_{enc}, account_\mathcal{D}||pwd)$.
(12) Generate $\tau_\mathcal{D} \leftarrow \mathsf{MAC}(\mathsf{k}_{mac}, U||n'_\mathcal{B}||c_\mathcal{D}||s_\mathcal{D}||C_\mathcal{D})$, and output $comm_{res} := (\tau_\mathcal{D}, U, n'_\mathcal{B}, c_\mathcal{D}, s_\mathcal{D}, C_\mathcal{D})$.

AEP-M Service saves $n'_\mathcal{B}$ and k_{mac} in Secure Memory as well as stores $blob_m$ in non-volatile storage. Then $\mathcal{D}$ switches back to NW and sends $comm_{res}$ to $\mathcal{B}$.

4. On input $comm_{res}$, $\mathcal{B}$ runs the following algorithm to generate a digital coin σ on m for $\mathcal{D}$:

$$(\sigma, \tau_\mathcal{B}) \leftarrow \mathsf{Gen\_DC}(comm_{res}, params', \mathsf{k}_{mac}, \mathsf{k}_{enc}, n_\mathcal{B}, \overline{sk}_\mathcal{B}).$$

The algorithm has seven steps:
(1) Verify $\tau_\mathcal{D} = \mathsf{MAC}(\mathsf{k}_{mac}, U||n'_\mathcal{B}||c_\mathcal{D}||s_\mathcal{D}||C_\mathcal{D})$, and check whether $n'_\mathcal{B} = n_\mathcal{B}$.
(2) Check whether U has not been used before by querying the database.
(3) Compute $R'_\mathcal{D} := g^{s_\mathcal{D}} \cdot U^{-c_\mathcal{D}}$ and $c'_\mathcal{D} := \mathsf{H}_1(g||U||R'_\mathcal{D}||C_\mathcal{B}||\alpha||n_\mathcal{B})$.
(4) Check whether $c'_\mathcal{D} = c_\mathcal{D}$.
(5) Decrypt $C_\mathcal{D}$ using Dec_{sym} and k_{enc}: $account_\mathcal{D}||pwd \leftarrow \mathsf{Dec}_{sym}(\mathsf{k}_{enc}, C_\mathcal{D})$, then check the plaintext's validness via communicating with the related commercial bank. If the account balance is enough, deduct money from the account and temporarily save it in $\mathcal{B}$.
(6) Choose a random number $a \overset{\$}{\leftarrow} \mathbb{Z}_p^*$, compute $A := g^a$, $B := A^y$, $C := g^{ax} \cdot U^{axy}$ and $D := U^{ay}$, and generate $\sigma := (A, B, C, D)$.
(7) Generate $\tau_\mathcal{B} \leftarrow \mathsf{MAC}(\mathsf{k}_{mac}, \sigma||n_\mathcal{D}||n_\mathcal{B})$.

In the above algorithm, if any check is failed, $\mathcal{B}$ aborts the process. If not, $\mathcal{B}$ sends $(\sigma, \tau_\mathcal{B})$ to $\mathcal{D}$, and sends $(U, \mathsf{dpk}, \mathsf{ID}_{bank}, \mathsf{ID}_{user})$ to $\mathcal{T}$ to backup for detecting possible double-spender. ID_{bank} is the identity of the commercial bank which the user account belongs to, and ID_{user} is the identity of the user.

5. Upon receiving $(\sigma, \tau_\mathcal{B})$, $\mathcal{D}$ switches into SW and verifies $\tau_\mathcal{B}$ using MAC, k_{mac} and $n'_\mathcal{B}$. Then, Data Handler calls Data_Seal() to seal σ and generates $blob_\sigma$. Finally, Logic Engine deletes $n_\mathcal{D}$, $n'_\mathcal{B}$ and k_{mac} from Secure Memory.

Pre-Compute. After the above step, $\mathcal{D}$ returns back to NW. AEP-M Proxy executes pre-computation in the background (off-line) to prepare for the following **Spend** phase. Preprocessing Engine calls AEPM_NW_PreCmpt() to generate a blinded coin:

$$(l, R, S, T, W) \leftarrow \mathsf{AEPM\_NW\_PreCmpt}(blob_{params}, blob_\sigma),$$

where the algorithm consists of the following steps.
(1) Get *params* and digital coin σ by directly reading the plaintext part of $blob_{params}$ and $blob_\sigma$ respectively.
(2) Parse σ as (A, B, C, D).
(3) Choose $l \xleftarrow{\$} \mathbb{Z}_p^*$ and compute $(R, S, T, W) := (A^l, B^l, C^l, D^l)$.
(4) Output (l, R, S, T, W).
Preprocessing Engine stores (l, R, S, T, W) together with $blob_\sigma$.

Spend. This is an interactive phase executed between a user with his mobile device $\mathcal{D}$ and a merchant $\mathcal{M}$, which enables $\mathcal{D}$ to anonymously pay some digital coins to $\mathcal{M}$.

1. App of $\mathcal{D}$ sends a nonce $\bar{n}_\mathcal{D} \xleftarrow{\$} \{0,1\}^\lambda$ to the merchant $\mathcal{M}$ for initiating a transaction.
2. Receiving $\bar{n}'_\mathcal{D}$, $\mathcal{M}$ chooses a nonce $n_\mathcal{M} \xleftarrow{\$} \{0,1\}^\lambda$ and generates a signature $\beta \leftarrow \mathsf{Sign}(\mathsf{sk}_\mathcal{M}, \texttt{"Spend"}\|info)$ where $info := (v, date, trans, \mathsf{pk}_\mathcal{M}, \bar{n}'_\mathcal{D}, n_\mathcal{M})$. info is the string collection containing the amount value v of coins to pay, transaction *date*, other necessary transaction information and the related nonce values. $\mathcal{M}$ sends $(info, cert_\mathcal{M}, \beta)$ to $\mathcal{D}$. In fact, issued by CA, $cert_\mathcal{M}$ is $\mathcal{M}$'s certificate, containing $\mathsf{ID}_\mathcal{M}$, $\mathsf{pk}_\mathcal{M}$ and the signature $\mathsf{Sign}_{\mathsf{CA}}(\mathsf{ID}_\mathcal{M}\|\mathsf{pk}_\mathcal{M})$, where $\mathsf{ID}_\mathcal{M}$ indicates the identitiy of $\mathcal{M}$.
3. When $\mathcal{D}$ receives the above data, AEP-M Proxy assembles the command to request AEP-M Service for payment. Without loss of generality, we assume that the user has a coin of value δ such that $\delta \geq v$. For the case that $\delta < v$, the user could spend another several coins in the same way in order that the sum amounts value of all coins equals v. On account of the request, $\mathcal{D}$'s environment is switched into SW. First, Logic Engine verifies β using Verify and $\mathsf{pk}_\mathcal{M}$ with $cert_\mathcal{M}$. Then, $\mathcal{D}$ enters the secure GUI after authenticating the user's inputted PIN (or fingerprint). Relying on Framebuffer, the secure GUI displays $\mathsf{ID}_\mathcal{M}$ and the content of v, *date* and *trans*. It is important for the user to confirm the exact $\mathsf{ID}_\mathcal{M}$ and transaction information in case an adversary falsifies the transaction. When the user presses the button of "OK", Logic Engine calls AEPM_SW_Spend() to create a master serial number $\boldsymbol{Z}$ of value v of coins together with a proof π of its validity, using the related pre-computation result as:

$$(\boldsymbol{Z}, \pi) \leftarrow \mathsf{AEPM\_SW\_Spend}(blob_{params}, blob_m, blob_\sigma\|(l, R, S, T, W), info),$$

where the detailed process is presented as follows:
(1) Unseal the blobs to get *params*, (m, CT, δ) and (A, B, C, D) by calling Data_Unseal().

(2) Check whether $\bar{n}'_{\mathcal{D}} = \bar{n}_{\mathcal{D}}$.

(3) Represent v by bits: $v = b_n b_{n-1}...b_0$ and set $\Phi := \{i|\ 0 \le i \le n \wedge b_i = 1\}$.

(4) For each $i \in \Phi$ from n to 0, based on CT, select uniformly at random an unspent node $s_i \in \mathcal{S}_n$ of level $n - i$ in the tree, and then mark it as the spent one.

(5) For each chosen node s_i, compute $t_{s_i} := g_{s_i}^m$, and form three sets: $\boldsymbol{s} := \{s_i | i \in \Phi\}$, $\boldsymbol{g_s} := \{g_{s_i} | i \in \Phi\}$ and $\boldsymbol{t_s} := \{t_{s_i} | i \in \Phi\}$. Set $\boldsymbol{Z} := (\boldsymbol{s}, \boldsymbol{t_s})$.

(6) Choose a random number $\bar{r} \xleftarrow{\$} \mathbb{Z}_p^*$, compute $L_i := g_{s_i}^{\bar{r}}$ for each $i \in \Phi$, form a set $\boldsymbol{L} := \{L_i | i \in \Phi\}$ and compute $\overline{L} := B^{l \cdot \bar{r}}$.

(7) Compute $\bar{c} := \mathsf{H}_2(\boldsymbol{g_s}||\boldsymbol{t_s}||R||S||T||W||\boldsymbol{L}||\overline{L}||\mathsf{info})$.

(8) Compute $\bar{z} := \bar{r} + \bar{c} \cdot m \pmod{p}$.

(9) Set $\pi := (R, S, T, W, \bar{c}, \bar{z})$.

(10) Delete (l, R, S, T, W) from the non-volatile storage.

(11) Update CT and $\delta := \delta - v$. If $\delta > 0$, call $\mathsf{Data\_Seal}()$ again to regenerate $blob_m$ using the updated CT and δ, else delete $blob_m$ and $blob_\sigma$.

After the API finally returns, $\mathcal{D}$ switches back into NW and sends $(\boldsymbol{Z}, \pi)$ to $\mathcal{M}$.

4. $\mathcal{M}$ sets $\mathsf{Tr} := (\mathsf{info}, \boldsymbol{Z}, \pi)$ and verifies Tr by the means of calling the specialized verification algorithm $\mathsf{Tr\_Verify}()$ as:

$$res \leftarrow \mathsf{Tr\_Verify}(params, \overline{\mathsf{pk}}_{\mathcal{B}}, \mathsf{Tr}),$$

where the algorithm runs in detail as follows:

(1) Parse Tr as $(\mathsf{info}, \boldsymbol{Z} = (\boldsymbol{s}, \boldsymbol{t_s}), \pi = (R, S, T, W, \bar{c}, \bar{z}))$.

(2) For any two nodes in $\boldsymbol{s}$, check that the one does not belong to the subtree rooted at the other one (i.e., each node is not a prefix of any other one).

(3) Compute $\overline{L}' := S^{\bar{z}} \cdot W^{-\bar{c}}$, $L'_i := g_{s_i}^{\bar{z}} \cdot t_{s_i}^{-\bar{c}}$ for each $s_i \in \boldsymbol{s}$, and set $\boldsymbol{L}' := \{L'_i | i \in \Phi\}$.

(4) Compute $\bar{c}' := \mathsf{H}_2(\boldsymbol{g_s}||\boldsymbol{t_s}||R||S||T||W||\boldsymbol{L}'||\overline{L}'||\mathsf{info})$.

(5) Check whether the relations $R \ne 1$, $W \ne 1$, $e(R, Y) = e(S, \tilde{g})$, $e(T, \tilde{g}) = e(R \cdot W, X)$ and $\bar{c}' = \bar{c}$ hold.

(6) If all the above checks are passed, then $res := true$, else $res := false$.

According to the verification result res, $\mathcal{M}$ decides whether to accept the payment from $\mathcal{D}$ and provide services or goods for the user. If $\mathcal{M}$ accepts the transaction, he sends $\mathcal{D}$ a receipt $\theta_{\mathcal{M}} \leftarrow \mathsf{Sign}(\mathsf{sk}_{\mathcal{M}}, \texttt{"receipt"}||\mathsf{Tr})$ as the proof of accepting digital coins.

Pre-$Compute$. After Step 3 above, AEP-M Proxy of $\mathcal{D}$ in NW executes pre-computation again in the background to generate a new tuple (l', R', S', T', W') w.r.t. some $blob_\sigma$, if exists, for the next **Spend** use.

Deposit. In this phase, $\mathcal{M}$ could deposit money from Tr to his preferable bank $account_{\mathcal{M}}$ through the central bank $\mathcal{B}$.

1. $\mathcal{M}$ generates a signature $\gamma \leftarrow \mathsf{Sign}(\mathsf{sk}_{\mathcal{M}}, \texttt{"Deposit"}||\mathsf{Tr}||account_{\mathcal{M}})$. Then he sends Tr, $account_{\mathcal{M}}$ and γ together with $cert_{\mathcal{M}}$ to $\mathcal{B}$.

2. $\mathcal{B}$ first verifies γ using Verify and $\text{cert}_{\mathcal{M}}$. Secondly, $\mathcal{B}$ retrieves $\text{pk}_{\mathcal{M}}$ from info and checks whether it is the same one inside $\text{cert}_{\mathcal{M}}$. Thirdly, $\mathcal{B}$ computes $H_3(\text{Tr})$ and queries database DB_{Tr} to check whether Tr has been used before. If not, $\mathcal{B}$ runs the verification algorithm Tr_Verify() to verify the validity of Tr. If it is valid, $\mathcal{B}$ immediately transfers the exact amount v of real money to $account_{\mathcal{M}}$ with the help of some commercial bank.

3. $\mathcal{B}$ detects double-spending off-line after the above step. The detection process is presented as follows:
   (1) Retrieve $\boldsymbol{s}$ and $\boldsymbol{t_s}$ from Tr, and load $params'$.
   (2) For each $t_{s_i} \in \boldsymbol{t_s}$ and each $f \in \mathcal{F}_n(s_i)$, compute $d_{s_i \mapsto f} := e(t_{s_i}, \tilde{g}_{s_i \mapsto f})$ and $d_{s_i,f} := H_4(d_{s_i \mapsto f})$.
   (3) Set $\boldsymbol{d} := \{d_{s_i,f} | s_i \in \boldsymbol{s} \wedge f \in \mathcal{F}_n(s_i)\}$.
   (4) Insert the item $(H_3(\text{Tr}), \text{Tr}, \boldsymbol{d})$ into DB_{Tr}.
   (5) For each $d_{s_i,f}$, query DB_{Tr} to check whether there exists a transaction Tr$'$ that has the same $d_{s_i,f}$. If exists, send both Tr and Tr$'$ to $\mathcal{T}$ through the secure channel for revealing the identity of the double-spender.

Identify. This phase endows $\mathcal{T}$ with the ability to reveal the identity of some double-spender.

1. When $\mathcal{T}$ receives the double-spending report (Tr, Tr') from $\mathcal{B}$, it executes the verification algorithm Tr_Verify() to verify the validity of Tr and Tr$'$. If both valid, $\mathcal{T}$ chooses one node $s_i \in \boldsymbol{s}$ from data of Tr and finds out the related r_{s_i} from its secret coin keys $\textsf{sck}$ to recover U by computing $U := t_{s_i}^{1/r_{s_i}}$ (i.e. g^m). Likewise, $\mathcal{B}$ recovers U' from Tr$'$.

2. If $U = U'$, it indicates that Tr and Tr$'$ lead to a double-spending. $\mathcal{T}$ would publish the spender's information $(\text{Tr}, \text{Tr}', U, \text{dpk}, \text{ID}_{\text{bank}}, \text{ID}_{\text{user}})$. Then, some possible penalties on the user ID_{user}, for example deducting money from the user's account or temporally prohibiting the user from using e-payment system, would be triggered.

4.4 Optional Defense Mechanisms and Security Analysis

AEP-M satisfies the desired security properties such as unlinkability, traceability, exculpability, confidentiality and authenticity. Additional defense mechanisms could further enhance our scheme's security. The detailed description of these properties, mechanisms and the analysis can be found in the full paper [21].

5 Implementation and Evaluation

In this section, we first present the prototype of AEP-M from both aspects of hardware and software. Afterwards, we show the efficiency of the proposed scheme. Finally, we give the performance evaluation and analysis based on our prototype system.

5.1   Implementation

Hardware Platform. To simulate real environment, we implement the role of merchant on one PC platform, and central bank as well as trusted authority on another one. For simulating mobile device, we leverage a development board Zynq-7000 AP Soc Evaluation Kit [18] to implement functions of AEP-M. It is TrustZone-enabled and equipped with ARM Cortex-A9 MPCore, 1 GB DDR3 memory and On-Chip Memory (OCM) module. We utilize an SRAM chip that is the type IS61LV6416-10TL [15] to act as our SRAM PUF. The processor can fetch the SRAM data in the RAM cache via the bus. In addition, the methods given in [13] are applied to fulfill Framebuffer for secure display.

Software Implementation. The software implementation on the development board for mobile device is divided into two parts. In *secure world*, we use Open Virtualization SierraTEE as the basic TEE OS which is compliant with GP's TEE Specifications [14]. For Crypto Library, we use OpenSSL-1.0.2 g for general cryptographic algorithms, and Pairing-Based Cryptography (PBC) 0.5.14 library for computations of elliptic curves and bilinear maps. The security parameter λ is set to 128 (bits), so we choose SHA256 for H_3 and H_4, HMAC-SHA256 for MAC, 3072-bit RSA for $Enc_{asym} - Dec_{asym}$, 256-bit ECDSA for Sign-Verify and 128-bit AES-CBC for $Enc_{sym} - Dec_{sym}$. 5268 lines of code (LOC) in C language totally make up our components and auxiliary functions in *secure world*. In *normal world*, we run a Linux as REE OS with kernel 3.8.6. AEP-M Proxy totally comprises 2879 LOC. Besides we program one test application that could execute upon AEP-M scheme. It contains 1268 LOC for App running in NW and 661 LOC for App Trustlet in SW. Furthermore, there are several tens of thousands of LOC for other entities.

5.2   Efficiency and Performance Evaluation

The specific analysis of AEP-M's efficiency also appears in the full paper [21]. Since the resource-constrained mobile device is the performance bottleneck as well as the focus of our attention, we measure the performance of AEP-M on the prototype system revolving around mobile device. We select BN curve with embedding degree 12. For testing the security level of 128-bit, we conduct the experiments using BN256. Each average experimental result is taken over 50 test-runs.

For coins of value 2^{10} and 2^{20} respectively, and spending 287 of them, Fig. 4 illustrates the average time overheads of critical processes including the computations of **Withdraw**, *Pre-Compute* and **Spend** on mobile device for user side and **Spend** on PC for merchant side. The results show that the frequent computations about either *Pre-Compute* or **Spend** only take less than 450 milliseconds (ms), while infrequent and time-consuming **Withdraw** spends less than 660 ms. Moreover, the time overhead is indeed low on PC platform.

Figure 5 shows the average time overheads of **Spend** phase on mobile device for user side using $n = 10$ and different v. $|\Phi|$ takes corresponding values from v's representations by bits. We can see that as the value of $|\Phi|$ increases, the time

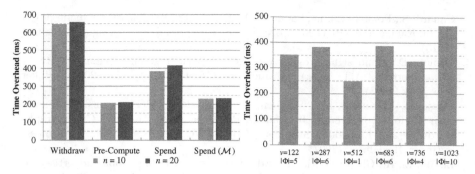

Fig. 4. Time overheads of the critical processes for coins of 2^n and $v = 287$.

Fig. 5. Time overheads of **Spend** phase for $n = 10$ with different values of v.

overheads of **Spend** have evident growth, which nearly has nothing to do with v itself, big or small. Encouragingly, under the worst-case scenario where $|\Phi| = 10$, the resulting overhead spends less than 500 ms, which is completely acceptable for a mobile user. According to our efficiency analysis and experimental results, AEP-M can be considered as a reasonably efficient scheme for mobile device.

6 Conclusion

In this paper, we propose AEP-M, a complete and practical anonymous e-payment scheme using TrustZone and divisible e-cash. AEP-M tackles both security and privacy issues specially for mobile electronic payers. The scheme allows users to withdraw a coin of value 2^n and spend it in several times by dividing it. Pre-computation and the bit-decomposition technique for coin's representation are carefully taken into our consideration to raise scheme's efficiency and flexibility. What is more, TrustZone provides data and execution protection for AEP-M. Our implementation and evaluation convince that AEP-M is quite practical for payers using resource-constrained mobile devices.

Acknowledgment. This work was supported in part by grants from the National Natural Science Foundation of China (No. 91118006 and No. 61402455).

References

1. Limited ARM: ARM security technology-building a secure system using TrustZone technology, April 2009
2. Camenisch, J.L., Hohenberger, S., Lysyanskaya, A.: Compact e-cash. In: Cramer, R. (ed.) EUROCRYPT 2005. LNCS, vol. 3494, pp. 302–321. Springer, Heidelberg (2005)
3. Canard, S., Gouget, A.: Divisible e-cash systems can be truly anonymous. In: Naor, M. (ed.) EUROCRYPT 2007. LNCS, vol. 4515, pp. 482–497. Springer, Heidelberg (2007)

4. Canard, S., Pointcheval, D., Sanders, O., Traoré, J.: Divisible e-cash made practical. In: Katz, J. (ed.) PKC 2015. LNCS, vol. 9020, pp. 77–100. Springer, Heidelberg (2015)
5. Chaum, D.: Blind signatures for untraceable payments. In: Chaum, D., Rivest, R.L., Sherman, A.T. (eds.) Advances in Cryptology, pp. 199–203. Springer, New York (1983)
6. Guajardo, J., Kumar, S.S., Schrijen, G.J., Tuyls, P.: FPGA intrinsic PUFs and their use for IP protection. In: Paillier, P., Verbauwhede, I. (eds.) Cryptographic Hardware and Embedded Systems - CHES 2007. LNCS, vol. 4727, pp. 63–80. Springer, Heidelberg (2007)
7. Li, W., Li, H., Chen, H., Xia, Y.: AdAttester: secure online mobile advertisement attestation using TrustZone. In: Proceedings of MobiSys 2015, pp. 75–88. ACM (2015)
8. Lim, A.S.: Inter-consortia battles in mobile payments standardisation. Electron. Commer. Res. Appl. **7**(2), 202–213 (2008)
9. Preibusch, S., Peetz, T., Acar, G., Berendt, B.: Purchase details leaked to PayPal (short paper). In: Böhme, R., Okamoto, T. (eds.) FC 2015. LNCS, vol. 8975, pp. 217–226. Springer, Heidelberg (2015)
10. Reaves, B., Scaife, N., Bates, A., Traynor, P., Butler, K.R.B.: Mo(bile) money, mo(bile) problems: analysis of branchless banking applications in the developing world. In: Proceedings of the 24th USENIX Conference on Security Symposium (2015)
11. Rial, A.: Privacy-preserving e-commerce protocols. Ph.D. thesis, Faculty of Engineering Science, KU Leuven, March 2013
12. Suh, G.E., Devadas, S.: Physical unclonable functions for device authentication and secret key generation. In: 44th ACM/IEEE DAC 2007, pp. 9–14 (2007)
13. Sun, H., Sun, K., Wang, Y., Jing, J.: Trust OTP: transforming smartphones into secure one-time password tokens. In: Proceedings of CCS 2015, pp. 976–988. ACM (2015)
14. GlobalPlatform: Tee client API specification version 1.0 (2010)
15. Integrated Silicon Solution Inc, IS61LV6416-10TL. http://www.alldatasheet.com/datasheet-pdf/pdf/505020/ISSI/IS61LV6416-10TL.html
16. ISO/IEC: 15946-5: 2009 Information Technology-Security Techniques: Cryptographic Techniques based on Elliptic Curves: Part 5: Elliptic Curve Generation (2009)
17. Proxama (2015). http://www.proxama.com/platform/. Accessed 15 Oct 2015
18. Xilinx: Zynq-7000 all programmable soc zc702 evaluation kit. http://www.xilinx.com/products/boards-and-kits/EK-Z7-ZC702-G.htm
19. Yang, B., Feng, D., Qin, Y.: A lightweight anonymous mobile shopping scheme based on DAA for trusted mobile platform. In: IEEE TrustCom 2014, pp. 9–17. IEEE (2014)
20. Yang, B., Yang, K., Qin, Y., Zhang, Z., Feng, D.: DAA-TZ: an efficient DAA scheme for mobile devices using ARM TrustZone. In: Conti, M., Schunter, M., Askoxylakis, I. (eds.) TRUST 2015. LNCS, vol. 9229, pp. 209–227. Springer, Heidelberg (2015)
21. Yang, B., Yang, K., Zhang, Z., Qin, Y., Feng, D.: AEP-M: practical anonymous e-payment for mobile devices using ARM Trust Zone and divisible e-cash (full version). ePrint (2016)
22. Zhao, S., Zhang, Q., Hu, G., Qin, Y., Feng, D.: Providing root of trust for ARM trust zone using on-chip SRAM. In: Proceedings of TrustED 2014, pp. 25–36. ACM (2014)

Universally Composable Two-Server PAKE

Franziskus Kiefer[1](✉) and Mark Manulis[2]

[1] Mozilla, Berlin, Germany
mail@franziskuskiefer.de
[2] Department of Computer Science, Surrey Center for Cyber Security,
University of Surrey, Guildford, UK
mark@manulis.eu

Abstract. Two-Server Password Authenticated Key Exchange
(2PAKE) protocols apply secret sharing techniques to achieve protection against server-compromise attacks. 2PAKE protocols eliminate the
need for password hashing and remain secure as long as one of the servers
remains honest. This concept has also been explored in connection with
two-server password authenticated secret sharing (2PASS) protocols for
which game-based and universally composable versions have been proposed. In contrast, universally composable PAKE protocols exist currently only in the single-server scenario and all proposed 2PAKE protocols use game-based security definitions.

In this paper we propose the first construction of an universally
composable 2PAKE protocol, alongside with its ideal functionality. The
protocol is proven UC-secure in the standard model, assuming a common reference string which is a common assumption to many UC-secure
PAKE and PASS protocols. The proposed protocol remains secure for
arbitrary password distributions. As one of the building blocks we define
and construct a new cryptographic primitive, called Trapdoor Distributed Smooth Projective Hash Function (TD-SPHF), which could be of
independent interest.

1 Introduction

Password Authenticated Key Exchange (PAKE) protocols have been extensively
researched over the last twenty years. They allow two protocol participants sharing a low-entropy secret (password) to negotiate an authenticated secret key.
Several PAKE security models are widely used such as the game-based PAKE
model, called BPR, by Bellare, Pointcheval and Rogaway [4,8] and the PAKE
model in the Universal Composability (UC) framework by Canetti [18]. PAKE
protocols are often considered in a client-server scenario where the client password is registered and stored in a protected way on the server side such that it
can be used later to authenticate the client. This approach however leads to an
intrinsic weakness of single-server PAKE protocols against server-compromise
attacks. An attacker who breaks into the server can efficiently recover client's
password and impersonate the client to the server as well as to other servers
if this password is used across many client accounts which is often the case. A

© Springer International Publishing Switzerland 2016
M. Bishop and A.C.A. Nascimento (Eds.): ISC 2016, LNCS 9866, pp. 147–166, 2016.
DOI: 10.1007/978-3-319-45871-7_10

number of approaches have been proposed to alleviate this threat. For instance, verifier-based PAKE [11,22,34], also known as augmented PAKE [9], considers an asymmetric setting in which the server uses a randomized password hash to verify a client holding the corresponding password. The crucial weakness of VPAKE protocols is that they do not protect against offline dictionary attacks on compromised password hashes, i.e. an attacker can still recover the password, which can often be done efficiently with current tools like [23,31].

Two-server PAKE (2PAKE) protocols solve this problem through secret sharing techniques. The client password is split into two shares and each server receives its own share upon registration. In order to authenticate the client both servers take part in the protocol execution. 2PAKE security typically holds against an active attacker who can compromise at most one server and thus learn the corresponding password share. 2PAKE protocols can be symmetric (e.g. [12,27,29,33]) where both servers compute the same session key and asymmetric (e.g. [27]) where each server can compute an independent session key with the client or assist another server in the authentication process [26,35] without computing the key. A potential drawback of symmetric protocols is that by corrupting one server the attacker may use learned key material to read communications between the client and the other server. Existing 2PAKE protocols were analysed using variants of the BPR model and do not offer compositional security guarantees. While 2PAKE can be seen as a special case of Threshold PAKE (TPAKE), e.g. [30,32], that adopt t-out-of-n secret sharing, existing TPAKE protocols do not necessarily provide solutions for 2PAKE, e.g. [32] requires $t < n/3$. Finally, we note that UC-security was considered for a class of Two-Server/Threshold Password Authenticated Secret Sharing (2/TPASS) protocols, e.g. [13,14,24], that address a different problem of sharing a chosen key across multiple servers and its subsequent reconstruction from the password.

In this paper we propose the first UC-secure (asymmetric) 2PAKE protocol where one of the two servers computes an independent session key with the client. We rely on a common reference string, which is a standard assumption for UC-secure PAKE protocols. As a consequence of UC modeling our protocol offers security for all password distributions, which is notoriously difficult to achieve in BPR-like models. One challenge in achieving UC security is that the protocol must remain simulatable against active attackers that play with a correctly guessed password (unlike in game-based models where simulation can be aborted). In order to achieve simulatability we introduce a new building block, called *Trapdoor Distributed Smooth Projective Hash Functions (TD-SPHF)*, offering distributed SPHF properties from [29] and the SPHF trapdoor property from [10]. While traditional SPHF were used in the design of single-server PAKE protocols, the 2PAKE protocol framework from [29], a generalisation of [27] that was proven secure in the BPR-like model, required an extension of SPHF to a distributed setting. Such distributed SPHF alone are not sufficient for achieving the UC security. Our TD-SPHF helps to achieve simulatability for 2PAKE protocols and could be of independent interest for other UC-secure constructions.

2   Preliminaries and Building Blocks

Our 2PAKE protocol is defined over bilinear groups $\mathbb{G}_1$ and $\mathbb{G}_2$ of prime order q with an efficiently computable map $e : \mathbb{G}_1 \times \mathbb{G}_2 \mapsto \mathbb{G}_T$. The following properties have to hold: (i) If g_1 is a generator of $\mathbb{G}_1$ and g_2 is a generator of $\mathbb{G}_2$, then $e(g_1, g_2)$ is a generator of $\mathbb{G}_T$. (ii) For generators g_1, g_2 and scalar $x \in_R \mathbb{Z}_q$ it holds that $e(g_1^x, g_2) = e(g_1, g_2^x) = e(g_1, g_2)^x$. We require further that the Symmetric External Diffie-Hellman assumption (SXDH) ([5,6] amongst others) holds in those groups. SXDH states that the DDH problem is hard in $\mathbb{G}_1$ and $\mathbb{G}_2$. All computations defined on a q-order group in the following are performed in $\mathbb{G}_1$. Let λ denote the security parameter throughout this work.

Commitments. By $\mathsf{C} = (\mathsf{CSetup}, \mathsf{Com})$ we denote an efficient commitment scheme and use Pedersen commitments in our constructions where $(g, h, q, \lambda) \leftarrow \mathsf{CSetup}(\lambda)$ and $C \leftarrow \mathsf{Com} = (x; r) = g^x h^r$ with g and h being generators of a cyclic group $\mathbb{G}$ of prime-order q with bit-length in the security parameter λ and where the discrete logarithm of h with respect to base g is not known. Pedersen commitments are *additively homomorph*, i.e. for all $(C_i, d_i) \leftarrow \mathsf{Com}(x_i; r_i)$, $i \in 0, \ldots, m$ we have $\prod_{i=0}^{m} C_i = \mathsf{Com}(\sum_{i=0}^{m} x_i; \sum_{i=0}^{m} r_i)$.

Committed Zero-Knowledge Proofs. We use committed Σ-protocols for security against malicious verifiers [21,25]. Note that we do not require extractability (proof of knowledge) here, which allows us to avoid the necessity of rewinding. A zero-knowledge proof ZKP is executed between a prover and a verifier, proving that a word x is in a language L, using a witness w proving so.[1] Let $P_1(x, w, r)$ and $P_2(x, w, r, c)$ denote the two prover steps of a Σ-protocol and $H : \{0, 1\}^* \mapsto \mathbb{Z}_q$ a collision-resistant hash function. A committed Σ-protocol is then given by the following four steps:

- The prover computes the first message $\mathsf{Co} \leftarrow P_1(x, w, r)$ and $m_1 \leftarrow \mathsf{Com}(H(x, \mathsf{Co}); r_1) = g^{H(x, \mathsf{Co})} h^{r_1}$, and sends m_1 to the verifier.
- The verifier chooses challenge $\mathsf{Ch} = c \in_R \mathbb{Z}_q$ and returns it to the prover.
- The prover computes the second message $\mathsf{Rs} \leftarrow P_2(x, w, r, c)$ and $m_2 \leftarrow \mathsf{Com}(H(\mathsf{Rs}); r_2) = g^{H(\mathsf{Rs})} h^{r_2}$, and sends m_2 to the verifier.
- Further, the prover opens the commitments m_1 and m_2 sending $(x, \mathsf{Co}, \mathsf{Rs}, r_1, r_2)$ to the verifier.
- The verifier accepts iff both commitments are valid and if the verification of the Σ-protocol $(x, \mathsf{Com}, \mathsf{Ch}, \mathsf{Rs})$ is successful.

Cramer-Shoup Encryption with Labels. Let $C = (\ell, \boldsymbol{u}, e, v) \leftarrow \mathsf{Enc}_{\mathsf{pk}}^{\mathsf{CS}}(\ell, m; r)$ (on label ℓ, message m, and randomness r) with $\boldsymbol{u} = (u_1, u_2) = (g_1^r, g_2^r)$, $e = h^r g_1^m$ and $v = (cd^\xi)^r$ with $\xi = H_k(\ell, \boldsymbol{u}, e)$ denote a labelled Cramer-Shoup ciphertext. We assume $m \in \mathbb{Z}_q$ and $\mathbb{G}$ is a cyclic group of prime order q

---

[1] Zero-knowledge languages L are independent from the smooth projective hashing languages introduced in Sect. 2.1.

with generators g_1 and g_2 such that $g_1^m \in \mathbb{G}$. The CS public key is defined as $\mathrm{pk} = (p, \mathbb{G}, g_1, g_2, c, d, H_k)$ with $c = g_1^{x_1} g_2^{x_2}, d = g_1^{y_1} g_2^{y_2}, h = g_1^z$ and hash function H_k such that $\tau = (x_1, x_2, y_1, y_2, z)$ denotes the decryption key. Decryption is defined as $g_1^m = \mathrm{Dec}_{\mathrm{dk}}^{\mathrm{CS}}(C) = e/u_1^z$ if $u_1^{x_1 + y_1 \cdot \xi'} u_2^{x_2 + y_2 \cdot \xi'} = v$ with $\xi' = H_k(\ell, \boldsymbol{u}, e)$.

2.1 Smooth Projective Hashing (SPHF)

First, we recall definitions for classical SPHF tailored to the PAKE use-case and cyclic groups $\mathbb{G}$ of prime-order q. We use languages of ciphertexts with the password as message and the randomness as witness. An SPHF language L for a given password pw from dictionary $\mathcal{D}$ is given by L_{pw}. The public parameter of the language is the common reference string crs containing the public key pk of the encryption scheme. By τ we denote the crs trapdoor, the secret key to pk. Let $\mathcal{L}$ be the encryption scheme used to generate words in L_{pw}. Unless stated otherwise we assume that $\mathcal{L}$ is a labelled CCA-secure encryption scheme, e.g. labelled Cramer-Shoup scheme.

Definition 1 (Languages of Ciphertexts). *Let $L_{\mathrm{pw}} \subseteq \{(\ell, C, \mathrm{pw}^*)\} = \mathcal{C}$ denote the language of labelled ciphertexts under consideration with ciphertext (ℓ, C) under pk and password $\mathrm{pw}^* \in \mathcal{D}$. A ciphertext C is in language L_{pw} iff there exists randomness r such that $C \leftarrow \mathrm{Enc}_{\mathrm{pk}}^{\mathcal{L}}(\ell, \mathrm{pw}; r)$.*

Smooth projective hashing for languages of ciphertexts where the projection key does not depend on the ciphertext is defined as follows (see also [10, 28]).

Definition 2 (KV-SPHF). *Let L_{pw} denote a language of ciphertexts such that $C \in L_{\mathrm{pw}}$ if there exists randomness r proving so. A smooth projective hash function for ciphertext language L_{pw} consists of the following four algorithms:*

- *$\mathrm{KGen}_{\mathrm{H}}(L_{\mathrm{pw}})$ generates a random hashing key $\mathrm{k_h}$ for language L_{pw}.*
- *$\mathrm{KGen}_{\mathrm{P}}(\mathrm{k_h}, L_{\mathrm{pw}})$ derives the projection key $\mathrm{k_p}$ from hashing key $\mathrm{k_h}$.*
- *$\mathrm{Hash}(\mathrm{k_h}, L_{\mathrm{pw}}, C)$ computes hash value h from hashing key $\mathrm{k_h}$ and ciphertext C.*
- *$\mathrm{PHash}(\mathrm{k_p}, L_{\mathrm{pw}}, C, r)$ computes hash value h from projection key $\mathrm{k_p}$, ciphertext C and randomness r.*

A SPHF has to fulfil the following three properties:

- *Correctness:* If $C \in L$, with r proving so, then $\mathrm{Hash}(\mathrm{k_h}, L_{\mathrm{pw}}, C) = \mathrm{PHash}(\mathrm{k_p}, L_{\mathrm{pw}}, C, r)$.
- *Smoothness:* If $\{(\ell, C, \mathrm{pw}^*)\} \ni C \notin L_{\mathrm{pw}}$, the hash value h is (statistically) indistinguishable from a random element.
- *Pseudorandomness:* If $C \in L_{\mathrm{pw}}$, the hash value h is (computationally) indistinguishable from a random element.

2.2 Trapdoor Smooth Projective Hashing

For efficient one-round UC-secure PAKE a new SPHF flavor, called Trapdoor SPHF (T-SPHF), was introduced in [10]. T-SPHF adds three additional functions to the classical SPHF definition allowing computation of the hash value from the projection key, ciphertext and trapdoor τ'.[2]

Definition 3 (Trapdoor SPHF). *Let L_{pw} denote a language of ciphertexts such that $C \in L_{\mathrm{pw}}$ if there exists randomness r proving so. A trapdoor smooth projective hash function for a ciphertext language L_{pw} consists of the following seven algorithms:*

- $\mathsf{KGen_H}$, $\mathsf{KGen_P}$, Hash *and* PHash *are as given in Definition 2*
- $\mathsf{TSetup(crs)}$ *generates a second* $\mathsf{crs'}$ *with trapdoor τ' on input of a* crs
- $\mathsf{VerKp(k_p}, L_{\mathrm{pw}})$ *returns 1 iff* $\mathsf{k_p}$ *is a valid projection key, 0 otherwise*
- $\mathsf{THash(k_p}, L_{\mathrm{pw}}, C, \tau')$ *computes the hash value h of C using the projection key* $\mathsf{k_p}$ *and trapdoor τ'*

We assume $\mathsf{crs'}$ is, like crs, made available to all parties.

2.3 Distributed Smooth Projective Hashing

Another flavor, called Distributed SPHF (D-SPHF), was introduced in [29] for use in (non-composable) 2PAKE protocols such as [27] where servers hold password shares pw_1 and pw_2 respectively, and the client holds $\mathrm{pw} = \mathrm{pw}_1 + \mathrm{pw}_2$. Due to the nature of the words considered in D-SPHF they produce two different hash values. One can think of the two hash values as h_0 for C_0 (from the client) and h_x for C_1, C_2 (from the two servers). The hash value h_0 can be either computed with knowledge of the client's hash key $\mathsf{k_{h0}}$ or with the server's witnesses r_1, r_2 that C_1, C_2 are in L_{pw_i}, $i \in \{1, 2\}$ respectively. The hash value h_x can be computed with knowledge of the server hash keys $\mathsf{k_{h1}}, \mathsf{k_{h2}}$ or with the client's witness r_0 that C_0 is in L_{pw}. The combined language is denoted by $L_{\widehat{\mathrm{pw}}}$.

Definition 4 (Distributed SPHF). *Let $L_{\widehat{\mathrm{pw}}}$ denote a language such that $C = (C_0, C_1, C_2) \in L_{\widehat{\mathrm{pw}}}$ if there exists a witness $r = (r_0, r_1, r_2)$ proving so, $\mathrm{pw} = \mathrm{pw}_1 + \mathrm{pw}_2$ and there exists a function $\mathsf{Dec'}$ such that $\mathsf{Dec'}(C_1 C_2) = \mathsf{Dec'}(C_0)$. A distributed smooth projective hash function for language $L_{\widehat{\mathrm{pw}}}$ consists of the following six algorithms:*

- $\mathsf{KGen_H}(L_{\widehat{\mathrm{pw}}})$ *generates a hashing key $\mathsf{k_{h}}_i$ for $i \in \{0, 1, 2\}$ and language $L_{\widehat{\mathrm{pw}}}$.*
- $\mathsf{KGen_P}(\mathsf{k_{h}}_i, L_{\widehat{\mathrm{pw}}})$ *derives projection key $\mathsf{k_{p}}_i$ from hashing key $\mathsf{k_{h}}_i$ for $i \in \{0, 1, 2\}$.*
- $\mathsf{Hash}_x(\mathsf{k_{h0}}, L_{\widehat{\mathrm{pw}}}, C_1, C_2)$ *computes hash value h_x from hashing key $\mathsf{k_{h0}}$ and two server ciphertexts C_1 and C_2.*
- $\mathsf{PHash}_x(\mathsf{k_{p0}}, L_{\widehat{\mathrm{pw}}}, C_1, C_2, r_1, r_2)$ *computes hash value h_x from projection key $\mathsf{k_{p0}}$, two ciphertexts C_1 and C_2, and witnesses r_1 and r_2.*

---

[2] Note that τ' is a different trapdoor than the CRS trapdoor τ.

- $\texttt{Hash}_0(\texttt{k}_{\texttt{h}1}, \texttt{k}_{\texttt{h}2}, L_{\widehat{\texttt{pw}}}, C_0)$ *computes hash value* h_0 *from hashing keys* $\texttt{k}_{\texttt{h}1}$ *and* $\texttt{k}_{\texttt{h}2}$ *and ciphertext* C_0.
- $\texttt{PHash}_0(\texttt{k}_{\texttt{p}1}, \texttt{k}_{\texttt{p}2}, L_{\widehat{\texttt{pw}}}, C_0, r_0)$ *computes hash value* h_0 *from projection keys* $\texttt{k}_{\texttt{p}1}$ *and* $\texttt{k}_{\texttt{p}2}$, *the ciphertext* C_0, *and witness* r_0.

A distributed SPHF protocol between three participants C, S_1, S_2 computing h_x and h_0 is described by three interactive protocols $\texttt{Setup}$, $\texttt{PHash}_x^D$ and $\texttt{Hash}_0^D$. Let Π denote D-SPHF as described above.

- $\texttt{Setup}(\text{pw}, \text{pw}_1, \text{pw}_2, C, S_1, S_2)$ initialises a new instance for each participant with (pw, C, S_1, S_2) for C, $(\text{pw}_1, S_1, C, S_2)$ for S_1 and $(\text{pw}_2, S_2, C, S_1)$ for S_2. Eventually, all participants compute and broadcast projection keys $\texttt{k}_{\texttt{p}_i}$ and encryptions $C_i \leftarrow \texttt{Enc}_{\texttt{pk}}^{\mathcal{L}}(\ell_i, \text{pw}_i; r_i)$ of their password (share) pw_i using $\Pi.\texttt{KGen}_\texttt{H}$, $\Pi.\texttt{KGen}_\texttt{P}$ and the associated encryption scheme $\mathcal{L}$. Participants store incoming $\texttt{k}_{\texttt{p}_i}, C_i$ for later use. After receiving $(\texttt{k}_{\texttt{p}1}, C_1, \texttt{k}_{\texttt{p}2}, C_2)$, the client computes $h_0 \leftarrow \Pi.\texttt{PHash}_0(\texttt{k}_{\texttt{p}1}, \texttt{k}_{\texttt{p}2}, L_{\widehat{\texttt{pw}}}, C_0, r_0)$ and $h_x \leftarrow \Pi.\texttt{Hash}_x(\texttt{k}_{\texttt{h}0}, L_{\widehat{\texttt{pw}}}, C_1, C_2)$.
- $\texttt{PHash}_x^D$ is executed between S_1 and S_2. Each server S_i performs $\texttt{PHash}_x^D$ on input $(\texttt{k}_{\texttt{p}0}, \text{pw}_i, C_1, C_2, r_i)$ such that S_1 eventually holds h_x while S_2 learns nothing about h_x.
- $\texttt{Hash}_0^D$ is executed between S_1 and S_2. Each server S_i performs $\texttt{Hash}_0^D$ on input $(\text{pw}_i, \texttt{k}_{\texttt{h}i}, C_0, C_1, C_2)$ such that S_1 eventually holds h_0 while S_2 learns nothing about h_0.

2.4   Ideal Functionalities

For our 2PAKE realisation we rely on some commonly used ideal functionalities within the UC framework. These are: $\mathcal{F}_{\texttt{crs}}$ for the common reference string from [17], $\mathcal{F}_{\texttt{CA}}$ for the CA from [16] to establish verified public keys for the servers, $\mathcal{F}_{\texttt{init}}$ from [7] to establish unique query identifiers between the parties in a protocol. We refer for their descriptions to the original sources.

3   Trapdoor Distributed Smooth Projective Hashing

T-SPHF enabled constructions of one-round UC-secure PAKE [10] because of simulatability even in presence of attackers who guess correct passwords. In order to use the trapdoor property for simulatability in 2PAKE protocols T-SPHF must first be extended to the distributed setting of D-SPHF (cf. Sect. 2.3). We denote this new flavor by TD-SPHF and describe it specifically for usage in our 2PAKE, i.e. using languages based on Cramer-Shoup ciphertexts. A more general description of TD-SPHF accounting for more servers and/or other languages can be obtained similarly to the general description of D-SPHF in [29].

Definition 5 (TD-SPHF). *Let* $L_{\widehat{\texttt{pw}}}$ *denote a language such that* $C = (C_0, C_1, C_2) \in L_{\widehat{\texttt{pw}}}$ *if there exists a witness* $r = (r_0, r_1, r_2)$ *proving so,* $\text{pw} = \text{pw}_1 + \text{pw}_2$ *and there exists a function* $\texttt{Dec}'$ *such that* $\texttt{Dec}'(C_1 C_2) = \texttt{Dec}'(C_0)$. *A trapdoor distributed smooth projective hash function for language* $L_{\widehat{\texttt{pw}}}$ *consists of the following ten algorithms:*

- $(\mathtt{crs}', \tau') \xleftarrow{R} \mathsf{TSetup}(\mathtt{crs})$ *generates* $\mathtt{crs}'$ *with trapdoor* τ' *from* $\mathtt{crs}$
- $\mathsf{KGen}_\mathsf{H}, \mathsf{KGen}_\mathsf{P}, \mathsf{Hash}_x, \mathsf{PHash}_x, \mathsf{Hash}_0, \mathsf{PHash}_0$ *behave as for D-SPHF*
- $b \leftarrow \mathsf{VerKp}(\mathsf{k_p}, L_{\widehat{\mathrm{pw}}})$ *returns* $b = 1$ *iff* $\mathsf{k_p}$ *is a valid projection key and* $b = 0$ *otherwise*
- $h_x \leftarrow \mathsf{THash}_x(\mathsf{k_{p0}}, L_{\widehat{\mathrm{pw}}}, C_1, C_2, \tau')$ *computes hash value* h_x *of ciphertexts* C_1 *and* C_2 *using projection key* $\mathsf{k_{p0}}$ *and trapdoor* τ'
- $h_0 \leftarrow \mathsf{THash}_0(\mathsf{k_{p1}}, \mathsf{k_{p2}}, L_{\widehat{\mathrm{pw}}}, C_0, \tau')$ *computes hash value* h_0 *of* C_0 *using projection keys* $\mathsf{k_{p1}}$ *and* $\mathsf{k_{p2}}$, *and trapdoor* τ'

Security of TD-SPHF can be derived from D-SPHF security and the extensions made on SPHF for T-SPHF. However, we do not consider security of TD-SPHF on its own but rather incorporate it in the security proof of the 2PAKE protocol in the following section. This is due to the fact that description of TD-SPHF is done only for this specific application such that a separate security definition is more distracting than giving any benefit. However, we define correctness and soundness of TD-SPHF since they differ from that of D-SPHF. In particular, *correctness* of TD-SPHF extends correctness of D-SPHF by the statement that for every valid ciphertext triple (C_0, C_1, C_2), generated by $\mathcal{L}$, and honestly generated keys $(\mathsf{k_{h0}}, \mathsf{k_{h1}}, \mathsf{k_{h2}})$ and $(\mathsf{k_{p0}}, \mathsf{k_{p1}}, \mathsf{k_{p2}})$, it holds not only that

$$\mathsf{Hash}_0(\mathsf{k_{h1}}, \mathsf{k_{h2}}, L_{\widehat{\mathrm{pw}}}, C_0) = \mathsf{PHash}_0(\mathsf{k_{p1}}, \mathsf{k_{p2}}, L_{\mathrm{pw},\mathrm{pw}_1,\mathrm{pw}_2}, C_0, r_0), \text{ and}$$
$$\mathsf{Hash}_x(\mathsf{k_{h0}}, L_{\widehat{\mathrm{pw}}}, C_1, C_2) = \mathsf{PHash}_x(\mathsf{k_{p0}}, L_{\mathrm{pw},\mathrm{pw}_1,\mathrm{pw}_2}, C_1, C_2, r_1, r_2)$$

but also that $\mathsf{VerKp}(\mathsf{k_{p_i}}, L_{\widehat{\mathrm{pw}}}) = 1$ for $i \in \{0, 1, 2\}$ and

$$\mathsf{Hash}_0(\mathsf{k_{h1}}, \mathsf{k_{h2}}, L_{\widehat{\mathrm{pw}}}, C_0) = \mathsf{THash}_0(\mathsf{k_{p1}}, \mathsf{k_{p2}}, L_{\mathrm{pw},\mathrm{pw}_1,\mathrm{pw}_2}, C_0, \tau') \text{ and}$$
$$\mathsf{Hash}_x(\mathsf{k_{h0}}, L_{\widehat{\mathrm{pw}}}, C_1, C_2) = \mathsf{THash}_x(\mathsf{k_{p0}}, L_{\mathrm{pw},\mathrm{pw}_1,\mathrm{pw}_2}, C_1, C_2, \tau').$$

To capture soundness of TD-SPHFs we define (t, ε)-*soundness*, complementing the previous correctness extension, as follows.

Definition 6 (TD-SPHF(t, ε)-soundness). *Given* $\mathtt{crs}$, $\mathtt{crs}'$ *and* τ, *no adversary running in time at most* t *can produce a projection key* $\mathsf{k_p}$, *a password* pw *with shares* pw_1 *and* pw_2, *a word* (C_0, C_1, C_2), *and valid witness* (r_0, r_1, r_2), *such that* $(\mathsf{k_{p0}}, \mathsf{k_{p1}}, \mathsf{k_{p2}})$ *are valid, i.e.* $\mathsf{VerKp}(\mathsf{k_{p_i}}, L_{\widehat{\mathrm{pw}}}) = 1$ *for* $i \in \{0, 1, 2\}$, *but*

$$\mathsf{THash}_x(\mathsf{k_{p0}}, L_{\widehat{\mathrm{pw}}}, C_1, C_2, \tau') \neq \mathsf{PHash}_x(\mathsf{k_{p0}}, L_{\widehat{\mathrm{pw}}}, C_1, C_2, r_1, r_2) \text{ or}$$
$$\mathsf{THash}_0(\mathsf{k_{p1}}, \mathsf{k_{p2}}, L_{\widehat{\mathrm{pw}}}, C_0, \tau') \neq \mathsf{PHash}_0(\mathsf{k_{p1}}, \mathsf{k_{p2}}, L_{\widehat{\mathrm{pw}}}, C_0, r_0)$$

with probability at least $\varepsilon(\lambda)$. *The perfect soundness states that the property holds for any* t *and any* $\varepsilon(\lambda) > 0$.

3.1   Cramer-Shoup TD-SPHF

In the following we present TD-SPHF for labelled Cramer-Shoup ciphertexts by extending the corresponding D-SPHF from [29] with the trapdoor property from [10] in the setting of bilinear groups. Let $C = (\ell, u_1, u_2, e, v)$ denote a Cramer-Shoup ciphertext as defined in Sect. 2.

- $\mathtt{TSetup}(\mathbf{crs})$ draws a random $\tau' \in_R \mathbb{Z}_q$ and computes $\mathbf{crs}' = \zeta = g_2^{\tau'}$
- $\mathtt{KGen_H}(L_{\widehat{pw}})$ returns $\mathbf{k}_{hi} = (\eta_{1,i}, \eta_{2,i}, \theta_i, \mu_i, \nu_i) \in_R \mathbb{Z}_p^{1 \times 5}$ for $i \in \{0,1,2\}$
- $\mathtt{KGen_P}(\mathbf{k}_{hi}, L_{\widehat{pw}})$ generates

$$\mathbf{k}_{P_i} = (\mathbf{k}_{P1,i} = g_{1,1}^{\eta_{1,i}} g_{1,2}^{\theta_i} h^{\mu_i} c^{\nu_i}, \mathbf{k}_{P2,i} = g_{1,1}^{\eta_{2,i}} d^{\nu_i}, \mathbf{k}_{P3,i})$$

with $\mathbf{k}_{P3,i} = (\chi_{1,1,i}, \chi_{1,2,i}, \chi_{2,i}, \chi_{3,i}, \chi_{4,i})$ and

$$\chi_{1,1,i} = \zeta^{\eta_{1,i}}, \chi_{1,2,i} = \zeta^{\eta_{2,i}}, \chi_{2,i} = \zeta^{\theta_i}, \chi_{3,i} = \zeta^{\mu_i}, \chi_{4,i} = \zeta^{\nu_i} \text{ for } i \in \{0,1,2\}$$

- $\mathtt{Hash}_x(\mathbf{k}_{h0}, L_{\widehat{pw}}, C_1, C_2)$ computes

$$h'_x = (u_{1,1} \cdot u_{1,2})^{\eta_{1,0} + (\xi_1 + \xi_2)\eta_{2,0}} (u_{2,1} \cdot u_{2,2})^{\theta_0} ((e_1 \cdot e_2)/g_{1,1}^{\mathrm{pw}})^{\mu_0} (v_1 \cdot v_2)^{\nu_0}$$

and returns $h_x = e(h'_x, g_2)$
- $\mathtt{PHash}_x(\mathbf{k}_{P0}, L_{\widehat{pw}}, C_1, C_2, r_1, r_2)$ computes $h'_x = \mathbf{k}_{P1,0}^{r_1 + r_2} \mathbf{k}_{P2,0}^{\xi_1 r_1 + \xi_2 r_2}$ and outputs $h_x = e(h'_x, g_2)$
- $\mathtt{Hash}_0(\mathbf{k}_{h1}, \mathbf{k}_{h2}, L_{\widehat{pw}}, C_0)$ computes

$$h'_0 = u_{1,0}^{\eta_{1,1} + \eta_{1,2} + \xi_0(\eta_{2,1} + \eta_{2,2})} u_{2,0}^{\theta_1 + \theta_2} (e_0/g_{1,1}^{\mathrm{pw}})^{\mu_1 + \mu_2} v_0^{\nu_1 + \nu_2}$$

and outputs $h_0 = e(h'_0, g_2)$
- $\mathtt{PHash}_0(\mathbf{k}_{P1}, \mathbf{k}_{P2}, L_{\widehat{pw}}, C_0, r_0)$ computes

$$h'_0 = (\mathbf{k}_{P1,1} \mathbf{k}_{P1,2})^{r_0} (\mathbf{k}_{P2,1} \mathbf{k}_{P2,2})^{r_0 \xi_0}$$

and outputs $h_0 = e(h'_0, g_2)$
- $\mathtt{VerKp}(\mathbf{k}_{P_i}, L_{\widehat{pw}})$ verifies that

$$e(\mathbf{k}_{P1,i}, \mathbf{crs}') \overset{?}{=} e(g_{1,1}, \chi_{1,1,i}) \cdot e(g_{1,2}, \chi_{2,i}) \cdot e(h, \chi_{3,i}) \cdot e(c, \chi_{4,i})$$

and

$$e(\mathbf{k}_{P2,i}, \mathbf{crs}') \overset{?}{=} e(g_{1,1}, \chi_{1,2,i}) \cdot e(d, \chi_{4,i}) \text{ for } i \in \{0,1,2\}$$

- $\mathtt{THash}_0(\mathbf{k}_{P1}, \mathbf{k}_{P2}, L_{\widehat{pw}}, C_0, \tau')$ computes

$$h_0 = \left[e(u_{1,0}, \chi_{1,1,1} \chi_{1,1,2} (\chi_{1,2,1} \chi_{1,2,2})^{\xi_0}) \cdot e(u_{2,0}, \chi_{2,1} \chi_{2,2}) \right.$$

$$\left. \cdot e(e_0/g_{1,1}^{\mathrm{pw}}, \chi_{3,1} \chi_{3,2}) \cdot e(v_0, \chi_{4,1} \chi_{4,2}) \right]^{1/\tau'}$$

- $\mathtt{THash}_x(\mathbf{k}_{P0}, L_{\widehat{pw}}, C_1, C_2, \tau')$ computes

$$h_x = \left[e(u_{1,1} u_{1,2}, \chi_{1,1,0} \chi_{1,2,0}^{\xi_1 + \xi_2}) \cdot e(u_{2,1} u_{2,2}, \chi_{2,0}) \cdot e((e_1 e_2)/g_{1,1}^{\mathrm{pw}}, \chi_{3,0}) \right.$$

$$\left. \cdot e(v_1 v_2, \chi_{4,0}) \right]^{1/\tau'}$$

Distributed computation of $\texttt{PHash}_x$ and $\texttt{Hash}_0$ is done as in D-SPHF with additional proofs for correctness and adding the pairing computation at the end to lift the hash value into $\mathbb{G}_T$. We formalise execution of the Cramer-Shoup TD-SPHF in the following paragraph. Necessary zero-knowledge proofs are described in the subsequent two paragraphs and only referenced in the description of the TD-SPHF. We describe the Σ protocol here, which we can use after transforming it to a committed Σ protocol (cf. Sect. 2). Note that we merge $\texttt{crs}$ and $\texttt{crs}'$ here for readability. Protocol participants are denoted C, S_1 and S_2 if their role is specified, or P, Q and R otherwise. Let further 0 denote the client's index and 1, 2 the indices of servers S_1, S_2, respectively. The session ID is given by $\texttt{sid} = C\|S_1\|S_2$ and the unique query identifier $\texttt{qid}$ is agreed upon start using $\mathcal{F}_{\texttt{init}}$.

All TD-SPHF participants have $\texttt{crs} = (q, g_{1,1}, g_{1,2}, h, c, d, \mathbb{G}_1, g_2, \zeta, \mathbb{G}_2, \mathbb{G}_T, e, H_k)$ as common input where $\tau = (x_1, x_2, y_1, y_2, z)$ is the $\texttt{crs}$ trapdoor, i.e. the according Cramer-Shoup secret key, and τ' the trapdoor, i.e. discrete logarithm to base g_2, of $\texttt{crs}' = \zeta$. Each server holds an ElGamal key pair $(\texttt{pk}_1, \texttt{dk}_1)$ and $(\texttt{pk}_2, \texttt{dk}_2)$ respectively such that $\texttt{pk}_1$ is registered with the CA for S_1 and $\texttt{pk}_2$ for S_2 and thus available to all parties (using $\mathcal{F}_{\texttt{CA}}$). An, otherwise unspecified, protocol participant P is initiated with $(\texttt{NS}, \texttt{sid}, \texttt{qid}, P, x)$. We further define $\texttt{pw}_0 = \texttt{pw}$.

CS TD-SPHF Computation

(a) Generate TD-SPHF keys $\texttt{k}_{hi} \in_R \mathbb{Z}_q^5$ and $\texttt{k}_{\texttt{P}i} = (\texttt{k}_{\texttt{P}1,i} = g_{1,1}^{\eta_{1,i}} g_{1,2}^{\theta_i} h^{\mu_i} c^{\nu_i},$
$\texttt{k}_{\texttt{P}2,i} = g_{1,1}^{\eta_{2,i}} d^{\nu_i}, \chi_{1,1,i} = \zeta^{\eta_{1,i}}, \chi_{1,2,i} = \zeta^{\eta_{2,i}}, \chi_{2,i} = \zeta^{\theta_i}, \chi_{3,i} = \zeta^{\mu_i}, \chi_{4,i} = \zeta^{\nu_i}).$
Encrypt $\texttt{pw}_i$ to $C = (\ell_i, u_{1,i}, u_{2,i}, e_i, v_i) \leftarrow (\ell, g_{1,1}^{r_i}, g_{1,2}^{r_i}, h^{r_i} g_{1,1}^{\texttt{pw}_i}, (cd^{\xi_i})^{r_i})$
with $\xi_i = H_k(\ell_i, u_{1,i}, u_{2,i}, e_i)$ for $\ell_i = \texttt{sid}\|\texttt{qid}\|\texttt{k}_{\texttt{P}i}$ and $r_i \in_R \mathbb{Z}_q$. If $P = S_1$, set $h_0 = h_x = \texttt{null}$. Output $(\texttt{sid}, \texttt{qid}, 0, P, C_i, \texttt{k}_{\texttt{P}i})$ to Q and R.

(b) When P, waiting for the initial messages, is receiving a message $(\texttt{sid}, \texttt{qid}, 0, Q, C_1, \texttt{k}_{\texttt{P}1})$ and $(\texttt{sid}, \texttt{qid}, 0, R, C_2, \texttt{k}_{\texttt{P}2})$ it proceeds as follows. P proceeds only if the projection keys $\texttt{k}_{\texttt{P}1}$ and $\texttt{k}_{\texttt{P}2}$ are correct, i.e. $\texttt{VerKp}(\texttt{k}_{\texttt{P}1}, L_{\widehat{\texttt{pw}}}) = 1$ and $\texttt{VerKp}(\texttt{k}_{\texttt{P}2}, L_{\widehat{\texttt{pw}}}) = 1$. If the verification fails, P outputs $(\texttt{sid}, \texttt{qid}, \perp, \perp)$ and aborts the protocol.

(i) If $P = C$, compute
$$h_x = e((u_{1,1} \cdot u_{1,2})^{\eta_{1,0} + (\xi_1 + \xi_2)\eta_{2,0}} (u_{2,1} \cdot u_{2,2})^{\theta_0}$$
$$((e_1 \cdot e_2)/g_{1,1}^{\texttt{pw}})^{\mu_0} (v_1 \cdot v_2)^{\nu_0}, g_2) \text{ and}$$
$$h_0 = e\left((\texttt{k}_{\texttt{P}1,1}\texttt{k}_{\texttt{P}1,2})^{r_0}(\texttt{k}_{\texttt{P}2,1}\texttt{k}_{\texttt{P}2,2})^{r_0\xi_0}, g_2\right), \text{ and outputs } (\texttt{sid}, \texttt{qid}, h_0, h_x).$$

(ii) If $P = S_2$, compute $h_{x,2} = (\texttt{k}_{\texttt{P}1,0} \cdot \texttt{k}_{\texttt{P}2,0}^{\xi_2})^{r_2}$ and $C_{h_{x,2}} = g_{1,1}^{H(h_{x,2}, \texttt{Co}_1)} h^{r_{c1}}$
with $r_{c1} \in_R \mathbb{Z}_q$ and send $(\texttt{sid}, \texttt{qid}, \texttt{PHash}_x, 0, S_2, C_{h_{x,2}})$ to S_1.

(iii) If $P = S_1$, compute $m_0 = \texttt{Enc}_{\texttt{pk}_1}^{\texttt{EG}}(g_{1,1}^{-\mu_1}; r)$ and $c_0 = \texttt{Enc}_{\texttt{pk}_1}^{\texttt{EG}}(g_{1,1}^{\texttt{pw}_1}; r')$ with $r, r' \in_R \mathbb{Z}_q$, and send $(\texttt{sid}, \texttt{qid}, \texttt{Hash}_0, 0, S_1, m_0, c_0)$ to S_2.

(c) On input $(\texttt{sid}, \texttt{qid}, \texttt{PHash}_x, 0, S_2, C_{h_{x,2}})$ S_1 in the correct state draws challenge $\mathfrak{c} \in_R \mathbb{Z}_q$ and returns $(\texttt{sid}, \texttt{qid}, \texttt{PHash}_x, 1, S_1, \mathfrak{c})$ to S_2.

(d) On input $(\text{sid}, \text{qid}, \text{PHash}_x, 1, S_1, c)$ S_2 in the correct state computes $C_{s_{h_{x,2}}} = g_{1,1}^{H(\text{Rs}_1)} h^{r_{c2}}$ with $r_{c2} \in_R \mathbb{Z}_q$ and sends $(\text{sid}, \text{qid}, \text{PHash}_x, 2, S_2, C_{s_{h_{x,2}}})$ to S_1. Subsequently, it sends $(\text{sid}, \text{qid}, \text{PHash}_x, 3, S_2, h_{x,2}, \text{Co}_1, \text{Rs}_1, r_{c1}, r_{c2})$ to S_1.

(e) On input $(\text{sid}, \text{qid}, \text{PHash}_x, 2, S_2, C_{s_{h_{x,2}}})$ S_1 in the correct state stores it and waits for the final PHash_x message.

(f) On input $(\text{sid}, \text{qid}, \text{PHash}_x, 3, S_2, h_{x,2}, \text{Co}_1, \text{Rs}_1, r_{c1}, r_{c2})$ S_1 in the correct state parses Co_1 as (t_1, t_2) and Rs_2 as $s_{h_{x,2}}$ and verifies correctness of commitments and the ZKP and computes $h_x = e\left(h_{x,2} \cdot (\text{k}_{\text{P0},1} \cdot \text{k}_{\text{P0},2}^{\xi_1})^{r_1}, g_2 \right)$ if the verifications are successful, $h_x \neq \bot$ and $h_0 \neq \bot$, or sets $h_0 = \bot$ and $h_x = \bot$ otherwise.

(g) On input $(\text{sid}, \text{qid}, \text{Hash}_0, 0, S_1, m_0, c_0)$ S_2 in the correct state retrieves pk_1 from $\mathcal{F}_{\text{CA}}$ and computes $C_{\text{Hash}_0,1} = g_{1,1}^{H(m_1, m_2, \text{Co}_2)} h^{r_{c3}}$ with $r_{c3} \in_R \mathbb{Z}_q$, $m_1 \leftarrow m_0^{\text{pw}_2} \times c_0^{-\mu_2} \times \text{Enc}_{\text{pk}_1}^{\text{EG}}(g_{1,1}^{-\mu_2 \cdot \text{pw}_2} \cdot u_{1,0}^{\eta_{1,2} + \xi_0 \eta_{2,2}} \cdot u_{2,0}^{\theta_2} \cdot e_0^{\mu_2} \cdot v_0^{\nu_2}; r'')$, and $m_2 \leftarrow \text{Enc}_{\text{pk}_1}^{\text{EG}}(g_{1,1}^{-\mu_2}; r''')$ with $r'', r''' \in \mathbb{Z}_q$, and sends $(\text{sid}, \text{qid}, \text{Hash}_0, 1, S_2, C_{\text{Hash}_0,1})$ back to S_1.

(h) On input $(\text{sid}, \text{qid}, \text{Hash}_0, 1, S_2, C_{\text{Hash}_0,1})$ S_1 in the correct state draws challenge $c \in_R \mathbb{Z}_q$ and returns $(\text{sid}, \text{qid}, \text{Hash}_0, 2, S_1, c)$ to S_2.

(i) On input $(\text{sid}, \text{qid}, \text{Hash}_0, 2, S_1, c)$ S_2 in the correct state computes $C_{\text{Rs}2} = g_{1,1}^{H(\text{Rs}_2)} h^{r_{c4}}$ with $r_{c4} \in_R \mathbb{Z}_q$ and sends $(\text{sid}, \text{qid}, \text{Hash}_0, 3, S_2, C_{\text{Rs}2})$ to S_1. Subsequently, it sends $(\text{sid}, \text{qid}, \text{Hash}_0, 4, S_2, m_1, m_2, \text{Co}_2, \text{Rs}_2, r_{c3}, r_{c4})$ to S_1.

(j) On input $(\text{sid}, \text{qid}, \text{Hash}_0, 4, S_2, m_1, m_2, \text{Co}_2, \text{Rs}_2, r_{c3}, r_{c4})$ S_1 in the correct state parses Co_2 as $(t_{\overline{m}1}, t_{\overline{m}2}, t_{e2}, t_{v2}, t_{\text{k}_p12}, t_{\text{k}_p22})$ and Rs_2 as $(s_{\text{pw}_2}, s_{\mu2}, s_{\eta12}, s_{\eta22}, s_{\theta2}, s_{\nu2}, s_{r2})$, verifies correctness of commitments and ZKP, and computes $h_0 = e\left(g_{1,1}^{-\mu_1 \cdot \text{pw}_1} \cdot \text{Dec}_{\text{dk}_1}^{\text{EG}}(m_1) \cdot u_{1,0}^{\eta_{1,1} + \xi_0 \eta_{2,1}} \cdot u_{2,0}^{\theta_1} \cdot e_0^{\mu_1} \cdot v_0^{\nu_1}, g_2 \right)$ if the verifications are successful, $h_x \neq \bot$ and $h_0 \neq \bot$, or sets $h_0 = \bot$ and $h_x = \bot$.

(k) Eventually S_1 outputs $(\text{sid}, \text{qid}, h_0, h_x)$ if $h_0 \neq \text{null}$ and $h_x \neq \text{null}$.

ZK Proof for PHash_x Correctness In order to ensure correct computation of h_x on S_1 server S_2 has to prove correctness of his computations. To this end S_2 sends, in addition to the PHash_x message $h_{x,2}$ the following zero-knowledge proof.

$$\text{ZKP}\{(r_2): \ h_{x,2} = (\text{k}_{\text{P1},0} \text{k}_{\text{P2},0}^{\xi_2})^{r_2} \wedge v_2 = (cd^{\xi_2})^{r_2}\} \tag{1}$$

where r_2 is the randomness used to create C_2, ξ_2 and v_2 are part of C_2, $\text{k}_{\text{P1},0}, \text{k}_{\text{P2},0}$ are part of C's projection key, and c, d are from the crs. The construction of the according zero-knowledge proof is straight-forward. The prover computes commitments

$$t_{hx2} = (\text{k}_{\text{P1},0} \text{k}_{\text{P2},0}^{\xi_2})^{k_{hx2}}; \quad t_{v2} = (cd^{\xi_2})^{k_{hx2}}$$

with fresh randomness $k_{hx2} \in_R \mathbb{Z}_q$, and response $s_{r2} = k_{hx2} - cr_2$ for verifier provided challenge c. This allows the verifier to check

$$t_{hx2} \stackrel{?}{=} h_{x,2}^c (\text{k}_{\text{P1},0} \text{k}_{\text{P2},0}^{\xi_2})^{s_{hx2}}; \quad t_{v2} \stackrel{?}{=} v_2^c (cd^{\xi_2})^{s_{hx2}}.$$

It is easy to see that this zero-knowledge proof is correct, sound and (honest-verifier) simulatable. We refer to the messages as $\mathsf{Co}_1 = (t_{hx2}, t_{v2})$, $\mathsf{Rs}_1 = s_{r2}$, and $\mathsf{Ch}_1 = \mathfrak{c}$.

ZK Proof for Hash_0 *Correctness* Let $\overline{m}_1$ and $\overline{m}_2$ denote the messages encrypted in m_1 and m_2 respectively and $m_{0,1}$ and $c_{0,1}$ the second part (e) of the ElGamal ciphertext m_0, c_1 respectively. In order to ensure correct computation of h_0 on S_1 server S_2 has to prove correctness of his computations. To this end S_2 sends, additionally to the Hash_0 messages $\overline{m}_1$ and $\overline{m}_2$ the following zero-knowledge proof

$$\mathsf{ZKP}\big\{ (x, \eta_{1,2}, \eta_{2,2}, \theta_2, \mu_2, \nu_2, r_2) : \ \overline{m}_1 = m_{0,1}^{\mathrm{pw}_2} c_{0,1}^{-\mu_2} g_{1,1}^{-\mu_2 x} u_{1,0}^{\eta_{1,2} + \xi_0 \eta_{2,2}} u_{2,0}^{\theta_2} e_0^{\mu_2} v_0^{\nu_2}$$

$$\wedge \ \overline{m}_2 = g_{1,1}^{-\mu_2} \ \wedge \ e_2 = h^{r_2} g_{1,1}^{\mathrm{pw}_2} \ \wedge \ v_2 = (cd^{\xi_2})^{r_2}$$

$$\wedge \ \mathsf{k}_{\mathsf{P}1,2} = g_{1,1}^{\eta_{1,2}} g_{1,2}^{\theta_2} h^{\mu_2} c^{\nu_2} \ \wedge \ \mathsf{k}_{\mathsf{P}2,2} = g_{1,1}^{\eta_{2,2}} d^{\nu_2} \big\},$$

$$(2)$$

where r_2 is the randomness used to create C_2, ξ_2 and v_2 are part of C_2, ξ_0 is part of C_0, $(\mu_2, \eta_{1,2}, \eta_{2,2}, \theta_2, \nu_2)$ is S_2's hashing key, pw_2 S_2's password share, and c, d are from the crs. The construction of the according Σ proof is straight-forward. The prover computes commitments

$$t_{\overline{m}1} = m_{0,1}^{\mathrm{pw}_2} c_{0,1}^{k_{\mu 2}} \overline{m}_2^{-k_x} u_{1,0}^{k_{\eta 12} + \xi_0 k_{\eta 22}} u_{2,0}^{k_{\theta 2}} e_0^{-k_{\mu 2}} v_0^{k_{\nu 2}}; \quad t_{\overline{m}2} = g_{1,1}^{k_{\mu 2}}; \quad t_{e2} = h^{k_{r2}} g_{1,1}^{\mathrm{pw}_2};$$

$$t_{v2} = (cd^{\xi_2})^{k_{r2}}; \quad t_{\mathsf{k}_{\mathsf{p}}12} = g_{1,1}^{k_{\eta 12}} g_{1,2}^{k_{\theta 2}} h^{k_{\mu 2}} c^{k_{\nu 2}}; \quad t_{\mathsf{k}_{\mathsf{p}}22} = g_{1,1}^{k_{\eta 22}} d^{k_{\nu 2}}$$

for $k_{\mathrm{pw}_2}, k_{\mu 2}, k_{\eta 12}, k_{\eta 22}, k_{\theta 2}, k_{\nu 2} \in_R \mathbb{Z}_q$

and responses

$$s_{\mathrm{pw}_2} = k_{\mathrm{pw}_2} - \mathfrak{c}\mathrm{pw}_2; \quad s_{\mu 2} = k_{\mu 2} + \mathfrak{c}\mu_2; \quad s_{\eta 12} = k_{\eta 12} - \mathfrak{c}\eta_{1,2}; \quad s_{\eta 22} = k_{\eta 22} - \mathfrak{c}\eta_{2,2};$$

$$s_{\theta 2} = k_{\theta 2} - \mathfrak{c}\theta_2; \quad s_{\nu 2} = k_{\nu 2} - \mathfrak{c}\nu_2; \quad s_{r2} = k_{r2} - \mathfrak{c}r_2$$

for verifier provided challenge $\mathfrak{c}$. This allows the verifier to check

$$t_{\overline{m}1} \overset{?}{=} \overline{m}_1^{\mathfrak{c}} m_{0,1}^{s_{\mathrm{pw}_2}} c_{0,1}^{s_{\mu 2}} \overline{m}_2^{s_{\mathrm{pw}_2}} u_{1,0}^{s_{\eta 12} + \xi_0 s_{\eta 22}} u_{2,0}^{s_{\theta 2}} e_0^{s_{\mu 2}} v_0^{s_{\nu 2}}; \quad t_{\overline{m}2} \overset{?}{=} \overline{m}_2^{\mathfrak{c}} g_{1,1}^{s_{\mu 2}}; \quad t_{e2} \overset{?}{=} e_2^{\mathfrak{c}} h^{s_{r2}} g_{1,1}^{s_{\mathrm{pw}_2}};$$

$$t_{v2} \overset{?}{=} v_2^{\mathfrak{c}} (cd^{\xi_2})^{s_{r2}}; \quad t_{\mathsf{k}_{\mathsf{p}}12} \overset{?}{=} \mathsf{k}_{\mathsf{P}1,2}^{\mathfrak{c}} g_{1,1}^{s_{\eta 12}} g_{1,2}^{s_{\theta 2}} h^{s_{\mu 2}} c^{s_{\nu 2}}; \quad t_{\mathsf{k}_{\mathsf{p}}22} \overset{?}{=} \mathsf{k}_{\mathsf{P}2,2}^{\mathfrak{c}} g_{1,1}^{s_{\eta 22}} d^{s_{\nu 2}}.$$

While this is mainly a standard zero-knowledge proof $t_{\overline{m}1}$ uses $\overline{m}_2$ instead of $g_{1,1}$ as base for the third factor and k_{pw_2} as exponent (s_{pw_2} in the verification). This is necessary due to the fact that the exponent $-\mu_2\mathrm{pw}_2$ of the third factor in $\overline{m}_1$ is a product of two values that have to be proven correct. The ZK proof uses the auxiliary message $\overline{m}_2$ to prove that $\log_{g_{1,1}}(\overline{m}_2) = -\mu_2$ such that it is sufficient to prove $\log_{\overline{m}_2}(\overline{m}_2^{\mathrm{pw}_2}) = \mathrm{pw}_2$. We refer to the messages as $\mathsf{Co}_2 = (t_{\overline{m}1}, t_{\overline{m}2}, t_{e2}, t_{v2}, t_{\mathsf{k}_{\mathsf{p}}12}, t_{\mathsf{k}_{\mathsf{p}}22})$, $\mathsf{Rs}_2 = (s_{\mathrm{pw}_2}, s_{\mu 2}, s_{\eta 12}, s_{\eta 22}, s_{\theta 2}, s_{\nu 2}, s_{r2})$, and $\mathsf{Ch}_2 = \mathfrak{c}$.

4   Universally Composable Two-Server PAKE

With TD-SPHF it is straight forward to build a 2PAKE protocol. We follow the general framework described in [29] to build 2PAKE protocols from distributed smooth projective hash functions. However, instead of aiming for key generation, where the client establishes a key with each of the two servers, we focus on a protocol that establishes a single key with one server, w.l.o.g. the first server. By running the protocol twice, keys can be exchanged between the client and the second sever. Note that UC security allows concurrent execution of the protocol such that round complexity is not increased by establishing two keys.

4.1   The Protocol

We obtain our 2PAKE protocol using the general 2PAKE framework from [29] yet using our TD-SPHF instead of original D-SPHF. Client C and both servers S_1 and S_2 execute a TD-SPHF protocol from Sect. 3 which provides C and S_1 with two hash values h_0 and h_x each. The session key is then computed by both as a product $\mathrm{sk} = h_0 \cdot h_x$.

4.2   Ideal Functionality for 2PAKE

Our ideal functionality for 2PAKE with implicit client authentication, $\mathcal{F}_{2\mathrm{PAKE}}$, is given in Fig. 1. Observe that implicit client authentication is sufficient for building UC-secure channels [19]. The ideal adversary can take control of any server from the outset of the protocol and learn the corresponding password share. The actual password remains hidden unless the adversary corrupts both servers. The use of static corruptions is motivated in the following. First, as explained in [18], PAKE security against static corruptions in the UC model implies security against adaptive corruptions in the BPR model. Second, existing single-server PAKE protocols that are UC-secure against adaptive corruptions, e.g. [1–3], rely on more complex SPHF constructions that are not translatable to the distributed setting of D-SPHF.

2PAKE Functionality. Our $\mathcal{F}_{2\mathrm{PAKE}}$ is very similar to single-server PAKE functionality but assumes two servers from which one generates a session key. The main difference is in the modelling of participants. We specify two initialisation interfaces **KEX Init**, one for the client and one for the servers. A client is initialised with a password pw while a server gets a password share α_b. The **TestPwd** interface allows the ideal world adversary to test client passwords. A tested session is marked `interrupted` if the guess is wrong, i.e. client and server in this session receive randomly chosen, independent session keys, or marked as `compromised` if the password guess is correct, i.e. the attacker is now allowed to set the session key. The attacker can only test client passwords but not password shares of the servers. Without knowledge of the password or any password share, a share is a uniformly at random chosen element and therefore not efficiently guessable. If the adversary corrupted server S_2, retrieving the second password

---

Functionality $\mathcal{F}_{2\text{PAKE}}$

The functionality $\mathcal{F}_{2\text{PAKE}}$ is parameterised by a security parameter λ. It interacts with an adversary, a client C and two servers S_1 and S_2 via the following interfaces. Without loss of generality the key is exchanged between C and S_1.

KEX Init$_C$: Upon input (KEXinit, sid, qid, pw) from client C, check that sid is (C, S_1, S_2) and that qid is unique (entries (KEX, sid, qid, S_1, α_1) or (KEX, sid, qid, S_2, α_2) may exist) and send (KEX, sid, qid, C) to SIM. If this is a valid request, create a *fresh* record (KEX, sid, qid, C, pw).

KEX Init$_S$: Upon input (KEXinit, sid, qid, α_b) from server S_b, $b \in \{1, 2\}$, check that sid is (C, S_1, S_2) and that qid is unique (entries (KEX, sid, qid, C, pw) or (KEX, sid, qid, S_{3-b}, α_{3-b}) may exist) and send (KEX, sid, qid, S_b) to SIM. If this is a valid request, create a fresh record (KEX, sid, qid, S_b, α_b).

TestPwd: Upon input (TP, sid, qid, pw′) from SIM check that a fresh record (KEX, sid, qid, C, pw) exists. If this is the case, mark (KEX, sid, qid, S_1, α_1) as **compromised** and reply with "correct guess" if pw = pw′, and mark it as **interrupted** and reply with "wrong guess" if pw ≠ pw′.

Failed: Upon input (FA, sid, qid) from SIM check that records (KEX, sid, qid, C, pw) and (KEX, sid, qid, S_1, α_1) exist that are not marked **completed**. If this is the case, mark both as **failed**.

NewKey: Upon input (NK, sid, qid, P, sk′) from SIM with $P \in \{C, S_1\}$, check that a respective (KEX, sid, qid, C, pw) or (KEX, sid, qid, S, α_1) record exists, sid = (C, S_1, S_2), $|\text{sk}'| = \lambda$, then:
  - If the session is **compromised**, or either C or S_1 and S_2 are corrupted, then output (NK, sid, qid, sk′) to P; else
  - if the session is *fresh* and a key sk was sent to P' with sid = (P, P', S_2) or sid = (P', P, S_2) while (KEX, sid, qid, $P', \cdot$) was fresh, then output (NK, sid, qid, sk) to P.
  - In any other case, pick a new random key sk of length λ, and send (NK, sid, qid, sk) to P.

In any case, mark qid as **completed** for P.

Fig. 1. Ideal functionality $\mathcal{F}_{2\text{PAKE}}$

share α_1 from S_1 is equivalent to guessing the password. Complementing the TestPwd interface is a **Failed** interface that allows the adversary to let sessions fail. This allows the attacker to prevent protocol participants from computing any session, i.e. failed parties do not compute a session key. Eventually the **NewKey** interface generates session keys for client C and server S_1. NewKey calls for S_2 are ignored. If client C or server S_1 and S_2 are corrupted, or the attacker guessed the correct password, the adversary chooses the session key. If a session key was chosen for the partnered party and the session was fresh at that time, i.e. not `compromised` or `interrupted`, the same session key is used again. In any other case a new random session key is drawn.

Instead of using a single session identifier `sid` we use `sid` and `qid`. The session identifier `sid` is composed of the three participants (C, S_1, S_2) (note that we use the client C also as "username" that identifies its account on the servers) and therefore human memorable and unique. To handle multiple, concurrent 2PAKE executions of one `sid`, we use a query identifier `qid` that is unique within `sid` and can be established with $\mathcal{F}_{\texttt{init}}$. In the multi-session extension $\widehat{\mathcal{F}}_{\texttt{2PAKE}}$ the `sid` becomes `ssid` and `sid` is a globally unique identifier for the used universe, i.e. server public keys (`CA`) and `crs`.

4.3 Security

The following theorem formalises the security of the proposed 2PAKE protocol. Note that we do not rely on any security of the TD-SPHF. Instead we reduce the security of our 2PAKE protocol directly to the underlying problem (SXDH). Thereby, we give an indirect security proof of the proposed TD-SPHF.

Theorem 1. *The 2PAKE protocol from Sect. 4.1 securely realises $\widehat{\mathcal{F}}_{\texttt{2PAKE}}$ with static corruptions in the $\mathcal{F}_{\texttt{crs}}\text{-}\mathcal{F}_{\texttt{CA}}$-hybrid model if the DDH assumption holds in both groups $\mathbb{G}_1$ and $\mathbb{G}_2$ and if H_k is a universal one-way hash function.*

Proof (Sketch). In the following we highlight changes in the sequence of games from the real-world execution in $\mathcal{G}_1$ to the ideal-world execution via $\mathcal{F}_{\texttt{2PAKE}}$ in $\mathcal{G}_{17}$ and describe the ideal-world adversary SIM. Due to space limitations the analysis of game hops is available in the full version.

$\mathcal{G}_1$: Game 1 is the real-world experiment in which $\mathcal{Z}$ interacts with real participants that follow, if honest, the protocol description, and the real-world adversary $\mathcal{A}$ controlling the corrupted parties.

$\mathcal{G}_2$: In this game all honest participants are replaced by a challenger $\mathcal{C}$ that generates `crs` together with its trapdoor τ and interacts with $\mathcal{A}$ on behalf of honest parties.

$\mathcal{G}_3$: When $\mathcal{C}$, on behalf of S_1, receives first messages $(C_0, \mathsf{k}_{\mathsf{p}_0})$ and $(C_2, \mathsf{k}_{\mathsf{p}_2})$, it decrypts C_0 to pw$'$ and checks if this is the correct password, i.e. pw$' = $ pw. If this is not the case, pw$' \neq$ pw, $\mathcal{C}$ chooses a random $h_0' \in_R \mathbb{G}_T$ if the subsequent `Hash`$_0$ computation with S_2 is successful, i.e. all zero-knowledge proofs can be verified, and aborts S_1 otherwise.

$\mathcal{G}_4$: In this game $\mathcal{C}$ chooses $\mathsf{sk} \in_R \mathbb{G}_T$ at random if h_0 was chosen at random (as in $\mathcal{G}_0$) and computation of sk on S_1 is successful.

$\mathcal{G}_5$: Upon receiving an adversarially generated C_1 or C_2 on behalf of client C, challenger $\mathcal{C}$ chooses $h_x \in_R \mathbb{G}_T$ uniformly at random instead of computing it with `Hash`$_x$ if C_1 or C_2 do not encrypt the correct password share pw$_1$ or pw$_2$ respectively.

$\mathcal{G}_6$: In this game $\mathcal{C}$ chooses $\mathsf{sk} \in_R \mathbb{G}_T$ at random if h_x was chosen at random (as in $\mathcal{G}_0$) and computation of sk on C is successful, i.e., projection keys $\mathsf{k}_{\mathsf{p}1}$ and $\mathsf{k}_{\mathsf{p}2}$ are correct.

$\mathcal{G}_7$: $\mathcal{C}$ replaces computation of hash values h_0 and h_x with a lookup table with index $(\mathsf{k}_{\mathsf{h}1}, \mathsf{k}_{\mathsf{h}2}, L_{\mathrm{pw},\mathrm{pw}_2,\mathrm{pw}_2}, C_0)$ for h_0 and $(\mathsf{k}_{\mathsf{h}0}, L_{\mathrm{pw},\mathrm{pw}_2,\mathrm{pw}_2}, C_1, C_2)$ for h_x. If no such value exists, it is computed with the appropriate Hash or PHash function and stored in the lookup table.

$\mathcal{G}_8$: Instead of computing Hash_0 for S_1 in case pw' decrypted from C_0 is the same as pw, $\mathcal{C}$ draws a random $h_0 \in_R \mathbb{G}_T$.

$\mathcal{G}_9$: In this game $\mathcal{C}$ chooses $\mathsf{sk} \in_R \mathbb{G}_T$ at random in case h_0 was chosen at random (as in $\mathcal{G}_0$) and computation of sk on S_1 is successful.

$\mathcal{G}_{10}$: Upon receiving correct C_1 or C_2, i.e. encrypting pw_1 and pw_2 respectively, on behalf of client C, challenger $\mathcal{C}$ chooses $h_x \in_R \mathbb{G}_T$ uniformly at random instead of computing it with Hash_x.

$\mathcal{G}_{11}$: In this game $\mathcal{C}$ chooses $\mathsf{sk} \in_R \mathbb{G}_T$ at random in case h_0 was chosen random (as in $\mathcal{G}_0$) and computation of sk on C is successful (projection keys $\mathsf{k}_{\mathsf{p}1}$ and $\mathsf{k}_{\mathsf{p}2}$ are correct).

$\mathcal{G}_{12}$: The entire crs including ζ is chosen now by challenger $\mathcal{C}$.

$\mathcal{G}_{13}$: Upon receiving C_1 and C_2, encrypting correct password shares, $\mathcal{C}$ uses THash_0 to compute h_0 on client C instead of PHash_0. This is possible because $\mathcal{C}$ now knows trapdoor τ'.

$\mathcal{G}_{14}$: Upon receiving C_0, encrypting correct password, $\mathcal{C}$ uses THash_x to compute h_x on server S_1 instead of PHash_x. This is possible because $\mathcal{C}$ now knows trapdoor τ'.

$\mathcal{G}_{15}$: Instead of encrypting the correct password pw in C_0 on behalf of client C, $\mathcal{C}$ encrypts 0 (which is not a valid password).

$\mathcal{G}_{16}$: Instead of encrypting the correct password share pw_i in C_i on behalf of server S_i with $i \in [1,2]$, $\mathcal{C}$ encrypts a random element $\mathrm{pw}'_i \in_R \mathbb{Z}_q$.

$\mathcal{G}_{17}$: This is the final game where instead of the challenger $\mathcal{C}$ the simulation is done by the ideal-world adversary (simulator) SIM that further interacts with the ideal functionality $\mathcal{F}_{\mathrm{2PAKE}}$. While this game is structurally different from $\mathcal{G}_0$ the interaction with $\mathcal{A}$ is indistinguishable from the latter. This combined with the following description of the simulator concludes the proof.

Simulator. We describe SIM for a single session $\mathsf{sid} = (C, S_1, S_2)$. The security then follows from the composition theorem [15] covering multiple sessions and from the joint-state composition theorem [20], covering creation of a joint state by $\mathcal{F}_{\mathrm{CA}}$ and $\mathcal{F}_{\mathrm{crs}}$ for all sessions and participants. As before, we assume that 0 is not a valid password.

First, SIM generates $\mathsf{crs} = (q, g_{1,1}, g_{1,2}, h, c, d, \mathbb{G}_1, g_2, \zeta, \mathbb{G}_2, \mathbb{G}_T, e, H_k)$ with Cramer-Shoup secret key as trapdoor $\tau = (x_{\mathsf{j}}, x_2, y_1, y_2, z)$ and second trapdoor

τ' for $\zeta = g_2^{\tau'}$ to answer all $\mathcal{F}_{crs}$ queries with crs. Further, SIM generates ElGamal key pairs (g^{z_1}, z_1) and (g^{z_2}, z_2), and responds to Retrieve(S_i) queries to $\mathcal{F}_{CA}$ from S_i with (Retrieve, S_i, (g^{z_i}, z_i)) for $i \in \{1, 2\}$ and with (Retrieve, S_i, g^{z_i}) to all other request.

When receiving (KEX, sid, qid, P) with sid $= (C, S_1, S_2)$ and $P \in \{C, S_1, S_2\}$ from $\mathcal{F}_{2PAKE}$, SIM starts simulation of the protocol for party P by computing $M_i = (C_i, k_{p_i})$ for $i \in \{0, 1, 2\}$ and encrypting a dummy value (0 for $P = C$ and a random value $\alpha_i' \in_R \mathbb{Z}_q$ for $P = S_i$, $i \in \{1, 2\}$). SIM outputs (C_i, k_{p_i}) to $\mathcal{A}$. The first round of messages is handled as follows.

(i) When a party receives an adversarially generated but well formed first message M_i, $i \in \{1, 2\}$ from uncorrupted S_i, i.e. VerKp on the projection key k_{p_i} is 1, SIM queries (FA, sid, qid), which marks the session failed for the receiving party and thus ensures that the party receives an independent, random session key (if any) on a NewKey query.

(ii) When a party receives an adversarially generated but well formed first message M_2 from a corrupted S_2 while S_1 is not corrupted, SIM decrypts C_2 to α_2'. If this value is not correct, $\alpha_2' \neq \alpha_2$ (the party is corrupted such that SIM knows the correct value), SIM queries (FA, sid, qid) to ensure independent session keys on NewKey queries.

(iii) When client C receives an adversarially generated but well formed first message M_1 from a corrupted S_1 while S_2 is not corrupted, SIM decrypts C_1 to α_1'. If this value is *not* correct, $\alpha_1' \neq \alpha_1$, SIM queries (FA, sid, qid) to ensure independent session keys on NewKey queries.

(iv) When a party receives adversarially generated but well formed first messages M_1, M_2 from corrupted S_1, S_2, SIM decrypts C_1 and C_2 to α_1', α_2' respectively, and verifies their correctness against α_1 and α_2. If they are correct, SIM computes $h_0 \leftarrow$ THash$_0(k_{p_1}, k_{p_2}, L_{pw, pw_1, pw_2}, C_0, \tau')$, $h_x \leftarrow$ Hash$_x(k_{p_0}, L_{\widehat{pw}}, C_1, C_2)$, and sk$_C = h_0 \cdot h_x$. Otherwise choose a random sk$_C \in \mathbb{G}_T$.

(v) When an honest S_1 or S_2 receives an adversarially generated but well formed first message M_0, i.e. VerKp on k_{p_0} is true, SIM extracts pw$'$ from C_0 and sends (TP, sid, qid, C, pw$'$) to $\mathcal{F}_{2PAKE}$. If $\mathcal{F}_{2PAKE}$ replies with "correct guess", SIM uses pw$'$, crs and τ' to compute $h_x \leftarrow$ THash$_x(k_{p_0}, L_{\widehat{pw}}, C_1, C_2, \tau')$, $h_0 \leftarrow$ Hash$_0(k_{h1}, k_{h2}, L_{pw, pw_1, pw_2}, C_0)$, and sk$_S = h_0 \cdot h_x$.

(vi) If verification of any k_{p_i} fails at a recipient, SIM aborts the session for the receiving participant.

If a party does not abort, SIM proceeds as follows. After C received all ciphertexts and projection keys and the previously described checks were performed SIM sends (NK, sid, qid, C, sk$_C$) to $\mathcal{F}_{2PAKE}$ if sk$_C$ for this session exists, or (NK, sid, qid, C, $\perp$) otherwise. After S_1 and S_2 received all ciphertexts and projection keys and the previously described checks were performed, SIM simulates PHash$_x$ and Hash$_0$ computations between S_1 and S_2 with random elements and simulated zero-knowledge proofs. If all messages received by S_1 are oracle generated, SIM sends (NK, sid, qid, S_1, sk$_S$) to $\mathcal{F}_{2PAKE}$ if this session is compromised and $(\mathtt{NK}, \mathtt{sid}, \mathtt{qid}, S_1, \perp)$ if not. If any $\mathtt{PHash}_x$ or $\mathtt{Hash}_0$ message received by S_1 can not be verified, $\mathtt{SIM}$ does nothing and aborts the session for S_1.

5 $\mathcal{F}_{2\mathrm{PAKE}}$ Discussion

$\mathcal{F}_{2\mathrm{PAKE}}$ and the BPR 2PAKE Model. While other security models for 2PAKE protocols where proposed [33], the BPR-like security model from [27] is the most comprehensible and (in its two-party version) established model. To compare security of a 2PAKE protocol Π in a game-based and UC setting we have to ensure that it supports session ids (necessary in the UC framework). We therefore assume that Π already uses UC compliant session ids. Before looking into the relation between the game-based model for 2PAKE and $\mathcal{F}_{2\mathrm{PAKE}}$ we want to point out that Π, securely realising $\mathcal{F}_{2\mathrm{PAKE}}$, offers "forward secrecy", i.e. even an adversary that knows the correct password is not able to attack an execution of Π without actively taking part in the execution. With this in mind it is easy to see that Π, securely realising $\mathcal{F}_{2\mathrm{PAKE}}$, is secure in the BPR-like model from [27]. This is because the attacker is either passive, which is covered by the previous observation, or is active and is therefore able tests one password. Those password tests ($\mathtt{TestPwd}$ in $\mathcal{F}_{2\mathrm{PAKE}}$ and Send in the game based model) give the attacker a success probability of $q/|\mathcal{D}|$, with q the number of active sessions and $|\mathcal{D}|$ the dictionary size, when considering a uniform distribution of passwords inside the dictionary $\mathcal{D}$.

$\mathcal{F}_{2\mathrm{PAKE}}$ and $\mathcal{F}_{\mathrm{PAKE}}$. While $\mathcal{F}_{\mathrm{PAKE}}$ and $\mathcal{F}_{2\mathrm{PAKE}}$ are very similar they contain some significant difference we want to point out here. First, the key-exchange is performed between all three participants, but only C and, w.l.o.g., S_1 agree on a common session key. The $\mathtt{role}$ is a technical necessity in $\mathcal{F}_{\mathrm{PAKE}}$ for correct execution. Since we have explicit roles in $\mathcal{F}_{2\mathrm{PAKE}}$ this is not necessary here. Due to the asymmetry in $\mathcal{F}_{2\mathrm{PAKE}}$ (a client negotiates with two servers) we assume that the client is always the invoking party. The asymmetric setting in $\mathcal{F}_{2\mathrm{PAKE}}$ further restricts $\mathtt{TestPwd}$ queries to the client since the servers hold high entropy password shares. While it is enough for the attacker to corrupt one party in $\mathcal{F}_{\mathrm{PAKE}}$ to control the session key, in $\mathcal{F}_{2\mathrm{PAKE}}$ he has to either corrupt or compromise the client, or corrupt both servers. As long as only one server is corrupted, the adversary has no control over the session keys and the parties receive uniformly at random chosen session keys In $\mathcal{F}_{2\mathrm{PAKE}}$ session ids are human memorisable, consisting of all three involved parties (C, S_1, S_2), and unique query identifier is used to distinguish between different (possibly concurrent) protocol runs of one account ($\mathtt{sid}$). This is a rather technical difference to $\mathcal{F}_{\mathrm{PAKE}}$ that uses only session identifiers.

Corruptions. The two-server extension of the BPR 2PAKE model used in [27] does not consider corruptions at all. While parties can be malicious in the model (static corruption), the attacker is not allowed to query a corrupt oracle to retrieve passwords or internal state of participants. In our model the attacker is allowed to corrupt parties before execution. This however implies security in the model from [27] even if the attacker is allowed to corrupt clients to retrieve their passwords. This is because the environment can provide the BPR attacker with the password. However, this does not increase his success probability. Dynamic corruptions in $\mathcal{F}_{2PAKE}$ on the other hand are much more intricate. While UC-secure two party PAKE protocols with dynamic corruptions exist, their approaches are not translatable to the 2PAKE setting. The challenge of dynamic corruptions is that the simulation has to be correct even if the attacker corrupts one party *after* the protocol execution has started. This is left open for future work.

6   Conclusion

This paper proposed the first UC-secure 2PAKE and introduced Trapdoor Distributed Smooth Projective Hashing (TD-SPHF) as its building block. The proposed 2PAKE protocol uses a common reference string and the SXDH assumption on bilinear groups and is efficient thanks to the simulatability of TD-SPHF.

References

1. Abdalla, M., Benhamouda, F., Blazy, O., Chevalier, C., Pointcheval, D.: SPHF-friendly non-interactive commitments. In: Sako, K., Sarkar, P. (eds.) ASIACRYPT 2013, Part I. LNCS, vol. 8269, pp. 214–234. Springer, Heidelberg (2013)
2. Abdalla, M., Benhamouda, F., Pointcheval, D.: Removing Erasures with Explainable Hash Proof Systems. Cryptology ePrint Archive, Report 2014/125 (2014)
3. Abdalla, M., Chevalier, C., Pointcheval, D.: Smooth projective hashing for conditionally extractable commitments. In: Halevi, S. (ed.) CRYPTO 2009. LNCS, vol. 5677, pp. 671–689. Springer, Heidelberg (2009)
4. Abdalla, M., Fouque, P.-A., Pointcheval, D.: Password-based authenticated key exchange in the three-party setting. In: Vaudenay, S. (ed.) PKC 2005. LNCS, vol. 3386, pp. 65–84. Springer, Heidelberg (2005)
5. Ateniese, G., Camenisch, J., Hohenberger, S., de Medeiros, B.: Practical group signatures without random oracles. Cryptology ePrint Archive, 2005:385 (2005)
6. Ballard, L., Green, M., de Medeiros, B., Monrose, F.: Correlation-resistant storage via keyword-searchable encryption. Cryptology ePrint Archive, 2005:417 (2005)
7. Barak, B., Lindell, Y., Rabin, T.: Protocol Initialization for the Framework of Universal Composability. Cryptology ePrint Archive, 2004:6 (2004)
8. Bellare, M., Pointcheval, D., Rogaway, P.: Authenticated key exchange secure against dictionary attacks. In: Preneel, B. (ed.) EUROCRYPT 2000. LNCS, vol. 1807, pp. 139–155. Springer, Heidelberg (2000)
9. Bellovin, S.M., Merritt, M.: Augmented encrypted key exchange: a password-based protocol secure against dictionary attacks and password file compromise. In: ACM CCS 1993, pp. 244–250. ACM (1993)
10. Benhamouda, F., Blazy, O., Chevalier, C., Pointcheval, D., Vergnaud, D.: New techniques for SPHFs and efficient one-round PAKE protocols. In: Canetti, R., Garay, J.A. (eds.) CRYPTO 2013, Part I. LNCS, vol. 8042, pp. 449–475. Springer, Heidelberg (2013)

11. Benhamouda, F., Pointcheval, D.: Verifier-based password-authenticated key exchange: New models and constructions. Cryptology ePrint Archive, 2013:833 (2013)
12. Brainard, J., Juels, A.: A new two-server approach for authentication with short secrets. In: USENIX03 (2003)
13. Camenisch, J., Enderlein, R.R., Neven, G.: Two-Server Password-Authenticated Secret Sharing UC-Secure Against Transient Corruptions. Cryptology ePrint Archive, 2015:006 (2015)
14. Camenisch, J., Lysyanskaya, A., Neven, G.: Practical yet universally composable two-server password-authenticated secret sharing, pp. 525–536. ACM (2012)
15. Canetti, R., Security, U.C.: A new paradigm for cryptographic protocols. In: FOCS 2001, p. 136. IEEE CS, Washington, DC, USA (2001)
16. Canetti, R.: Universally composable signature, certification, and authentication. In: CSFW 2004, p. 219. IEEE CS (2004)
17. Canetti, R., Fischlin, M.: Universally composable commitments. In: Kilian, J. (ed.) CRYPTO 2001. LNCS, vol. 2139, pp. 19–40. Springer, Heidelberg (2001)
18. Canetti, R., Halevi, S., Katz, J., Lindell, Y., MacKenzie, P.: Universally composable password-based key exchange. In: Cramer, R. (ed.) EUROCRYPT 2005. LNCS, vol. 3494, pp. 404–421. Springer, Heidelberg (2005)
19. Canetti, R., Krawczyk, H.: Analysis of key-exchange protocols and their use for building secure channels. In: Pfitzmann, B. (ed.) EUROCRYPT 2001. LNCS, vol. 2045, pp. 453–474. Springer, Heidelberg (2001)
20. Canetti, R., Rabin, T.: Universal composition with joint state. In: Boneh, D. (ed.) CRYPTO 2003. LNCS, vol. 2729, pp. 265–281. Springer, Heidelberg (2003)
21. Damgård, I.B.: Efficient concurrent zero-knowledge in the auxiliary string model. In: Preneel, B. (ed.) EUROCRYPT 2000. LNCS, vol. 1807, pp. 418–430. Springer, Heidelberg (2000)
22. Gentry, C., MacKenzie, P.D., Ramzan, Z.: A method for making password-based key exchange resilient to server compromise. In: Dwork, C. (ed.) CRYPTO 2006. LNCS, vol. 4117, pp. 142–159. Springer, Heidelberg (2006)
23. hashcat. hashcat - advanced password recovery (2014). http://hashcat.net/. Accessed 1 Dec 2014
24. Jarecki, S., Kiayias, A., Krawczyk, H.: Round-optimal password-protected secret sharing and T-PAKE in the password-only model. In: Sarkar, P., Iwata, T. (eds.) ASIACRYPT 2014, Part II. LNCS, vol. 8874, pp. 233–253. Springer, Heidelberg (2014)
25. Jarecki, S., Lysyanskaya, A.: Adaptively secure threshold cryptography: introducing concurrency, removing erasures (Extended Abstract). In: Preneel, B. (ed.) EUROCRYPT 2000. LNCS, vol. 1807, pp. 221–242. Springer, Heidelberg (2000)
26. Jin, H., Wong, D.S., Xu, Y.: An efficient password-only two-server authenticated key exchange system. In: Qing, S., Imai, H., Wang, G. (eds.) ICICS 2007. LNCS, vol. 4861, pp. 44–56. Springer, Heidelberg (2007)
27. Katz, J., MacKenzie, P.D., Taban, G., Gligor, V.D.: Two-server password-only authenticated key exchange. In: Ioannidis, J., Keromytis, A.D., Yung, M. (eds.) ACNS 2005. LNCS, vol. 3531, pp. 1–16. Springer, Heidelberg (2005)
28. Katz, J., Vaikuntanathan, V.: Round-optimal password-based authenticated key exchange. In: Ishai, Y. (ed.) TCC 2011. LNCS, vol. 6597, pp. 293–310. Springer, Heidelberg (2011)

29. Kiefer, F., Manulis, M.: Distributed smooth projective hashing and its application to two-server password authenticated key exchange. In: Boureanu, I., Owesarski, P., Vaudenay, S. (eds.) ACNS 2014. LNCS, vol. 8479, pp. 199–216. Springer, Heidelberg (2014)
30. MacKenzie, P., Shrimpton, T., Jakobsson, M.: Threshold password-authenticated key exchange. In: CRYPTO 2002, p. 141 (2002)
31. Openwall. John the Ripper password cracker (2014). http://www.openwall.com/john/. Accessed 1 Dec 2014
32. Raimondo, M.D., Gennaro, R.: Provably secure threshold password-authenticated key exchange. In: EUROCRYPT 2003, p. 507523 (2003)
33. Szydlo, M., Kaliski, B.: Proofs for two-server password authentication. In: Menezes, A. (ed.) CT-RSA 2005. LNCS, vol. 3376, pp. 227–244. Springer, Heidelberg (2005)
34. Wu, T.: RFC 2945 - The SRP Authentication and Key Exchange System, September 2000
35. Yang, Y., Deng, R., Bao, F.: A practical password-based two-server authentication and key exchange system. IEEE TDSC 3(2), 105–114 (2006)

Yet Another Note on Block Withholding Attack on Bitcoin Mining Pools

Samiran Bag[1]([⊠]) and Kouichi Sakurai[2]

[1] Newcastle University, Newcastle upon Tyne, UK
samiran.bag@newcastle.ac.uk
[2] Kyushu University, Fukuoka, Japan
sakurai@csce.kyushu-u.ac.jp

Abstract. In this paper we provide a short quantitative analysis of Bitcoin Block Withholding (BWH) Attack. In this study, we investigate the incentive earned by a miner who either independently or at the diktat of a separate mining pool launches Block Withholding attack on a target mining pool. The victim pool shares its earned revenue with the rogue attacker. We investigate the property revenue function of the attacker and find parameters that could maximize the gain of the attacker. We then propose a new concept that we call "special reward". This special rewarding scheme is aimed at discouraging the attackers by granting additional incentive to a miner who actually finds a block. A BWH attacker who never submits a valid block to the pool will be deprived from this special reward and her gain will be less than her expectation. Depending upon the actual monetary value of the special reward a pool can significantly reduce the revenue of a BWH attacker and thus can even ward off the threat of an attack.

1 Introduction

Bitcoin [1] has been the most successful cryptocurrency ever. It was proposed by an anonymous person who identified himself as 'Satoshi Nakamoto' but did not ever disclose his real identity. In this cryptocurrency system users make online payments by creating digital transactions signed by the payer which is then broadcast across the entire Bitcoin network. Since, there is no centralized authority to validate a transaction, the Bitcoin network cleverly bestows this task on the most resourceful parties of the Bitcoin network who are called miners. The miners demonstrate their computing resources through solving a hard computational puzzle, which cannot be solved beforehand and thus present a proof of work thereof. This proof of work scheme is based on Back's Hashcash [2] technique. The miners create a block of transactions created by Bitcoin users and find an appropriate proof of work that can be associated with the block. This block of transactions along with the proof of work is then broadcast over the

---

S. Bag—A member of the faculty of Kyushu University, Fukuoka, Japan at the time of submission of this paper.

M. Bishop and A.C.A. Nascimento (Eds.): ISC 2016, LNCS 9866, pp. 167–180, 2016.
DOI: 10.1007/978-3-319-45871-7_11

entire Bitcoin network. Every node in the Bitcoin network receives it and verifies the PoW associated with the block. If the verification is successful and the block is constructed flawlessly, the Bitcoin nodes append it to the 'blockchain' which is a data structure containing chain of blocks of transactions which are linked to each other by means of cryptographic hash function. Since, construction of a Bitcoin block requires solving a proof of work puzzle, which in turn requires a huge amount of computation, it is very unlikely for a small miner having limited amount of computing resources to be able to mine a single block until a long time. Hence, such a small miner will need to mine patiently for a long time (which may be as long as few years) before she is able to mine her first block and earn her first Bitcoin. This is the reason why small miners join hands to form large pools owning sizable amount of computing resources. This computing powerhouse has higher probability of winning the mining race than a small miner. So, the mining pool earns Bitcoins more often and distributes the earned revenue among its members following a fair policy keeping in view the contribution of each miner towards solving the proof of work puzzle. Thus, the purpose of mining pools is to allow every small miners to earn small incentives frequently rather than winning 25BTC once after patiently mining for few years. But the pools do not blindly trust their members. Since, many pools are open and allow untrusted miners to join and mine on behalf of the pool, there should be a strategy in place for assessment of the performance of the miners. For this purpose, the pool administrator requires all the pool members to submit 'pool shares'. These pool shares are partial proofs of work with a limited difficulty, usually lower than the difficulty level of the full proof of work associated with a valid Bitcoin block. Hence, pool members tend to find them more frequently than they find a full proof of work. The pool administrator sets a difficulty level for the partial PoWs so that it does not cause a high overhead on the administrator to check every pool share submitted by the miners and at the same time the frequency of a small miner to find a pool share remains as high as possible making it easy for the pool administrator to judge the performance of a miner on the basis of the number of pool shares the latter submitted. In block withholding attack [3], a rogue miner secretly discards all full proofs of work computed by her and only submits partial proofs of work that cannot be used by the pool administrator to construct a valid Bitcoin block worth 25BTC. Thus, a block withholding attacker reduces the revenue of a mining pool, breaching the pool's trust on her. This paper deals with analyzing block withholding attack in Bitcoin mining pools and also suggests a simple trick that could be used to repel BWH attackers to some extent. We discuss this attack in details in Sect. 3.1.

1.1   Contribution

The contribution of this paper is two fold. Firstly, we analyze Block Withholding Attack with respect to a simple though realistic model of Bitcoin mining pool. We consider two pools and an attacker. The attacker uses a part of her computing power for mining honestly in one pool and uses the rest of her computing power to attack the second pool. We show that the attacker can expend upto $\frac{p}{2}$ fraction of her entire computing power to attack the pool beyond which her earned incentive will tend to decrease, p being the computing power of the victim pool. We show this in Lemma 1. Then, we consider a k-BWH attacker who does not withhold all her valid blocks found while mining for the victim pool but instead, selects a fraction of blocks for withholding. This attacker withholds a block with a fixed probability p_B and submits a block to the pool with probability $1 - p_B$. We show that for such an attacker her gain will be maximized when she uses a certain fraction of computing power for attacking the victim pool.

Then we discuss one simple technique for countering block withholding attack. In this technique a Bitcoin mining pool gives extra reward to a miner who actually find the winning block on behalf of the pool. We study the long term viability of this technique for a Bitcoin mining pool. Our main results from Sect. 5 is given in Lemmas 4 and 5. In Lemmas 4 and 5 we compute the minimum amount of special reward that would make a Bitcoin pool resistant to BWH attack and will repulse unwanted attackers.

2 Related Work

In Block Withholding attack (BWH) [3], a malicious miner submits to the pool administrator, as pool shares, all those PoWs that do not constitute a full PoW which can be used to generate a revenue of 25BTC, and withholds all pool shares that represent a valid full PoW for the Bitcoin network. That is the miner submits all partial proofs of work that allows her to convince the pool administrator that she is indeed trying to mine for the pool. However, when by chance she ends up finding a full proof of work, she withholds it. The pool protocol does not allow her to submit the full proof of work to the Bitcoin network and claim the entire reward [3,4]. So, she chooses to conceal her find causing a loss of at least 25BTC to the pool for which she has been mining. The pool, however remains oblivious of this act and she, satisfied with the pool shares submitted by the attacker, considers her to be a valuable asset and shares her earned revenue with the attacker. One can argue that by launching BWH attack on a pool, a miner decreases its own revenue. But, the efficacy of BWH attack lies in the fact that by reducing one pool's revenue the miner actually increases the revenue of other pools in the Bitcoin network as the rate at which Bitcoins are generated in Bitcoin network remains the same for ever. Saha et al. [5] discussed a scenario where a miner can use BWH attack for increasing her revenue by splitting her computing power and attacking one pool using a fraction of her computing power and mining independently with the rest of her computing power. Saha et al. showed that in their model the BWH attack always increases the incentive of a rational attacker. Eyal et al. [6] analyzed a scenario where mining pools send their miners to attack other pools and reduce their revenue and thereby increasing the selfish pool's incentive. They have defined and analyzed a game where identical mining pools attack each other. They have showed that in a scenario where the mining pools attack each other, there exists a Nash equilibrium where all of them earn less than what they should have if none had attacked. Laszka et al. [7] defined and analyzed a game theoretic model of Bitcoin block withholding attack and showed interesting results about the long term viability of a Nash equilibrium between attacking pools in the Bitcoin network.

3   Preliminaries

3.1   Bitcoin Mining and Block Withholding Attack

Bitcoin block withholding attack [3,5–7] has long been discussed on Bitcoin forums. In this attack, a rogue miner joins an open Bitcoin mining pool pretending to be an honest miner. She then demonstrates her work to the pool administrator by regularly submitting 'pool shares' which are partial proofs of work having an associated difficulty level which is generally lower than the difficulty level of the proof of work acceptable for the Bitcoin network. This partial proofs are such that the set of all partial proofs of work for any pool contain within itself, the set of full Bitcoin proofs. Hence, these partial proofs are in reality potential candidates for Bitcoin proofs of work. The mining pool scrutinizes all partial proofs submitted by its miners. This serves as a check for the performance of the miners as the number of partial proofs submitted by an individual miner faithfully reflect the amount of computing power that particular miner has indeed invested for mining. Again, since, the partial proofs are potential candidates for Bitcoin proof of work, the pool administrator may find a full proof from the set of partial proofs computed by the members of the pool. So, the pool administrator tasks itself with checking all partial proofs until it finds a proof of work that matches the difficulty level of the Bitcoin system. If it finds such a proof of work, it uses the same to claim the mining reward from the Bitcoin system. So, with this strategy, the pool administrator's job reduces to examining all partial proofs of work submitted by the miners until it finds a full proof of work that will allow it to earn 25BTC by submitting the same to the Bitcoin network. If someone outside the pool finds the correct proof of work, the particular pool refreshes its mining parameters and starts mining with a fresh set of parameters. In BWH attack, a malicious member submits to the administrator only those partial proofs that cannot be associated with a valid Bitcoin block and withholds all partial proofs that represent a full proof of work and which can be used to earn 25BTC from the system. However, the pool administrator remains oblivious to the fact and considers the miner to be an asset to the pool. Hence, the pool administrator shares the earned revenue of the pool with the malicious miner even though she does not contribute even a single block. Thus, a BWH attacker incurs loss to an unsuspecting mining pool. One may argue that a malicious miner thus causes loss to herself whenever she withholds a block as the pool would have shared with her, the revenue generated with the block if she had submitted a valid block to the pool. But, the key to success of BWH attack is that whenever a miner launches BWH attack on a pool, its revenue decreases. The decreased revenue gets distributed to all other miners of the Bitcoin network depending upon there computing power.

So, the malicious miner who attacks one pool causes the revenue of other miners to increase as the total amount of Bitcoins generated in a fixed amount of time is nearly constant. Thus, BWH attack is launched generally by other pools who send their miners to infiltrate victim mining pools [6]. When the mining pool comes under BWH attack, its revenue goes down which in turn results in the increase of the revenues of other pools including the one that sent its agents to attack the victim pool. BWH attack can also be carried out profitably by solo miners who expend part of their hashpower to attack some other pool [5].

3.2   Notations and Terminologies

We, in our analysis assume that the total computing power of the Bitcoin network is 1, so that the computing power of each entity can be expressed as a fraction. Thus, when we write α to be the computing power of any miner, we mean that the computing power of that entity is α fraction of that of the entire Bitcoin network. Similarly, we express the earned incentive of miners as a fraction of the entire revenue generated by Bitcoin network(which is 25BTC plus transaction fees per block). We use the term 'gain' to imply the earned revenue of any Bitcoin miner. If g_m be the incentive earned by a miner and g_B be the incentive earned by the entire Bitcoin network, we say the gain of the miner is $\frac{g_m}{g_B}$. This is the same definition of gain of a miner as was used in [5].

4   Analysis of BWH Attack

We start our analysis with a network model comprising only two pools and an attacker. It is obvious that this model is too simplistic as there are multiple mining pools in today's Bitcoin network. However this study can build the basis for substantial studies of BWH attack taking multiple pools into account. Let, $\mathcal{A}$ be an attacker with computing power α. We consider two pools P and P' with computing power p and p' respectively. The attacker $\mathcal{A}$ splits her computing power into two parts. She uses β fraction of her computing power to launch block withholding attack against pool P. She joins the open pool P pretending to be an honest miner having computing power α and upon joining, she regularly submits partial PoW to the pool to demonstrate that she, like every other miner of the pool, is striving to find an appropriate PoW that could allow the pool to earn some revenue. However, she never submits a complete PoW. Whenever, by chance, she computes one full PoW, she secretly discards it and continues to compute partial PoWs. The pool P' has no way to detect this treacherous act of $\mathcal{A}$ and remains oblivious about the deception. The partial PoWs submitted by the attacker $\mathcal{A}$ allow her to mislead the pool P' into trusting her. The miner uses the rest of her computing power to mine honestly in the second pool P'. We assume the Bitcoin network comprises of these two pools only. Hence, $p + p' = 1$.

Lemma 1. *When the attacker launches BWH attack on P and mines honestly in pool P', her gain will be an increasing function as long as she expends upto $\frac{p}{2}$ fraction of her computing power for attacking P.*

Proof. When $\mathcal{A}$ launches BWH attack on the pool P, the computing power of the pool reduces from p to $p - \alpha\beta$ and the computing power of the entire Bitcoin network reduces to $1 - \alpha\beta$. The incentive earned by $\mathcal{A}$ from the pool P is $G_P = \frac{p-\alpha\beta}{1-\alpha\beta}\frac{\alpha\beta}{p} = \frac{\alpha\beta(p-\alpha\beta)}{p(1-\alpha\beta)}$. Similarly, the incentive earned from pool P' is $G_{P'} = \frac{p'}{1-\alpha\beta}\frac{\alpha(1-\beta)}{p'} = \frac{\alpha(1-\beta)}{1-\alpha\beta}$. So, the total amount of incentive earned by $\mathcal{A}$ is $G = \frac{\alpha\beta(p-\alpha\beta)}{p(1-\alpha\beta)} + \frac{\alpha(1-\beta)}{1-\alpha\beta} = \frac{p\alpha - \alpha^2\beta^2}{p(1-\alpha\beta)}$. Taking partial derivative with respect to β,

$$\frac{\partial G}{\partial \beta} = \frac{(p\alpha - \alpha^2\beta^2)p\alpha - 2p(1-\alpha\beta)\alpha^2\beta}{p^2(1-\alpha\beta)^2}$$

That is $\frac{\partial G}{\partial \beta} = \frac{\alpha}{p(1-\alpha\beta)^2}(p\alpha + \alpha^2\beta^2 - 2\alpha\beta)$.

At $\beta = 0$, $\frac{\partial G}{\partial \beta}\big|_{\beta=0} = \alpha^2 > 0$. Hence, the gain of the attacker $\mathcal{A}$ is increasing at $\beta = 0$. Now, the value of G will reach a maxima/minima at some $\beta = \beta_0$ such that $\frac{\partial G}{\partial \beta} = 0$. This will happen if $p\alpha + \alpha^2\beta^2 - 2\alpha\beta = 0$. Now, this quadratic equation will have a solution at $\beta = \frac{1-\sqrt{1-p\alpha}}{\alpha}$. Now, if $\alpha \ll 1$, then $p\alpha \ll 1$. Hence, $\beta \approx \frac{1-(1-0.5p\alpha)}{\alpha} = \frac{p}{2}$. Hence, the gain of the attacker is an increasing function in the range $(0, \frac{p}{2})$. $\square$

The attacker $\mathcal{A}$ may not use all the β fraction of her computing power for attacking the pool P as the pool administrator may grow skeptical about the intention of $\mathcal{A}$ if she does not submit even a single valid PoW in the long run. So, the attacker can spare some of her computing power to mone honestly in the pool P, while investing a major fraction to launch BWH attack. So, we consider a scenario where the attacker honestly mines in P with δ fraction of her computing power, attacks the same pool P using β fraction of her computing power and mines honestly in pool P' with the rest of her computing power and prove the following lemma.

Lemma 2. *When the attacker uses some k fraction of her computing power invested in the pool P to mine honestly, and mines honestly in P' with the rest of her computing power, her gain will be an increasing function as long as she expends $\frac{p}{2(k+1)}$ fraction of her computing power for attacking P.*

Proof. The incentive she gains for mining pool P is $G_P = \frac{p-\alpha\beta}{1-\alpha\beta}\frac{\alpha\beta+\alpha\delta}{p}$. Similarly, the incentive obtained from P' will be $G_{P'} = \frac{1-\alpha(\beta+\delta)}{1-\alpha\beta}$. The total gain of the attacker will be $G = \frac{\alpha(p-\alpha\beta)(\beta+\delta)+p-p\alpha(\beta+\delta)}{p(1-\alpha\beta)}$. Now, in order to deceive the pool P, δ should be proportional to β. Let, $\delta = k\beta$. Now, the value of the earned incentive will be $G = \frac{\alpha(p-\alpha\beta)\beta(1+k)+p-p\alpha(1+k)\beta}{p(1-\alpha\beta)}$. Now, $\frac{\partial G}{\partial \beta} = \frac{\alpha^2(\alpha(1+k)\beta^2-2(1+k)\beta+p)}{p(1-\alpha\beta)^2}$. Thus $\frac{\partial G}{\partial \beta}\big|_{\beta=0} = \frac{\alpha^2}{(1-\alpha\beta)^2} > 0$. Hence, G is an increasing function of β at $\beta = 0$. Also, G will have a local maxima at $\beta = \beta_0$ if $\alpha(1+k)\beta_0^2 - 2\alpha(1+k)\beta_0 + p = 0$. Solving this quadratic equation we get, $\beta_0 = \frac{2(1+k)-\sqrt{4(1+k)^2-4\alpha(1+k)p}}{2\alpha(1+k)} = \frac{1-\sqrt{1-\frac{\alpha p}{1+k}}}{\alpha}$. Now, if $\alpha \ll 1$, then $\frac{\alpha p}{1+k} \ll 1$. Then, $\beta_0 \approx \frac{p}{2(1+k)}$. $\square$

Next, we consider a scenario where the attacker acts as a liaison of the smaller pool P'. That is, she attacks pool P but transfers all the incentive received from P to P'. P' then distributes the total incentive to all the members. So, the attacker does not gain anything directly from P. Instead, she receives all her incentives from P'. The amount of incentive the attacker gets from P(that she transfers to P') is $g_1 = \frac{p-\alpha\beta}{1-\alpha\beta} \frac{\alpha\beta}{p} = \frac{\alpha\beta(p-\alpha\beta)}{p(1-\alpha\beta)}$. Now, the total incentive P' receives through mining honestly is $g_2 = \frac{p'}{1-\alpha\beta}$. As the attacker transfer her ill-gotten incentive to P', the total incentive earned by P' reaches $g = g_1 + g_2 = \frac{\alpha\beta(p-\alpha\beta)}{p(1-\alpha\beta)} + \frac{p'}{1-\alpha\beta}$. Now, the incentive P' gives to the attacker is $G = g * \frac{\alpha}{p'} = \frac{\alpha}{1-\alpha\beta} + \frac{\alpha^2\beta(p-\alpha\beta)}{pp'(1-\alpha\beta)}$. Now, replacing p' by $1-p$, we get, $G = \frac{\alpha}{1-\alpha\beta} + \frac{\alpha^2\beta(p-\alpha\beta)}{p(1-p)(1-\alpha\beta)} = \frac{\alpha}{p(1-p)} \frac{p-p^2+\alpha\beta p-\alpha^2\beta^2}{1-\alpha\beta}$. Differentiating partially with respect to β,

$$\frac{\partial G}{\partial \beta} = \frac{\alpha^2}{p(1-p)} \frac{(p-\alpha\beta)(2-\alpha\beta-p)}{(1-\alpha\beta)^2}.$$

Since, $p \gg \alpha\beta$, $\frac{\partial G}{\partial \beta}$ is always positive. Hence, the attacker always gains more by attacking a pool on behalf on another pool.

If the attacker was honest, she could have earned only α by honestly mining in either pool. The ratio of increase of her earned incentive with respect to the incentive for honest mining is $\frac{G-\alpha}{\alpha} = \frac{1}{1-\alpha\beta} + \frac{\alpha\beta(p-\alpha\beta)}{p(1-p)(1-\alpha\beta)} - 1$.

We now consider another scenario, where there are three pools, namely P, P' and P'' with computing power $p + a_p, p'$ and $1 - p - p'$. P'' is a closed pool. Out of the p computing workforce, a_p computing power is already invested for attacking P, reducing the active computing power of P to p. Now, the pool P' sends the attacker $\mathcal{A}$ to attack P using β fraction of her computing power α.

As such, the total expected incentive of P will be $g_1 = \frac{p-\alpha\beta}{1-\alpha\beta}$. So, the incentive given to the attacker is $\frac{p-\alpha\beta}{1-\alpha\beta} \frac{\alpha\beta}{p+a_p}$. The incentive obtained from the pool P' is $\frac{\alpha(1-\beta)}{1-\alpha\beta}$. So, the total incentive earned by the attacker will be $G = \frac{p-\alpha\beta}{1-\alpha\beta} \frac{\alpha\beta}{p+a_p} + \frac{\alpha(1-\beta)}{1-\alpha\beta}$. Partially differentiating with respect to β we get, $\frac{\partial G}{\partial \beta} = \frac{\alpha}{p+a_p} \left(1 - \frac{1-p}{(1-\alpha\beta)^2}\right) - \frac{\alpha(1-\alpha)}{(1-\alpha\beta)^2}$. Simplifying we get, $\frac{\partial G}{\partial \beta} = \frac{\alpha(\alpha^2\beta^2-2\alpha\beta+\alpha p+\alpha a_p-a_p)}{(p+a_p)(1-\alpha\beta)^2}$. Now, if $a_p > \frac{\alpha p}{1-\alpha}$, $\alpha^2\beta^2 - 2\alpha\beta + (\alpha p + \alpha a_p - a_p)$ will always be negative and hence, $\frac{\partial G}{\partial \beta}$ will be negative. So, the gain of the attacker will decrease as she invests more computing resources for attacking the pool which is already under attack by an attacker having sufficiently high computing power.

5   Proposed Remedy to BWH Attack

Here, we propose a new strategy to defeat BWH attack. The key idea here is to reduce the incentive of an attacker to such an extent that she does not find it profitable to launch block withholding attack on a pool. In the existing pool protocol, a miner gets entitled to the share of the revenue generated by the entire pool by submitting partial proofs to the pool. The number of partial proofs a miner submits to the pool is directly proportional to her computing power. So, a mining pool judges the computing power of the miner by counting the partial proofs submitted by a miner. It can be noted that a BWH attacker does not submit any full proof but that does not decrease her count of partial proofs as she finds a full PoW quite often. Thus, even though she withholds all full proofs, the number of partial proofs computed by her does not decrease significantly and the dividend she gets from the pool thus becomes directly proportional to her computing power thanks to the high number of partial proofs computed by her. Now, we propose to introduce a new notion of rewarding a miner. We call this a "special reward" that is to be disbursed to a miner who actually solves the PoW puzzle and constructs a full proof of work that will eventually be used by the pool to earn a revenue from the Bitcoin system. This special reward will be paid from the pool's revenue. Hence, every time the pool wins the mining game a fixed amount of special reward will be given to the miner who actually computed a full PoW and constructed a valid Bitcoin block. The rest of the revenue will be shared among all the miners of the pool (including the winner of the special reward) depending upon their contribution to the pool which can be calculated on the basis of the number of partial proofs submitted by them. Any miner who launches BWH attack will never receive a special reward and will only be entitled to the normal share of revenue for submitting the partial proofs. As such the incentive earned by the attacker will be less as she will never receive a special reward. This should discourage the attacker from launching this attack. The Lemma 3 shows that this strategy does not decrease the gain of an honest miner of the pool in the long run. Thus, this strategy only reduces the long term revenue of an attacker while the revenue of all the honest miners remains the same as before in the long run.

Lemma 3. *The proposed strategy does not alter the earned revenue of an honest miner of a pool in the long run.*

Proof. Let, there be a pool P with computing power p. Also, let A be a miner of the pool with computing power a. Assume that the computing power of the entire Bitcoin network is 1. If the pool does not adopt the special rewarding strategy, the earned incentive of A will always be a, which is the fraction of computing power of A with respect to the computing power of the entire Bitcoin network, which we take as 1 in this study. Let W_P be the event that pool P wins the mining game on a certain epoch. Also, let W_A be the event that A has computed the winning block. Let $I(A)$ be the incentive function of A that corresponds to the expected amount of incentive earned by miner A by mining within the pool P. Hence, the expected incentive of A will be $E(I) = P[W_P] * I(A|W_P) = P[W_P] * [I(A|W_A)P[W_A|W_P] + I(A|\bar{W}_A \cap W_P)P[\bar{W}_A|W_P]] = I(A|W_A) * P[W_P] * P[W_A|W_P] + I(A|W_P \setminus W_A) * P[W_P] * P[\bar{W}_A|W_P] = I(A|W_A) * P[W_A \cap W_P] + I(A|W_P \setminus W_A) * P[W_P \setminus W_A]$. Now, $I(A|W_A)$ is the incentive earned by A when A constructs the winning block. Hence, $I(A|W_A) = \gamma + \frac{(1-\gamma)a}{p}$.

Similarly, $I(A|W_P\backslash W_A) = \frac{(1-\gamma)a}{p}$. Now, $P[W_A \cap W_P] = a$. Also, $P[W_P\backslash W_A] = p - a$. Hence, $E(I) = a(\gamma + \frac{(1-\gamma)a}{p}) + (p - a)\frac{(1-\gamma)a}{p} = a$. □

We consider a situation where the pool P gives extra reward to a miner that solves a PoW. Let this fixed amount of incentive be denoted as γ. Whenever some miner of the pool P solves a proof of work and successfully constructs a valid Bitcoin block worth 25BTC, the pool P offers this special reward to her. The rest of the revenue is then distributed fairly among the miners of the pool depending upon their invested computing power which is measured on the basis of the number of pool shares submitted by them. Therefore, the revenue of the miner who constructs a valid block is two fold, she gets the special reward from the pool and also her regular share from the pool's generated revenue. Like before, here too we consider an attacker with computing power α who uses β fraction of her computing power for attacking the pool P. So, the computing power dedicated to attacking the pool P is $\alpha\beta$. The attacker uses the rest of her computing power to mine privately. While the attack is being carried out, the revenue of the pool P gets reduced from p to $\frac{p-\alpha\beta}{1-\alpha\beta}$. The pool P will spend an expected γ fraction of this amount for paying the special reward to the miners who really constructed Bitcoin blocks. The rest of the incentive amounting $\frac{p-\alpha\beta}{1-\alpha\beta}(1 - \gamma)$ will be distributed among all the miners depending upon the exact quantity of computing power they have contributed to the pool. The miner who gets the special reward will also be entitled to a share of this incentive. The attacker does not contribute any block and hence will not ever get the special reward. What she will get is $\frac{\alpha\beta}{p}$ fraction of the remaining incentive after the special reward is paid. Thus, she will receive an amount given by $G_{\mathcal{A}}^{BWH} = \frac{\alpha\beta(p-\alpha\beta)(1-\gamma)}{p(1-\alpha\beta)}$. She also mines privately utilizing $\alpha(1 - \beta)$ computing power and the expected incentive gained from this private mining will be $G_{\mathcal{A}}^{PRIV} = \frac{\alpha(1-\beta)}{1-\alpha\beta}$. Thus, her total generated revenue will be $G_{\mathcal{A}}^{Total} = \frac{\alpha\beta(p-\alpha\beta)(1-\gamma)}{p(1-\alpha\beta)} + \frac{\alpha(1-\beta)}{1-\alpha\beta} = \frac{p\alpha-\alpha^2\beta^2-\gamma\alpha\beta(p-\alpha\beta)}{p(1-\alpha\beta)}$. The attacker will be discouraged against launching attack on the pool P if $G_{\mathcal{A}}^{Total}$ is less than the incentive that she could make through mining honestly. Since, the attacker's computing power is α, she could earn an incentive of $G_{\mathcal{A}}^{Hons} = (p-\gamma p)\frac{\alpha}{p} + (\gamma p)\frac{\alpha}{p} = \alpha$ if she had mined honestly. So, the attacker will not be interested in the execution of the attack if $\frac{p\alpha-\alpha^2\beta^2-\gamma\alpha\beta(p-\alpha\beta)}{p(1-\alpha\beta)} < \alpha$. This inequality will hold if $\gamma > \alpha(p-\beta)$. From this, we can state the following lemma;

Lemma 4. *In the above scenario, the attacker can be prevented from launching BWH attack if the pool chooses a special reward γ higher than $\alpha p - \alpha\beta$.*

Lemma 5. *Let there be two pools P and P' having computing power p and p' respectively such that $p + p' = 1$ i.e. all the computing power of the Bitcoin network is held by both the pools. The pool P' tasks some of its members with infiltrating the pool P and carrying out BWH attack on the pool P. Whatever incentive the rogue infiltrators make by joining the pool P is transferred to the pool P'. The pool P' then distributes them to all of its members. Let δ be the total amount of computing power of all the miners of pool P' who are sent to* *infiltrate and attack the pool P. As such the attack could be defeated by choosing a special reward $\gamma > 1 - p - \delta$.*

Proof. As such, the computing power of the entire Bitcoin network will get reduced to $1 - \delta$. Thus, when the infiltrators start attacking P, the gain of pool P will become $G_P^{BWH} = \frac{p}{1-\delta}$. The miner who finds a block by solving the PoW will get an extra amount γ. The rest of the incentive amounting $\frac{p}{1-\delta}(1-\gamma)$ will be given away to all the miners of the pool P including the selfish infiltrators. So, the infiltrators of the pool will get $G_{INF}^{BWH} = (\frac{p}{1-\delta}(1-\gamma))\frac{\delta}{p+\delta}$ amount of incentive totally. This amount will be transferred to the pool P'. The gain of the pool P' from its own mining will be $G_{P'}^{priv} = \frac{p'-\delta}{1-\delta}$. So, the total gain of P' will be $G_{P'}^{BWH} = G_{P'}^{priv} + G_{INF}^{BWH} = \frac{p'-\delta}{1-\delta} + \frac{p(1-\gamma)}{1-\delta}\frac{\delta}{p+\delta}$. Substituting p' by $1 - p$, we get $G_{P'}^{BWH} = G_{P'}^{priv} + G_{INF}^{BWH} = \frac{1-p-\delta}{1-\delta} + \frac{p\delta(1-\gamma)}{(1-\delta)(p+\delta)}$. If P' had mined honestly, its total gain would be $G_{P'}^{HONS} = p' = 1 - p$. In order to discourage P' from attacking the pool P, $G_{P'}^{HONS}$ must be higher than $G_{P'}^{BWH}$, that is the gain of the miner from honest mining should be higher than the gain from selfish mining. This can hold only if $\frac{1}{1-\delta}(1-p-\delta+\frac{p\delta(1-\gamma)}{p+\delta}) < 1 - p$. Hence, the attacker will stop attacking P if $\gamma > 1 - p - \delta$.   □

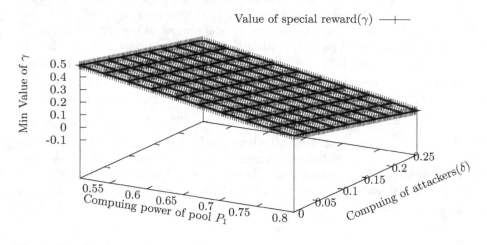

Fig. 1. Graphical presentation of the minimum value of γ to counter BWH attack for different value of p and δ.

Lemma 5 shows that in order to discourage the second pool from launching BWH attack on the pool P, the pool P should make provision for a special reward amounting $(1 - \text{fraction of gross hashpower of } P) \times 25\text{BTC}$, to be awarded to the miner who constructs a block by solving the PoW puzzle. As such, the other pool P' won't be interested to attack pool P and will stick to honest mining in its own pool or in the other pool P. So, the pool P's computing power will drop down to p. Every time the pool P wins the mining race, the pool administrator spends $\gamma = 1 - p$ fraction of the reward (25BTC) for giving the special reward to the miner who constructed the block with the appropriate PoW. Now, every time the pool P wins 25BTC, a miner who did not find the block will get $\frac{\zeta}{p}p = \zeta$ fraction of 25BTC, ζ being the computing power of that miner. Without the provision of the special reward, the miner would have got $\frac{\zeta}{p}$ fraction of 25BTC. So, this special reward system can cause a reduction of $\zeta \frac{1-p}{p}$ of incentive to the miner every time the pool wins but the miner does not construct the block for that epoch. However, as we have shown in Lemma 3, the long term incentive of an honest miner will be same as what she would have earned without having the special reward scheme in place. So, this special reward scheme will be viable in long term as it does not deprive honest miners in long term. Figure 1 shows the minimum values of the special reward γ for different values of p and δ. Figure 2 shows the difference between the earned incentives of honest miners and attacker in the pool P per unit hashpower. The value of the earned incentives of both honest miners and the attacker are normalized by dividing the gain by the actual computing power invested by the miners/attacker. Figure 2 shows that when the computing power of the attacker is low, the gain of honest miners is much higher than the attackers. However, as more attackers pour in and the attackers hashpower in pool P increase, the gain of attackers per unit hashpower tends to be close to that of the honest miners. So, our technique can significantly reduce the earned incentive of attackers when the computing power of all the attackers is low. In other words, this special rewarding scheme is effective when the combined hashpower of all the attackers expended in this attack is less than a certain limit.

Consider two mining pools, P and P', each with computing power p and p' respectively. The computing power of the rest of the Bitcoin network is $1 - p - p'$. The pool P' sends miners having δ computing power to infiltrate and attack pool P. As such, the computing power of P goes up to $p + \delta$. The computing power of P' gets reduced to $p' - \delta$. The incentive earned by the attackers sent by pool P' is transferred back P'. This amount is given by $G_{INF}^{BWH} = \frac{p}{1-\delta}(1 - \gamma)\frac{\delta}{p+\delta}$. The incentive earned through P''s own mining is given by $G_{P'}^{priv} = \frac{p'-\delta}{1-\delta}$. Thus, the total earned incentive of pool P' is given by $G_{P'}^{BWH} = G_{P'}^{priv} + G_{INF}^{BWH} = \frac{p'-\delta}{1-\delta} + \frac{p(1-\gamma)}{1-\delta}\frac{\delta}{p+\delta}$. If P' had mined honestly, it would have generated an incentive given by $G_{P'}^{Hons} = p'$. Now, if P' has to be discouraged against attacking the pool P, $G_{P'}^{Hons}$ should be higher than $G_{P'}^{BWH}$. So, P' will be discouraged if $\frac{p'-\delta}{1-\delta} + \frac{p\delta(1-\gamma)}{(p+\delta)(1-\delta)} < p'$ that is $\gamma > p' - \frac{\delta(1-p')}{p}$.

Corollary 1. *Let $P_1, P_2, \ldots, P_n$ be n mining pools all having trusted miners as members. The hashpower of pool i is $p_i, \forall i \in \{1, 2, \ldots, n\}$. If a pool $P_i, i \in \{1, 2, \ldots, n\}$ wants to accept an external miner having hashpower δ from any other pool, she should choose a special reward γ, given by $\max_{j=1, j\neq i}^n (p_j - \frac{\delta(1-p_j)}{p_i})$ in order to repel all attackers.*

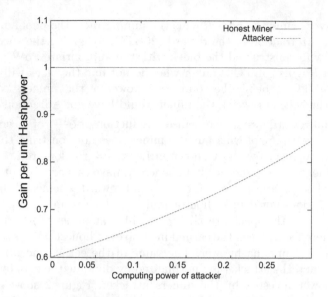

Fig. 2. Graphical presentation of the gain of an honest miner and an attacker per unit computing power. Here the computing power of pool P is $p = 0.6$ of the entire Bitcoin network. We choose a γ equal to the numeric value of $1 - p - \delta$, where δ is the total hashpower of all the attackers in the pool P.

Let there be n pools in the Bitcoin network represented as $P_1, P_2, \ldots, P_n$. The pools have computing powers $p_1, p_2, \ldots p_n$ respectively, satisfying $\sum_{j=1}^{n} p_j = 1$. The pool P_j spends α_j computing power to attack other pools and is itself attacked by miners from other pools having a total of β_j computing power. So, $\sum_{j=1}^{n} \alpha_i = \sum_{j=1}^{n} \beta_j$. Now, we assume that a particular pool $P_i, i \in \{1, 2, \ldots, n\}$, uses a special reward γ to discourage all traitors against block withholding. Now, since P_i uses α_i amount of computing power to attack other pools, its computing power reduces to $p_i - \alpha_i$. We assume that P_i is honest and hence, $\alpha_i = 0$. The expected amount of revenue generated by P_i is $\frac{p_i}{1 - \sum_{j=1}^{n} \beta_j}$. The pool P_i uses γ fraction of this to give special reward to the miners who found the winning block(s). The rest of the revenue amounting to $\frac{(p_i - \alpha_i)(1-\gamma)}{1 - \sum_{j=1}^{n} \beta_j}$ is distributed among all the miners. So, the block withholding attackers will earn a total revenue of $\frac{(p_i - \alpha_i)(1-\gamma)}{1 - \sum_{j=1}^{n} \beta_j} \frac{\beta_i}{p_i + \beta_i}$. We have assumed P_i to be honest and hence $\alpha_i = 0$. Therefore, the earned incentive of all honest miners of P_i will be $\frac{p_i}{1 - \sum_{i=1}^{n} \beta_i} \gamma + \frac{(p_i)(1-\gamma)}{1 - \sum_{j=1}^{n} \beta_j} \frac{p_i}{p_i + \beta_i}$. The miners of pool P_i won't be deprived if $\frac{p_i}{1 - \sum_{i=1}^{n} \beta_i} \gamma + \frac{(p_i)(1-\gamma)}{1 - \sum_{j=1}^{n} \beta_j} \frac{p_i}{p_i + \beta_i} \geq p_i$. Simplifying this relation we get,

$$\gamma \geq 1 - \sum_{j=1}^{n} \beta_j - \frac{p_i}{\beta_i} \sum_{j=1}^{n} \beta_j. \tag{1}$$

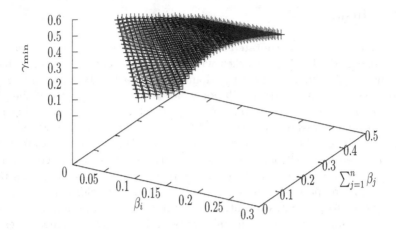

Fig. 3. The value of $\gamma_{\min} = 1 - \sum_{j=1}^{n} \beta_j - \frac{p_i}{\beta_i} \sum_{j=1}^{n} \beta_j$ for different values of the parameters. We choose $p_i = 0.25$.

It can be seen from Eq. 1 that the minimum value of γ decreases as the computing power of P_i increases. So, the special reward can be low for a big mining pool. In other words, a big pool can ward off the threat of BWH attack by affording a small special reward. Figure 3 gives a pictorial depiction of how the minimum value of the special reward varies with other parameters. We fixed the computing power of P_i to be 0.25 in Fig. 3.

Now, let us consider another scenario where there are n pools represented as $P_1, P_2, \ldots, P_n$. These pools have computing power denoted by $p_1, p_2, \ldots, p_n$. Each pool $P_i, i \in \{1, 2, \ldots, n\}$ uses α_i computing power to attack other pool. Also, each pool P_i is itself attacked by other pools' members having an aggregate of β_i computing power. Since, there are only n pools, $\sum_{i=1}^{n} \alpha_i = \sum_{i=1}^{n} \beta_i$. Now, a pool P_i's effective computing power will come down to $p_i - \alpha_i$ after it has sent few miners to attack other pools. The revenue earned by the pool P_i from mining within the pool is given by $G_{P_i}^{self} = \frac{p_i - \alpha_i}{1 - \sum_{j=1}^{n} \beta_j}$. We assume that $\beta_i = 0$, i.e. the pool P_i is not attacked. Let, P_i attacks a pool $P_j, 1 \leq j \leq n, j \neq i$ with a computing power α_{ij}. So, if a pool P_j follows the special rewarding policy, the incentive the attackers from pool P_i could earn can be written as $G_{ij}^{BWH} = \frac{(1-\gamma)(p_j - \alpha_j)\alpha_{ij}}{(1 - \sum_{i=1}^{n} \beta_j)(p_j - \alpha_j + \beta_j)}$. Hence, the total earned incentive of pool P_i will be $G_{P_i} = \frac{p_i - \alpha_i}{1 - \sum_{j=1}^{n} \beta_j} + \sum_{j=1, j \neq i}^{n} \frac{(1-\gamma)(p_j - \alpha_j)\alpha_{ij}}{(1 - \sum_{i=1}^{n} \beta_j)(p_j - \alpha_j + \beta_j)} = \frac{p_i - \alpha_i}{1 - \sum_{j=1}^{n} \beta_j} + \frac{(1-\gamma)}{(1 - \sum_{i=1}^{n} \beta_j)} \sum_{j=1, j \neq i}^{n} \frac{(p_j - \alpha_j)\alpha_{ij}}{(p_j - \alpha_j + \beta_j)}$. Now P_i will be benefited from the attack only if $\frac{p_i - \alpha_i}{1 - \sum_{j=1}^{n} \beta_j} + \frac{(1-\gamma)}{(1 - \sum_{j=1}^{n} \beta_j)} \sum_{j=1, j \neq i}^{n} \frac{(p_j - \alpha_j)\alpha_{ij}}{(p_j - \alpha_j + \beta_j)} \geq \frac{p_i}{1 - \sum_{j=1, j \neq i}^{n} \beta_j}$. Now, while the attack is put on execution, the gain of P_i will be maximized if $\frac{(p_j - \alpha_j)\alpha_{ij}}{p_j - \alpha_j + \beta_j} = c, \forall j \in \{1, 2, \ldots, n\} \backslash \{i\}$. This will hold if $\alpha_{ij} = \frac{c(p_j - \alpha_j + \beta_j)}{p_j - \alpha_j}$ that is $\alpha_{ij} = c(1 + \frac{\beta_j}{p_j - \alpha_j})$ for all c.

6 Conclusion

In this paper, we have incorporated few results that deal with the optimal strategy of a block withholding attacker who, with an intention to maximize her revenue uses her computing power rationally to attack a Bitcoin mining pool. We also showed that this attack will not yield the desired outcome if the victim pool is already under attack by one or more attackers having computing power beyond a specific threshold. In other words, an overwhelmingly invaded mining pool may not be good destination for a BWH attacker who wants to utilize her computing power most efficiently in order to earn incentives at the cost of unsuspecting miners of an open mining pool. We also propose and analyze a strategy that honest mining pools can use to repulse all block withholding attackers. This paper proposes to pay extra incentive to a miner who submits a valid Bitcoin block to the mining pool which the mining pool can use to earn incentives from the Bitcoin network. We have discussed the exact amount of the special reward that can eliminate the risk of BWH attack in a pool for some chosen models of Bitcoin mining pools.

Acknowledgement. The authors were partially supported by the Japan Society for the Promotion of Science (Japan) and the Department of Science and Technology (India) under the Japan-India Science Cooperative Program of research project named: "Computational Aspects of Mathematical Design and Analysis of Secure Communication Systems Based on Cryptographic Primitives".

The second author was partially supported by JSPS Grants-in-Aid for Scientific Research named: "KAKEN-15H02711".

References

1. Nakamoto, S.: Bitcoin: a peer-to-peer electronic cash system. Consulted **1**(2012), 28 (2008)
2. Back, A.: Hashcash - a denial of service counter-measure. Technical report, August 2002
3. Rosenfeld, M.: Analysis of bitcoin pooled mining reward systems. CoRR, abs/1112.4980 (2011)
4. Courtois, N.T., Bahack, L., On subversive miner strategies, block withholding attack in bitcoin digital currency. arXiv preprint arXiv:1402.1718 (2014)
5. Luu, L., Saha, R., Parameshwaran, I., Saxena, P., Hobor, A., On power splitting games in distributed computation: the case of bitcoin pooled mining. In: proceedings of IEEE Computer Security Foundations Symposium, CSF 2015, Verona, Italy (2015)
6. Eyal, I.: The miner's dilemma. In: IEEE Symposium on Security and Privacy, San Jose (2015)
7. Laszka, A., Johnson, B., Grossklags, J.: When bitcoin mining pools run dry. In: Brenner, M., Christin, N., Johnson, B., Rohloff, K. (eds.) FC 2015 Workshops. LNCS, vol. 8976, pp. 63–77. Springer, Heidelberg (2015)

Network and Systems Security and Access Control

Cyber Security Risk Assessment
of a DDoS Attack

Gaute Wangen[1(✉)], Andrii Shalaginov[1], and Christoffer Hallstensen[2]

[1] NISlab, Norwegian Security Laboratory,
Center for Cyber and Information Security, Gjøvik, Norway
{gaute.wangen2,andrii.shalaginov}@NTNU.no
[2] IT Services, NTNU, Trondheim, Norway
christoffer.hallstensen@NTNU.no

Abstract. This paper proposes a risk assessment process based on distinct classes and estimators, which we apply to a case study of a common communications security risk; a distributed denial of service attack (DDoS) attack. The risk assessment's novelty lies in the combination both the quantitative (statistics) and qualitative (subjective knowledge-based) aspects to model the attack and estimate the risk. The approach centers on estimations of assets, vulnerabilities, threats, controls, and associated outcomes in the event of a DDoS, together with a statistical analysis of the risk. Our main contribution is the process to combine the qualitative and quantitative estimation methods for cyber security risks, together with an insight into which technical details and variables to consider when risk assessing the DDoS amplification attack.

1 Introduction to InfoSec Risk Assessment

To conduct an information security (InfoSec) risk analysis (ISRA) is *to comprehend the nature of risk and to determine the level of risk* [2]. InfoSec risk comes from applying technology to information [6], where the risks revolve around securing the confidentiality, integrity, and availability of information. InfoSec risk management (ISRM) is the process of managing these risk while maximizing long-term profit in the presence of faults, conflicting incentives, and active adversaries [19]. Risks for information systems are mainly analyzed using a probabilistic risk analysis [3,17], where risk is defined by estimations of consequence for the organization (e.g. financial loss if an incident occurred) and the probability of the risk occurring within a time interval. ISRA is mostly conducted using previous cases and historical data. Depending on statistical data (quantitative) alone for risk assessments will be too naive as the data quickly become obsolete [18] and is limited to only previously observed events [16]. While the subjective (qualitative) risk assessment is prone to several biases [11] (Part II) [16]. ISRM methods claim to be mainly quantitative [6,8] or qualitative [7], but the quantitative versus qualitative risk situation is not strictly either-or. There are degrees of subjectivity and human-made assumptions in any risk assessment, and the intersection of these two approaches remains largely unexplored. The goal of

© Springer International Publishing Switzerland 2016
M. Bishop and A.C.A. Nascimento (Eds.): ISC 2016, LNCS 9866, pp. 183–202, 2016.
DOI: 10.1007/978-3-319-45871-7_12

this paper is to explore this intersection and discuss the benefits and drawbacks from each approach, and how they can complement each other. Moreover, we will discuss alternative ways of expressing uncertainty in risk assessment.

The remainder of the paper is structured as follows: The two following subsections introduces the reader to Distributed Denial of Service attacks and discusses the related work in ISRA. The Sect. 2 provides a brief description of the DDoS attack and development trend. Also, we present the method applied for ISRA and statistical analysis of the DDoS attack. Later in the Sect. 3 we give an insight into the qualitative ISRM approach together with results and the quantitative risk assessment in the Sect. 4 based on statistical methods. Lastly, we discuss and conclude the results, the relationship between this work and previous ISRA work, limitations and propose future work in the Sect. 5.

1.1 Distributed Denial of Service Attacks

A denial of service (DoS) occurs when an ICT (Information and Communication Technology) resource becomes unavailable to its intended users. The attack scenario is to generate enough traffic to consume either all of the available bandwidth or to produce enough traffic on the server itself to prevent it from handling legitimate requests (resource exhaustion). The attacker needs to either exploit a vulnerable service protocol or to exploit network device(s) to generate traffic, or to amplify his requests via a server to consume all of the bandwidth. The DoS attack is distributed (DDoS) when the attacker manages to send traffic from multiple vulnerable devices. The attacker can achieve amplification through the exploitation of vulnerable protocols or through using botnets.

The increase of Internet throughput capacity has also facilitated the growth in traffic volume for DDoS-attacks. According to Arbor Networks, the largest observed attack in 2002 was less 1 Gbps (Gigabit per second). While the biggest observed attack until now targeted a British television channel and reportedly generated ≈ 600 Gbps of traffic. That is an approximate 60x development in capacity for DDoS attacks over the course of about 14 years, see Fig. 1.

1.2 Related Work in ISRA

The ISRA approach presented in this paper primarily builds on two previous studies; firstly, Wangen et al.'s [17] Core Unified Risk Framework (CURF), which is a bottom-up classification of nine ISRA methods. The motivation behind CURF, was that there are several ISRA methods which conduct similar tasks, but there is no common way to conduct an ISRA. The approach ranked as most complete in CURF was ISO27005 [3] (from this moment referred to as ISO27005), while ISO27005 has many strengths, such as the process descriptions and taxonomies, one of the primary deficits of the ISO27005 is the lack of variables to consider and risk estimation techniques. The proposed approach in this paper builds on ISO27005 and addresses the outlined issues by defining classes and estimations for each step. Second, the probabilistic model presented in this paper builds on the feasibility study conducted by Wangen and

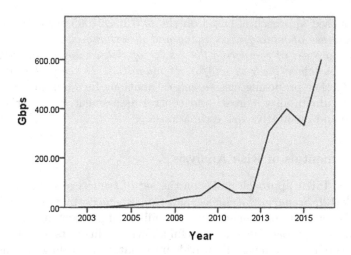

Fig. 1. The development of bandwidth consumption (Gbps) of DDoS-attacks during the last 15 years. *Data source: Arbor Networks and media reports*

Shalaginov [18], which discusses statistics and *Black Swan* (see Taleb [16]) issues in ISRA. The Authors [18] found that there are Black Swan related aspects of the ICT domain that may render past observations (Statistics) inappropriate for probability, such as for novel and unique attacks, and the fast development of ICT, for example, Fig. 1. However, the authors also found that quantifying and modeling InfoSec risks have utility as long as the risk assessor is aware of the properties of the risk and the domain we are modeling. The Single and Annual Loss Expectancy (SLE/ALE) represent the most developed area of statistics in ISRA, where risk is described as the probability of a loss occurring [6]. Yet, risk must be considered as more than an expected loss [5]. Knowledge-based probabilities represent the main approach in ISRA [17], as previously discussed, there is utility in statistical data. The combination of these two approaches to probability has remained relatively unexplored in ISRA. So, this study proposes to combine a statistical and a qualitative ISRA to address the research gap.

Thus, this paper proposes a step-by-step process model for an ISRA of a distributed denial of service (DDoS) attack, and we apply the model to a real-world case as a proof of concept and feasibility study. The proposed ISRA approach is compliant with ISO27005.

2   Choice of Methods

This section outlines the core risk assessment concepts applied in this paper. First, we present the fundamentals of our risk analysis approach, then the qualitative ISRA method, and, lastly, discuss the statistical methods employed for quantitative analysis. Our overarching approach to validation is case study.

The proposed approach is based on the two ISO27005 steps (i) *Risk Identification -process of finding, recognizing and describing risks* [2], and (ii) *Risk Estimation - process of comparing the results of risk analysis with risk criteria to determine whether the risk and/or its magnitude is acceptable or tolerable* [2]. We go further proposing classes and estimations for qualitative asset evaluation, and vulnerability, threat, and control assessment, together with both quantitative and qualitative risk estimations.

2.1 Fundamentals of Risk Analysis

Our proposed ISRA approach builds on the *set of triplets* as defined by Kaplan and Garrick [12], *Scenario, Likelihood,* and *Consequences.* In which we define the scenario as a combination of assets, vulnerability, threat, controls, and outcome. Each step in the approach generates useful knowledge in on its own, for example, a thorough threat assessment will provide information regarding opponents that are also useful in other risk-related activities and decision-making.

We combine the two approaches to risk and probability proposed by Aven [5]: (i) the frequentist (*"the fraction of times the event A occurs when considering an infinite population of similar situations or scenarios to the one analyzed"*), and (ii) the subjective knowledge-based probability (*"assessor's uncertainty (degree of belief) of the occurrence of an event"*). In terms of risk analysis, the key components of a risk (R) related to an activity for discussion and calculation are as follows [4] (p. 229): R is described as a function of events (A), consequences (C), associated uncertainties (U), and probabilities (P). U and P calculations rely on background knowledge (K) which captures the qualitative aspect of the risk, for example, low K about a risk equals more U. Model sensitivities (S) display the underlying dependencies on the variation of the assumptions and conditions. Thus, $R = f(A, C, U, P, S, K)$ allows for a comprehensive output and incorporates the most common components of risk.

In the following section, we define the classes and estimators for each of the key elements of InfoSec risk as subjective knowledge, where the classes describe and categorize the risk components, and the estimators represent qualitative estimations based on expert knowledge and collected data. We do not define the scales for each estimator in this paper as this is individual for each organization.

2.2 Proposed Methodology for Qualitative Risk Analysis

The proposed qualitative methodology is based on descriptions, classes, and estimators. Based on ISO27005 we defined these for Assets evaluation, Vulnerability assessment, Threat assessment, and Control Assessment.

Asset Identification and Evaluation. To start, the Institution needs to identify and know its assets. We define *Asset Identification* as the process of identifying assets, while asset *Evaluation* assess their value, importance, and criticality. According to ISO27005 [3] Annex B, there are two primary assets, (i) Business

Processes & activities and (ii) Information. While *Asset Container* identifies where assets are stored, transported, and processed [7].

As a part of the process, we map the organizational goals and objectives for risk assessment, as these are important in deriving security goals for the InfoSec program. Also, we consider these when determining the risk event outcome.

- Assets - Something of value to the organization, person, or entity in question.
- Asset type - Description of the asset class, E.g. sensitive information.
- Asset Container - refers to where and how the asset is stored [7].
- Asset value - Estimated, either monetary or some intangible measurement of value
- Importance in Business Process is an estimation of the criticality of the asset in daily operations
- Asset criticality is the comprehensive assessment of the asset value and role in business process estimations.

Vulnerability Assessment. *Vulnerability Identification* is the process of identifying vulnerabilities of an asset or a control that can be exploited by a threat [2]. Vulnerability *Assessment* is the process of identifying, quantifying, and prioritizing (or ranking) the vulnerabilities in a system. Vulnerabilities can be discovered through many activities, such as automated vulnerability scanning tools, security tests, security baselining, code reviews, and penetration testing. In the case of network penetration from a resourceful attacker, the analyst should also consider the *attacker graph*: how compromising one node in the network and establishing a foothold in the network can be exploited to move laterally inside the network and compromising additional nodes.

- Vulnerability type - A classification and description of vulnerability, *weakness of an asset or control that can be exploited by one or more threats* [2].
- Attack description - description of the attack for single attacks such as DDoS, or *attacker graph* where the adversary obtains access to an asset or asset group. The attacker graph is a visual representation of how the attacker traverses the network and gains access to an asset or a group of assets.
- Attack difficulty - Estimation, how difficult is it to launch the attack?
- Vulnerability severity - Estimation of the seriousness of the vulnerability
- System Resilience - How well will the system function under and after an assault, especially important for availability related risk
- Robustness - is the measure of how strong an attack will the system absorb.
- Exposure assessment - Determines exposure of entity's assets through the vulnerability and attack

Threat Identification and Assessment. *Threat identification* is the process of identifying relevant threats for the organization. A Threat is a potential cause of an unwanted incident, which may result in harm to a system or organization [2]. Besides mother nature, the threat is always considered as a human. For example, the threat is not the computer worm, but the worm's author. While the threat *Assessment* comprises of methods and approaches to determine the credibility and seriousness of a potential threat. The assessment process relies on the quality of threat intelligence and understanding of the adversary. For each threat, we propose to consider the following classes and estimators:

- Threat actor - Describes the human origin of the threat. There are several classes of threat agents in InfoSec, for example, malware authors, Cyberspies, and hackers.
- Intention - Defines what the threat actor's objectives with the attack, for example, unauthorized access, misuse, modify, deny access, sabotage, or disclosure.
- Motivation - Defines the primary motivation for launching the attack, previous work on malicious motivations [13] suggests Military or Intelligence, Political, Financial, Business, Grudge, Amusement, Self-assertion, Fun, and Carelessness.
- Breach type - which type of security breach is the threat actor looking to make; either confidentiality, integrity, availability, non-repudiation, or accountability.
- Capacity - Estimation of the resources he/she has at their disposal to launch the attack. For example, if an attack requires a lengthy campaign against your systems to succeed, the threat actor must have the resources available to launch such an attack.
- Capability - Estimation the threat's *know how and ability* for launching the attack.
- Willingness to attack - Estimation of how strong the motivation is to attack. For example, historical observations of the threat actor's frequency attacking the system is a good indicator.
- Threat severity is the comprehensive assessment of the above variables and the main output of the process.

Control Efficiency Estimation. Existing controls are measures already in place in the organization to modify risk [2]. *Control identification* is the activity of identifying existing controls for asset protection. *Control (efficiency) Assessment* are methods and approaches to determine how effectively the existing controls are at mitigating an identified risk.

The important issue to consider here is if the control sufficiently mitigates the risk in question. If the control is considered adequate, the risk can be documented for later review.

- Control Objectives - a written description or classification of what the control is in place to achieve.
- Control domain - Addresses in what domain the identified control is, either in the physical, technical, or administrative [9] (pp. 166–167).
- Control class - Addresses what the control is supposed to achieve; either prevent, detect, deter, correct, compensate or recovery [9] (pp. 166–167).
- Risk Event components - Consists of the *Asset Criticality*, *Exposure Assessment*, and *Threat Severity* for the identified risk event.
- Control efficiency - Estimation, addresses how efficient the control is at modifying the identified threat event and how well it achieves the control objectives.

2.3   Methodology for Statistical Risk Analysis

The main statistical approaches considered in this paper are for theoretical analysis of the supplied historical data to run calculations. The motivation is to use conventional statistical methods to extract particular characteristics that are suitable for Quantitative ISRA. Additionally, we make hypotheses about an applicability of each particular method concerning available data. The calculations in this article are based on DDoS attacks data from the Akamai Technology's *State of the Internet* Reports (duration and magnitude) [1] and data gathered from the assessed case study institution on occurrence. These data are considered as quantitative observation of metrics of selected events, for example, some DDoS attacks over time. We utilize several community-accepted methods to deal with the historical data when it is necessary to make predictions in numbers. In particular, these are *Conditional Probability* and *Bayes Theorem*. First, the probabilistic model $p(x)$ is suggested and the corresponding set of parameters are estimated from the data to fit suggested distribution. In sequence, we apply statistical testing, which is an important part of our work since further for the DDoS case study we will justify the usage of a specific statistical method and make a hypothesis about their applicability. By testing, we can make a quantitative analysis of different statistical models quality. However, this is based only on pure analysis of the case's data and deducing the most applicable model that can describe the data and fit the purposes. The testing is suitable for determining whether the data follow a particular distribution model with some degree of defined beforehand confidence interval measured in %. The tests evaluate the actual observed data O with the expected data E from the hypothesized distribution. This is done with a help of QQ-PLOT or Quantile-Quantile plot representing a probability plot by depicting expected theoretical quantiles E and observed practical quantiles O against each other. The quality of hypothesized data distribution can be evaluated using linearity in this plot. It means that if the expectations match observations, even with some minor outliers, then the null hypothesis can be rejected, and data fit selected distribution. Second, the probabilistic model can be used to estimate the probability of similar events in this very period or later on. We observe the following well-known shortcomings of the probabilistic modeling. First, very few data points from history may cause a wrong decision. Second, very rare events have negligibly small probabilities which might cause trouble in predicting corresponding outcomes. The authors have applied the statistical analysis software IBM *SPSS*, GNU *PSPP* and *RapidMiner*. Later on, we also discuss the application of this methodology and possible ways of its improvement.

3   Case Study: Qualitative Risk Assessment of a DDoS Attack

The case data together with relevant available statistics was collected from an institution whose IT-operations delivers services to about 3,000 users. The Case study Institution (hence referred to as "The Institution") is a high-availability organization delivering a range of services to the employees and users, mainly within research and development. The objectives of the IT-operations is to deliver reliable services with minimal downtime. The target of this study has a 10 Gbps main fiber optics connection link, which is the threshold of a successful DDoS attack. Figure 2 displays the institution's network capacity and average traffic during regular weekdays, this case study considers attacks on the main link. During the five previous years, the Institution has had an average annual occurrence of two DDoS attempts, whereas none has been successful thus far. The goal of this assessment is to derive the qualitative risk of the Institution experiencing a successful attack by applying the proposed method.

The case study starts with asset identification and evaluation, further, considering vulnerabilities, threat assessment, control efficiency, and outcomes. Our contribution in this section is the application of the classes and qualitative estimators for each step of the risk assessment process.

Case Asset Evaluation. A DDoS attack is primarily an attack on the availability of the organization's Internet connection. We compare the Internet connection capacity with a pipeline; it's capacity limits the pipe's throughput. Once the capacity is filled, no additional traffic can travel through the pipe. The attacker's goal is to fill the pipeline with traffic and effectively block all legitimate traffic from traveling through the pipe.

In the considered case, a successful DDoS attack will lock the users out of the network and prevent them from conducting their connectivity-dependent tasks. Most of the organization's value chain is dependent on some level of connectivity, which makes the availability of services and assets the top priority when considering DDoS attacks. For simplicity, we consider "Service" as the main asset. As the institution is high availability and has up-time as one of the top priorities, service delivery is seen as crucial for production. Table 1 shows the classification and estimation considered for protection in the case study.

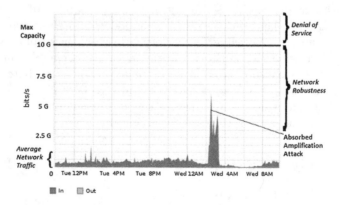

Fig. 2. Illustration of network robustness with an absorbed amplification attack. Network capacity at 10 Gbps, everything above constitutes a DoS.

Table 1. Asset considerations for the DDoS attack

Asset	Container	Protection attribute	Importance in business process	Asset value	Asset criticality
Service delivery	Infrastructure - Internet Pipeline	Availability	Essential (70–100)	Very high (50–85)	Essential (70–100)

Case Vulnerability Assessment Results. The Institution is exposed to several attack vectors for achieving DoS; for example resource starvation, application layer-based, and volumetric/flood. We provide a technical description of one attack, together with a vulnerability assessment. These estimations assume a 10 Gbps connection and the current security level in the Institution.

We measure the robustness in the DDoS-case in the gap between maximum network capacity and average traffic, illustrated in Fig. 2. A narrow gap between average load and maximum capacity is an indicator of fragility towards traffic generating attacks. To describe the network robustness we look at the maximum load versus the average load and measure the gap. The average load on the network is ≈ 1 Gbps; the system can absorb DDoS attacks up to ≈ 9 Gbps before the users experience denial of service, Fig. 2.

On resilience, the network will continue to function within acceptable service delivery up to traffic of about approximately 6–9 Gbps, depending on several variables such as weekday and hours, before users start to experience a degradation in service. Although attacks in this vicinity do not entirely cause a DoS, they reduce the latency in the network and efficiency of the workforce.

Based on our assessment of the network, we define four events (A) for further assessment:

1. Attacks less than 6 Gbps which will be absorbed by the network robustness and will go by unnoticed by the users (A1).
2. Attacks ranging 6–9 Gbps can cause reduction of service in the network (A2).
3. Attacks ranging above 9 Gbps will cause DoS together with day-to-day use (A3).
4. Attacks ranging from approximately 50 Gbps carry the potential for causing damage at the Internet Service Provider (ISP) level but carry the same consequences for the institution (A4).

Attacks need to be able to generate a traffic within the ranges of scenarios A2–A4 to be considered a threat potential threat in the case study, for illustration purposes, we only considered volumetric and flood-based attacks. The Institution's vulnerability is then the generic network capacity; we assume that no vulnerable services are running on the Institution's internal network. *Volumetric and flood based* attacks aims to saturate the amount of connections to the Link, through UDP (User Datagram Protocol) amplification generating a small amount of data from the attacker resulting in a lot of data traffic to the victim. UDP DDoS attacks exploit the fact that the UDP does not require a handshake to transmit data, and requires the service to return more bytes than the attacker sent with spoofed source IP. Hilden [10] provides the following example, *services running a vulnerable CharGen (Character generator protocol) can be exploited to generate traffic: the attacker sends a 1-byte sized packet with a spoofed IP (the target's IP) to the vulnerable servers. Due to no handshake, the servers immediately responds with a 1024 byte large packet to the target IP. The attacker can amplify his traffic (bytes sent) with 1024x (bytes received by the target) by exploiting one vulnerable server.* The Table 2 represents the attacker's bandwidth limits the attack.

The UDP amplification attack requires access to either a botnet or vulnerable service, both of which are readily available on the Internet, the former for hire and the latter for exploitation. The technical expertise required to launch an attack is low, where the trick is to locate vulnerable services through scans. The attacker can create traffic volumes in the ranges A2–A4, whereas attacks within ranges A2 and A3 are easily achieved with a low number of vulnerable services, Table 2. The A4 scenario requires more resources regarding bandwidth and services, but is still easily achieved for the technically skilled.

With a 10 Gbps connection, the Institution is inherently vulnerable to DDoS attacks, and since this is an attack on availability, the duration of the attack is also important to consider. We have defined the following downtime scenarios according to the Institution's risk tolerance:

1. Attack ranging between 0–10 min are considered negligible (B1).
2. 11–30 min will produce a slight loss in production (B2).
3. 31–120 min will produce a moderate loss in production, it is also likely that employees will seek out the helpdesk and cause extra overhead (B3).
4. 2–24 h will produce a critical loss in production, at this point everyone will have exhausted their tasks that can be solved without connectivity (B4).
5. >24 h will qualify as a catastrophe (B5).

The Institution is exposed to volumetric and flood-based attacks due to ease of exploitation and effective amplification. Attacks ranging within A2–A3 are easily achievable with an initial technical insight, while ability to maintain the attack up to scenarios B3–B4 depend on a number of externalities that have a high level of uncertainty related to them, such as internal reaction time, threat capacity, and ISP capabilities. We address uncertainty related to the threat actor in the next section.

Case Threat Assessment Results. Based on the exposure assessment, we identify and assess one threat actor in the position to trigger the attacks. For the threat actor, we consider the motivation, intention, willingness, capacity, and capability, to determine threat severity. The amplification attacks in question are easy to implement as long as vulnerable services are running, so, the analyst should consider less able attackers. However, for the case study we consider only one threat actor based on the estimated properties regarding the specifically analyzed DDoS attack:

Actor 1 is the politically motivated hacktivist whose weapon of choice is commonly the DDoS attack. Due to some of the research conducted in the Institution being controversial, they are the a potential target of Actor 1. We estimate the capacity for maintaining a lengthy attack (B3–B4) as *Moderate* and the capability for launching the attacks A2–A5 as *Very high*. It is uncertain whether this actor has been observed attacking their networks in the past, Table 3.

Table 2. Examples of approximate amplifications by exploiting vulnerable UDP, including possible amplification of the 100 Mbps connection. *Data source: Hilden* [10], *Norwegian Security Authority (NSM)*

Protocol	Amplification Ratio	100 Mbit/s ⇒
NTP	1:556	55.6 Gbit/s
CharGen	1:358	35.8 Gbit/s
QOTD	1:140	14 Gbit/s
Quake (servers)	1:63	6.3 Gbit/s
DNS (open resolver)	1:28–54	2.8–5.4 Gbit/s
SSDP	1:30	3 Gbit/s
SNMP	1:6	600 Mbit/s
Steam (Servers)	1:6	600 Mbit/s

Table 3. Threat assessment for DDoS attack, K represents confidence in the estimates

Threat actor	Motivation	Intention	Capacity	Capability	Willingness	K	Threat severity
Actor 1	Political	Disruption	Moderate	Very high	Moderate	Low	High
Actor 2	Military or Intelligence	Access	Very high	Very high	Very low	Medium	Medium
Actor 3	Self-assertion	Deny access	Low	Medium	Very high	High	Medium

Control Assessment Case Results. We provide a description of countermeasures for the considered attack, together with an estimation of efficiency which, for reactive controls, can be measured in time until the attack is mitigated.

In the case organization, the first and primary control strategy is to filter vulnerable UDP protocols on ingress network traffic. This control limits the attack surface of the organization's network and limits the effectiveness of exploiting vulnerable UDP based protocols. This control does not completely mitigate the possibility of attack because there is still network nodes that need to respond to UDP like Network Time Protocol and Domain Name System, but these are

Table 4. Control efficiency estimation. K represents confidence in the estimates

Control objectives	Control domain	Control class	K	Control efficiency
1. Filter UDP traffic	Logical	Preventive	Medium	*Medium*
2. Agreement with upstream ISP	Organizational	Reactive	High	*High*

configured to provide low possibility for amplification values so that threat actors cannot effectively use them for attacking other systems on the Internet.

The second available mitigation strategy is to have a close cooperation with the Internet service provider's CSIRT. This control is vital because of the ISP's capabilities to blackhole (null-routing), rate-limit or even block network traffic that originates outside of their own network, or the country itself. For large DDoS attacks, the ISP is the only one capable of filtering away this traffic efficiently. On a day-to-day basis and within normal work hours, to involve the ISP CSIRT to start shaping or blocking traffic is highly effective and possible to implement within 1 to 2 h. After working hours, 2 to 5 h is estimated.

3.1   Events and Results

The *Event outcomes* describes the range of outcomes of the event, consisting of asset, vulnerability, threat, and control, and how it affects the stakeholders and the organization. The process consists of identifying and describing the likely outcome(s) of the event regarding breaches of confidentiality, integrity, and availability, which does not entail calculations of consequence, as this is performed in the risk analysis. For example, an event outcome can have a financial impact or an impact on reputation.

The qualitative risk assessment shows that the most severe risk facing the organization is a DDoS campaign in the ranges A3–A4 (>9 Gbps) and lasting longer than 2 h (B4–B5). The Institution is currently vulnerable to such attacks due to the dependency on connectivity for running business processes. There is currently one politically motivated threat actor with a high capability of launching such an attack, but a moderate capacity for maintaining a lengthy campaign. We estimate the existing controls to be quite efficient to mitigate UDP amplification attacks, although the upstream ISP option includes third party dependencies which the institution does not control and introduces another layer of uncertainty. We continue the ISRA with the quantitative assessment of available real-case data from Akamai in the next section.

4   Quantitative Risk Analysis

The Risk analysis phase consists of estimating risk concerning $R = f(A, C, U, P, S, K)$. We assign the identified adverse outcomes, Sect. 3.1, probability according to previous observations and subjective knowledge. *A* (event) is the result of the risk identification process and in the analysis described as a range of adverse outcomes based on the consequence calculations. There are primarily two approaches to probability, frequentist or subjective knowledge-based assessments (quantitative and qualitative). This section starts with the quantitative risk approach, before combining it with the qualitative results to obtain the risk.

4.1   Risk Calculations

The goal of the risk estimation is to reduce *U* related to risk occurring. For *P&C* calculations, we suggest merging the objective data gathered through observations and statistics with the subjective knowledge-based probabilities. We define the following:

- *Quantitative Assessment (Objective data)* - prior frequencies of occurrence, including past observations of the risk and generic risk data used to derive objective measurements of probability. Together with the gathering of relevant metadata through observations made by others.
- *Qualitative Assessment (Knowledge-based data)* - a combination of knowledge that is specific to the organization and the threat it is facing. Primarily derived from the *risk event components*, Sect. 3.
- *Risk Estimate* - The final estimate of the probability for the risk, derived from quantitative and qualitative data.

The consequence estimation is derived primarily from two factors, monetary loss and intangible losses such as loss of reputation. Besides, the consequence estimation should consider the organizational objectives and goals [3]. The loss calculation is challenging as complex systems may fail in unpredictable ways. Possible data sources and input for consequence/impact considerations: prior loss data, monetary losses, consequences for organizational goals and objectives, and risk specific factors such as response time and attack duration.

Observed Frequencies of DDoS Attacks. By monitoring activity, we can obtain reliable numbers on how large the average DDoS attack and generate corresponding reports. The data applied in this article was provided by Akamai [1], and is based on 4,768 valid observations from 2014–2015, shown in the Table 5. There was no observed attack magnitudes over 255 Gbps in the data set. The observed frequencies of attacks towards the case study institution averaged two annual attacks during the last five years, $P_{occ} = \frac{1}{6} \approx 17\%$ of monthly occurrence, none of which have succeeded in attaining the necessary magnitude to achieve DoS. One of which managed to cause instability in the wireless network, thus, classifying as an A2 scenario.

Further, to test our hypothesis about the distribution of the data we used Q-Q plot, depicted in the Fig. 3. The plot shows the dependency between the observed data and expected data according to **Gamma distribution** prediction. Also,

Table 5. Frequencies of DDoS magnitude observations from Akamai dataset [1].

Characteristic	Valid	Missing	Mean	Median	Std. Dev.	Minimum	Maximum
Duration	4768	0	154,931.00	48,180.00	622,073.00	600	29,965,740.00
Gbps	4768	0	6.09	1.50	15.63	10^{-5}	249.00

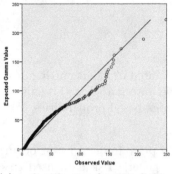

(a) γ-distribution for DDoS Gbps

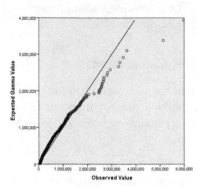

(b) γ-distribution for DDoS Duration

Fig. 3. Fitting DDoS Magnitude and duration data set by means of Q-Q Plot using γ-distribution. Two outliers are evident at the high end of the range for both distributions.

one can see two outliers at the high bandwidth interval indicating either unusual events or possible error in logging the characteristics of the events.

Observed Values for Impact Estimation. By monitoring activity, we can also obtain reliable numbers on the duration of DDoS attacks and generate distributions. Our data provides us with Table 5, the data shows that the documented DDoS durations observed in this period were in the range from 600 up to $29 \cdot 10^6$ s, the longest lasting attack lasting approximately 347 days with magnitudes reaching about 4 Gbps. Removing two outliers from the data set gives a new mean value equal to $1.4 \cdot 10^5$ s. The Fig. 4 displays the data clustering in the area around the mode and median. The majority of the data are distributed in this particular interval. In the case of probabilistic estimation, it means that the data located far from this region are going to have a negligible level of occurrence.

Our tests showed that there is no correlation between the variables "attack duration" and "attack magnitude". There is a small difference between the mean attack durations in the considered outcomes, but it is not statistically significant, Table 6. The A3 attacks seem to have shorter durations than the other; the one-way ANOVA (Analysis of variance model) shows that these two groups of observations are similar only to significance P = 85 %. Yet, if we combine the A3 and A4 attacks this mean duration rises, and there is no significance.

Figure 5 depicts the correlation between duration and magnitude, where the attacks from the *A1* and *A2* scenarios are distributed nearly uniformly across the

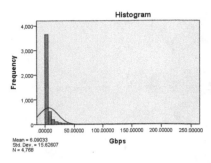

(a) Histogram for DDoS Gbps

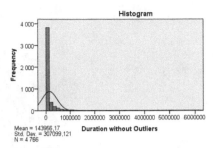

(b) Histogram for DDoS duration

Fig. 4. Histogram of DDoS magnitudes and durations with normal curve, without two largest outliers. *Data Source: Akamai* [1]

Table 6. Frequencies for the defined events, *A*. *Data Source: Akamai* [1]

Scenario	Magnitude Gbps	Mean	Median	N	Std. Dev	% of attacks	$P(P_{occ} \wedge A)$
A1	<6	159,956.64	48,900	3,713	682,039.967	77.9	13.2 %
A2	6–8.9	162,124.35	44,700	331	450,382.579	6.9	2.6 %
A3	9–49.4	117,437.50	46,080	624	259,646.272	13.1	1.8 %
A4	>49.5	178,485.20	52,380	100	284,012.424	2.1	0.4 %

duration scale. It means that the nature of such attacks is more random and non-deterministic, which was also confirmed by our correlation tests. Going further, one can see that the majority of the attacks from the range of *A3* are located in the duration range around $10^3 \cdots 10^6$ s. Finally, same stands for the scenario *A4*, where the dispersion of possible magnitudes is large in comparison to *A3*. However, much higher frequency in case of probabilist model suppresses less frequent cases, while fuzzy logic describes data independently from the frequency of its appearance, only taking into consideration its possibility as described before by Shalaginov et al. [14].

4.2 Probabilistic Modeling for Risk Estimation

Unplanned downtime is an adverse event for which most ICT-dependent organizations need to have contingencies. The Institution considered in this paper have defined the severity metrics in Table 7, ranging from "Negligible" to "Catastrophe", together with the distribution of duration within the defined intervals. Losses are considered to be moderate up to two hours downtime, as most employees will be able to conduct tasks that do not require connectivity for a short period. Losses are estimated to start to accumulate after 2 h of downtime. The analysis shows that the defined events *B3–B5* are over 99 % likely to last more than 2 h, which falls well outside of the Institutions risk tolerance. The conditional probability that the institution will suffer DDoS events in a given month

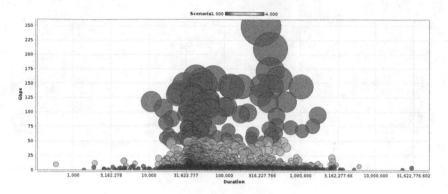

Fig. 5. Bubble plot of the attack bandwidth depending on the duration for each scenario. Size of the bubble also denote magnitude of the attack. Scenarios are depicted with different colours (Color figure online).

Table 7. Overview attack severity for the case study and duration frequencies. *Data Source: Akamai* [1]

Outcome	Interval (min)	Seconds	Severity	Frequency	% of Attacks
B1	0–10 min	0–600	Negligible	1	0.0
B2	11–30 min	601–1,800	Slight	1	0.0
B3	31–120 min	1,800–7,200	Moderate	28	0.6
B4	2–24 h	7,201–86,400	Critical	3,346	70.2
B5	> 24 h	> 86400	Catastrophe	1,392	29.2

is described in Table 6, right column. The risk estimation is modeled as an *Event tree*, Fig. 6, based on conditional probabilities $P(P_{occ} \wedge A \wedge B)$.

Sensitivity. The most sensitive numbers for the risk calculation is the P_{occ}, which is based on approximately ten observations from the last five years. The low amount of observations makes the mean sensitive to changes and one can capture this aspect in the analysis by assigning ranges to P_{occ} instead of concrete numbers. A probability range will help to make the assessment more robust, by for example adjusting for a range of 1–6 (or more) occurrences of DDoS attacks every year.

5   Discussion and Conclusion

In this section, we discuss the possibility of adjusting the risk model with additional qualitative input and propose an expanded model. We then discuss the limitations of the work and the potential future directions for the work.

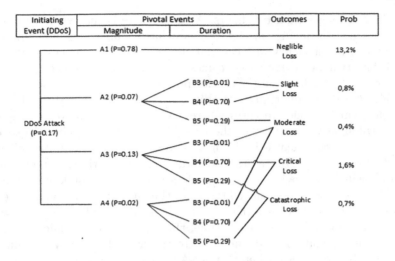

Fig. 6. Event tree displaying probability of monthly DDoS occurrence for the case study.

5.1   Adjusting for Knowledge-Based Probability Estimations

The primary objective of the ISRA process is to provide the decision-maker with as good a decision basis as possible. The benefit of the quantitative analysis is that the results are grounded in reality and defensible in a risk communication process. From the other side, the advantage of the qualitative risk assessment is that it allows more dynamic risk assessments. The main fragility of quantitative approaches is the dependence on the data quality and quantity of observations. We know about the fast-paced developments in ICT, for example, Fig. 1, showed the progress in capacity for DDoS attacks, and that attack trends may vary which have implications for the annual occurrence (discussed in [18]). The duration and magnitude of γ distributions should be more stable although the observed values are likely to increase according to the trend. However, the limitation of quantitative risk assessments is that attacks may not be present in the dataset, which makes the probabilistic approach less flexible as conducted in Sect. 4.2. It means that there is a need to have a control or introduce an additional factor that may indicate the possibility of the attacks.

One specific finding is the *Control efficiency*, Table 4, in which we have identified one proactive and one reactive control in place to mitigate an attack. For this discussion, we disregard the proactive control *Filter UDP traffic* as attacks have been occurring at a regular rate even with this control in place. We consider the reactive control, *Agreement with upstream ISP*, as a part of the risk assessment, where, during the workday we can expect an attack to be mitigated within 1–2 h, and after working hours the handling time is between 2–5 h. Although our quantitative analysis, Fig. 6, shows the combined risk of a monthly DDoS attack ranging from critical to a catastrophic loss at $\approx 2,3\%$. Further, if we include the control efficiency assessment we can adjust down the risk estimate for DDoS attacks lasting longer than two hours. A caveat here is that we must consider the event of control failure, in this case, we have a high degree of knowledge about the control efficiency and can put more trust in its functionality. However, third party dependency always comes with uncertainties due to information asymmetry problems between the service provider and the institution.

We also have the opportunity to adjust P_{occ} estimates based on the threat assessment, which applies to cases where the attacker attributes changes, for example, willingness to attack in the case of controversial political events. A thorough threat assessment is likely the best data source for more technical and rarer attacks than the DDoS. An understanding of the threats intention and motivation will also provide a better understanding of possible consequences. The qualitative risk assessment shows that the Institution is facing one serious threat actor who both has the capacity, capability, and moderately willing to launch an attack. At the current time, the UDP-based amplification attack vector is easily exploitable and can generate traffic far beyond system limits to achieve all adverse scenarios between A2-A4. Which means that threat actors with less capacity and capability will be able to produce more powerful attacks. For a more technical and resource intensive attack, it would make sense to consider the threat assessment where the more resourceful threats are linked to the more advanced attacks, for example, Threat *Actor 2* (Table 3) is more likely to be behind attacks in the critical to catastrophic loss events. *Actor 3* will be responsible for most attacks, but due to his limitations in capacity and capability; attacks will primarily be limited to short lasting and small magnitude attacks. While *Actor 2* is rarely observed, but can launch the catastrophic range attacks.

Taking into account both the threat and control assessments, we modify the Event tree to accommodate the qualitative assessment. For the combined assessment, we consider control efficiency concerning subjective ranges for P of a successful attack with Control 2 in place. To operationalize the threat assessment in the model, we have visualized our estimated attack ranges assigned to the identified threat actors in the left column, Fig. 7.

5.2   Limitations and Future Work

Our work has proposed an approach on how to combine quantitative and qualitative risk estimates. However, there is a limitation in our model due to the combination of the subjective and statistical assessments. We believe that application of possibilistic models such that Fuzzy Logic may help to understand the reasoning of statistical models better when the probabilities of two events are nearly equal and are very small. It means that the difference between two similar events can be below the limit of computing error because the event falls under the category of what Taleb defines as *Extremistan* (see [15,18]). Therefore, applying a combination of subjective and objective estimators, we will be able to achieve better generalization of the model. Another way to improve the methodology is to use hierarchical models that ensemble inference of human-understandable Fuzzy Rules (also used for decision support) into a comprehensive framework.

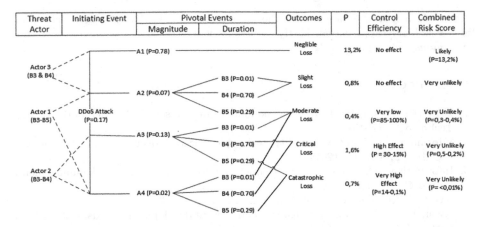

Threat Actor	Initiating Event	Pivotal Events		Outcomes	P	Control Efficiency	Combined Risk Score
		Magnitude	Duration				

Fig. 7. Expanded event tree also including subjective estimates of threat actors and control efficiency.

We propose to apply our approach to model other cyber risks for further validation. The risk considered in this paper is a very technical communications risk, and the risk model would benefit from testing in areas where historical data is less available. Another limitation is the limited generalization of our case study; the ISRA approach should also be applied to other types of organizations.

5.3 Conclusion

In this paper, we have proposed and applied classes and estimators for qualitative ISRA, which should contribute towards making the overall risk assessment process easier and more comprehensive. Our work shows that applying statistical methods for a cyber risk is feasible as long as there is data available. Moreover, with more accurate data there are possibilities for even more accurate and better quality models. Also, we adjusted the quantitative risk estimates with qualitative findings, for example, the definitions of scenario events (A and B) were based on qualitative measures of vulnerability and applied to categorize objective data. This paper also took the merging further by implementing the findings from the qualitative threat and control efficiency assessments into the probabilistic model. The control estimation is crucial to the risk estimation as it directly affects the estimation result, which in our case study made the most severe outcomes very unlikely. Thus, the conclusion is that combination of both the qualitative and quantitative aspects of ISRA is both feasible and beneficial. Defining an ISRM method as either-or in this manner may cause the risk analyst to miss out on valuable information for the assessment.

Acknowledgements. The authors acknowledge Professors Einar Snekkenes, Katrin Franke, and Dr. Roberto Ferreira Lopes from NTNU, Anders Einar Hilden from the Norwegian Security Authority (NSM), Karine Gourdon-Keller, David Fernandez, and

Martin McKeay from Akamai. Also, the support from the COINS Research School for InfoSec is highly appreciated. Lastly, we acknowledge the contributions made by the anonymous reviewers.

References

1. 2014–2015 DDoS attack duration and magnitude dataset. Technical report, Akamai Technologies (2015)
2. Information technology, security techniques, ISMS, overview and vocabulary, ISO/IEC 27000:2014 (2014)
3. Information technology, security techniques, information security risk management, ISO/IEC 27005:2011 (2011)
4. Aven, T.: Misconceptions of Risk. Wiley, New York (2011)
5. Aven, T.: The risk concept - historical and recent development trends. Reliab. Eng. Syst. Saf. **99**, 33–44 (2012)
6. Blakley, B., McDermott, E., Geer, D.: Information security is information risk management. In: Proceedings of the 2001 Workshop on New Security Paradigms, pp. 97–104. ACM (2001)
7. Caralli, R.A., Stevens, J.F., Young, L.R., Wilson, W.R.: Introducing octave allegro: Improving the information security riskassessment process. Technical report, DTIC Document (2007)
8. Freund, J., Jones, J.: Measuring and Managing Information Risk: A FAIR Approach. Butterworth-Heinemann, Newton (2014)
9. Gregory, P.H.: All in One - CISA - Certified Information Systems Auditor - Exam Guide. McGraw-Hill Companies, New York (2012)
10. Hilden, A.E.: UDP-Based DDoS Amplification Attacks. Norwegian Security Authority (NSM). Lecture held at NTNU (Gjøvik), 7 October 2015
11. Kahneman, D.: Thinking, Fast and Slow. Macmillan, New York (2011)
12. Kaplan, S., Garrick, B.J.: On the quantitative definition of risk. Risk Anal. **1**(1), 11–27 (1981)
13. Pipkin, D.L.: Halting the Hacker: A Practical Guide to Computer Security, 2nd edn. Pearson Education, New York (2003)
14. Shalaginov, A., Franke, K.: A new method of fuzzy patches construction in neuro-fuzzy for malware detection. In: IFSA-EUSFLAT. Atlantis Press (2015)
15. Taleb, N.N.: Errors, robustness, and the fourth quadrant. Int. J. Forecast. **25**(4), 744–759 (2009)
16. Taleb, N.N., Swan, T.B.: The Impact of the Highly Improbable, 2nd edn. Random House LLC, New York (2010)
17. Wangen, G., Hallstensen, C., Snekkenes, E.: A framework for estimating information security risk assessment method completeness - core unified risk framework. Submitted for Review (2016)
18. Wangen, G., Shalaginov, A.: Quantitative risk, statistical methods and the four quadrants for information security. In: Lambrinoudakis, C., Gabillon, A. (eds.) CRiSIS 2015. LNCS, vol. 9572, pp. 127–143. Springer, Heidelberg (2016). doi:10.1007/978-3-319-31811-0_8
19. Wangen, G., Snekkenes, E.A.: A comparison between business process management and information security management. In: Paprzycki, M., Ganzha, M., Maciaszek, L. (ed.) Proceedings of the 2014 Federated Conference on Computer Science and Information Systems, vol. 2, pp. 901–910. IEEE (2014). Annals of Computer Science and Information Systems

Moving Target Defense Against Network Reconnaissance with Software Defined Networking

Li Wang and Dinghao Wu$^{(\boxtimes)}$

College of Information Sciences and Technology,
The Pennsylvania State University, University Park, PA 16802, USA
{lzw158,dwu}@ist.psu.edu

Abstract. Online hosts and networks are easy targets of network attacks due to their static nature, which creates an information asymmetry and makes them easy to attack and hard to defend. To break the asymmetry, Moving Target Defense was proposed to bring uncertainties to computer systems. It can be applied to all levels of protections, covering applications, system software, operating systems, and networks. In this paper, we present, Sniffer Reflector, a new method to practice Moving Target Defense against network reconnaissance, which is usually considered as the very first step of most attacks. Sniffer Reflector employs Software-Defined Networking to disturb network reconnaissance. We use virtualization to provide an obfuscated reconnaissance result for attackers. Our method can be easily combined with existing security tools for network forensics as well. We have developed a prototype in a virtual local area network. Our experiment results show that Sniffer Reflector is effective and efficient in blurring various network reconnaissance.

Keywords: Network reconnaissance · Network reflector · Software-Defined Networking · Moving Target Defense · Shadow Networks

1 Introduction

Online hosts and networks are easy targets of various attacks due to their static nature. Under the current Internet architecture, it is not easy for networked computer systems to change their network parameters once being established. Most networked services are deployed with a set of well-devised computing infrastructure and serve in a stable network environment. For example, typically, a web server open to public visit will be deployed with a fixed domain name and connected to a physical network device, router or switch, and assigned a public IP address locatable on the Internet. Once deployed, the web server's network parameters will not be changed frequently. This is a good practice because users can easily get online service through the server's domain name or IP address. However, this also exposes valuable network information to attackers for malicious use. Theoretically, attackers have unlimited time to study the server's network

© Springer International Publishing Switzerland 2016
M. Bishop and A.C.A. Nascimento (Eds.): ISC 2016, LNCS 9866, pp. 203–217, 2016.
DOI: 10.1007/978-3-319-45871-7_13

environment and find out a method to finally take over it. Although existing security tools, like firewalls and intrusion detection systems, can prevent most common attacks, these protections are essentially static and cannot change the static nature of the online servers. The static nature of online hosts and networks leads to an asymmetry between the attackers and defenders, and the attack and defense game is always unfair.

To change the attack and defense game, Moving Target Defense (MTD) was proposed to break the asymmetry between the attacker and defender. By introducing uncertainties and diversifications, MTD provides a dynamic defense environment for adversaries. Adversaries are forced to reprobe, reassess and restudy the protected environment. For example, an MTD strategy can periodically change a part of an operating system that makes it harder for attackers to find a known vulnerability for a specific OS. With MTD, it is hard for adversaries to decide and verify the authenticity of the obtained information. Therefore, adversaries can hardly start effective attacks in an MTD protected environment. Inspired by the promising defense philosophy of MTD, a lot of research has been proposed [8] to seek adaptive, dynamic, and practical security solutions for modern computer systems.

Software-Defined Networking (SDN) is a rising network technology which offers sufficient control flexibilities for users to modify network behaviors on the fly. Consider the fact that more and more critical services are moved from offline to online, network naturally becomes the top attack vector in most attack scenarios. Attackers have to make use of network infrastructures to achieve their attack goals. Typically, there are four attack phases for an attack over network [13]: network reconnaissance[1], targeting vulnerable machines, finding exploits, and conducting effective attacks. Each phase represents an attack step. For example, network reconnaissance is used to collect effective information in a target network, like how many nodes are alive, what are the versions, and so on. It provides valuable information for an attacker to conduct the rest attack steps. Existing research results show that, port scan should be regarded as the initial step of the cyber attack routine [2,6], and more than 70 % of the network scans [16] are connected with attack activities. We manage to use MTD to mitigate network reconnaissance with the SDN technologies.

Scan is a special network activity, which can be used both for protectors and attackers. Protectors, like security engineers and network administrators, use network scans to assess the security status of a target network. In most cases, scan activity is regarded highly dangerous in real networks. In our work, unless with a special mention, we recognize network scan as an attack activity.

In this paper, we present Sniffer Reflector, a new MTD method against network scans with SDN. Different from traditional protection mechanisms, Sniffer Reflector does not block or drop network scan traffic. Instead, it reflects scan traffic to a shadow network where scan replies are generated and obfuscated. Shadow network can simulate arbitrary network structures and services with an

---

[1] Technically, the terms *network reconnaissance* and *network scan* are exchangeable when describing network probe activities. We use them equally in this paper.

acceptable overhead. Therefore, attackers can only obtain obfuscated network views from a shadow network instead of a target network. With our method, attackers can no longer collect effective network information through network reconnaissance. Consequently, it is hard for attackers to continue the rest three attack phases and finish desired attacks.

In summary, the main contributions of Sniffer Reflector are as follows:

- *Provide a new MTD method against network reconnaissance.* As far as we know, it is the first work that employs MTD on network scan. We try to prevent attackers from collecting effective network information through network scan. Consider attackers rely on scan responses to collect vulnerability information, our method fail attackers by obfuscating the returned scan responses.
- *Obfuscate attackers' view with shadow network.* We use shadow network to provide forged responses to scan traffic. Shadow network can establish any desired network environment to obfuscate attackers' view, which is an invisible and isolated environment, and implemented with small overhead through the virtualization technologies.
- *Achieve stealthy protection.* Sniffer Reflector provides "stealthy" protection against network scan. Here, "stealthy" means an attacker cannot detect the fact that the responses he obtained are from a shadow network instead of the target network. Our method finishes at link layer and does not give any hints about its existence. As a result, Sniffer Reflector is invisible to the network layer and above.

Sniffer Reflector can be easily combined with other security tools for network forensic purposes. We have developed a Sniffer Reflector prototype by using virtualization technologies. It mainly has three components, Scan Sensor, Reflector and Shadow Network, and the three components cooperate together to protect a target network. Our implementation can be easily deployed in real productive network environments. The experimental results show that Sniffer Reflector is effective and efficient in defending various network scans. We tested our prototype in a local area network. The experiment results show attackers can only receive obfuscated scan responses from the shadow network.

2 Background

2.1 Moving Target Defense

For a long time, the cyber defenses are mainly static. Security analysts follow the conventional process to deploy protections in a productive network, which includes accessing information properties, planning defense strategies, deploying defense technologies and conducting penetration tests. Once the defense is established, it will keep running statically as deployed for a long time. Consequently, adversaries can take time to systematically study the network environment, plan malicious activities, find out a break point and conquer the protected system finally.

Moving Target Defense tries to increase attack bar for adversaries by introducing uncertainties and diversifications to computer systems. It forces adversaries to reprobe, reassess and restudy the target systems. Providing dynamic defense makes MTD a promising research topic in academia. The existing MTD researches can be categorized at two levels: system level and network level. At system level [5], researchers tried MTD on operating system, processor architecture, program runtime environment, application source code and binary code, and so on. At network level [1,21], MTD was practiced with frequent IP address reshuffling, network port remapping, network configurations adaptation, network topology mutation, dynamic changes on routing information and IP hopping. More MTD research details will be discussed in Sect. 6.

2.2   Network Scan

Network scan is composed of a set of network activities systematically collecting network information from a target network. Based on different purposes, network scan can be divided into three phases, which are illustrated as follows:

Host Detection. The first phase of network scan is host detection. Host detection tries to determine the accessible hosts and their IP addresses in an unknown network. The most used method for host detection is sending an ICMP echo request. If the remote host is alive and the request arrives unblocked, an ICMP echo reply will be answered. When the ICMP reply is received, we know the remote host is on.

Port Discovery. Port discovery is the second phase of a network scan, which puts efforts on searching all open network ports of a live host. In general, there are two scans to discover open ports of a host, UDP scan and TCP scan. The most dominant one is TCP scan because most valuable services are implemented in TCP protocol. In this paper, we mainly introduce TCP scan. There are several ways to perform TCP scan: (1) *Full TCP Scan.* scanning host tries to finish the classic TCP *three-way handshake* and establish a full TCP connection with the target host; (2) *Half-Open TCP Scan.* Scanning host sends a SYN segment to a selected port on the target host. If a SYN/ACK reply is received, scanning host knows the selected port is open and the target host is on; if a RST reply is received, that means the port is closed. It does not establish a complete TCP connection; (3) *Stealth Scan.* The word "illegal"means the standard TCP three-way handshake does not consist any of these segments. Attackers forge illegal TCP segments (mainly on control bits) to start a *Stealth Scan* [20], which includes FIN scan, Xmas scan, NULL scan and so on.

Vulnerability Assessment. Vulnerability assessment is the last phase of network scan. After identifying the live hosts and open ports, an attacker needs to know further details about a target system, such as OS version, service version, and configuration, to make further attack strategy. Due to most security vulnerabilities are OS dependent, it is necessary to fingerprint the OS information. Also, the protocol and service versions are valuable to attackers as well, for choosing attack exploits. By figuring out the specific OS and protocol information, the attacker can find out corresponding vulnerabilities and start effective attacks.

2.3   Software-Defined Networking

Software-Defined Networking (SDN) is a hot topic both in academia and industry. The key innovation of SDN is changing the static nature of network devices and making them programmable [11]. SDN separates the control function and forwarding function of a network device, which provides network users abilities to change network behaviors dynamically. Prior to SDN, network devices are designed like "dead" boxes. Each network device will be installed with the same chips and firmware in the factory. Once delivered, network devices can only work in a predefined way. If users want to make some changes to their networks, they have to start over and rebuild the physical network environment, which proves a time-consuming task. With SDN, users can easily modify network packets, change traffic flow directions, form a new network topology, and so on.

Inspired by the defense philosophy of MTD, we consider putting MTD practice into network reconnaissance protection. Since SDN provides the power to modify network behaviors during running, we try to make use of SDN's flexibilities to modify scan traffic flow and provide obfuscated scan replies. If network scan is obfuscated, the following attack steps will not succeed. We believe our method can greatly raise the attack bar for attackers.

3   System Architecture

In this section, we present Sniffer Reflector architecture, which seeks to employ SDN technologies and shadow networks to provide forged scan responses to attackers. Figure 1 shows the Sniffer Reflector architecture.

As can be seen in Fig. 1, Sniffer Reflector has three main components, *Scan Sensor*, *Reflector*, *Shadow Network*. The wholly protected network environment is called a protection domain. Each protection domain will have one or more protected objects. Each object is connected at least one Reflector, and each Reflector has one or more protected objects connected. In the figure, we have two protected objects in a protection domain, one target host and one target network. The green node stands for a Scan Sensor, which is responsible for monitoring network traffic. The red node represents an attacker node and sends out scan traffic, which is represented by a red curved arrow. The dash line in red shows the scan traffic is visible to the Scan Sensor when going through the Reflector. If scan traffic is detected, the Scan Sensor will send out a reflection message to the Reflector. After receiving the reflection message, the Reflector will manipulate the scan traffic and reflect it to a Shadow Network, where scan responses will be generated. The working process for the Sniffer Reflector architecture could

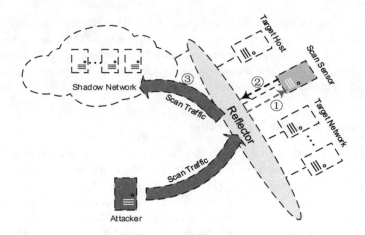

Fig. 1. The Sniffer Reflector architecture (Color figure online)

be illustrated as following steps: (1) Scan Sensor will be monitoring the coming traffic for target hosts and target networks. (2) Once scan traffic is detected, Scan Sensor will send a reflection message to the corresponding Reflector, and the Reflector will follow the message and redirect the scan traffic to Shadow Network. (3) Shadow Network receives redirected traffic from Reflector and disturbs attackers' view by making obfuscated scan responses.

3.1   Scan Sensor

Scan Sensor is a scan detection engine in our architecture, which is responsible for detecting scan activities and generating reflection messages. To perceive scan activities, it is required to observe all the traffic happening in a protection domain. Once scan activities are detected, Scan Sensor will notify Reflector, and Reflector and Shadow Network will cooperate together to reflect scan traffic. Scan Sensor finishes two functions: (1) Detecting scan traffic. Scan Sensor monitor the network traffic and detect the scan activities. It uses scan detection algorithms to find out possible scan traffic. (2) Generating reflection message. Once scan activities are confirmed, Scan Sensor needs to generate a reflection message to notify Reflector. The message contains scan type, scan source, scan target, port info, sensitive level, and action. An example message could be like this: *(TCP Xmas scan, source IP 192.168.2.12, target IP 192.168.1.105, target port 80, high density, reflection flag)*. Reflector will receive the message and execute the reflection action.

3.2   Reflector

Reflector is in charging of reflecting scan traffic. It communicates with the Scan Sensor and the Shadow Network, and executes reflection actions to redirect the scan traffic to the Shadow Network. In a protection domain, each Reflector will have at lease one protected host or protected network connected. For these connected objects, Reflector should be invisible to them. When there is no scan activity, Reflector acts as a regular network device and provides traffic flow functions to the connected nodes. It maintains the packets switching and traffic management like most network devices do. Consider the scan types may cover ICMP, TCP, UDP and other protocols, when reflecting, Reflector should be able to conduct fine-grained traffic control to identify scan traffic by protocol. Besides, to reflect scan traffic, the Reflector needs to provide at least two routes to change the flow of scan traffic. One route is for normal traffic, which leads to the target host; the other route is for scan traffic, which goes to the Shadow Network.

3.3 Shadow Network

All the scan traffic will be redirected to a Shadow Network. The Shadow Network is an isolated and invisible network, which is composed of shadow nodes. Except for the Reflector, no node in the protection domain is aware of the existence of the Shadow Network. Unless receiving scan traffic from the Reflector, the Shadow Network does not create any network traffic to the protection domain. A protection domain may have multiple Shadow Networks. And, these Shadow Networks can cooperate together to simulate a complex network structure. A typical strategy could be to simulate a replica of the protection domain. The forged network environment has the same look of the protection domain, such as the same number of nodes, the same network topology, and configurations. It can be used to cheat attackers to believe the reached network is the target network. Shadow Network can provide further responses if attackers keep attacking.

4   Design and Implementation

4.1   Design Principles

We have three design principles. First, all three components of Sniffer Reflector should be trusted and not disturbed by attackers. Moreover, the communications within Sniffer Reflector should be invisible to attackers and target nodes. Second, in the protection domain, no network traffic can escape from being monitored, and Scan Sensor can observe any communication happened in the protection domain. Last, once being detected and reflected, attackers cannot bypass the reflection mechanisms of Sniffer Reflector.

4.2   Prototype Implementation

Based on our three design principles, we implemented a prototype of Sniffer Reflector to provide scan protections in a virtual network. The prototype runs a protection domain which consists of two parts, a target network and a Sniffer Reflector framework. As shown in Fig. 2, the target network has two virtual

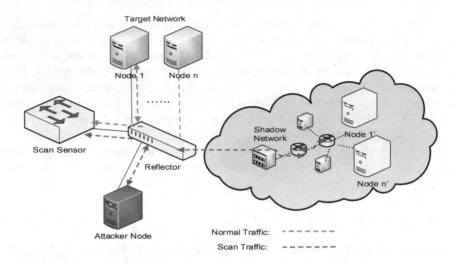

Fig. 2. A prototype of Sniffer Reflector

machine nodes and both of them are connected to Reflector. Accordingly, the Sniffer Reflector framework has three components, just as we discussed in Sect. 3, Scan Sensor, Reflector and Shadow Network.

We choose to implement the prototype on a virtualization platform, for more and more organizations are transplanting their services to cloud and data center. We employ the Kernel Virtual Machine (KVM) [10] virtualization platform. On KVM platform, 80 % of instructions on guest machine can be directly executed on physical CPU, which provides a high efficiency of resource utilization. We customized a virtual switch, VDE_Switch, to run as Reflector. Both KVM guest machines and VDE_Switch are running as user processes at host OS. We illustrate the three components' implementation details as follows.

We modify Snort to implement a Scan Sensor. There are three working modes in Snort, sniffer mode, packet logger mode and NIDS mode. To fit our purpose, we take advantage of NIDS mode. Under NIDS mode, Snort allows security engineers install a set of security rules to detect suspicious events over network. When a suspicious event is detected, an alert information will be generated and written to the log file. We located the source code where the alert information is generated and inserted an additional module. The module is responsible for looking scan alert patterns and generating reflection messages. The message includes scan type, scan source, scan target, port info, sensitivity level and action. Moreover, we developed a communication module sending reflection messages to Reflector. More details about message content and scan detection policies will be given in Sect. 5.

The implementation of Reflector is based on a virtual switch, VDE_Switch [3]. The basic functionalities of VDE_Switch are receiving, processing and forwarding network packets for the connected nodes. Each connected node will be assigned a dedicated port number for packets switching purpose. We modified VDE_Switch and inserted it an extra layer for packet checking function. When receiving reflection messages, VDE_Switch will translate the messages to reflection rules. Reflection rules contain the characteristics of scan traffic, such as scan source IP and protocol information. All the reflection rules will be stored in a *Refl_Rules_List*. All the packets going through VDE_Switch will be checked by the list. If any traffic got a match, it will be reflected. The modified VDE_Switch is capable of verifying packet headers from link layer to transport layer, which enables VDE_Switch to identify traffic flows by protocol. That helps us achieve a fine-grained traffic control. To implement secret channels for communications happened within Sniffer Reflector, we designed two reserved port numbers in VDE_Switch. These two reserved port numbers are dedicated to Scan Sensor and Decoy Network. For each reserved port, we set a switch flag. When there is no scan traffic, the switch flag will be set off and the reserved port is blocked; when there is scan traffic detected, the switch flag will turn on and scan traffic will be reflected through reserved ports. The detailed reflection decision is demonstrated in Algorithm 1.

Algorithm 1. Scan traffic reflection decision

Require: The current packet p; The switch flag sf; The set of reflection rules: *Refl_Rules_List*; The port used by Shadow Network: SN_port;
1: **for** every packet p **do**
2:   **if** $((p.source\_IP, p.protocol) \in [Refl\_Rules\_List])$ &&(sf is on) **then** send p to SN_port;
3:   **end if**
4: **end for**

Shadow Network is implemented with virtualization technologies as well. We use KVM virtual machines to simulate nodes of Shadow Network. These virtual machines are connected with virtual network devices to construct a network structure. On each virtual machine, we deploy the corresponding OS and network services, which are intended to provide obfuscated scan responses to attackers. The virtual machines are in full control and can be configured with arbitrary network parameters to simulate desired network behaviors. The reason why we use real virtual machines instead of virtual honeypots/honeynets in Shadow Network is virtual machine can provide a fully responsible OS and TCP/IP stack for attackers. Some light-weight virtual honeynets, like honeyd [17], can be used to simulate a network, but their responses can be easily detected by attackers.

5   Evaluation

We evaluate our prototype in a virtual LAN 192.168.1.0/24. The entire virtual network facility is deployed on a physical machine, and all other network nodes and devices are running as virtual machines. Our testbed was deployed on Intel Core i7-3370 3.4 Ghz processor with 16 GB RAM. The host machine is running

```
alert tcp any any -> $HOME_NET any (msg:"TCP SYN"; flow:stateless; flags:S;
detection_filter:track by_dst, count 100, seconds 5; sid:1000001;rev:1)

alert tcp $EXTERNAL_NET any -> $HOME_NET any (msg:"SCAN NULL"; flow:stateless;
ack:0; flags:0; seq:0; reference:arachnids,4; classtype:attempted-recon; sid:623; rev:6;)

alert tcp $EXTERNAL_NET any -> $HOME_NET any (msg:"SCAN SYN FIN";
flow:stateless; flags:SF,12; reference:arachnids,198; classtype:attempted-recon; sid:624; rev:7;)

alert tcp $EXTERNAL_NET any -> $HOME_NET any (msg:"SCAN XMAS"; flow:stateless;
flags:SRAFPU,12; reference:arachnids,144; classtype:attempted-recon; sid:625; rev:7;)

alert tcp $EXTERNAL_NET any -> $HOME_NET any (msg:"SCAN nmap XMAS";
flow:stateless; flags:FPU,12; reference:arachnids,30; classtype:attempted-recon; sid:1228;
rev:7;)
```

Fig. 3. Scan detection rules in Snort

CentOS 6.0 with kernel 2.6.32 x86_64 and qemu-kvm-0.15.1. Similarly, all other guest OSes are running CentOS Linux as well. The Scan Sensor virtual machine is running Snort 2.9.7.5 for scan detection. The Shadow Network is deployed with two virtual machines as shadow nodes to provide forged scan responses. The Reflector employs VDE_Switch 2.3.2.

As can be seen in Fig. 2, we have five nodes in prototype. The host is running with 192.168.1.107. Two other nodes, 192.168.1.153 and 198.168.1.154, run Linux OS as normal nodes of the protection domain. Node 192.168.1.153 is running with open ports 22, 23, 80, 111 and 443; node 192.168.1.154 is running with open ports 22, 80 and 111. Shadow Network has two shadow nodes, which both run Windows XP systems with the same IP addresses as normal nodes, 192.168.1.153 and 192.168.1.154. Both two shadow nodes are configured to open ports, 53, 135, 139, 445 and 3389. An attacker node runs with IP 192.168.2.1 in subnet 192.168.2.0/24. The attacker node is used to send scan traffic and collect scan results. Besides, Snort and VDE_Switch are running as user-space processes on host OS.

Scan detection rules were configured to detect TCP SYN flood and TCP stealth scans on Snort. The content of rules is shown in Fig. 3. The first rule detects TCP SYN flood scan, which is defined as X TCP SYN requests in Y time period. In our evaluation, we define TCP SYN flood as any TCP connection requests sent more than 100 times in 5 s (this assumption can be modified to accommodate different detection scenarios). Then, the following detection rules give the details of TCP stealth scans, including FIN scan, NULL scan, and XMAS scan.

We employ nmap to generate scan traffic. To demonstrate the effectiveness of our protection, we design three types of scans, SYN scan, Xmas scan and version detection scan. See the nmap commands as follows:

1. nmap -sS 192.168.1.0/24
2. nmap -sX 192.168.1.153/154
3. nmap -A -T4 -F 192.168.1.153/154

The first nmap command executes a TCP SYN scan for LAN 192.168.1.0/24. TCP SYN scan probes remote ports through a full TCP three-way handshake. Since TCP SYN scan is the most typical scan over the internet, most system logs will default capture TCP SYN scan. Then, the second scan command starts a TCP Xmas scan on two hosts 192.168.1.153/154. TCP Xmas scan is one of the TCP stealth scans, which is usually blocked by most firewalls. The third scan in our evaluation is a composite scan, which contains a series of scan activities. The -A option means the scan command will run OS detection, version detection, script scanning, and traceroute. nmap will match the scan results with its fingerprint database and estimate the OSes and version information of the scanned hosts. We execute the scan commands in two rounds. The first scan

Table 1. Scan results comparisons on 192.168.1.0/24

	Without Sniffer Reflector Protection	With Sniffer Reflector Protection
TCP SYN scan nmap -sS 192,168.1.0/24	192.168.1.107 is up	192.168.1.107 is up
	0.00015 s latency	0.00032 s latency
	Not shown: 998 closed ports	Not shown: 998 closed ports
	Port State Service	Port State Service
	22/tcp open ssh	22/tcp open ssh
	111/tcp open rpcbind	111/tcp open rpcbind
	192.168.1.153 is up	192.168.1.153 is up
	0.00029 s latency	0.015 s latency
	Not shown: 995 closed ports	Not shown: 995 closed ports
	Port State Service	Port State Service
	22/tcp open ssh	53/tcp open domain
	23/tcp open telnet	135/tcp open msrpc
	80/tcp open http	139/tcp open netbios-ssn
	111/tcp open rpcbind	445/tcp open microsoft-ds
	443/tcp open https	3389/tcp filtered ms-term-serv
	192.168.1.154 is up	192.168.1.154 is up
	0.00023s latency	0.015s latency
	Not shown: 997 closed ports	Not shown: 995 closed ports
	Port State Service	Port State Service
	22/tcp open ssh	53/tcp open domain
	80/tcp open http	135/tcp open msrpc
	111/tcp open rpcbind	139/tcp open netbios-ssn
		445/tcp open microsoft-ds
		3389/tcp filtered ms-term-serv

round is executed without the protection of Sniffer Reflector; the second round is scanned with the protection of Sniffer Reflector. Then, we compared the scan results, which show our prototype is effective and efficient in defending various network scans.

Table 1 shows the scan results on subnetwork 192.168.1.0/24. The first column is the nmap scan command. The second and third columns show the scan results without and with the protection of Sniffer Reflector. As can be seen from the second column, scan results return all the live nodes in 192.168.1.0/24. For each live node, the results show scan latency, how many ports are closed, and a list of open port/state/service information. When there is no Sniffer Reflector, the scanner is able to receive network information from the protection domain. From the returned IP and open ports information, we can see the attacker can easily collect network information from the target network. The third column lists the scan results under the protection of Sniffer Reflector. The listed open ports match with the virtual machines we configured as shadow nodes, which shows the scanner actually obtains the network information from shadow network. Comparing the scan results of two columns, we can tell there is a difference in scan latency. With Sniffer Reflector, the average latency of simulated decoy service is 0.015 s, which is higher than the latency of real nodes (0.00015 s). The delay is probably because the reflection is implemented in software, and VDE_Switch needs to reflect scan traffic packet one by one.

Table 2 illustrates the TCP Xmas scan results on two hosts 192.168.1.153/154. The nmap scan command is in column one, and the rest columns demonstrate the scan results received with/without the protection of Sniffer Reflector. Different from SYN scan, the port state information in Xmas scan results shows as "o/f", representing "open or filed". The scan results show, when Sniffer Reflector is on, both two scans are returning simulated results from shadow nodes. That presents, the scan traffic is reflected and the scan results come from shadow network. And, we also got the scan results from the nmap composite scan. The results show that, with the protection of Sniffer Reflector, the composite scan can also be detected and reflected in our framework.

Table 2. Scan result comparisons on a single host

	Without Sniffer Reflector Protection		With Sniffer Reflector Protection	
	for 153	for 154	for 153	for 154
TCP Xmas scan	192.168.1.153 is up 0.00029s latency 995 closed ports Port State Service 22/tcp o/f ssh	192.168.1.154 is up 0.00028s latency 997 closed ports Port State Service 22/tcp o/f ssh	192.168.1.153 is up 0.015s latency 999 closed ports Port State Service 3389/tcp o/f m-t-s	192.168.1.154 is up 0.0020 latency 999 closed ports Port State Service 3389/tcp o/f m-t-s
nmap -sX 192.168.1.X	23/tcp o/f telnet 80/tcp o/f http 111/tcp o/f rpcbin 443/tcp o/f https scanned in 14.26 sec	80/tcp o/f http 111/tcp o/f rpcbin scanned in 14.25 sec	scanned in 14.28 sec	scanned in 14.25 sec

6 Related Work

MTD researches are quite popular in recent years. Many MTD methods have been proposed to mitigate security threats. The existing researches on MTD can be divided into two levels, system level and network level. The network level MTD solutions [1,4] include IP address reshuffling, network configuration randomization, and so on. These methods all tried to obfuscate attackers at network level. The system level MTD methods cover platform [15], runtime environment [19], and software applications [18]. Kewley et al. [9] proposed to reduce network attacks using dynamic network reconfigurations. They tried to force attackers to use the outdated network configurations to prepare an attack. A live IPv6 version MTD was implemented by Groat et al. [7]. By using DHCPv6 protocol, a hidden connection is built between IPv6 address and DHCP identity, which could be used to protect sensitive communications in government or confidential organizations. Unlike our method, these methods are still in the phase of prototype and far from mature for deployment.

Our work is also motivated by SDN researches. OpenFlow [14] is the most widely accepted SDN protocol. It is designed to dynamically change network behaviors by using controllers and switches. Controllers are responsible for managing the attached switches and deciding the flow tables of switches, and switches are using flow tables to forward network traffic. Lara et al. [12] provided a survey on innovations of OpenFlow, which has been used in network management, traffic analysis, fault tolerance, security, and many other areas. SnortFlow [22] is proposed to integrate intrusion prevention systems (IPS) with OpenFlow on a cloud. The basic idea is using the IPS alerts to change cloud behaviors. However, due to the heavy structure of cloud, It is impractical to provide timely reactions on security events. In comparison, Sniffer Reflector provides instant changes on reflection actions and can be deployed with no difficulties in a real network.

7 Conclusion

The relatively static nature of today's computing and network systems has led to an information asymmetry between the attackers and defenders. To break this asymmetry, in this paper, we present Sniffer Reflector, a new Moving Target Defense method against network reconnaissance with Software-Defined Networking, to obfuscate the attackers' view of the target network information. We have designed and implemented a prototype in a virtual local area network. Our experimental results with various scan activities show that Sniffer Reflector is effective and efficient.

Acknowledgements. This work was supported in part by the NSF Grants CCF-1320605 and CNS-1223710, and ONR Grants N00014-13-1-0175 and N00014-16-1-2265.

References

1. Al-Shaer, E.: Toward network configuration randomization for moving target defense. In: Jajodia, S., Ghosh, A.K., Swarup, V., Wang, C., Wang, X.S. (eds.) Moving Target Defense, vol. 54, pp. 153–159. Springer, New York (2011)
2. Allman, M., Paxson, V., Terrell, J.: A brief history of scanning. In: Proceedings of the 7th ACM SIGCOMM Conference on Internet Measurement, pp. 77–82 (2007)
3. Davoli, R.: VDE: virtual distributed ethernet. In: 1st International Conference on Testbeds and Research Infrastructures for the Development of Networks and Communities, pp. 213–220. IEEE (2005)
4. Dunlop, M., Groat, S., Urbanski, W., Marchany, R., Tront, J.: MT6D: A moving target IPv6 defense. In: Military Communications Conference. IEEE (2011)
5. Evans, D., Nguyen-Tuong, A., Knight, J.: Effectiveness of moving target defenses. In: Jajodia, S., Ghosh, A.K., Swarup, V., Wang, C., Wang, X.S. (eds.) Moving Target Defense, vol. 54, pp. 29–48. Springer, New York (2011)
6. Gadge, J., Patil, A.A.: Port scan detection. In: 16th IEEE International Conference on Networks, ICON 2008, pp. 1–6. IEEE (2008)
7. Groat, S., Dunlop, M., Urbanksi, W., Marchany, R., Tront, J.: Using an IPv6 moving target defense to protect the smart grid. In: 2012 IEEE PES Innovative Smart Grid Technologies (ISGT), pp. 1–7, January 2012
8. Jajodia, S., Ghosh, A.K., Swarup, V., Wang, C., Wang, X.S. (eds.): Moving Target Defense - Creating Asymmetric Uncertainty for Cyber Threats. Advances in Information Security, vol. 54. Springer, New York (2011)
9. Kewley, D., Fink, R., Lowry, J., Dean, M.: Dynamic approaches to thwart adversary intelligence gathering. In: Proceedings of DARPA Information Survivability Conference and Exposition II, pp. 176–185 (2001)
10. Kivity, A., Kamay, Y., Laor, D., Lublin, U., Liguori, A.: KVM: the Linux virtual machine monitor. Proc. Linux Symp. **1**, 225–230 (2007)
11. Lantz, B., Heller, B., McKeown, N.: A network in a laptop: rapid prototyping for software-defined networks. In: Proceedings of the 9th ACM SIGCOMM Workshop on Hot Topics in Networks, p. 19 (2010)
12. Lara, A., Kolasani, A., Ramamurthy, B.: Network innovation using openflow: a survey. IEEE Commun. Surv. Tutorials **16**, 493–512 (2014)
13. Liao, H.J., Lin, C.H.R., Lin, Y.C.: Intrusion detection system: a comprehensive review. J. Netw. Comput. Appl. **36**, 16–24 (2013)
14. McKeown, N., Anderson, T., Balakrishnan, H., Parulkar, G., Peterson, L., Rexford, J., Shenker, S., Turner, J.: Openflow: enabling innovation in campus networks. ACM SIGCOMM Comput. Commun. Rev. **38**(2), 69–74 (2008)
15. Okhravi, H., Comella, A., Robinson, E., Haines, J.: Creating a cyber moving target for critical infrastructure applications using platform diversity. Int. J. Crit. Infrastruct. Prot. **5**(1), 30–39 (2012)
16. Panjwani, S., Tan, S., Jarrin, K.M., Cukier, M.: An experimental evaluation to determine if port scans are precursors to an attack. In: Proceedings on International Conference on Dependable Systems and Networks, DSN 2005, pp. 602–611. IEEE (2005)
17. Provos, N.: Honeyd-a virtual honeypot Daemon. In: 10th DFN-CERT Workshop, Hamburg, Germany, vol. 2, p. 4 (2003)
18. Rinard, M.: Manipulating program functionality to eliminate security vulnerabilities. In: Jajodia, S., Ghosh, A.K., Swarup, V., Wang, C., Wang, X.S. (eds.) Moving Target Defense, vol. 54, pp. 105–115. Springer, New York (2011)

19. Shacham, H., Page, M., Pfaff, B., Goh, E.J., Modadugu, N., Boneh, D.: On the effectiveness of address-space randomization. In: Proceedings of the 11th ACM Conference on Computer and Communications Security, pp. 298–307 (2004)
20. Staniford, S., Hoagland, J.A., McAlerney, J.M.: Practical automated detection of stealthy portscans. J. Comput. Secur. **10**(1/2), 105–136 (2002)
21. Wang, H., Jia, Q., Fleck, D., Powell, W., Li, F., Stavrou, A.: A moving target DDoS defense mechanism. Comput. Commun. **46**, 10–21 (2014)
22. Xing, T., Huang, D., Xu, L., Chung, C.J., Khatkar, P.: Snortflow: a openflow-based intrusion prevention system in cloud environment. In: Research and Educational Experiment Workshop (GREE), Second GENI, pp. 89–92. IEEE (2013)

Uni-ARBAC: A Unified Administrative Model for Role-Based Access Control

Prosunjit Biswas, Ravi Sandhu$^{(\boxtimes)}$, and Ram Krishnan

Institute for Cyber Security,
University of Texas at San Antonio, San Antonio, USA
prosun.csedu@gmail.com, {ravi.sandhu,ram.krishnan}@utsa.edu

Abstract. Many of the advantages of Role Based Access Control (RBAC) accrue from the flexibility of its administrative models. Over the past two decades, several administrative models have been proposed to manage user-role, permission-role and in some cases role-role relations. These models are based on different administrative principles and bring inherent advantages and disadvantages. In this paper, we present a unified model, named Uni-ARBAC, for administering user-role and permission-role relations by combining many of the administrative principles and novel concepts from prior models. For example, instead of administering individual permissions Uni-ARBAC combines permissions into tasks which are assigned to roles as a unit. Slightly differently, users are assigned to user-pools from where individual users are assigned to roles. The central concept of Uni-ARBAC is to integrate user-role and task-role administration into a more manageable unit called an Administrative Unit (AU). AUs partition roles, tasks and user-pools and they are organized in a rooted tree hierarchy. Administrative users are assigned to AUs with possibility of restricting their authority to user-role assignment or task-role assignment. While most existing models assume existence of administrative roles for managing regular roles, we present an approach for engineering AUs based on structured partitioning of roles and tasks.

1 Introduction

Role Based Access Control (RBAC) [6,17] is one of the most widely deployed and studied access control models. Instead of directly assigning permissions to users, RBAC assigns permissions to roles and users to roles. While the number of roles in a large organization might vary from dozens to thousands, the number of users or permissions could vary from tens of thousands to hundreds of thousands and even millions. Thus, maintaining the user-role and permission-role relations are the most commonly carried out administrative actions in RBAC. While some models also speak to administering the role-role hierarchy [5,15], it is evident that modifications to the role-role relationship can have significant impact, so it might be advisable to keep this authority relatively centralized. Hence, we limit our scope in this paper to decentralized administration of the user-role and permission-role relations.

© Springer International Publishing Switzerland 2016
M. Bishop and A.C.A. Nascimento (Eds.): ISC 2016, LNCS 9866, pp. 218–230, 2016.
DOI: 10.1007/978-3-319-45871-7_14

To a large degree the advantages of RBAC accrue from the flexibility of administering the permission-role and user-role relations. In this regard, several administrative models have been proposed in the literature (see Sect. 2). These models are based on different administrative principles and offer inherent advantages and disadvantages. Each one incorporates some novel and putatively useful concepts relative to the others. To our knowledge, there has been no effort so far to comprehensively consolidate the various novel concepts introduced in different administrative models into a coherent unified model that potentially brings together the inherent advantages of the individual models.

In this paper, we present a novel unified model, named Uni-ARBAC, for administering user-role and permission-role relations by combining many of the existing administrative principles and novel concepts. For example, instead of administering individual permissions, Uni-ARBAC combines permissions into tasks and assigns tasks to roles. For administrative purposes, Uni-ARBAC decouples users and tasks from roles following the decoupling principle of ARBAC02 [13]. Uni-ARBAC utilizes user-pools as sets of candidate users who can be assigned to a role, while tasks act as permission-pools. One advantage of using tasks as permission-pools is that tasks can be designed during role engineering according to some top-down approaches (e.g. [11]). User-pools on the other hand can be designed via the organization structure.

Uni-ARBAC integrates user-role and task-role administration into a more manageable unit, we call Administrative Unit (AU). AUs partition roles, tasks and user-pools, and are organized in a rooted tree hierarchy. Administrative users are assigned to AUs with possibility of restricting their authority to user-role assignment or task-role assignment. The partitioning of roles and tasks across AUs leads us to propose an engineering process for AUs for given role hierarchy and/or task hierarchy. The potential for engineering AUs in this manner is a significant advantage of Uni-ARBAC.

This paper makes the following contributions. We have presented a unified model (Uni-ARBAC) for administering user-role and permission-role relation for RBAC. Uni-ARBAC combines several novel concepts and administrative principles from prior models into a more powerful and manageable unit called administrative unit (AU). We proposed an engineering approach for developing AUs. While most other administrative models assume the existence of separate administrative roles, we relax this assumption and our approach for engineering administrative units can also be used for engineering administrative roles.

The remainder of this paper is organized as follows. We discuss related work in Sect. 2 highlighting the concepts we have adopted from prior administrative models. We present our model in Sect. 3 and some variations of the model in Sect. 4. Section 5 discusses our approach for engineering AUs. We conclude the paper in Sect. 6.

2   Background and Related Work

In this section, we review prior models for administering RBAC, emphasizing those of their driving principles which have been incorporated in Uni-ARBAC. These concepts and principles are summarized in Table 1.

The value of grouping permissions into a higher level abstraction has often been recognized in the literature. Task-role based access control (TRBAC) [12] proposes the notion of a task as a group of permissions which constitute a fundamental unit of business work in an enterprise. Similar to TRBAC, two-sorted RBAC [9] and scenario-based role engineering [11] organize tasks into another higher level abstract.

One of the central notions of RBAC administration is to separate user-role and permission-role assignments. Introduced in ARBAC97 [15], this notion is adopted by many other models including [13,14,16]. Uni-ARBAC accepts this separation to be at the core of the model.

Another essential concept of ARBAC97 is to keep administration of roles separate from regular roles. To this end, ARBAC97 introduced the concept of administrative roles. Uni-ARBAC adopts the former separation principle, but eschews the use of administrative roles for this purpose. Instead, Uni-ARBAC introduces a more sophisticated construct of Administrative Units to achieve the desired separation.

Table 1. Concepts motivating Uni-ARBAC (* denotes source of the concept)

Concepts and principles	ARBAC97 [15]	ARBAC02 [13]	SARBAC [4,5]	UARBAC [10]	Role graph model [18]	Uni-ARBAC
Task and task hierarchy						✓
Separation of user and permission administration	✓ *	✓	✓		✓	✓
Separation of regular roles from administration	✓ *	✓			✓	✓
User pools and user pool hierarchy		✓ *				✓
Administrative structure design			✓ *		✓	✓
Reversibility and administrative structure flexibility				✓ *		✓
Senior most administrators					✓ *	✓

ARBAC02 [13] is another influential model for administrative RBAC. It documents a number of problems with ARBAC97 and introduces the notions of user-pools and permission-pools. Uni-ARBAC adopts the user-pool and user-pool hierarchy from ARBAC02, while on the permission side it adopts the task and task hierarchy from [12] as discussed above.

Crampton et al. developed a model called SARBAC [4,5] based on the concept of administrative scope, which confines the side effects of role hierarchy modification in a highly disciplined manner relative to ARBAC97. Notably, administrative scope becomes a means to define administrative roles which are otherwise assumed to be given in ARBAC97 and ARBAC02. Administrative scope is mathematically defined based on the given role hierarchy. Uni-ARBAC incorporates the general notion that the role hierarchy should influence the administrative structure. However, it departs from the strict mathematical definition of SARBAC to accommodate a heuristic top-down approach in designing administrative units, based on the role-hierarchy and task-role allocation.

The UARBAC [10] model proposes a number of principles for RBAC administration, such as scalability and flexibility, psychological acceptability and economy of mechanism. As noted earlier Uni-ARBAC departs from UARBAC on the question of whether or not administrative permissions should be assigned to regular roles. However, all the other principles of UARBAC are considered similarly desirable in Uni-ARBAC. As it stands some of the UARBAC principles, such as psychological acceptability and scalability, are qualitative and difficult to convincingly claim for a given model. Here we confine our attention to the two principles of reversibility and administrative structure flexibility, which have been explicitly adopted from UARBAC into Uni-ARBAC. The reversibility principle requires that administrative operations should be reversible. This is incorporated in Uni-ARBAC by coupling grant and revoke operations for user-role or task-role assignment in a single administrative unit. The principle of administrative structure flexibility (called policy neutrality in [10]) argues against the tight coupling of administrative structure to role hierarchy, such as in SARBAC.

It remains to consider the Role-Graph Administration Model [18]. It partitions roles into units called administrative domains. The model explicitly includes a single highest administrative domain which includes the MaxRole and MinRole from the underlying Role-Graph model. Uni-ARBAC adopts this concept embodying it in the highest administrative unit at the root of the administrative unit tree hierarchy.

In addition to the administrative models discussed above, there are other notable models developed in various applied contexts, especially in temporal/location aware RBAC [1,2], Enterprise RBAC [7,8]), event driven RBAC [3], administration of cryptographic RBAC [19] etc.

3   The Uni-ARBAC Model

In this section, we describe the Uni-ARBAC model, along with formal definitions. The overall structure of Uni-ARBAC is illustrated in Fig. 1. We consider Uni-ARBAC in two parts: the operational model for RBAC with respect to regular

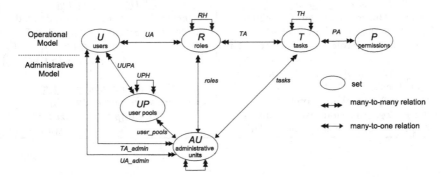

Fig. 1. The **Uni-ARBAC** model for user-role and task-role administration

roles and permissions, and the administrative model for administering the user-role and task-role relations of the former. These are respectively discussed in the following two subsections.

3.1 Uni-ARBAC Operational Model

The sets and relations in the top part of Fig. 1 represent the Uni-ARBAC operational model, which is slightly different from the standard RBAC model [6]. The most salient difference is that there is a level of indirection in role-permission assignment, so permissions are assigned to tasks and tasks are assigned as a unit to roles. As discussed in Sect. 2, this additional indirection has emerged in several different administrative models in the literature. Additionally tasks are organized in a partial order $\succeq_t$, whereby a senior task inherits all permissions from its juniors. For example, in Fig. 2(b), task t1 is senior to tasks t2 and t3, so it inherits permissions from both of them, and so on. User-role assignment remains unchanged from standard RBAC, so individual users are assigned to and deassigned from roles. For simplicity, we have not considered the standard RBAC concepts of sessions and role activation.

The Uni-ARBAC operational model is formalized in Table 2. Item I specifies the familiar components carried over from traditional RBAC. Item II specifies the additional components which effect the additional indirection between permissions and roles via tasks. Item III formalizes the interaction between the role hierarchy, task hierarchy, and permission-task and task-role assignments. The interaction is schematically depicted in Fig. 3 and formally expressed in the *authorized_perms* function. The authorization function in item IV specifies the authorization required for a user to exercise a permission, which is that the permission must be authorized to at least one role assigned to the user. A familiar role hierarchy from the literature and an example task hierarchy are shown in Figs. 2(a) and (b), respectively.

Table 2. Uni-ARBAC operational model

I. Traditional RBAC Sets & Relations	III. Derived Function
- U, R and P (users, roles and permission)	- $authorized\_perms(r : R) \rightarrow 2^P$, defined as
- $RH \subseteq R \times R$, partial order on roles $\geq$	$authorized\_perms(r) = \{p \mid (\exists t, t' \in T)$
- $UA \subseteq U \times R$, user-role assignment relation	$(\exists r' \in R)[r \geq r' \wedge t \succeq_t t' \wedge$
	$(t, r') \in TA \wedge (p, t') \in PA]\}$
II. Added RBAC Sets & Relations	
- T, set of tasks	**IV. Authorization Function**
- $TH \subseteq T \times T$, partial order on tasks $\succeq_t$	- $can\_exercise\_permission(u : U, p : P) =$
- $PA \subseteq P \times T$, permission-task assignment rel.	$(\exists r \in R)[p \in authorized\_perms(r)$
- $TA \subseteq T \times R$, task-role assignment relation	$\wedge (u, r) \in UA]$

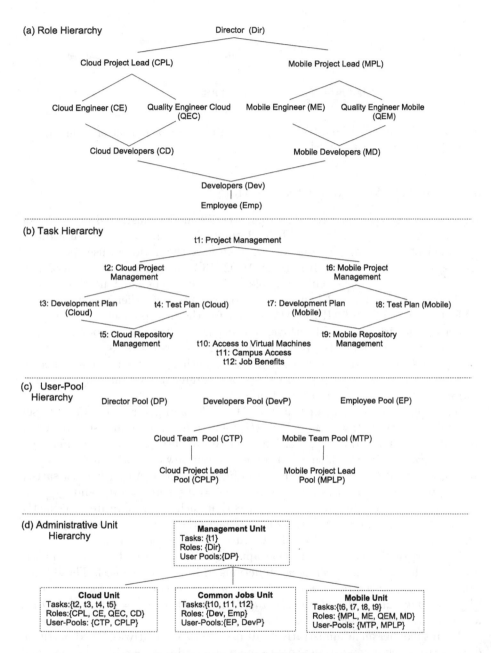

Fig. 2. Examples of Uni-ARBAC hierarchies

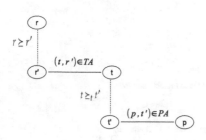

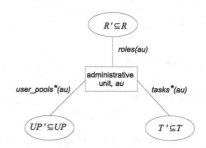

Fig. 3. Interaction of role and task hierarchies in operational model

Fig. 4. Scope of control of an AU

3.2 Uni-ARBAC Administrative Model

We now turn to the Uni-ARBAC administrative model illustrated in the lower part of Fig. 1, and formalized in Table 3. The administrative model introduces a number of additional components. First we have the notion of user-pools and user-pool hierarchy adopted from ARBAC02 [13]. An example user-pool hierarchy is shown in Fig. 2(c). This example has three independent user-pools DP, DevP and EP, with DevP being senior to a number of other user-pools, i.e., CTP, CPLP, MTP and MPLP. Motivation for the user-pool hierarchy in this instance is by virtue of qualifications, so every user in the CPLP pool is also eligible to be a developer in the DevP pool. The hierarchy obviates the need to do multiple assignments of a user to both pools in such cases. Users are assigned to user-pools via the $UUPA$ user to user-pool assignment relation. The user-pool notion is formally specified in item I of Table 3.

The central mechanism in Uni-ARBAC is the administrative unit. The set of administrative units is denoted as AU, while individual administrative units are indicated as au, au_i, au_j, etc. Uni-ARBAC requires that each au manages an exclusive set of roles which is not under the purview of another au. The $roles$ function in item II of Table 3 is a partitioning assignment in that it must satisfy the requirements that $roles(au_i) \cap roles(au_j) = \emptyset$ for $au_i \neq au_j$, and $\bigcup_{au \in AU} roles(au) = R$. The effect of partitioning is that each role is allocated to exactly one au for administration. Each au only manages the roles it is directly assigned. The effect of the role hierarchy is limited to the operational model.

The partitioning concept is further applied in Uni-ARBAC to tasks and user-pools via the $tasks$ and $user\_pools$ functions in item II of Table 3. These functions must satisfy the requirements $tasks(au_i) \cap tasks(au_j) = \emptyset$ for $au_i \neq au_j$, $\bigcup_{au \in AU} tasks(au) = T$, $user\_pools(au_i) \cap user\_pools(au_j) = \emptyset$ for $au_i \neq au_j$, and $\bigcup_{au \in AU} user\_pools(au) = UP$.

In this manner an administrative unit manages a explicitly assigned partition of roles, to which it can assign users from an assigned partition of user-pools and tasks from an assigned partition of tasks. Unlike for roles, Uni-ARBAC extends the authority of an au to junior tasks and user-pools, for which purpose we define the $tasks^*$ and $user\_pools^*$ functions in item III of Table 3. The net effect

Table 3. Uni-ARBAC administrative model

I. User-Pools Sets & Relations	*V. Authorization Functions*
- UP, set of user-pools - $UPH \subseteq UP \times UP$, partial order $\succeq_{up}$ - $UUPA \subseteq U \times UP$, user to user-pool assignment relation	- $can\_manage\_task\_role(u : U, t : T, r : R) =$ $(\exists au_i, au_j)[(u, au_i) \in TA\_admin \wedge$ $au_i \succeq_{au} au_j \wedge r \in roles(au_j) \wedge t \in tasks^*(au_j)]$ - $can\_manage\_user\_role(u_1 : U, u_2 : U, r : R) =$ $(\exists au_i, au_j)[(u_1, au_i) \in UA\_admin \wedge$
II. AU and Partitioned Assignments	$au_i \succeq_{au} au_j \wedge r \in roles(au_j) \wedge (\exists up \in UP)$ $[(u_2, up) \in UUPA \wedge up \in user\_pools^*(au_j)]$

- AU, set of administrative units
- $roles(au : AU) \rightarrow 2^R$, assignment of roles
- $tasks(au : AU) \rightarrow 2^T$, assignment of tasks
- $user\_pools(au : AU) \rightarrow 2^{UP}$,
      assignment of user-pools
	VI. Administrative Actions
	- $assign\_task\_to\_role(u : U, t : T, r : R)$
	 Authorization:
	  $can\_manage\_task\_role(u, t, r) = True$

III. Derived Functions
- $tasks^*(au : AU) \rightarrow 2^T$, defined as
 $tasks^*(au) = \{t' \mid (\exists t \in tasks(au))t \succeq_t t'\}$
- $user\_pools^*(au : AU) \rightarrow 2^{UP}$, and
 $user\_pools^*(au) = \{up' \mid (\exists up \in$
 $user\_pools(au))up \succeq_{up} up'\}$
	Effect: $TA' = TA \cup \{(t, r)\}$
	- $revoke\_task\_from\_role(u : U, t : T, r : R)$
	 Authorization:
	  $can\_manage\_task\_role(u, t, r) = True$
	 Effect: $TA' = TA \backslash \{(t, r)\}$
	- $assign\_user\_to\_role(u_1 : U, u_2 : U, r : R)$
	 Authorization:
	  $can\_manage\_user\_role(u_1, u_2, r) = True$
	 Effect: $UA' = UA \cup \{(u, r)\}$

IV. Administrative User Assignments
- $TA\_admin \subseteq U \times AU$
- $UA\_admin \subseteq U \times AU$
- $AUH \subseteq AU \times AU$, rooted tree
      partial order $\succeq_{au}$
	- $revoke\_user\_from\_role(u_1 : U, u_2 : U, r : R)$
	 Authorization:
	  $can\_manage\_user\_role(u_1, u_2, r) = True$
	 Effect: $UA' = UA \backslash \{(u, r)\}$

is illustrated in Fig. 4, and further discussed in context of item V of Table 3. An example partitioning of roles, tasks and user-pools across four administrative units is shown in Fig. 2(d). We also note that for a given au it is permissible to assign an empty partition of roles, tasks or user-pools. While, such situations may be unusual, the model does not prohibit them.

Next we consider assignment of users to administrative units (item IV of Table 3). Users can be assigned via the TA_admin or the UA_admin relation. The former authorizes the task-role assigment power of an au, while the latter authorizes the user-role assignment power. In this way, these two capabilities can be separately assigned to users, even though they are coupled in the au. This embodies the separation of user and permission assignment principle of Sect. 2. A user assigned to any au via TA_admin or UA_admin is said to be an administrative user.

For convenience in maintaining the TA_admin and UA_admin relations, Uni-ARBAC also defines a hierarchy $\succeq_{au}$ on administrative units. Assignment of user u to a senior au_i for task-role administration, i.e., $(u, au_i) \in TA\_admin$, effectively also assigns u for task-role administration to all au_j for $au_i \succeq_{au} au_j$. Likewise for $(u, au_i) \in UA\_admin$. For simplicity, Uni-ARBAC requires $\succeq_{au}$ to

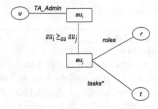

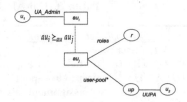

Fig. 5. Task-role authorization   **Fig. 6.** User-role authorization

be a rooted tree hierarchy. One effect of this is that there is a seniormost au. An example administrative units tree is shown in Fig. 2(d).

The authorization functions of Uni-ARBAC are specified in item V of Table 3 as boolean functions that return true or false. The function $can\_manage\_task\_role(u : U, t : T, r : R)$ specifies the conditions for user u to assign/revoke task t to/from role r. The requirement is schematically depicted in Fig. 5. User u must be assigned as a TA_Admin to the unique administrative unit au_j which has jurisdiction over role r, or alternately so assigned to an administrative unit $au_i \succeq_{au} au_j$. In either case task t must be assigned to au_j or be junior to a task assigned to au_j.

The $can\_manage\_user\_role(u_1 : U, u_2 : U, r : R)$ similarly specifies the conditions for user u_1 to assign/revoke user u_2 to/from role r, and is schematically depicted in Fig. 6. User u_1 must be assigned as UA_Admin to the unique administrative unit au_j which has jurisdiction over role r, or alternately to an administrative unit $au_i \succeq_{au} au_j$. In either case user u_2 must be assigned via $UUPA$ to a user-pool which is directly assigned to au_j or to some user-pool junior to a user-pool directly assigned to au_j.

Item VI of Table 3 formalizes the four administrative actions of Uni-ARBAC. Assigning and revoking have the same authorization and the effect is self-explanatory. The alignment of authorization for assign and revoke embodies the principle of reversibility of administrative actions discussed in Sect. 2.

3.3 Uni-ARBAC Invariants

Invariants are properties that hold for the lifetime. For the moment assume we start with an initial state in which both TA and UA are empty, i.e., the roles have no tasks or users assigned. Let us denote TA where all possible TA assignments have been made as TA_{max}. It is evident that,

$$TA_{max} = \bigcup_{au \in AU} \{(t, r) | t \in tasks^*(au) \wedge r \in roles(au)\}.$$

Because of reversibility of assign and revoke, and the independence of each assignment from another, in any state TA must satisfy

$$\emptyset \subseteq TA \subseteq TA_{max}. \tag{1}$$

It is further evident that any value of TA bounded as in Eq. 1 is realizable. We can either build up from an empty TA or build down from TA_{max}. In fact we can take the system from any value of TA compliant with Eq. 1 to any other compliant value. Further, we can relax our assumption of an empty TA in the initial state. Any initial state with TA compliant with Eq. 1 will ensure that TA is maintained within these bounds. Finally, let TA_0 denote TA in the initial state. Any $(t, r) \in TA_0/TA_{max}$ cannot be revoked and will persist in all subsequent states. These observations can be proved formally but are quite evident. As an example, the upper bound of TA for the administrative unit hierarchy of Fig. 2(d) does not contain the pair of $(t1, CPL)$. Thus the task $t1$ cannot be assigned to the role CPL using the instance of the AUs in Fig. 2(d).

We can make similar observations with respect to the maximal possible values of UA as follows,

$$UA_{max} = \bigcup_{au \in AU} \{(u, r) | (\exists up \in UP)[(u, up) \in UUPA \land r \in roles(au) \land up \in user\_pools^*(au)]\}$$

so UA is bounded as follows.

$$\emptyset \subseteq UA \subseteq UA_{max} \tag{2}$$

4   Variations of Uni-ARBAC

In this section we discuss some variations of Uni-ARBAC which materially alter the characteristics of the model. Uni-ARBAC has a rich structure so it is not surprising that many variations are possible. Some are relatively incremental, such as allowing the administrative hierarchy to be a general partial order rather than a rooted tree. Here we discuss a few variations that raise some substantial policy issues.

4.1   Aggressive Inheritance Model

In this variation we allow a senior au to do more than simply the sum of what the au itself is authorized to do plus what each of the junior au's are allowed. To be concrete consider the AU hierarchy of Fig. 2(d), and an administrative user u assigned as TA_admin and UA_admin to the Management Unit administrative unit. This user also inherits membership in the Cloud Unit and Mobile Unit administrative units. User u is thereby authorized, for example, to assign task t2 to role CPL and users from user-pool CTP to role CPL. However, u cannot make assignments across the junior administrative units, such as task t2 to role MPL and users from user-pool MTP to role CPL. We denote this form of inheritance in the AU hierarchy as membership inheritance.

With the alternate aggressive inheritance we allow cross assignments across junior administrative units by a senior administrator. There is clearly a major policy difference between the two forms of inheritance. The senior au effectively serves as a single consolidated au with freedom to assign any task to any role, and users from any user-pool to any role. With aggressive inheritance the user

Table 4. Uni-ARBAC with aggressive inheritance

V'. Modified Authorization Functions
- $can\_manage\_task\_role(u : U, t : T, r : R) = (\exists au_i, au_j, au_k)[(u, au_i) \in TA\_admin \land$
$au_i \succeq_{au} au_j \land au_i \succeq_{au} au_k \land r \in roles(au_j) \land t \in tasks^*(au_k)]$
- $can\_manage\_user\_role(u_1 : U, u_2 : U, r : R) = (\exists au_i, au_j, au_k)[(u_1, au_i) \in UA\_admin$
$\land au_i \succeq_{au} au_j \land au_i \succeq_{au} au_k \land r \in roles(au_j) \land (\exists up \in UP)[(u_2, up) \in UUPA \land$
$up \in user\_pools^*(au_k)]$

u discussed above will be able to assign t2 to role MPL and users from user-pool MTP to role CPL. Perhaps more dangerously, user u will be able to assign task t1 to MPL. The effect of aggressive inheritance is formally stated in the modified authorization functions of Table 4. Everything else from Table 3 applies unchanged to Uni-ARBAC with aggressive inheritance. The senior most au at the root of the AU hierarchy can assign any task to any role and any user (assuming every user is in at least one user-pool) to any role, so $TA_{max} = T \times R$ and $UA_{max} = U \times R$. Equations 1 and 2 continue to hold. Other less aggressive variations can also be considered.

4.2   No Self-administration Model

Consider the administrative actions $assign\_user\_to\_role(u_1 : U, u_2 : U, r : R)$ and $revoke\_user\_from\_role(u_1 : U, u_2 : U, r : R)$ of Table 3 item VI. In general, u_1 can equal u_2, so it is permissible for u_1 to assign and revoke himself to and from roles. In some contexts, this may be considered as a conflict of interest. To avoid this, an additional check that $u_1 \neq u_2$ can be added to the $can\_manage\_user\_role$ authorization function of Table 3.

5   Engineering Administrative Units

There can be different meaningful AU hierarchies for a given set of roles. For example, for the roles in Fig. 2(a), two different instances of AUs are given in Figs. 2(d) and 7, based on different partitioning of the roles. Crampton and Loizou partition roles in defining 'administrative scopes' in [5] to confine the side effects of role hierarchy modification in a highly disciplined manner. The AU structure in Fig. 2(d) is based on the partitioning of roles defined in administrative scope.

As we have argued, one particular partition is not suitable for all requirements. UARBAC [10] argues that role partitioning according to administrative scope does not work well for all different types of role hierarchies and there should be flexibility in the administration structure.

We develop the following simple and flexible partitioning heuristics, which are applicable to many different types of hierarchies and can be configured to produce different partitions for a given set of roles or tasks.

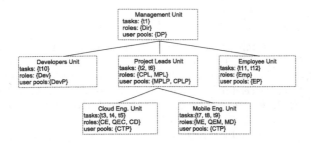

Fig. 7. Alternate AUs for roles, tasks and user-pools as given in Fig. 2

1. Most senior roles in the role hierarchy contain critical tasks. They should be administered separately. Similarly, most junior roles contain tasks that most other roles inherit. They should also be administered separately.
2. For rest of the roles, iteratively select the most senior and most junior roles until all roles are partitioned.

After partitioning and specifying role set for each AU, we can populate tasks in the AUs using given task-role allocation. Scenario based top-down approach for engineering roles [11] derives tasks in an intermediate process and assign them to roles. Thus, we believe, the process of engineering roles can be utilized to derive task-role allocation. On the other hand, the process of engineering user-pools and allocating them in AUs, assigning administrative users into AUs should also be carried out to develop working AUs. We do not further elaborate these issues here.

6   Conclusion

In this paper, we present Uni-ARBAC, a unified model for administering user-role and task-role relations in Role Based Access Control. It combines various novel concepts and administrative principles from prior works. It integrates user-role and permission-role administration into a manageable unit, we call Administrative Unit. While most of the previous models assume existence of administrative roles for managing regular roles, we relax this assumption and our approach for engineering administrative units can be used for engineering administrative roles.

Nonetheless, Uni-ARBAC has limitations. It uses several sets and relations of which it administers only task-role and user-role relations. Further research in this area is needed to realize a complete model.

Acknowledgement. This research is partially supported by NSF Grants CNS-1111925 and CNS-1423481.

References

1. Bertino, E., Bonatti, P.A., Ferrari, E.: TRBAC: a temporal role-based access control model. TISSEC **4**(3), 191–233 (2001)
2. Bertino, E., Catania, B., Damiani, M.L., Perlasca, P.: GEO-RBAC: a spatially aware RBAC. In: Proceedings of 10th SACMAT, pp. 29–37. ACM (2005)
3. Bonatti, P., Galdi, C., Torres, D.: ERBAC: event-driven RBAC. In: Proceedings of 18th SACMAT, pp. 125–136. ACM (2013)
4. Crampton, J.: Understanding and developing role-based administrative models. In: Proceedings of 12th ACM CCS, pp. 158–167 (2005)
5. Crampton, J., Loizou, G.: Administrative scope: a foundation for role-based administrative models. ACM TISSEC **6**(2), 201–231 (2003)
6. Ferraiolo, D.F., Sandhu, R., Gavrila, S., Kuhn, D.R., Chandramouli, R.: Proposed NIST standard for role-based access control. ACM TISSEC **4**(3), 224–274 (2001)
7. Kern, A.: Advanced features for enterprise-wide role-based access control. In: Proceedings of 18th ACSAC, pp. 333–342. IEEE (2002)
8. Kern, A., Schaad, A., Moffett, J.: An administration concept for the enterprise role-based access control model. In: Proceedings of 8th ACM SACMAT, pp. 3–11 (2003)
9. Kuijper, W., Ermolaev, V.: Sorting out role based access control. In: Proceedings of 19th ACM SACMAT, pp. 63–74 (2014)
10. Li, N., Mao, Z.: Administration in role-based access control. In: Proceedings of 2nd ACM ASIACCS, pp. 127–138 (2007)
11. Neumann, G., Strembeck, M.: A scenario-driven role engineering process for functional RBAC roles. In: Proceedings of 7th ACM SACMAT, pp. 33–42 (2002)
12. Oh, S., Park, S.: Task-role-based access control model. Inf. Syst. **28**(6), 533–562 (2003)
13. Oh, S., Sandhu, R.: A model for role administration using organization structure. In: Proceedings of 7th ACM SACMAT, pp. 155–162 (2002)
14. Sandhu, R.: The ASCAA principles for next-generation role-based access control. In: Proceedings of 3rd ARES (2008)
15. Sandhu, R., Bhamidipati, V., Munawer, Q.: The ARBAC97 model for role-based administration of roles. ACM TISSEC **2**(1), 105–135 (1999)
16. Sandhu, R., Munawer, Q.: The ARBAC99 model for administration of roles. In: Proceedings of 15th Annual ACSAC, pp. 229–238. IEEE (1999)
17. Sandhu, R.S., Coyne, E.J., Feinstein, H.L., Youman, C.E.: Role-based access control models. Computer **29**(2), 38–47 (1996)
18. Wang, H., Osborn, S.L.: An administrative model for role graphs. In: De Capitani di Vimercati, S., Ray, I., Ray, I. (eds.) Data and Applications Security XVII. IFIP, vol. 142, pp. 302–315. Springer, New York (2004)
19. Zhou, L., Varadharajan, V., Hitchens, M.: Secure administration of cryptographic role-based access control for large-scale cloud storage systems. JCSS **80**(8), 1518–1533 (2014)

SKALD: A Scalable Architecture for Feature Extraction, Multi-user Analysis, and Real-Time Information Sharing

George D. Webster$^{(\boxtimes)}$, Zachary D. Hanif, Andre L.P. Ludwig,
Tamas K. Lengyel, Apostolis Zarras, and Claudia Eckert

Technical University of Munich, Garching, Germany
webstergd@sec.in.tum.de

Abstract. The inability of existing architectures to allow corporations to quickly process information at scale and share knowledge with peers makes it difficult for malware analysis researchers to present a clear picture of criminal activity. Hence, analysis is limited in effectively and accurately identify the full scale of adversaries' activities and develop effective mitigation strategies. In this paper, we present SKALD: a novel architecture which guides the creation of analysis systems to support the research of malicious activities plaguing computer systems. Our design provides the scalability, flexibility, and robustness needed to process current and future volumes of data. We show that our prototype is able to process millions of samples in only few milliseconds per sample with zero critical errors. Additionally, SKALD enables the development of new methodologies for information sharing, enabling analysis across collective knowledge. Consequently, defenders can perform accurate investigations and real-time discovery, while reducing mitigation time and infrastructure cost.

1 Introduction

Cyber crime has evolved to the point where teams of criminals command sophisticated tools and infrastructures, while possessing the resources required to stay ahead of the defender [4,18]. For instance, in 2012, McAfee received over 100 thousand samples per day [3], yet on one day in 2015, VirusTotal received over a million unique samples [25]. Disproportionately, malware analysis systems are struggling to scale to meet this challenge and present a clear picture of criminal activity [22,23,26]. A solution to this problem is twofold: (*i*) how can corporations extract features and retain a central repository at scale and (*ii*) how can industry peers collaborate in a timely manner without exposing sensitive data and retain essential context.

Simply put, the rise in malware, has strained the ability of the security teams to run analytics and maintain a central repository of generated information. The reason why is because current analytic tools fall short when being used together

© Springer International Publishing Switzerland 2016
M. Bishop and A.C.A. Nascimento (Eds.): ISC 2016, LNCS 9866, pp. 231–249, 2016.
DOI: 10.1007/978-3-319-45871-7_15

in an automated fashion and enabling a collaborative environment [22,23,26]. As stated by MITRE, this causes a situation where analysts often regenerate information and duplicate the work of their peers—a huge waste of time and resources [23]. With respect to collaboration with industry peers, the second problem is that once information is received, it is difficult to be shared with security partners in a manner that is timely, retains context, and protects the collection methods. Although, rapid information sharing is an essential element of effective cybersecurity, and will lessen the volume companies need to process, companies are weary of sharing data for fear of tarnishing their business reputation, loosing market share, impairing profits, privacy violations, and revealing internal sources and methods [6]. As a result, it is now common practice to share only with trusted groups large sets of data with minimal context or select post-processed information. Recognizing the criticality of this issue, the US Government recently issued an Executive Order to promoting the sharing of cybersecurity information in private sector [2].

In an attempt to alleviate the burden on analysts, a number of solutions has been developed to help triage data through feature extraction and create a central repository of the collected information [9,23,28]. Unfortunately, many of these tools struggle to scale and provide the fault-tolerance required to support the sheer volume of data needed to be processed in part due to the linear, monolithic, and tightly coupled processing pipeline. For instance, analysis tools, like CRITs [23] and MANTIS [9], are not separated from the core Django/Apache system and are executed on the same physical host, while VIPER [28] has been developed for a single user with the intention of being deployed on a workstation. Consequently, when these system becomes overloaded, a bottleneck occurs that prevents the feature extraction tools from executing properly. To make things even worse, when one of the aforementioned tools fails, it is difficult to perform a graceful exit or cleanup, and the system becomes overwhelmed with a load of only a few thousand malware samples. This results in a situation in which feature extraction cannot be performed quickly and at scale using current technologies and architectures. Furthermore, none of these tools address the issue of how to make assessments on the extracted features.

In this paper we present SKALD, a novel architecture to create systems that can perform feature extraction at scale and provide a robust platform for analytic collaboration and data-sharing. In essence, SKALD provides the required infrastructure to extract features at a scale that can: (i) cope with the growing volume of information, (ii) be resilient to system failures, and (iii) be flexible enough to incorporate the latest technology trends. In addition, SKALD takes a new approach in terms of how data is shared by providing a platform that grants analysts' tools access to the entire extracted feature set without requiring the analysts, and their tools, to have direct access to the raw malware samples or other primary analytic object, such as a domain or an IP address. This enables correlations, clustering, and data discovery over the entire set of collected knowledge, while still protecting the raw objects, the sources, and the methods used to obtain the data. To this end, we develop an open-source prototype and conduct extensive experiments that demonstrate that our architecture has a near linear growth rate and is able to eliminate critical failures when extracting features across millions of entries. Furthermore, we show major performance gains with the ability to conduct feature extraction at a rate of 3.1 ms per sample, compared to 2.6 s when using existing systems. Finally, we discuss how our methodology provides a platform for analysis on a collective set of raw data and extracted features, which enables more accurate clustering of malicious information and real-time data discovery while minimizing the need for redundant feature extraction and thereby reducing analysis time and infrastructure cost.

In summary, we make the following main contributions:

- We develop a framework for feature extraction that is scalable, flexible, and resilient, while it demonstrates near linear growth with zero critical errors over millions of samples.
- We display major speed improvements over traditional techniques using only 3.1 ms per sample when utilizing 100 workers.
- We exhibit the ability of our approach to allow partner organizations to submit new raw analysis objects in real-time leveraging the infrastructure from multiple organizations on different continents.
- We show SKALD's capacity to share resultant extracted features and analysis with geographically and organizationally diverse partners who are not sufficiently trusted to have unrestricted access to the raw analysis objects.

2   System Overview

SKALD is an architecture for developing analytic platforms for teams working to thwart cyber crime. At its core, SKALD dictates the required structure to perform asynchronous feature extraction against submitted objects. It scales these actions horizontally, at a near linear rate, across millions of objects while remaining resilient to failures and providing the necessary flexibility to change analytic methods and core components. SKALD additionally provides the necessary infrastructure to perform advanced analytics over the extracted features while empowering analysts to retrieve and share information using their preferred tools and scripting interface. Furthermore, SKALD's design allows the sharing of data with partners without requiring the release of raw data. This is because SKALD's *Access Control Layer* (ACL) and intelligent core components allow analytics to be executed across a combination of central instance and in-house replicas. Ergo, the information can be easily shared, enabling analysis over a more complete and collective set of data, which overcomes biases caused by informational gaps.

SKALD's achievements are primarily due to the approach of logically abstracting the system into "loose coupled" themes and core components, as depicted in Fig. 1; allowing the creation of systems that are scalable, flexible, and resilient. This level of abstraction between system elements is critically missing in widely-used systems such as CRITs, VIPER, and MANTIS. This has lead these systems to become monolithic and tightly coupled in design; creating a major hindrance in allowing them to evolve and scale by leveraging distributed computing techniques. However, SKALD apart from the scalability it offers, it enables systems to evolve so they can meet the challenges of future cyber criminals by allowing components to be easily exchanged, added, or subtracted. Thus, if a newer, better, or simply different method is discovered, this can be easily incorporated alongside existing methods or simply replace the old ones. This also allows system components to be outsourced to a company, institution, or organization specializing in that work. Finally, this design allows SKALD to orchestrate tasking which create efficient system by substantially reducing the infrastructure and network overhead for transmitting data to and from multiple Services.

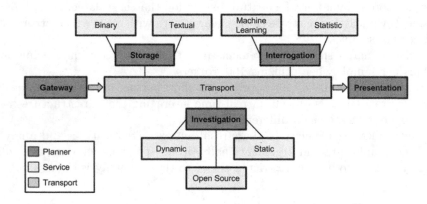

Fig. 1. Organization of SKALD's components and core themes.

The structure of SKALD is composed of three main components: *Transport, Planners*, and *Services*. As Fig. 1 illustrates, Transport is the main orchestrator and moves data and tasking to the Planners. Then, Planners allocate infrastructure, enforce security, and oversees the execution of Services. Services in turn perform the requested work and provide the resultant response along with pertinent meta information, such as error messages, back to Planners. This is further described in the following sections.

2.1 Planner

The Planner's primary purpose is to serve as an intelligent orchestrator for Services. At its core, it manages tasking, allocates resources, enforces the ACL, and provides an abstraction between the Services and other aspects of SKALD. The Planner also informs the Transport what Services are available for tasking and provides status information back to the system core. As previously mentioned, Planners are loosely coupled with other parts of SKALD. In this way, they provide flexibility by ensuring that changes to the core aspects of a Planner will not affect other parts of the system. This helps to improve resiliency by allowing the Transport to delegate tasking to redundant Planners during system failures [5, 19]. Furthermore, this allows the system to horizontally scale by permitting the Transport element to instantiate additional Planners under heavy load while also allowing the Planners to schedule the tasking of Services and allocated additional resources on internal servers and cloud infrastructure.

Communication with Services. The Planner communicates with Services through the HTTP over TLS protocol. We select this method for three main reasons. First, HTTP is widely understood and is capable of transferring a variety of data-types. Second, the wide adoption of the HTTP protocol allows analysts to be able to add new analytics in the language they feel most comfortable. Third, HTTP communication allows Services to be deployed either locally or across network partitions. However, SKALD does not dictate the format of messages transmitted over HTTP over TLS. That said, our prototype provides developers with interfaces for typed JSON message parsing for stronger message safety and a loosely-typed Map data-structure when message safety is not a concern. We choose this because when evaluating other formats in the past, such as Google's Protocol Buffers [8], we found the overall speed increase did not justify the greater level of difficulty.

SKALD provides two methods for the Planner to communicate objects to Services. For local analytics, the object is delivered via a RAM disk, with fail-over to local disks based on size. This creates a fast and easily-available data-store for analytic file reads. For external Services, the Planner will deliver the object via HTTP. When interacting with Services that do not require a local file, the Planner will attach the pertinent meta-data to the tasking message. We select this method for transmitting objects as many existing analytics require a local file to be read. As such, this allows existing tools to maintain relevance through leveraging SKALD's ability to distribute and scale workloads. Furthermore, this improves performance since each Service does not require a costly network transmission to access an object.

Service Orchestration and Management. The Planner leverages configuration management and containers to package Services together in order to manage the tasking of Services and optimize their execution. This provides three core benefits with respect to speed, flexibility, and resiliency [7]. Regarding speed, the packaging of multiple isolated Services on one worker reduces the volume and frequency of data passing through the network. This in turn is a major advantage with distributed and cloud based systems because it eliminates the need for multiple large file transfers. Furthermore, packaging Services increases the flexibility of SKALD-based systems by allowing the rapid deployment of new Services without concern for complex dependency management while ensuring discrete versions of Services and configurations. Finally, containers easily allow Quality of Service (QoS) operations to automatically re-instantiate critically failed Services.

Access Control Layer Enforcement. The Planner is the primary element responsible for managing the ACL by ensuring that tasking is authorized. The Planner also limits potential exposure caused by analyzing an object by enforcing

Service execution restrictions through the use of ACL meta-tags. This is done by allowing tasking to state that the execution should only run, for example, on internal hardware, without Internet access, and restrict DNS lookups. This is critically missing in current system designs.

2.2   Planner Themes

SKALD breaks down Planners into five discrete themes as Fig. 1 illustrates. This prevents systems from becoming monolithic through the separation of core parts of the system into categories based on their area of influence. Planners are categorized as members of one of the following: Gateway, Investigation, Storage, Interrogation, and Presentation.

Gateway. The Gateway's primary purpose is to receive taskings and push it to the Transport. When tasking is received, the Gateway first performs an ACL check to guarantee that the tasking is authorized. If authorized, Gateway then ensures taskings are valid, scheme compliant, and that the pipeline can handle the requested work. The Gateway can also automatically assign tasking based on an object type. Together this ensures that pipeline resources are not wasted and provides the first level of system security.

Investigation. This theme is responsible for performing feature extraction against objects. When tasking is received, it schedules the execution of its Services which are capable of performing static and dynamic analysis as well as gather data from third parties. During scheduling it optimizes the execution of Services by packaging them together and directly provides them the data. As taskings are executed, it performs the two-fold QoS strategy by monitoring the health of Services and validating received results. Additionally, it enforces the ACL and ensures Services adhere to the meta-tags configured restrictions.

To help illustrate the goals of the Investigation Planner, we describe an ideal execution flow. The Transport layer T2 (see Sect. 2.4) submits an object for Investigation along with a set of taskings and ACL tags. The Planner first identifies which Services are available for tasking and configures them according to the tasking request and ACL meta-tags. Next, it either packages an object together with a set of Services for execution on one node, or sends the object to a preexisting node dedicated to a Service. The Planner will then monitor the health of the Service and perform any remediation action as needed. For instance, if a Service is unable to gather data from a source due to a query cap, the Planner will reschedule the Service's execution once the cap has expired. When the results of a Service are received, the Planner passes them to the Transport layer which then commands the Storage Planner to archive them. Additionally, if a Service returns new objects, the Planner will submit the object back to the Transport layer with the pertinent ACL tags for storage and tasking to the Gateway.

Storage. This Planner controls how data is stored and retrieved in SKALD. At its core, it is an abstraction layer for the database Services. It enforces a standard storage scheme and passes the requests to the appropriate database elements, identified by UUID4. This enables SKALD to be storage system agnostic and utilize a single or hybrid data storage scheme for resultant data and objects of analysis. The benefit of this approach is that data can be stored in databases optimized for the data type while also easing the inclusion of legacy archives. For example, objects can use Amazon's Simple Storage Service (S3) while the features can be stored in a system optimized for textural data such as Cassandra [16]. This approach additionally has a major benefit of allowing industry partners to perform in-house replication of selected sets of data while enforcing restrictions on more restricted datasets. For instance, in our prototype, raw objects are stored in restricted datasets hosted by the originator while the extracted features are replicated across all partners. When restricted data is required, the Planner provides contact information to the requester and once approved the Planner will automatically configure access.

The design also enabled the Storage Planner to perform storage-based optimizations, for example, performing deduplication, compression, and deconflict updates and deletions of temporally sensitive datasets. Finally, the Planners overarching view of the system allows it to automatically perform QoS-based operations such as automatically instantiating additional storage shards and coordinating multi-datacenter replication.

Interrogation. The Interrogation Planner focuses on how to turn the extracted features into intelligence in two distinct forms. Aside these forms, the Planner is responsible for scaling the number of Services to balance system load. In the first form, Interrogation Services process system data for retrieval through mechanisms such as an API, plugin, or website. This deviates from previous systems by separating the feature extraction process from the rendering of data. Thus, displaying information is not bound to the extraction method and can change based on what a user's desire. For example, a Service can display the VirusTotal score along side any Yara rule matches. In the second form, an Interrogation Service orchestrates the execution of advanced analytics, such as Machine Learning. To do this, the Planner distributes the work load across available Services to complete the task at hand. This load balancing allows the integration with mini-batch training techniques to achieve large scale model training and generation and serve as a distribution layer for pre-trained models.

Presentation. This Planner provides a standard mechanism for interacting with stored data and the data the Interrogation layer generates. When requests are received, it first ensures that the request accessing information is authenticated and scheme compliant. As this Planner theme is the most likely to vary based on individual use cases, we opt to keep its definition as minimal as possible. That being said, the Presentation layer can be imagined as a microservice fog surrounding the datastores queried by the Interrogation layer.

2.3 Service

Services perform the work being orchestrated by Planners. In SKALD, Services are "loosely coupled" and only interact with their parent Planner. As discussed by Papazoglou et al. [19], this highly decentralized model allows services to be platform independent and scale as needed. Additionally, this improves fault-tolerance as no Service is reliant on another Services. Furthermore, the atomic nature of Services provides a great level of flexibility by allowing them to be exchanged as new technology emerges and requirements change. In essence, a Service is attached to a Planner and performs work associated with that Planner's theme. For example, under the Investigation Planner, Services can gather data from VirusTotal, generate a PEHash [27], and perform dynamic analysis through Cuckoo [10] and DRAKVUF [17]. An Interrogation Service, on the other hand, will perform an action across a set or subset of the data through data gathering, machine learning, or other data mining and exploration techniques and provide a mechanism to display the resulting information [14].

2.4   Transport

Transport's main task is to move data among the Planners. In addition, the Transport layer monitors the health of the Planners and performs remediation actions to support QoS. This enables a robust level of resilience by ensuring that requested work is always stored in a queue and that results and taskings are never lost. Additionally, the Transport layer improves SKALD's ability to scale by reducing adverse effects of system overloads by allowing the distribution of work across multiple Planners. This is a huge benefit over current available cybersecurity systems that are only design to scale vertically.

The Transport consists of four main sections (T1, T2, T3, and T4) as Fig. 2 illustrates. This permits the selection of optimized technology to handle the interaction among Planners. In the following we introduce these sections.

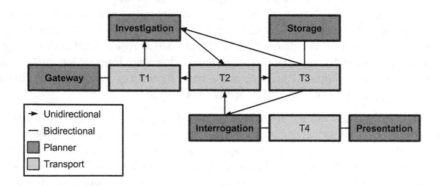

Fig. 2. Interaction between the Transports and Planners.

Transport - T1. T1 is focused on moving data to the Investigation Planner for analysis. To do this, T1 is required to perform three primary actions. The first action is to receive tasking from the Gateway Planner and schedule their transmission to the Investigation Planner. The second is to receive tasking from T2 and submit them to the Gateway Planner for validation. The final action is to monitor the health of the Interrogation Planner and perform QoS management. When implementing T1, we recommend the utilization of a distributed message broker such as Kafka or RabbitMQ.

Transport - T2. T2 is focused on receiving objects and results from the Investigation and Interrogation Planners. To do this, T2 has two primary actions. The first action is to receive data from the Investigation and Interrogation Planners and schedule their submission to the T3 for Storage. This is separated from T3 to allow a message queue service to be implemented to help throttle the storage of data during peak loads as storage operations can be costly but are often not time critical. The secondary action is to receive objects from the Investigation and Interrogation Planners and pass them to T1 for further analysis. Like T1, we recommend the use of a distributed message broker.

Transport - T3. T3 is focused on submitting and retrieving data from the Storage Planner. As such, T3 is responsible for three primary actions: (i) provide data from the Storage Planner directly to the Investigation and Interrogation Planner, (ii) receive information from T2 and pass the data on to the Storage Planner, and (iii) manage the QoS of the Storage Planner. We recommend the first two actions to be implemented with no message queues between the Storage Planner and the databases. This permits database Services to rely on their own optimization frameworks during the retrieval of data; databases are often heavily optimized for retrieval of data and implement their own form of message queues.

Transport - T4. T4 handles the exchange of information between the Interrogation and Presentation Planners. The first action is to provide a conduit for communication between the Presentation and Interrogation Planners. The second is to monitor the health of the Investigation Planner and perform QoS management. As the Interrogation Planner will typically provide data through HTTP calls, we recommend implementing T4 as HTTP load balancers.

3 System Wide Aspects

In this section we present system wide aspects of SKALD. We first introduce the QoS strategy we follow and then discuss the ACL requirements.

3.1 Quality of Service

Malware authors are incentivized to thwart analysis and as a result the failure of Services should be expected. To counter this, SKALD automatically recovers from issues arising from the execution of Services by leveraging a robust QoS pattern. It does this through a two-fold QoS philosophy of monitoring the actual Service and the resultant response from a Service. Thus, SKALD accounts for the scenario of when the returned work has failed or even if the actual Service has failed and accounts for them differently.

To implement the QoS strategy, the Planner monitors the health of the Service worker using container status messages and HTTP response codes. The Planner then evaluates the availability, response time, and throughput of each attached Service. If the evaluation responds by stating the Service is operating within normal bounds, the Planner will then send the results to the second stage to evaluate the returned work. This allows Service authors to specify deep level checks and perform automated remediation actions that are Service specific. If a failure occurs at either step, the Planner will determine if the Service has entered a failed state and perform remediation action, such as restarting the Service container. If the Service appears to be healthy, the Planner will re-queue the work that has demonstrated failure while saving otherwise successful Service results.

The unique aspect of this strategy is that SKALD views the tasking of each Service as single elements and process them individually. When the Service or returned work fails, SKALD will only discard failed work as opposed to abandoning combined tasking and queued work. This is a key difference between SKALD and previously systems. For example, the methodology used by systems which rely on the HDFS and MapReduce model for their data and task distribution, such as BinaryPig [11] and BitShred [13], will cease processing or discard successful results when percentages of work fail. However in SKALD, even if there is a high number of failed jobs, any successful result will be saved and failed work will be reattempted. Furthermore, in the event of pathological failure, SKALD will store the tasking in a separate queue for human intervention. Our evaluation showed this new approach provides significant performance benefits and was outright required when executing large, historical, analytic tasks across cybersecurity data as the data is often designed to confound investigations and cause failures.

SKALD's QoS system also accounts for congestion during times of peak load as well as Service shutdown and instantiation. To do this, the Planner enforces throttling of Services using a pull-based pattern with an "at least once" message delivery scheme [12]. In this scheme, the Planner is aware of the system's current message load and pulls a configurable number of messages, in which it is capable of completing, from the Transport layer. Each requested message is then tracked within the Planner as a discrete entity. Upon completion of work, the Planner notifies the Transport of the work status, pushes the results to the queue, and pulls additional tasking. This allows SKALD to prevent the overburdening of Planners while also ensuring that no work is lost due to component failure. This approach also reduces the chance that a failed Service state will replicate to other parts of the system and cause a work stoppage due to being overburdened.

3.2   Access Control Layer

The nature of handling unique and sensitive data within SKALD requires the incorporation of a complex ACL system. Unfortunately, the standard ACL model used in cybersecurity only allows for isolation based on the source of the raw data or a user's role. However, the problem is that this does not address access to sensitive capabilities, differentiate Services from users, or separate raw data from features and analytics; creating an "all or nothing" approach to access. Therefore, the systems cannot be designed to easily allow analysts or Services to derive intelligence across a collective set of information without granting the analysts access to all sources and sensitive capabilities.

To overcome this, SKALD creates a new ACL model by granting access based on *User*, *Capability*, *Source*, and *Meta-tags*. The *User* defines the users and components of the system. While *Capabilities* map to Services and their derived information, *Source* maps the origin of the raw data. Finally, *Meta-tags* provide Planners with Service execution restrictions. For example, these tags can specify that dynamic analysis can only execute without Internet access. While implemented in our prototype, the full definition of the ACL is left to future work.

4 Evaluation

To evaluate SKALD, we created an open-source prototype and performed a series of experiments. We acknowledge that the SKALD focuses on the structure of a system and does not prescribe implementation methods. While this can create varying performance metrics, we feel it is prudent to present a baseline implementation to demonstrate the significant improvements afforded by the SKALD architecture. Throughout this section, we use our prototype to evaluate the architecture's (i) scalability, (ii) resiliency, and (iii) flexibility.

4.1 Experimental Environment

We leveraged three hardware profiles for our evaluation. The first profile was used as a control and deployed CRITs using their recommended setup with the following nodes: (i) *Ingest VM*: 2 cores 4 GB RAM, (ii) *MongoDB VM*: 10 cores 32 GB RAM, and (iii) *CRITs VM*: 6 cores 32 GB RAM. CRITs was selected for our control as it is an industry standard for performing multi-user analytics. Additionally, we made the assumption that CRITs performs similarly to other systems such as MANTIS [9], as the architectures are remarkably similar and Django-based. The second profile deploys our prototype using a similar hardware profile as the control. In this profile we deployed the prototype in a cloud-based environment using the following nodes: (i) *Workers*: three AWS EC2 M3 large instances, (ii) *Transport*: One AWS M3 xlarge instance, and (iii) *Storage*: One AWS M3 medium instance. Finally, the third profile was used to evaluate the horizontal scalability of the SKALD architecture. In this profile we used the following nodes: (i) *Workers*: 100 AWS EC2 M3 large instances, (ii) *Transport*: One AWS M3 xlarge instance, and (iii) *Storage*: One AWS M3 medium instance.

In all experiments we used a diverse set of malicious PE32 samples from Virus Share, Maltrieve, Shadowserver, and private donations. These binaries encompasses traditional criminal malware, highly advanced state-sponsored malware, and programs which are not confirmed as malicious but are suspicious.

4.2 Scalability

In order to provide a meaningful evaluation of SKALD's scalability, we studied the ability to ingest PE32 samples and then execute a series of three Investigation Services on our profiles. We selected PE32s as this provides a direct mapping to CRITs and our prototype performs at a near identical level when processing Domain, PCAP, IP, and other objects. To this end, we did not perform any experiments related to the storage of Service results because these experiments would vary depending on the data-store used. Furthermore, we omitted experiments on the Interrogation Services as this architectural model is relatively similar to Investigation and it is difficult to perform clear correlations among existing systems.

During our experiments, we used four sets of data: 1000, 5000, 10,000 and 50,000 randomly selected samples. As SKALD is intended to scale to support large datasets, we additionally ran SKALD with a set of one million samples. In both cases we pushed the samples into each system through a linear sequence of RESTful calls with no delays. We did this to evaluate the ability for the systems to queue work and simulate batch queues that we regularly encounter in our work. Additionally, to ensure that the evaluation was as fair as possible, we levered existing CRITs Services and added a RESTful wrapper around Services to make them compatible with our prototype. These Services gather PEInfo data, check the file against the VirusTotal private API, and run each sample against 12,431 Yara [1] signatures provided by Yara Exchange [20].

Table 1. Average time to process samples in seconds.

Framework	1K	5K	10K	50K
CRITs	2.8000	3.1774	3.3781	1.1929
3 Workers	0.0502	0.0558	0.0616	0.1303
100 Workers	0.0032	0.0032	0.0032	0.0025

Our first scalability experiment revealed that SKALD outperforms existing systems, as shown in Table 1. Our prototype, which was deployed with 100 workers, was able to process each sample at an approximate rate of 3.1 ms. Additionally, this hardware setup began to show speed improvements, due to the scheduler caching, at about 50,000 samples.

An interesting finding is that even with similar hardware, the design still greatly outperformed existing systems and was able to process each sample at an average rate of 74.5 ms. In terms of speed, this is a significant improvement over CRITs' average rate of 2.64 s per sample. Furthermore, these results highlight one of the critical issues plaguing existing systems. The perceived rate increase with CRITs at 50,000 samples was caused by an overload of the system. When this occurred, the operating system began to randomly kill processes before completion, producing a false appearance of speed improvements.

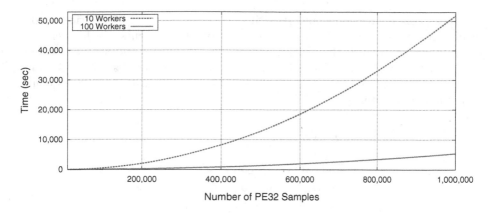

Fig. 3. SKALD's speed in processing one million samples.

Having established SKALD's performance against existing systems, we then studied the prototypes ability to scale to meet the demand of very large sample sets, as Fig. 3 depicts. Unfortunately, a direct comparison with CRITs was not possible because CRITs could not maintain stability beyond 50,000 samples. Nevertheless, this experiment clearly showed that our prototype was able to scale to meet the demand of processing a million samples and performed at an approximate rate of 7.5 ms per sample using 100 workers. When analyzing the results, we identified that the slowdown, with respect to the previous experiment, was caused by the Transport requiring disk reads due to tasking size. This however, was a constant rate and we are confident that SKALD performs at a similar rate irrespective of the volume of samples.

4.3   Resilience

Next, we executed two experiments to study the resiliency of SKALD. For the first experiment, we wrote 26 KB worth of random bytes across 20 % of the samples. This was done to generate files that represent work that will potentially confuse and fail analysis tasks. This allowed us to evaluate SKALD's ability to cope with failed, long-running, and troublesome work. For the second experiment, we studied how SKALD-based systems perform while core components of the infrastructure are unavailable or entered a failed state. We did this by running a script that randomly killed and restarted the machines, in isolation and in unison, hosting the Planner, Transport, and Service components of the system. During both experiments, we reran newly generated file sets through each hardware profile using the same method as used in the scalability experiments.

It was immediately apparent that SKALD's design and QoS paradigm greatly outperformed current systems. In the first experiment, as Table 2 shows, our prototype encountered zero critical errors, defined as a Service failing to complete tasking. While the Services did fail, the Planners successfully recovered from all errors encountered by Services and continued processing other unaffected

Services. This is in direct contrast with CRITs reporting 17,012 critical errors using a set of 50,000 samples. In an investigation of the results, the high critical error rate was because the system was unable to handle the load caused by failed Services and, as a result, these Services entered a locked state, consuming valuable resources, or were killed by the operating system before results could be submitted for storage.

Table 2. Critical failures in sample processing.

Framework	1K	5K	10K	50K
CRITs	0	0	151	17,012
3 Workers	0	0	0	0
100 Workers	0	0	0	0

During the second experiment, SKALD remained tolerant of faults. While the overall processing speed was decreased, our prototype's QoS paradigm was able to identify failed states and re-queue the tasking without any operator interaction. The final outcomes revealed that no work was lost and 100 % of the tasking were completed with no critical errors.

These results showed the benefits of SKALD. During these experiments, our prototype was not only resilient to handling poorly performing Services, but also easily handled failures to critical components of the system. Furthermore, these results revealed the benefits of SKALD's QoS structure in that work was never lost and the system never entered a failed state. By contrast, CRITs' failed Services remained in a failed state and human intervention would be required to clean faulty results from the database, reset service, and re-task the system.

4.4   Flexibility

In order to examine SKALD's flexibility, we first evaluated its ability to incorporate existing feature extraction methods. To do so, we created an Interrogation Service by modifying the existing CRITs PEInfo Service. This required the modification of less than 50 lines of code in order to remove CRITs specific commands and provide a RESTful HTTP wrapper. During evaluation, we noted that the wrapped Service performed with no discernible difference when compared to natively written SKALD Services. The prototype was able to seamlessly incorporate the Service to include performing QoS operations. In summary, this demonstrated SKALD's ability to provide flexibility by showing a minimal level or required work to incorporate a none native Service.

Next, we wanted to study SKALD's flexibility in changing a core component. We did this by testing the ability of our prototype to directly work with the CRITs database. This experiment was selected because the original implementation of SKALD's Storage Planner used Cassandra and Amazon S3 as the database back-end. In contrast, the CRITs framework uses MongoDB Documents and GridFS. Thus, we were not only required to change the underlying scheme for storing data but also the core database technologies. To do this, we made three modifications to the original prototype: (*i*) we created a Storage Service that parsed the results into the CRITs database scheme, (*ii*) we introduced an additional Storage Service that queried the existing CRITs database system to retrieve raw objects, and (*iii*) we included logic in the Storage Planner to identify which Storage Services to select when handling tasking. In total, we were able to make these changes using less than 100 lines of code. Furthermore, no modifications to the other parts of SKALD were required.

In summary, we are confident that SKALD provides the flexibility required to relatively easily change Services as well as core components of a system. We are confident that these results are applicable to other parts of the system.

5   Use Cases

In this section we present use cases to demonstrate how SKALD can overcome many of the current limitations in sharing data among industry partners. These use cases are based on our experience in using our framework to manage a collective set of millions of objects, including their associated analysis, across three globally distributed organizations.

5.1   Sharing Resources with Geographically Distributed Partners

Security partners rarely share infrastructure for fear that this will lead to informational exposure. This fear is well founded as it has in the past tarnished business reputations, reduced market share, impaired profits, caused privacy violations, and revealed internal sources and methods [6]. This poses a problem where processing resources become duplicated and large capital is required for their maintenance. While we acknowledge that SKALD cannot overcome the reasons behind why data is restricted, our framework can overcome the problem of how to develop a shared infrastructure.

This is because SKALD allows each partner to leverage their own Storage Planner to restrict the transfer of raw and restricted data and configure the Transport to acknowledge internal and external Planners. This in turn allows partners to leverage a collective set of hardware resources while allowing the creation of Investigation Services capable of performing data discovery, clustering, and colorations over the entire set of collective knowledge.

5.2   Sharing Derived Information with Partners

Cybersecurity analysis is mostly retrospective in nature and based on identifying previously observed attack patterns [6]. Sharing information is vital in this process as it allows analysts to make better correlations and see a more accurate picture of what is happening. Unfortunately, current methods for sharing information among partners require too much time to be truly effective. This is because information first needs to be identified as important, processed, validated as sharable, and then transmitted to partners often in the form of *Indicators of Compromise*. This is an inherently slow process but an even more critical issue is that feature extraction methods are not uniform and vary among industry partners [21]. This causes the receiving party to rerun extraction methods before the information can be leveraged by their analysts.

SKALD overcomes the above described issue by providing a platform that supports breaking the traditional paradigm of how information is shared. This is accomplished by providing the infrastructure to allow partners to directly share a central repository of extracted features and jointly perform analysis against the information. This approach allows partners: (*i*) to have a common set of extraction methods that is understood by all parties, (*ii*) to have real-time access to current information, and (*iii*) to have a wider view of the malicious activity during the investigation.

SKALD makes the aforementioned setup an easy task. In our deployment, we created this setup with globally distributed partners by adding a Storage Service that contained the logic required to replicate features between industry peers while restricting sensitive raw data. Thus, this created a uniformed platform that allowed all partners to leverage a collective pool of information. Furthermore, when working with Investigation tasking, the taskings can leverage the ACL meta-tags, discussed in Sect. 2.1, to inform the system if it should only use in-house resources or use collective resources.

6 Related Work

The need for a large scale malware analysis framework has been well discussed in prior work. Bitshred is a prime example and can perform malware correlation based on hashes of features, thus greatly increasing the throughput of the analysis system [13]. However, it only addresses the problem of dealing with already extracted data through clustering correlations on these hashed features. It does not address the full analysis lifecycle and is reliant on other systems for feature extraction. Hanif et al. [11] attempted to solve the problem of performing scalable feature extraction in their work on BinaryPig, by performing feature extraction in a distributed fashion through the use of a Hadoop based back-end. Unfortunately, BinaryPig only performs static analysis of malware binaries, it does not support deriving intelligence over the features, and the QoS schedule cannot appropriately handle failed work.

Significant research has been put forward in distributed setups addressing the issues of scaling, flexibility, and resilience. Service Oriented Architectures (SOA) breaks systems down into business processes, i.e., Services, and integrates them in an asynchronous event-driven manner through an Enterprise Service Bus (ESB) [15]. By establishing disjoint components, SOA enables distributed systems by defining how Services communicate versus their implementation. The xSOA architecture expands upon traditional SOA by allowing multiple Services to be combined under a single composite Service [19]. This composite server then provides a management layer which can perform Service orchestration, routing, provisioning, as well as integrity checking. Verma et al. [24] has taken xSOA a step further in their work on large-scale cluster management with Borg. They developed a xSOA like system and optimized it by introducing a BorgMaster. The BorgMaster serves as the master for Services and schedules their execution across Borglets. Together, the BorgMaster and Borglets intelligently manage the execution of Services and perform necessary actions to improve resilience, flexibility, and scalability. Additionally, Borg packages Services together for execution to improve efficiency by cutting down on transmission time, network bandwidth, and resource utilization. However, these architectures are not meant to deal with cybersecurity work as the data is often designed to cause component failures and SKALD's additional abstraction is required.

7   Conclusions

In this paper, we designed SKALD: an architecture to support the entire lifecycle of analysis against the ever growing volume of malicious activities plaguing computer systems. SKALD enables the design of a large-scale, distributed system that is applicable in the security domain. To this end, it supports multiple users and scales horizontally to support analysis of millions of objects. SKALD provides mechanisms for implementors to incorporate static and dynamic feature extraction techniques and apply advanced analytic across these features to derive intelligence. Furthermore, it breaks the paradigm that the automated extraction is the ultimate goal, opting to provide a flexible architecture to empower human analysts. In a similar manner, SKALD's design overcomes the limitation of current sharing models. It does this by providing the infrastructure needed to allow industry peers to perform analysis across collective knowledge while protecting sensitive data. Empirical results confirm that our solution scales horizontally, at near linear growth, and is able to process objects at a rate of 3.1 ms with zero critical error in contrast to existing system's rate of 2.6 s and thousands of critical errors. Additionally, SKALD allows analysis across a collaborative set of raw data and extracted features. Thus, it enables more accurate clustering of malicious information and real-time data discovery while minimizes the need for redundant feature extraction systems and thereby reduces analysis time and infrastructure cost.

Availability

The prototype used to evaluate the SKALD framework is open-source under the Apache2 license. It can be accessed at: http://holmesprocessing.github.io/.

Acknowledgments. We would like to thank the Technical University of Munich for providing ample infrastructure to support our prototype development. We would also like to thank the United States Air Force for sponsoring George Webster in his academic pursuit. In addition, we thank the German Federal Ministry of Education and Research for providing funding for hardware under grant 16KIS0328 (IUNO). Lastly, we would like to thank the members of VirusTotal, Yara Exchange, and DARPA for their valuable discussions and support.

References

1. Alvarez, V.M.: Yara 3.3.0. VirusTotal (Google Inc.) (2015). http://plusvic.github.io/yara/
2. Barack, O.: Executive Order No. 13691. Promoting Private Sector Cybersecurity Information Sharing (2015)
3. Bu, Z., Dirro, T., Greve, P., Lin, Y., Marcus, D., Paget, F., Pogulievsky, V., Schmugar, C., Shah, J., Sommer, D., et al.: McAfee Threats Report: Second Quarter 2012 (2012)
4. Choo, K.-K.R.: The cyber threat landscape: challenges and future research directions. Comput. Secur. **30**(8), 719–731 (2011)
5. Cristian, F.: Understanding fault-tolerant distributed systems. Commun. ACM **34**(2), 56–78 (1991)
6. DARPA: Cyber Information Sharing - DARPA Cyber Forum, October 2015
7. Estublier, J.: Software configuration management: a roadmap. In: Conference on the Future of Software Engineering (2000)
8. Google: Protocol Buffers, November 2015. https://developers.google.com/protocol-buffers/
9. Grobauer, B., Berger, S., Göbel, J., Schreck, T., Wallinger, J.: The MANTIS Framework: Cyber Threat Intelligence Management for CERTs, Boston, USA, June 2014
10. Guarnieri, C., Tanasi, A., Bremer, J., Schloesser, M.: The Cuckoo Sandbox (2012). http://cuckoosandbox.org
11. Hanif, Z., Calhoun, T., Trost, J.: BinaryPig: scalable static binary analysis over Hadoop. In: Black Hat USA (2013)
12. HiveMQ: MQTT Essentials Part 6: Quality of Service 0, 1 & 2 (2015). http://www.hivemq.com/blog/mqtt-essentials-part-6-mqtt-quality-of-service-levels
13. Jang, J., Brumley, D., Venkataraman, S.: BitShred: feature hashing malware for scalable triage and semantic analysis. In: Conference on Computer and Communications Security, CCS (2011)
14. Kolosnjaji, B., Zarras, A., Lengyel, T., Webster, G., Eckert, C.: Adaptive semantics-aware malware classification. In: Caballero, J., Zurutuza, U., Rodríguez, R.J. (eds.) DIMVA 2016. LNCS, vol. 9721, pp. 419–439. Springer, Heidelberg (2016). doi:10.1007/978-3-319-40667-1_21
15. Krafzig, D., Banke, K., Slama, D.: Enterprise SOA: Service-Oriented Architecture Best Practices. Prentice Hall Professional, Indianapolis (2005)
16. Lakshman, A., Malik, P.: Cassandra: a decentralized structured storage system. ACM SIGOPS Oper. Syst. Rev. **44**(2), 35–40 (2010)
17. Lengyel, T.K., Maresca, S., Payne, B.D., Webster, G.D., Vogl, S., Kiayias, A.: Scalability, fidelity and stealth in the DRAKVUF dynamic malware analysis system. In: Annual Computer Security Applications Conference, ACSAC (2014)
18. Ollmann, G.: Behind todays crimeware installation lifecycle: how advanced malware morphs to remain stealthy and persistent. Technical report, Damballa (2011)
19. Papazoglou, M.P., Van Den Heuvel, W.-J.: Service oriented architectures: approaches, technologies and research issues. VLDB J. **16**(3), 389–415 (2007)

20. Parkour, M., DiMino, A.: Deepend Research - Yara Exchange, May 2015. http://www.deependresearch.org/2012/08/yara-signature-exchange-google-group.htm
21. Shields, W.: Problems with PEHash Implementations, September 2014. https://gist.github.com/wxsBSD/07a5709fdcb59d346e9e
22. Stamos, A.: The Failure of the Security Industry, April 2015. http://www.scmagazine.com/the-failure-of-the-security-industry/article/403261/
23. The MITRE Corporation: Collaborative Research Into Threats (CRITs), June 2014. http://www.mitre.org/capabilities/cybersecurity/overview/cybersecurity-blog/collaborative-research-into-threats-crits
24. Verma, A., Pedrosa, L., Korupolu, M.R., Oppenheimer, D., Tune, E., Wilkes, J.: Large-scale cluster management at Google with Borg. In: European Conference on Computer Systems, EuroSys (2015)
25. VirusTotal: File Statistics, May 2015. https://www.virustotal.com/en/statistics/
26. Vixie, P.: Internet Security Marketing: Buyer Beware, April 2015. http://www.circleid.com/posts/20150420_internet_security_marketing_buyer_beware/
27. Wicherski, G.: PEHash: a novel approach to fast malware clustering. In: USENIX Workshop on Large-Scale Exploits and Emergent Threats, LEET (2009)
28. Zeltser, L.: SANS - Managing and Exploring Malware Samples with Viper, June 2014. https://digital-forensics.sans.org/blog/2014/06/04/managing-and-exploring-malware-samples-with-viper

Privacy and Watermarking

Leveraging Internet Services
to Evade Censorship

Apostolis Zarras(✉)

Technical University of Munich, Garching, Germany
`zarras@sec.in.tum.de`

Abstract. Free and uncensored access to the Internet is an important
right nowadays. Repressive regimes, however, prevent their citizens from
freely using the Internet and utilize censorship to suppress unwanted
content. To overcome such filtering, researchers introduced circumvention systems that avoid censorship by cloaking and redirecting the censored traffic through a legitimate channel. Sadly, this solution can raise
alerts to censors, especially when it is mistakenly used. In this paper, we
argue that relying on a *single* channel is not sufficient to evade censorship since the usage pattern of a circumvention system differs compared
to a legitimate use of a service. To address this limitation of state-of-the-
art systems, we introduce CAMOUFLAGE, an approach to combine multiple *non-blocked* communication protocols and dynamically switch among
these tunnels. Each protocol is only used for a limited amount of time and
the Internet connection is transparently routed through instances of different censorship circumvention systems. We prototype CAMOUFLAGE by
using applications which are based on these protocols and also offer end-
to-end encryption to prevent censors from distinguishing circumvention
systems from regular services. We evaluate CAMOUFLAGE in countries
that impose censorship and demonstrate that our approach can successfully bypass existing censorship systems while remaining undetected.

1 Introduction

The Internet has become the medium of choice for people to search for information, conduct business, and enjoy entertainment. It offers an abundance of
information accessible to anyone with an Internet connection. Additionally, the
explosion of social media plays a central role in shaping political debates, as
for example the coordination of movements in "Arab Spring", where citizens
used Facebook, Twitter, and YouTube to put pressure on their governments.
However, free communication threatens repressive regimes as it, in many cases,
exposes concealed truths and government corruption. Personal communication
is then becoming subject to pervasive monitoring and surveillance, and various
state and corporate actors are trying to block access to controversial information. In detail, regimes can trace, monitor, filter, and block data flows using
sophisticated technologies such as IP address blocking, DNS hijacking, and deep
packet inspection [12,27].

© Springer International Publishing Switzerland 2016
M. Bishop and A.C.A. Nascimento (Eds.): ISC 2016, LNCS 9866, pp. 253–270, 2016.
DOI: 10.1007/978-3-319-45871-7_16

With the use of censorship technologies to be more timely than ever, researchers developed a number of different systems to retain the freedom of the Internet [5,16,19], which are widely-known as *censorship circumvention systems* and most of the time try to deploy a redirection proxy that provides access to blocked websites. Nevertheless, censors can locate such proxies and instantly block them [20,33]. The root cause of most of the systems' identification is the differentiation they exhibit from regular Internet traffic. To overcome this limitation, *unobservable circumvention systems* were introduced. These systems try to impersonate popular applications to blend with the *allowed* Internet traffic [35,44,45]. Although these systems sound promising, they fail to raise the bar against censorship mostly because they only implement the imitating protocol partially and thus fall into discrepancies that censors can locate [25].

Consequently, unobservability by imitation is a fundamentally flawed approach [25]. For this reason, researchers introduced new systems that operate higher in the protocol stack. These systems avoid censorship by executing the actual protocol instead of trying to impersonate it [1,2,26,50] and thus can protect the users from various Internet restrictions. However, they suffer from being shut down if they got recognized, mostly due to their users' inexperience. In general, the average end-user is not familiar with the proper configuration and utilization of a circumvention system. Hence, the erroneous usage of such a system can create traffic that may appear suspicious to censors. Additionally, such systems support only one protocol and therefore are susceptible to loosing their functionality in case a government decides to completely block this protocol.

In this paper, we present CAMOUFLAGE, a novel approach that protects users from Internet censorship, while requiring limited expertise from the users' side compared to existing systems. More specifically, CAMOUFLAGE is a framework to which existing or future censorship-resistant systems can be plugged in, cooperate, and provide increased resistance against censorship. The main idea of our approach is to tunnel Internet traffic inside multiple non-blocked communication protocols and dynamically switch among them. Many of the existing systems can protect users from being subjects of censorship by tunneling the traffic through various protocols [1,2,26,50], yet they all work independently from each other. Thus, users who want to avoid censors' surveillance should install and configure as many of these systems on their computers, and manually rotate the forwarded traffic among them, which is likely an error-prone process. In our approach, the censorship-resistant tools can be attached to CAMOUFLAGE as plugins and the framework itself decides when and for how long they will be used.

To demonstrate the functionality of our framework, we built a prototype implementation, which supports four different protocols used by four widely-used applications. To evaluate CAMOUFLAGE in real-world, we chose countries that impose censorship and browsed the web with censored terms from inside these countries. The experimental results exhibit that our prototype can be successfully used for web browsing, while resisting censors' blocking efforts. To the best of our knowledge, our approach is the first attempt to combine a variety of different censorship circumvention systems under one solid framework.

In summary, we make the following main contributions:

- We propose CAMOUFLAGE, a novel approach for censorship circumvention that combines the advantages of existing systems while makes it easier for users to employ them. Its *plug-and-play* architecture can be used by existing or future censorship-resilient systems.
- We build a prototype based on widely-used applications and evaluate its performance and security. We show that the produced overhead is related to the implementation of each circumvention system.
- We evaluate our prototype in existing censored networks and show its ability to bypass censorship in real-world, concealing at the same time its presence from the censors.

2   Threat Model

Throughout this paper we use the following threat model. We assume that a user connects to the Internet through an Internet Service Provider (ISP) that utilizes some kind of censorship system. In fact, governments can control and regulate ISPs, which can be forced to monitor and block users' access to certain Internet destinations. For example, China can filter IP packets [9], or even can censor services such as blog platforms [52], chat programs [31], and search engines [51]. In our scenario, we assume that the user operates in an inhospitable network where ISPs can trace, monitor, filter, and block data flows using sophisticated technologies [12]. Additionally, we assume that the users are restricted from using proxies or other circumvention systems to evade censorship. Therefore, we take for granted that ISPs can identify and block the traffic that is forwarded through a censorship circumvention system using a variety of different features and strategies [22,25] and notify the authorities for any incident. Then, the violators might face severe punishments varying from payment of fines to even imprisonment [46].

We also assume that regimes do not want to jeopardize the usability of the Internet due to political and economical implications. For instance, services such as email and VoIP constitute important parts of today's communications among businesses, which benefit from these services to reduce their operational costs. Furthermore, VoIP and chat communications are also widespread among individuals due to the level of convenience and flexibility they offer. Finally, another type of Internet services that is popular, mostly among young people, is online games, which are used for entertainment purposes. Thus, we speculate that censorship regulations do not interfere with fundamental Internet services such as email communications, VoIP, file sharing, and even entertainment services including online games. In the extreme case in which a censor might decide to block some of these services, we believe that this decision will not affect all of them, but only a fraction of the services.

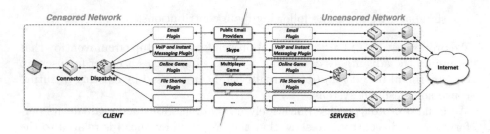

Fig. 1. Abstract architecture of CAMOUFLAGE.

3   System Design

CAMOUFLAGE acts as a pluggable framework for different circumvention systems that evade censorship by leveraging existing protocols and services. Thus, in order to gain access to the framework, users need to install the CAMOUFLAGE *client* on their computers along with the appropriate plugins (i.e., circumvention systems). On the other hand, administrators should install the CAMOUFLAGE *server* as well as the supported plugins. Note that a server can support multiple circumvention systems that operate in parallel and are synchronized. Both clients and servers consist of key components necessary for converting and forwarding the traffic. In the remainder of this section, we introduce the main components of CAMOUFLAGE and describe how do they contribute to the design and operation of a secure and easy-to-use framework.

3.1   Abstract Architecture

CAMOUFLAGE is composed of two key components: a client and a server. The client usually runs in a censored environment in which all the communications are monitored by censors, while the server operates in an uncensored and secure environment where the network traffic flows unrestricted. In addition, our framework transfers all the data transparently and thus any software which can be configured to use a proxy, such as a web browser, is able to use the client. On the server side, we set up a proxy (HTTP or SOCKS), which is responsible to communicate with the outside world and access the censored content. Figure 1 illustrates the framework's abstract architecture.

The client consists of the following components: a connector, a dispatcher, and different plugins. The connector receives the traffic from the user's application (e.g., the web browser), transforms it from multiple connections' traffic to a serialized data stream by adding suitable headers, and then encrypts it. Next, it forwards the traffic to the dispatcher, which decides through which plugin the data stream will be transferred. Finally, the plugin running in the client's side is responsible for transmitting the data stream. Accordingly, when a plugin receives an encrypted data stream transfers it to the connector through the dispatcher. In this component, the data stream is decrypted, is split into multiple connections, and the traffic is forwarded to the appropriate application.

The server, on the other hand, consists of the proxy, the connector and the supported plugin. When encrypted data is received, the plugin transfers it to the connector. The server's connector operates in a similar way to the one presented for the client. More specifically, when a data stream is received, it checks the connection identifier and forwards the data through the corresponding connection with the proxy, or opens a new connection if the identifier is currently unused (i.e., the connection was newly established on the client's side). Similar, when the server transmits data back to the client, the proxy sends the data sequences to the connector in order to transform them into an encrypted data stream and then the plugin transfers it through the suitable channel. As we previously mentioned, a server can support more than one circumvention system. In this case, the existence of a dispatcher is necessary in order to combine the different plugins and forward the traffic to the same proxy.

3.2   Connector

Modern applications, such as web browsers, use HTTP/1.1 protocol that allows multiple network connections to run in parallel. In detail, when a user visits a web page that has many different objects on it (e.g., images, JavaScript files, frames, data feeds, etc.), the browser tries to download several of them at once in different parallel sessions to obtain better performance. Most HTTP servers and browsers use an HTTP protocol feature called *keep-alive* that does not close the TCP connection when the client is done with it; the connection closes either after an idle connection timeout or after a maximum number of allowed requests. This makes sense since opening a remote connection is expensive due to the three-way TCP handshake, so it is faster to open one connection and then download n items compared to open and close a connection n times. However, the plugins provide only one data channel for forwarding the data. To overcome this obstacle, we need a middleware that combines the multiple connections generated from applications with the single data channel provided by the plugins. In CAMOUFLAGE, the connector plays the role of this middleware.

In detail, the connector bundles the traffic from all connections to one data stream by adding a small header in front of the payload. This header contains the identifier of the connection and the payload's length. Using the identifiers, the connector can forward the data to the correct connection within the user's application, while the length of the payload ensures that the appropriate amount of data will be read. The transformed data stream is then encrypted and forwarded to plugins. The connector also receives a similar data stream back from the plugins with the contents of the censored web pages. Additionally, it ensures that the state of each connection is consistent at both client and server side. Especially if one side closes a connection, this information is transmitted to the other side which in turn closes the connection as well.

3.3   Dispatcher

The dispatcher acts as an intermediate component between the connector and the actual plugins. More precisely, the use of the dispatcher is twofold: (i) forward the traffic from a client to a server and vice-versa, and (ii) schedule the orchestration of activities by planning and monitoring the plugins. For the first part, the dispatcher forwards the outgoing traffic from the connector to a server over the currently active plugins. Accordingly, it receives the incoming traffic from a server and delivers it to the connector. To this end, the dispatcher should know which plugins are supported by the current instance of CAMOUFLAGE. If a connection to a plugin is not feasible anymore, the dispatcher will temporarily disable this plugin and will try to connect with an alternative one. Nevertheless, the plugin will be enabled again after a timeout period.

For the second part, the dispatcher administers the plugins by implementing the schedule in which a plugin is selected. The implementation of the schedule could be created either manually by the user or automatically by integrating the manifest files of plugins. The former solution is recommended for advanced users, while inexperienced ones can always use the latter. The dispatcher can decide at any time if it wants to execute the plugins on parallel, or one at a time. This way, makes it difficult for the censors to create detection patterns.

3.4   Plugins

The data channels (circumvention systems) that used for tunneling the traffic are the lifeblood of CAMOUFLAGE. Each circumvention system is implemented in our framework as a standalone plugin. CAMOUFLAGE supports an abstract architecture, in which new circumvention systems can be easily plugged in. It is mandatory that the implementation of each system to leverage an actual protocol, or service, and not try to imitate it. Additionally, each plugin must accompanied by a manifest file. This file contains information such as the plugin's unique identifier, the recommended operation time used by the dispatcher, the suggested timeout period after which the dispatcher will try to reconnect to the plugin if a previous connection attempt failed, and the list of the required software. Both the usage time duration and the timeout period are only recommended values by developers and can be voluntarily modified by users.

Each plugin is responsible for concealing its tunneled data stream. Therefore, it is highly recommended the utilization of a two-layer encryption strategy. First, each plugin should leverage applications that use encrypted communications by default. This will prevent regimes that monitor the network traffic to have a direct access to unencrypted data. However, this may not be always the case. For instance, a company that wants to operate in a country that imposes censorship must accept the country's requirements in order to be allowed to enter the country's market. These requirements could permit, among the others, the government's censors to have access to unencrypted network traffic. Thus, no matter how strong is the encryption that the application uses, the plugin should implement a second layer of encryption as well. More specifically, it must encrypt the data stream before being encrypted by the application itself. This way, the censors that monitor the network traffic will be prevented from accessing the plaintext data if an agreement with the application provider is established. Overall, the two layers of encryption ensure that even if the application gets *compromised*, the content of the data stream will remain hidden.

4   Circumvention Systems

As we mentioned, the lifeblood of our system is the different plugins it supports and each of them implements a circumvention system. To prove the feasibility of our approach, we implemented a prototype of CAMOUFLAGE using four different circumvention systems that are attached to our framework as plugins. These plugins leverage four different services and protocols to evade censorship and do not rely on emulation or any other imitation technique, but each of them utilizes the actual implementation of a protocol. To test each service, we selected an application that supports a utilized protocol and implemented the censorship-resistant system on top of this application. In brief, we chose the following applications:

1. *Public email providers:* to implement a circumvention system based on SMTP protocol.
2. *Skype:* to leverage VoIP and instant messaging communication protocols.
3. *Runes of Magic:* to benefit from communication protocols in multiplayer online game platforms.
4. *Dropbox:* to conceal traffic within a file sharing service.

Note that these applications are selected only as proof-of-concept scenarios and therefore the circumvention systems can operate in the same way with differently selected applications (e.g., utilization of *Google Voice* as VoIP application). Same principle applies in the case that an application is forbidden in a country. It is worth to mention here that some of these applications were proposed in previous works [26,50], however, it was not possible to find their actual implementations and in order to evaluate our prototype we had to design and implement them from scratch. In addition, for the purpose of this paper, the communication between the client and the server is already initialized, which means that there is no need for any kind of registration. Nevertheless, sophisticated registration strategies could be applied in real-world circumvention systems to prevent attacks, such as denial-of-service against the system, but these strategies are outside the scope of this paper. In general, the framework itself allows developers to design their own registration strategies without any restrictions.

4.1   Email

Electronic mail (email) is a widely-used method of exchanging digital messages from an author to one or more recipients. Modern email systems are based on a *store-and-forward* architecture, which permits servers to accept, forward, deliver, and store messages. The wide acceptance of email allows us to create a circumvention system that utilizes the email delivery system to evade censorship. More precisely, the system uses publicly available email providers to hide the data stream inside email messages. Private email providers could be used as well to perform this task, but we believe they are more prone to manipulation by repressive regimes compared to public email services provided by international corporations. We assume that censors do not have access to the users' mailboxes hosted outside of the regime's geographical boundaries. In contrast, it is easier for a censor to access mailboxes of email providers that are hosted inside the regime's borders. Moreover, we consider as suitable email providers those which by default offer email encryption. This way, an encrypted email will not be considered suspicious by censorship authorities, if all the emails sent by this email provider are encrypted as well.

With regards to the design of the email circumvention system, both client and server plugins share the same mailbox that is hosted at a public service. The email servers of this service must reside outside the censors' jurisdiction (e.g., Gmail, Yahoo Mail, Outlook.com). The client communicates with the server, and vice-versa, by sending emails to the same mailbox. The data stream is divided into blocks, which are encoded in *Base64* format. Additionally, we use encryption to protect users from revealing the content of the visited web pages to a compromised email provider. The body of the email contains the data stream, while the header contains a message counter to retain the correct sequence of data blocks and a flag that indicates the sender of this message. Both client and server implementations use the *IMAP IDLE* feature to get notified on new emails and recreate the data stream from the received messages. This feature allows IMAP email users to immediately receive any mailbox changes without having to automatically and repeatedly ask the server for new messages [28].

4.2   VoIP and Instant Messaging

Voice over Internet Protocol (VoIP) is a technology used for the delivery of voice communications and multimedia sessions over Internet Protocol (IP) networks. Instant messaging, on the other hand, is a type of online chat that offers real-time text transmission over the Internet. Short messages are typically transmitted bi-directionally between two parties, where each party composes a sentence and sends it to the other. Modern applications that support VoIP often support instant messaging as well. One such application is Skype, in which users communicate with each other both with voice and text. For the prototype of this circumvention system we use Skype and implement it to transmit data either as audio or as encrypted text messages. However, any other VoIP software will work in a similar way. In the following paragraphs we analyze the operation of each service in more detail.

Nowadays, many companies switch from traditional telephony to VoIP to reduce their telephone costs [10,17]. Similarly, individuals communicate with each other using publicly available and usually free VoIP services. Hence, VoIP is difficult to be manipulated or even blocked by censorship authorities. We exploit this censorship weakness and create a system that transfers uncensored data over VoIP. More specifically, our system modulates binary data into audio signals, which are transmitted over the voice channel of a VoIP software. This concept is similar to the operation of modems (i.e., a device that modulates an analog carrier signal to encode digital information and also demodulates such a carrier signal to decode the transmitted information). The goal is to produce a signal, which can be transmitted and decoded, to reproduce the original digital data. For our prototype we used a 1200 baud Audio Frequency Shift Keying modem without sophisticated error correction. However, in a real-world application a more advanced approach should be applied.

We mentioned that for the prototype implementation of the system we selected Skype. Therefore, the data stream is converted to audio signals and transmitted from one edge to the other using *Skype call*. For sending the generated audio signal over Skype we use a virtual audio device. This device is installed as a driver and behaves like a real sound card. In Skype it can be configured as the microphone device. Additionally, a second virtual audio device is set as Skype's output and all signals pointing to this device are recorded by the system, and eventually demodulated back to the initial data stream.

Alongside with VoIP services, some companies leverage instant messaging as a complementary method to the customer support services they offer. In addition, instant messaging is one of the most popular Internet activities among individuals. According to studies, a vast majority of the Internet's population uses instant messaging as its main communication tool [29]. Consequently, if a repressive regime blocks a major application such as Skype, which provides instant messaging communication services to its users, this will cause a severe impact to a significant portion of the regime's inhabitants and companies. Hence, in some cases is preferable for a repressive regime to monitor this means of communication than completely forbid it. With this in mind, we created a circumvention system that utilizes the text channels of instant messaging services. To our benefit, Skype automatically encrypts the transmitted messages, fact that offers us a two-layer encryption similar to email circumvention system. In general, we used a similar approach to the one presented for the email circumvention system.

4.3 Online Gaming

Online games are video games which are played over the Internet. These games are divided in single-player where input from only one player is expected throughout the course of the gaming session and multi-player that allow more than one person to play in the same game environment at the same time. The latter provide their users (players) with a form of social communication channel in which the players can interact with each other. These communication services in multi-player games can be used as covert channels for the secret transfer of data streams. Modern multi-player games offer various communication channels such as public chat rooms, private instant messaging, and voice chat. We leverage the services offered by multi-player games to create a circumvention system.

We prototype a circumvention system, by using the voice chat communication channel of *Runes of Magic*, a massively multiplayer online role-playing game. We create two characters (avatars) which are connected to the same server. The client owns the first avatar, while the server possesses the second. The avatars create a voice chat room where they communicate. We utilize a similar approach to the one presented for the VoIP circumvention system. More accurately, we convert the digital data stream to audio signals and transmit them from one side to the other through the voice chat room. The other side demodulates the signals back to their original form and processes the requests.

This approach is only a proof-of-concept. We want to show that a strategy that leverages online games to evade censorship is feasible. Thus, we only use the voice chat rooms to transmit the data, which imposes limitations in the amount of data being transferred. To increase the amount of transmitted data, we can use in parallel approaches that benefit, for example, from the movements of avatars to transmit information over the online game engine [47, 48].

4.4   File Sharing

File sharing is a private or public distribution of data or resources, such as documents, multimedia, graphics, computer programs, images, or e-books, in a network with different levels of sharing privileges. More specifically, it allows to a number of people to simultaneously modify the same files. File sharing is known for quite a long period. It is massively used, from companies and universities, among people who work on the same projects and need to share and modify the same documents and data. Recently, as social media has become increasingly popular with hundreds of millions of users, file sharing became attractive among individuals as well. Nowadays, people use it to upload and share pictures and videos among friends. File sharing is so massively used in today's world, which offers a unique opportunity for a circumvention system based on this service.

We create a circumvention system on top of Dropbox. Dropbox is a personal cloud storage service frequently used for file sharing. Our censorship-resistant system uses the official Dropbox client to create a private communication channel between the client and the server. After configuring and starting Dropbox on both sides, the tool uses a shared folder for data exchange. The sent data is split into blocks which are then stored as encrypted files in the shared folder. The folder is automatically synchronized with the other party. Both plugins can monitor the folder for new files, read the data from them, and recreate the data stream. For additional security we can apply solutions that revoke the access to files after a certain period of time [6, 21, 37, 49].

5   Evaluation

We implemented a prototype of CAMOUFLAGE and evaluated the effectiveness of our approach. We first measured its performance and then studied the users' behavior to properly configure the plugins. Finally, we evaluated CAMOUFLAGE's ability to evade censorship on countries that impose censorship to their citizens.

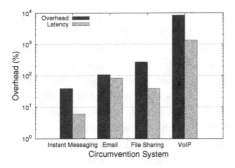
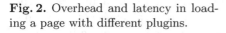

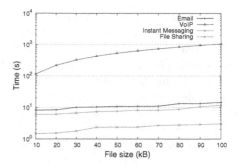

Fig. 2. Overhead and latency in loading a page with different plugins.

Fig. 3. Downloading time for files size from 10 to 100 kB.

5.1 Performance

Using a circumvention system is the only way to access a censored web page, especially if services such as Virtual Private Networks (VPNs) are forbidden. Thus, to show that a circumvention system must only be used for accessing censored web pages, and not for web surfing, we measured its performance. Therefore, all the performance experiments were conducted by using broadband connections for accessing the Internet on both client and server side. We believe that individuals can contribute to the Internet's freedom by hosting at least one circumvention system (Tor [16] uses a similar infrastructure). Hence, we chose to use DSL instead of a university's connection to make our experiments more realistic. For instance, individuals who want to contribute in the fight against censorship can run a CAMOUFLAGE server during the night when their bandwidth is usually idle. During our experiments, the utilized DSL connections remained idle and the only traffic sent over the network was the traffic which was created by the circumvention systems. Obviously, using broadband connections had a huge impact in the performance of CAMOUFLAGE, compared to previous works [26], but at the same time it provided us with realistic results instead of the ideal results we would have gotten in a sterile laboratory environment.

In our first experiment, we measured the overhead and the latency added by each circumvention system when a web page is downloaded. First, we visited the main page of popular websites with a browser without using any intermediate channel and captured all the incoming traffic which constitutes the bottom line of the experiment. Then, we visited the same web pages with different circumvention systems. Between the measurements we cleared the browser's cache so all the contents of the web pages were loaded directly from the network. Figure 2 shows the average values for the overhead and the latency of each system. As we can see, instant messaging increased the incoming data only by 39 % compared to VoIP, which increased the traffic by a factor of 84. This is caused because VoIP can conceal a smaller amount of information in its audio signals than the transferred information through raw text. Regarding email, it increased the overhead by 107 %, while file sharing increased it by 272 %.

The latency on the other hand is caused by different factors for the various systems. For instant messaging, the latency is low because this service is designed to be real-time and transferring data this way does not create much overhead. In the case of the email, the latency is mainly caused by the time email servers spend for processing the mails; the traffic overhead is considered negligible here. Same principle applies for the file sharing. The data is first sent to a file sharing server, processed there, and then synchronized with the file sharing client. Finally, the latency of the VoIP system is clearly caused by its limited bandwidth.

Expanding the previous experiment, we conducted downloads of files of various sizes and measured the time required for a complete download. The experiment used the same configurations described earlier. Figure 3 illustrates the results. The outcomes of this experiment help us to create a version of our framework in which all the supported circumvention systems can operate in parallel. In detail, we notice that a circumvention system that uses instant messaging is the most effective approach no matter what the size of the downloaded file is. Therefore, this approach could be used for web elements that require more bandwidth such as videos and high-resolution images. Email and file sharing systems behave in similar way and the offered bandwidth by each system is rather close to the other. Thus, these systems could be used to download medium and low quality images and medium size Flash applications. Finally, VoIP systems offer a rather small bandwidth for transferring data. Hence, VoIP censorship-resistant tools can be used to transmit a limited amount of data, such as a text entry in a microblogging service like Twitter, or to download the mobile versions of the websites. Although, in theory all circumvention systems can be used for all the tasks, in practice is not always feasible.

5.2   Traffic Patterns

CAMOUFLAGE offers unobservable connections. By unobservability we mean the ability of a circumvention system to hide its existence from censorship authorities. In other words, the censors should not be able to identify whether a user is using a circumvention system. This is essential for the user's safety because authorities can prohibit the use of technologies that evade censorship. To prevent a censor from detecting CAMOUFLAGE, the behavior of the protocols used as tunnels should be as close as possible to a user's normal use of those protocols.

For this, we monitored the users' behavior when they utilize the proposed applications. Additionally, we considered reports that studied traffic characteristics [8, 40, 41]. The outcomes allowed us to create generic traffic patterns that match the behavior of the majority of users. Table 1 depicts an overview of our exported results. Keep in mind that these results are affected by many factors, such as the age and cultural influences, and may look different in separate regions.

Table 1. Suggested time values for different applications.

Application	Duration (min)
Email	1–3
VoIP	20–30
Instant messaging	15–20
File sharing	5–10

Unfortunately, there exist individuals whose behaviors do not match with our proposed patterns. This allows to a well-trained censor to detect these discrepancies. With that in mind, we designed our prototype to liberally allow its users to configure it based on their demands. More precisely, during the installation process of a circumvention system, users get informed about the typically-used traffic pattern and asked for any modifications. Note that users are not bind to their initial decisions and can modify these traffic patterns any time they want. However, we recommend only the advanced users to manually modify these values because a misconfigured system can cause exactly the opposite results and the existence of CAMOUFLAGE to be revealed to censors.

Nevertheless, our experiments showed that is not always feasible for users to know their unique traffic patterns as they may do not have neither the experience nor the suitable tools to measure it. Simultaneously, users that use the very same services leveraged by our plugins in their everyday lives can raise suspicions when these services used with and without CAMOUFLAGE. This results in two different traffic patterns which can be observed by anomaly-based detection systems. To overcome this limitation we enhanced CAMOUFLAGE with the ability to monitor and decipher the network traffic patterns of a user, when the plugins are not used. This way, the generated network patterns will be unique and based on user's typical behavior. Consequently, a censor will not be able to distinguish between the real traffic and the artificial one, and thus will not be able to detect the existence of our framework, as the results of our experiments revealed. More precisely, the traffic patterns with and without the use of CAMOUFLAGE looked almost identical which makes it impossible for censors to spot the differences.

5.3   Real-World Deployment

To explore how CAMOUFLAGE performs in a real-world scenario, we evaluated it in countries that impose censorship to their citizens. Therefore, we acquired access to servers hosted in these countries and imposed to the same censorship as these countries' inhabitants. Then, we tried to access forbidden websites, which are usual websites that contain known forbidden keywords. We repeated our experiments in different time periods to capture any possible changes in the detection capabilities of the censors. Our early results demonstrate that our framework can successfully evade censorship, while it remains undetectable for a large period. In detail, without CAMOUFLAGE it was not possible to render a plethora of web pages that contained one or more forbidden keywords, which became possible once we utilized our framework. To be certain that our results are accurate, we repeated these experiments for a period of one month. During this period we were able to evade censorship that was imposed in different countries.

A perfect example of such a country is China. It is well known that China has the world's most complex Internet censorship system [23]. However, its censors are very prudent to perform DNS hijacking nowadays due to the risk of affecting the network in other countries [34]. Chinese censors impose strict restrictions on international Internet traffic and the most effective filtering mechanism is the keyword filtering. These factors make China an ideal candidate to evaluate CAMOUFLAGE. For this purpose, we used a list of known forbidden keywords. We ran our experiments in daily basis for a period of one month and monitored if during this period (*i*) we were able to access censored web pages and (*ii*) CAMOUFLAGE got detected and its services were banned. The outcomes of this experiment revealed that with the utilization of CAMOUFLAGE we could access web pages that otherwise would not be possible. Additionally, during the period of our experiment we were able to continuously access these forbidden web pages which shows that our framework was not detected by censors. Therefore, we believe that CAMOUFLAGE can assure unobservability as it blends with the real network traffic. It is worth to mention that not all the applications were available in China, for instance, we could not use Dropbox. Therefore, we performed our experiments only with the allowed plugins. Nevertheless, even if an application is not available to a country, it can easily be replaced by another. Therefore, by using more plugins we increase our chances to access the data we want in a heavy censored environment.

Overall, CAMOUFLAGE was able to evade censorship when applied. However, the main goal of the framework is to provide access to censored web pages. Thus, it should explicitly be used only for this specific scenario and not for everyday tasks such as web surfing, e-radio listening, or any other activities that require a high bandwidth connection and could raise suspicions due to highly-produced network traffic over services that are not designed to produce so.

6   Related Work

Circumvention systems try to ensure anonymity to their users. Anonymity is an old idea. Chaum proposed a technique based on public key cryptography that allows an electronic mail system to hide both the participants and the content of the communication [7]. There exist systems that provide anonymity by following a high-latency network design [13,24]. These systems can resist against strong adversaries, but introduce too much lag for interactive tasks such as web browsing. On the opposite side, systems with low-latency network design [3,4,11] can anonymize interactive network traffic, but it is difficult to prevent an attacker that eavesdrops both ends of the communication from correlating the timing and volume of traffic [38]. Finally, there is the peer-to-peer network design in which the participants both generate and relay traffic for others [30,39].

On the other hand, the oldest technique to evade censorship and surveillance is the use of open proxy servers [18,42]. Toward this direction, INFRANET [19] improves the proxies infrastructure by leveraging a tunnel protocol that provides a covert communication channel between the clients and the servers. Similarly, COLLAGE [5] uses user-generated content on social-networking and image-sharing websites such as Facebook and Flickr to embed hidden messages into cover traffic. To this end, researchers presented an obfuscation-based approach that enables users to follow privacy-sensitive channels, while makes it difficult for the censors to discover the users' actual interests [36]. However, these designs are susceptible to inside attacks where censors pretend to be ordinary users to locate and block the infrastructure of censorship resilient systems [20]. Over the years, researchers have designed better proxy distribution strategies that protect proxies from Sybil attacks [20,33]. Additionally, reputation systems might be used to detect censors who have infiltrated in the proxies' network [43]. These strategies are adequate against individual users, but are insufficient in thwarting censors that rule a significant amount of untrustworthy users.

One of the most effective circumvention tools is Tor [16], which is a circuit-based anonymous communication system that uses encryption to conceal the network packets. More specifically, it interposes at least three relays between each user and the website the user visits. The transferred packets through these relays are encrypted, and each relay can decrypt only the necessary information that leads the packet to the next relay. Although the relationship between the user and the visited website through Tor is secure, repressive governments can block the Tor itself [14]. To make Tor resilient to such attacks, developers have proposed a centralized discovery service to disseminate a restricted set of relay identities to requesting users [15]. OBFSPROXY [32] is the first Tor pluggable transport, which adds an additional layer of encryption to Tor's traffic to obfuscate its identifiers. However, albeit the developers' modifications, the problem is that the traffic generated by Tor remains recognizable by its characteristic patterns and content signatures.

Unobservable circumvention systems, on the other hand, instead of encrypting the web content and access it through proxies, imitate applications and blend with the authorized Internet traffic. SKYPEMORPH [35] is a system designed to encapsulate the Tor's traffic into a connection that resembles Skype video traffic. CENSORSPOOFER [44] is a framework for censorship-resistant web browsing that exploits the asymmetric nature of web browsing traffic. STEGOTORUS [45] is a tool that comprehensively disguises Tor from protocol analysis. Although these approaches sound promising, Houmansadr et al. [25] demonstrate that these systems fail to achieve unobservability because they implement only partially the imitating protocol and fall into discrepancies.

An alternative to unobservable circumvention systems by imitation is to run the actual protocols and tunnel the hidden content inside their traffic. FOE [2] and MAILMYWEB [1] are two systems that can download a requested website and send it as an email attachment to the requesting user. These systems can evade censorship, however, the users cannot interact with the actual website and they can only leverage these systems for accessing static websites. SWEET [50] encapsulates a censored user's traffic inside email messages. In detail, the client tunnels its network traffic inside a series of email messages that are changed between the client and an email server operated by SWEET's server. The server acts as an Internet proxy by forwarding the encapsulated traffic to the requested blocked destinations. FREEWAVE [26] operates by tunneling Internet traffic inside non-blocked VoIP communications by modulating them into acoustic signals that are carried over VoIP connections. These systems appear to work flawless. Nevertheless, their main limitation is that they only support one protocol and thus their overuse by users can trigger alerts on censors.

7   Conclusions

In this paper, we presented CAMOUFLAGE, a novel approach that protects users from Internet censorship. The key idea of CAMOUFLAGE is a framework where different circumvention systems can be plugged in and help users to access an uncensored Internet. The framework operates one layer below the circumvention system and thus the design and the implementation of each system lies in the hands of each developer. To demonstrate the feasibility of our approach, we built a proof-of-concept prototype on widely-used applications and evaluate its performance and security. We showed that different systems can co-exist with each other, while a central framework can synchronize them. Finally, we evaluated CAMOUFLAGE in countries that impose censorship. The outcomes of our experiments revealed that by using CAMOUFLAGE is possible to access an uncensored Internet while at the same time the existence of our framework remained hidden from the deployed censors.

Acknowledgments. The research was supported by the German Federal Ministry of Education and Research under grant 16KIS0328 (IUNO).

References

1. MailMyWeb, June 2013. http://www.mailmyweb.com
2. The Foe Project, November 2013. https://code.google.com/p/foe-project
3. Berthold, O., Federrath, H., Köpsell, S.: Web MIXes: a system for anonymous and unobservable internet access. In: Federrath, H. (ed.) Designing Privacy Enhancing Technologies. LNCS, vol. 2009, pp. 115–129. Springer, Heidelberg (2001)
4. Boyan, J.: The anonymizer - protecting user privacy on the web. Comput. Mediated Commun. Mag. **4**(9) (1997)
5. Burnett, S., Feamster, N., Vempala, S.: Chipping away at censorship firewalls with user-generated content. In: USENIX Security Symposium (2010)
6. Castelluccia, C., De Cristofaro, E., Francillon, A., Kaafar, M.-A.: EphPub: toward robust Ephemeral Publishing. In: IEEE International Conference on Network Protocols (2011)
7. Chaum, D.L.: Untraceable electronic mail, return addresses, and digital pseudonyms. Commun. ACM **24**(2), 84–90 (1981)
8. Chui, M., Manyika, J., Bughin, J., Dobbs, R., Roxburgh, C., Sarrazin, H., Westergren, M.: The Social Economy: Unlocking Value and Productivity Through Social Technologies. McKinsey, New York (2012)
9. Crandall, J.R., Zinn, D., Byrd, M., Barr, E.T., East, R.: ConceptDoppler: a weather tracker for internet censorship. In: ACM Conference on Computer and Communications Security, CCS (2007)
10. Crispin, J.: The importance of VoIP and a business continuity plan for business survival, June 2011. http://www.tech2date.com/the-importance-of-voip-and-a-business-continuity-plan-for-business-survival.html
11. Dai, W.: Pipenet 1.1. Usenet post (1996)
12. Dainotti, A., Squarcella, C., Aben, E., Claffy, K.C., Chiesa, M., Russo, M., Pescapé, A.: Analysis of country-wide internet outages caused by censorship. In: ACM SIGCOMM Conference on Internet Measurement, IMC (2011)

13. Danezis, G., Dingledine, R., Mathewson, N.: Mixminion: design of a type III anonymous remailer protocol. In: IEEE Symposium on Security and Privacy (2003)
14. Dingledine, R.: Tor and circumvention: lessons learned. In: Rogaway, P. (ed.) CRYPTO 2011. LNCS, vol. 6841, pp. 485–486. Springer, Heidelberg (2011)
15. Dingledine, R., Mathewson, N.: Design of a blocking-resistant anonymity system. The Tor Project, Technical report, 11:15–16 (2006)
16. Dingledine, R., Mathewson, N., Syverson, P.: Tor: the second-generation onion router. In: USENIX Security Symposium (2004)
17. Douglas, T.L.: The importance of VoIP, May 2010. http://ezinearticles.com/? The-Importance-of-VoIP&id=4278231
18. Dynamic Internet Technology: Dynaweb, November 2013. http://www.dit-inc.us/dynaweb
19. Feamster, N., Balazinska, M., Harfst, G., Balakrishnan, H., Karger, D.R.: Infranet: circumventing web censorship and surveillance. In: USENIX Security Symposium (2002)
20. Feamster, N., Balazinska, M., Wang, W., Balakrishnan, H., Karger, D.R.: Thwarting web censorship with untrusted messenger discovery. In: Dingledine, R. (ed.) PET 2003. LNCS, vol. 2760, pp. 125–140. Springer, Heidelberg (2003)
21. Geambasu, R., Kohno, T., Levy, A.A., Levy, H.M.: Vanish: increasing data privacy with self-destructing data. In: USENIX Security Symposium (2009)
22. Geddes, J., Schuchard, M., Hopper, N.: Cover your ACKs: pitfalls of covert channel censorship circumvention. In: ACM Conference on Computer and Communications Security, CCS (2013)
23. Global Internet Freedom Consortium (GIFC): The great firewall revealed, December 2002. http://www.internetfreedom.org/files/WhitePaper/ChinaGreatFirewallRevealed.pdf
24. Gulcu, C., Tsudik, G.: Mixing e-mail with Babel. In: ISOC Network and Distributed System Security Symposium, NDSS (1996)
25. Houmansadr, A., Brubaker, C., Shmatikov, V.: The parrot is dead: observing unobservable network communications. In: IEEE Symposium on Security and Privacy (2013)
26. Houmansadr, A., Riedl, T., Borisov, N., Singer, A.: I want my voice to be heard: IP over Voice-over-IP for unobservable censorship circumvention. In: ISOC Network and Distributed System Security Symposium, NDSS (2013)
27. Leberknight, C.S., Chiang, M., Poor, H.V., Wong, F.: A taxonomy of internet censorship and anti-censorship, December 2012. http://www.princeton.edu/chiangm/anticensorship.pdf
28. Leiba, B.: RFC 2177: IMAp. 4 IDLE Command, June 1997
29. Leskovec, J., Horvitz, E.: Planetary-scale views on a large instant-messaging network. In: International Conference on World Wide Web, WWW (2008)
30. Levine, B.N., Shields, C.: Hordes: a multicast based protocol for anonymity. J. Comput. Secur. **10**(3), 213–240 (2002)
31. MacKinnon, R.: Race to the bottom-corporate complicity in Chinese internet censorship. In: Human Rights Watch, HRW (2009)
32. Mathewson, N.: The Tor project - a simple obfuscating proxy, November 2013. https://gitweb.torproject.org/obfsproxy.git
33. McCoy, D., Morales, J.A., Levchenko, K.: Proximax: a measurement based system for proxies dissemination. In: International Conference on Financial Cryptography and Data Security, FC (2011)
34. McMillan, R.: China's great firewall spreads overseas, March 2010. http://www.networkworld.com/news/2010/032510-chinas-great-firewall-spreads.html

35. Moghaddam, H.M., Li, B., Derakhshani, M., Goldberg, I.: SkypeMorph: protocol obfuscation for Tor bridges. In: ACM Conference on Computer and Communications Security, CCS (2012)
36. Papadopoulos, P., Papadogiannakis, A., Polychronakis, M., Zarras, A., Holz, T., Markatos, E.P.: K-subscription: privacy-preserving microblogging browsing through obfuscation. In: Annual Computer Security Applications Conference, ACSAC (2013)
37. Perlman, R.: The ephemerizer: making data disappear. J. Inf. Syst. Secur. (JISSec) 1, 51–68 (2005)
38. Serjantov, A., Sewell, P.: Passive attack analysis for connection-based anonymity systems. In: Snekkenes, E., Gollmann, D. (eds.) ESORICS 2003. LNCS, vol. 2808, pp. 116–131. Springer, Heidelberg (2003)
39. Sherwood, R., Bhattacharjee, B., Srinivasan, A.: P^5: a protocol for scalable anonymous communication. In: IEEE Symposium on Security and Privacy (2002)
40. Statista: Daily time spent playing video games per capita in the United States, September 2014. http://www.statista.com/statistics/186960/time-spent-with-videogames-in-the-us-since-2002
41. Statistic Brain: Skype statistics, September 2012. http://www.statisticbrain.com/skype-statistics
42. Ultrareach Internet Corporation: Ultrasurf, November 2013. http://www.ultrasurf.us
43. Walsh, K., Sirer, E.G.: Experience with an object reputation system for peer-to-peer filesharing. In: USENIX Symposium on Networked Systems Design and Implementation, NSDI (2006)
44. Wang, Q., Gong, X., Nguyen, G.T., Houmansadr, A., Borisov, N.: CensorSpoofer: asymmetric communication using IP spoofing for censorship-resistant web browsing. In: ACM Conference on Computer and Communications Security, CCS (2012)
45. Weinberg, Z., Wang, J., Yegneswaran, V., Briesemeister, L., Cheung, S., Wang, F., Boneh, D.: StegoTorus: a camouflage proxy for the Tor anonymity system. In: ACM Conference on Computer and Communications Security, CCS (2012)
46. Wilkins, B.: 25 shocking facts about Chinese censorship, July 2009. http://www.onlinecollege.org/2009/07/05/25-shocking-facts-about-chinese-censorship
47. Zander, S., Armitage, G., Branch, P.: Covert channels in multiplayer first person shooter online games. In: IEEE Conference on Local Computer Networks, LCN (2008)
48. Zander, S., Armitage, G., Branch, P.: Reliable transmission over covert channels in first person shooter multiplayer games. In: IEEE Conference on Local Computer Networks, LCN (2009)
49. Zarras, A., Kohls, K., Dürmuth, M., Pöpper, C.: Neuralyzer: flexible expiration times for the revocation of online data. In: ACM Conference on Data and Application Security and Privacy, CODASPY (2016)
50. Zhou, W., Houmansadr, A., Caesar, M., Borisov, N.: SWEET: serving the web by exploiting email tunnels. In: Privacy Enhancing Technologies Symposium, PETS (2013)
51. Zhu, T., Bronk, C., Wallach, D.S.: An analysis of Chinese search engine filtering. arXiv preprint arXiv:1107.3794 (2011)
52. Zhu, T., Phipps, D., Pridgen, A., Crandall, J.R., Wallach, D.S.: The velocity of censorship: high-fidelity detection of microblog post deletions. In: USENIX Security Symposium (2013)

Analyzing Randomized Response Mechanisms Under Differential Privacy

Atsushi Waseda[✉] and Ryo Nojima

National Institute of Information and Communications Technology, Cybersecurity
Research Institute, 4-2-1, Nukui-Kitamachi, Koganei, Tokyo, Japan
{a-waseda,ryo-no}@nict.go.jp

Abstract. The randomized response technique was first introduced by
Warner in 1965 [27] as a technique to survey sensitive questions. Since
it is considered to protect the respondent's privacy, many variants and
applications have been proposed in the literature. Unfortunately, the
randomized response and its variants have not been well evaluated from
the privacy viewpoint historically. In this paper, we evaluate them by
using differential privacy. Specifically, we show that some variants have
a tradeoff between the privacy and utility, and that the "negative" survey
technique obtains negative results.

1 Introduction

The randomized response technique was first introduced by Warner in 1965 [27]
as a technique to solve the following survey problem: estimating the ratio of people in a population that has attribute $\mathcal{A}$ for example he/she is aged between 20
and 29. Intuitively, the randomized response technique works as follows: the questioner asks the responder P_i whether he has an attribute $\mathcal{A}$, then P_i answers honestly with some probability or otherwise answers randomly, and finally the questioner estimates the ratio by collecting all the answers from $\{P_1, \ldots, P_n\}$. Since
Warner developed it, many variants [1,3,4,12,15,21,23,24] and their applications [6,10,13,22] have been proposed. Due to the "randomized" (or the "noisy")
response, the questioner cannot confidently determine whether the responder
has $\mathcal{A}$, which means the responder has some kind of *deniability*. With this property, the randomized response technique is employed in many privacy preserving
applications, for instance location privacy [17,26], sensor network [16,18], and
data mining [7,19]. The weakness among the research related to the randomized
response is that how much privacy is leaked is not measured with the same measurement, and hence the application designers cannot decide which is the best
from the privacy perspective. In this paper, we consider employing *differential
privacy* as the common measurement.

Differential privacy [8] is a quantitative notion of privacy that bounds how
much a single individual's private data can contribute to a public output and is
employed in many privacy preserving applications [2,5,11,14,20]. In this notion,
intuitively, two neighboring databases D_1, D_2 are applied to some randomized

© Springer International Publishing Switzerland 2016
M. Bishop and A.C.A. Nascimento (Eds.): ISC 2016, LNCS 9866, pp. 271–282, 2016.
DOI: 10.1007/978-3-319-45871-7_17

algorithm M, which is called a *mechanism*, and the difference between their outputs $M(D_1)$ and $M(D_2)$ is measured. However, this setting is somewhat different from the randomized response. Let tA be a true answer of the respondent and $M(\text{tA})$ be an actual noisy response. Then what we want to measure is privacy loss of tA_i rather than that of the database D_i. Hence, we regard tA_i as a singleton set, i.e., $D_i = \{\text{tA}_i\}$, and analyze the privacy with the differential privacy. This is not the first time for kind of analysis [9]. However, Dwork and Roth only evaluated very simplified and tiny randomized response.

In this paper, we evaluate the randomized response techniques and those applications with the differential privacy. First, we evaluate the simplified randomized response technique and show a useful lemma (Lemma 1). By using this lemma, we evaluate several well studied randomized response techniques, Warner's mechanism [27], Kuk's mechanism [21], and negative survey [12] and its variants [3]. We show that the randomized response technique has a trade-off between the privacy and utility. Finally, the applications of the randomized response are evaluated. Specifically, we evaluate the location privacy proposed by Quercia et al. [26] and private collaborative recommendation algorithm proposed by Polat and Du [25].

This paper is organized as follows. In Sect. 2, we introduce the overview of differential privacy and the randomized response techniques. In Sect. 3, we evaluate the randomized response techniques under differential privacy. In Sect. 4, the variants of several randomized response and those applications are evaluated. Finally, we conclude this paper in Sect. 5.

2 Background

2.1 Differential Privacy

Differential privacy [8] is a quantitative notion of privacy that bounds how much a single individual's private data discloses. The standard setting involves a *database* of private information and a *mechanism* that computes an output given in the database. Formally, a database D is a multiset of records belonging to some *data universe* $\mathcal{X}$, where a record corresponds to one individual's private data. We say that the two databases are *neighbors* if they are identical except for a single record. A mechanism M is a randomized function that takes the database as input and outputs an element of the range $\mathcal{R}$.

Definition 1 ([8]). *Given $\epsilon \geq 0$, a mechanism M is ϵ-differentially private if, for any two neighboring databases D and D' and for any subset $\mathcal{S} \subseteq \mathcal{R}$ of outputs,*

$$\Pr[M(D) \in \mathcal{S}] \leq \exp(\epsilon) \cdot \Pr[M(D') \in \mathcal{S}]. \tag{1}$$

If $\mathcal{S}$ is a countable set, then we can modify the inequation (1) as

$$\Pr[M(D) = s] \leq \exp(\epsilon) \cdot \Pr[M(D') = s], \tag{2}$$

where we consider $\mathcal{S}$ as a singleton set. We use the inequation (2) in this paper and assume that $\mathcal{R} = \mathcal{X}$.

2.2   Randomized Response Mechanisms

The randomized response mechanism was first introduced by Warner [27] in 1965 to solve the following survey problem: to estimate the ratio of people in a population that has attribute $\mathcal{A}$. That is, the randomized response mechanism has two types of participants, the *questioner* Q and the *respondents* $\mathcal{P} = \{P_1, P_2, \cdots, P_n\}$, and Q wants to estimate

$$|\{P_i \in \mathcal{A} \mid P_i \in \mathcal{P}\}|.$$

To solve this, the mechanism works as follows:

- Q sends (the description of) $\mathcal{A}_0 := \mathcal{A}$, and $\mathcal{A}_1 := \mathcal{P} \setminus \mathcal{A}_0$.
- Then, P_i flips a coin b such that $\Pr[b = 0] = p$ and $\Pr[b = 1] = 1 - p$, and returns 1 if $P_i \in \mathcal{A}_b$ or 0 otherwise.

Since Q does not know b, he/she cannot decide whether $P_i \in A$ or not. After the proposal of the above mechanism, many extensions have been studied. For example, Abul-Ela et al. [1] considered the estimation of

$$|\{P_i \in \mathcal{A}_0 \mid P_i \in \mathcal{P}\}|, \ldots, |\{P_i \in \mathcal{A}_t \mid P_i \in \mathcal{P}\}|.$$

To analyze the differential privacy of the various randomized mechanisms, we provide a simple definition. The randomized response mechanism consists of the following algorithms $(\mathsf{S}, \mathsf{Res}, \mathsf{Eval})$, where

- S is a randomized algorithm that generates a question q.
- Res is a randomized algorithm that takes q and true answer tA as input and output the noisy answer nA. We often omit q and denote $\mathsf{Res}(\mathsf{tA})$ instead of $\mathsf{Res}(q, \mathsf{tA})$.
- $\mathsf{Eval}(\mathsf{nA}_1, \ldots, \mathsf{nA}_n)$ is a randomized algorithm that takes noisy answers $\mathsf{nA}_1, \ldots, \mathsf{nA}_n$ as input and outputs the estimation, for example $|\{P_i \in \mathcal{A} \mid P_i \in \mathcal{P}\}|$.

We formalize the privacy requirement of the randomized response mechanism as follows:

Definition 2. *Given* $\epsilon \geq 0$, *a randomized response mechanism* $M = (\mathsf{S}, \mathsf{Res}, \mathsf{Eval})$ *is* ϵ-*differentially private if, for any two answers* $\mathsf{tA}_0 \in \mathcal{X}$ *and* $\mathsf{tA}_1 \in \mathcal{X}$, *and for any element* $s \in \mathcal{X}$,

$$\Pr[\mathsf{Res}(\mathsf{tA}_0) = s] \leq \exp(\epsilon) \cdot \Pr[\mathsf{Res}(\mathsf{tA}_1) = s]. \tag{3}$$

If there exists a distance function d for every two answers $\mathsf{tA}_0 \in \mathcal{X}$ and $\mathsf{tA}_i \in \mathcal{X}$, then we can modify Eq. (3) as

$$\Pr[\mathsf{Res}(\mathsf{tA}_0) = s] \leq \exp(d(\mathsf{tA}_0, \mathsf{tA}_1) \cdot \epsilon) \cdot \Pr[\mathsf{Res}(\mathsf{tA}_1) = s],$$

which is similar to Andrés et al. [2]. We can consider d as Euclid distance, Manhattan distance, etc.

3   Analyzing the Randomized Response Mechanisms

We begin with the simplified variance that Warner proposed [27]. Let us consider two randomized functions Rand_0 and Rand_1 such that

- Rand_0 outputs 0 with probability p_0 and 1 with $p_1 = 1 - p_0$; and
- $\mathsf{Rand}_1(x)$ is a randomized algorithm whose domain is $\mathcal{X}$ such that for every $x \in \mathcal{X}$, there exists $i \in \mathcal{X}$ satisfying

$$\Pr[\mathsf{Rand}_1(x) = i] \geq 0,$$

and $\sum_{i \in \mathcal{X}} p_{x,i} = 1$, where $\Pr[\mathsf{Rand}_1(x) = i] = p_{x,i}$.

The algorithm $\mathsf{Res}(\mathsf{tA})$ works as follows:

1. Run Rand_0 to obtain $b_0 \in \{0, 1\}$
2. If $b_0 = 0$ then set $\mathsf{nA} = \mathsf{tA}$, or otherwise run Rand_1 to obtain b_1 and set $\mathsf{nA} = b_1$.
3. Output nA.

The case

$$p_0 = p_1 = 1/2, \mathcal{X} = \{0, 1\}, p_{0,0} = p_{0,1} = p_{1,0} = p_{1,1} = 1/2, \tag{4}$$

was evaluated by Dwork, which results in $\ln 3$-differential privacy. The more general case is proven as follows:

Lemma 1 (Key Lemma). *If for any* $\mathsf{tA}, s \in \mathcal{X}$, $\Pr[\mathsf{Res}(\mathsf{tA}) = s] > 0$, *then the simplified randomized response mechanism has* ϵ-*differential privacy, where*

$$\epsilon = \ln \max \left\{ \frac{1 + p_1(p_{\mathsf{tA}_0,\mathsf{tA}_0} - 1)}{p_1 p_{\mathsf{tA}_1,\mathsf{tA}_0}}, \frac{p_1 p_{\mathsf{tA}_0,s}}{p_1 p_{\mathsf{tA}_1,s}}, \frac{p_1 p_{\mathsf{tA}_0,\mathsf{tA}_1}}{1 + p_1(p_{\mathsf{tA}_1,\mathsf{tA}_1} - 1)} \right\}, \tag{5}$$

and the maximum is taken over $\mathsf{tA}_0 \in \mathcal{X}, \mathsf{tA}_1 \in \mathcal{X} \setminus \{\mathsf{tA}_0\}, s \in \mathcal{X} \setminus \{\mathsf{tA}_0, \mathsf{tA}_1\}$.

Proof. Let us denote the respondent's noisy output as $\mathsf{nA} \in \mathcal{X}$ and true answer as $\mathsf{tA} \in \mathcal{X}$. Modifying inequality (3), we have

$$\max_{\mathsf{tA}_0, \mathsf{tA}_1 \in \mathcal{X}, \mathsf{tA}_0 \neq \mathsf{tA}_1, s \in \mathcal{X}} \left\{ \frac{\Pr[\mathsf{Res}(\mathsf{tA}_0) = s]}{\Pr[\mathsf{Res}(\mathsf{tA}_1) = s]} \right\} \leq \exp(\epsilon), \tag{6}$$

where the randomness is taken over the choice of Res, and what we want to estimate is ϵ. To do so, we consider three cases:

Case 1: $s = \mathsf{tA}_0$,
Case 2: $s = \mathsf{tA}_1$,
Case 3: $s \neq \mathsf{tA}_0, s \neq \mathsf{tA}_1$.

Here, the probability of $\mathsf{Res}(\mathsf{tA})$ producing $\mathsf{nA} = \mathsf{tA}$ is

$$\Pr[\mathsf{nA} = \mathsf{tA}] = p_0 + p_1 p_{\mathsf{tA},\mathsf{tA}} = 1 + p_1(p_{\mathsf{tA},\mathsf{tA}} - 1),$$

and for every $\mathsf{nA} \in \mathcal{X}$ such that $\mathsf{tA} \neq \mathsf{nA}$,

$$\Pr[\mathsf{nA} \neq \mathsf{tA}] = p_1 p_{\mathsf{tA},\mathsf{nA}}.$$

Hence, for Case 1,

$$\frac{\Pr[\mathsf{Res}(\mathsf{tA}_0) = \mathsf{tA}_0)]}{\Pr[\mathsf{Res}(\mathsf{tA}_1) = \mathsf{tA}_0)]} = \frac{1 + p_1(p_{\mathsf{tA}_0,\mathsf{tA}_0} - 1)}{p_1 p_{\mathsf{tA}_1,\mathsf{tA}_0}}.$$

Case 2 is similar to Case 1. Finally, for Case 3,

$$\frac{\Pr[\mathsf{Res}(\mathsf{tA}_0) = s]}{\Pr[\mathsf{Res}(\mathsf{tA}_1) = s]} = \frac{p_1 p_{\mathsf{tA}_0,s}}{p_1 p_{\mathsf{tA}_1,s}}.$$

Putting these together with (6) and taking the logarithm we have proved the lemma. $\qquad\square$

Applying the case of (4), where Dwork analyzed, in the above lemma, we obtain

$$\epsilon = \ln \max \left\{ \frac{1 + 1/2(1/2 - 1)}{1/2 \cdot 1/2}, \frac{1/2}{1/2}, \frac{1/2 \cdot 1/2}{1 + 1/2(1/2 - 1)} \right\}$$
$$= \ln \max \{3, 1, 1/3\}$$
$$= \ln 3.$$

Hence, the result contains Dwork's analysis.

4   Privacy Analysis

4.1   Analyzing the Variants

Warner's Original Mechanism [27]**:** The aim is to estimate the proportion π_A of people who have some attribute $\mathcal{A}$. More formally, each person has some attribute $\mathcal{A}$ under Bernoulli distribution with π_A. The questioner estimates π_A. However, if $\mathcal{A}$ represent some sensitive attribute, the responder has a high possibility of responding untruthfully. Thus, Warner proposed the randomized response technique below.

The questioner prepares red cards, non-red cards and the box, where the ratio of red cards among all the cards is q with $0 < q < 1$ and $q \neq \frac{1}{2}$. The questioner puts those cards in the box and sends it to respondent P. Respondent P draws one card from the box. If the card is "red", then P truthfully replies "True" or "False" to the question "I am a member of $\mathcal{A}$," or otherwise truthfully replies to the question "I am *not* a member of $\mathcal{A}$". Let $\hat{T}$ be the proportion to which the respondents reply "True." It is easy to see that the expectation of $\hat{T}$ is

$$E(\hat{T}) = q\pi_A + (1-q)(1-\pi_A).$$

Then, the estimation of π_A is calculated as

$$\pi'_A = \frac{\hat{T} - (1-q)}{2q-1},$$

with sampling variance

$$V(\pi'_A) = \frac{E(\hat{T})(1-E(\hat{T}))}{n(2q-1)^2} = \frac{\pi_A(1-\pi_A)}{n} + \frac{q(1-q)}{n(2q-1)^2},$$

where n is the number of respondents. This variance will be small when q approaches 0 or 1. We can regard this mechanism as a special case of the simplified randomized response by setting

- $|\mathcal{X}| = 2$,
- (the definition of Rand_0) $p_0 = q$,
- (the definition of Rand_1)

$$p_{\mathsf{tA},x} = \begin{cases} 0 \text{ if } \mathsf{tA} = x, \\ 1 \text{ otherwise.} \end{cases}$$

By applying Lemma 1, we obtain the following result.

Lemma 2. *The randomized response proposed by Warner [27] satisfies the ϵ_W-differential privacy, where*

$$\epsilon_W = \max\left\{\ln\frac{1-q}{q}, \ln\frac{q}{1-q}\right\}.$$

This ϵ_W is small when q approaches $1/2$. It is desirable that ϵ_W and sampling variance are small. Hence, q must be chosen carefully since it controls $V(\pi'_A)$ and ϵ_W.

Kuk's Mechanism [21]: Kuk's proposed another kind of randomized response mechanism to estimate the proportion π_A, the same as Warner's mechanism. The questioner prepares two boxes, Box_1 and Box_2. These boxes contain red cards and non-red cards. The ratio of the red cards among all the cards in Box_i is q_i, where $0 < q_1, q_2 < 1, q_1 \neq q_2$. The questioner sends these boxes to respondent P. Then, respondent P takes one card from each box.

- If P is member of $\mathcal{A}$, then P replies "red card" or "non-red card" in accordance with the card taken from Box_1
- Otherwise, he does the same as above except that he takes a card from Box_2.

Let $\hat{R}$ be the proportion to which the respondents say "red card." It is easy to see that the expectation of $\hat{R}$ is

$$E(\hat{R}) = q_1 \pi_A + q_2(1 - \pi_A).$$

Then, the estimation of π_A is calculated as

$$\pi'_A = \frac{\hat{R} - q_2}{q_1 - q_2},$$

with sampling variance,

$$\begin{aligned} V(\pi'_A) &= \frac{E(\hat{R})(1 - E(\hat{R}))}{n(q_1 - q_2)^2} \\ &= \frac{((-q_1 + q_2)\pi_A + (1 - q_1))((q_1 - q_2)\pi_A + q_1)}{n(q_1 - q_2)^2}. \end{aligned}$$

We can also regard Kuk's mechanism as a special case of the simplified randomized response by setting

- $f : \{\text{"True"}, \text{"False"}\} \mapsto \{1, 2\}$,
- $|\mathcal{X}| = 2$,
- (the definition of Rand_0) $p_0 = 0$,
- (the definition of Rand_1)

$$p_{\text{tA},x} = \begin{cases} q_{f(\text{tA})} & \text{if } x = red\ card, \\ 1 - q_{f(\text{tA})} & \text{otherwise.} \end{cases}$$

By applying Lemma 1, we obtain the following result.

Lemma 3. *The randomized response proposed by Kuk [21] satisfies the ϵ_K-differential privacy, where*

$$\epsilon_K = \ln \max \left\{ \frac{q_1}{q_2}, \frac{q_2}{q_1}, \frac{1 - q_1}{1 - q_2}, \frac{1 - q_2}{1 - q_1} \right\}.$$

$V(\pi'_A)$ is small when $|q_1 - q_2|$ is large, but ϵ_K becomes large.

Thus, Warner's mechanism and Kuk's mechanism have a tradeoff between privacy ϵ and utility $V(\pi'_A)$.

Negative Survey Mechanism [12]: A negative survey mechanism is a special case of the randomized response mechanism. In this mechanism, the respondent always gives a non-true answer, i.e., always answering $\text{nA} \neq \text{tA}$. For example, if $\mathcal{X} = \{a_0, a_1, a_2, a_3\}$ and $\text{tA} = a_0$, then $\text{nA} = a_i$ such that $1 \leq i \leq 3$ and $\Pr[\text{nA} = a_i] = 1/3$ for all i. Since the respondent does not answer truthfully, the questioner never receives the true answer. By obtaining the noisy answers in this way, the number of people $X(i)$ whose true answer is a_i can be estimated as

$$X(i) = n - (\alpha - 1)Y(i),$$

where n is the number of respondents and α is the number of choices, i.e. $\alpha = |\mathcal{X}|$. $Y(i)$ is the number of people whose response nA is a_i. We consider the following setting

- $|\mathcal{X}| = \alpha$,
- (the definition of Rand_0) $p_0 = 0, p_1 = 1$,
- (the definition of Rand_1)

$$p_{\mathsf{tA},x} = \begin{cases} 0 \text{ if } \mathsf{tA} = x, \\ \frac{1}{\alpha-1} \text{ otherwise.} \end{cases}$$

Considering the case $s = \mathsf{tA}_1$ in inequation (3), to have the ϵ-differential privacy,

$$\Pr[\mathsf{Res}(\mathsf{tA}_0) = \mathsf{tA}_1] \leq \exp(\epsilon) \Pr[\mathsf{Res}(\mathsf{tA}_1) = \mathsf{tA}_1].$$

must be satisfied. However,

$$\Pr[\mathsf{Res}(\mathsf{tA}_0) = \mathsf{tA}_1] = \frac{1}{\alpha-1} > \exp(\epsilon) \Pr[\mathsf{Res}(\mathsf{tA}_1) = \mathsf{tA}_1] = 0,$$

for any finite ϵ. Hence we can conclude as follows:

Lemma 4. *The negative survey mechanism does not satisfy ϵ-differential privacy for any finite ϵ.*

Intuitively, if nA is given to the questioner, then he/she can know that the respondent's true answer is not nA. This makes the mechanism very weak.

Limited Negative Survey Mechanism [3]: Aoki et al., proposed the *limited negative survey* mechanism as an extension of the negative survey [3]. In limited negative survey mechanism, the privacy information is protected as follows. Let $\mathsf{tA} = (x_1, x_2, \ldots, x_D)$ be original data, where each x_i is chosen from $\mathcal{X}_i$, i.e., $\mathsf{tA} \in \mathcal{X} = \mathcal{X}_1 \times \cdots \times \mathcal{X}_D$. Then the noisy data $\mathsf{nA} = (x_1', x_2', \ldots, x_D')$ satisfies the following:

$$\forall i : x_i' < x_i - \beta_i \text{ or } x_i + \beta_i < x_i'.$$

for some positive β_i. For the same reason as the negative survey mechanism, we can conclude the following:

Lemma 5. *The limited negative survey mechanism does not satisfy ϵ-differential privacy with finite ϵ.*

t-times Negative Survey: Aoki and Sezaki [4] considered using the negative survey twice to make the time complexity of the estimation small. The privacy analysis is shown in the following theorem. Note that the analysis is given in a more general form:

Theorem 1. *Let* $|\mathcal{X}| = \alpha$ *and the respondent makes the response by running the negative survey mechanism t-times. Then, the mechanism satisfies the* ϵ-*differential privacy, where*

$$\epsilon = \begin{cases} \ln \frac{(1-\alpha)((1-\alpha)^{t-1})-1}{(1-\alpha)^t - 1)} & if\ t\ is\ even, \\ \ln \frac{(1-\alpha)^t - 1}{(1-\alpha)((1-\alpha)^{t-1}-1)} & if\ t\ is\ odd. \end{cases}$$

Proof. Let s_t be the response of t times negative survey mechanism. We construct the following recurrence relation:

$$\begin{cases} \Pr[s_0 = \mathsf{tA_0}] = 1, \\ \Pr[s_0 \neq \mathsf{tA_0}] = 0, \\ \Pr[s_t = \mathsf{tA_0}] = \Pr[s_{t-1} = \mathsf{tA_0}] \times 0 + (\alpha - 1) \times \Pr[s_{t-1} \neq \mathsf{tA_0}] \times \frac{1}{\alpha-1}, \\ \Pr[s_t \neq \mathsf{tA_0}] = \Pr[s_{t-1} = \mathsf{tA_0}] \times \frac{1}{\alpha-1} + (\alpha - 2) \times \Pr[s_{t-1} \neq \mathsf{tA_0}] \times \frac{1}{\alpha-1}. \end{cases}$$

Solving this recurrence relation, we obtain the following

$$\Pr[s_t = \mathsf{tA_0}] = \frac{1}{\alpha} \left(1 - \frac{1}{(1-\alpha)^{t-1}} \right),$$

$$\Pr[s_t \neq \mathsf{tA_0}] = \frac{1}{\alpha} \left(1 - \frac{1}{(1-\alpha)^t} \right).$$

By applying Lemma 1, we obtain this theorem. □

Note that ϵ becomes closer to 0 as t tends to ∞.

4.2 Analyzing Applications

We analyze the application to which the randomized response mechanism was applied.

Location Obfuscation Algorithm SpotMe [26]: SpotMe, proposed by Quercia et al., is a mechanism for aggregating user locations in real-time. They realized this mechanism by a randomized response. Furthermore, they implemented SpotMe on mobile phones and conducted experiments in Zurich and London. As a result, they reported that SpotMe works with reasonable communication, computational time and storage overheads.

Let k be the number of locations on a map. Then, the main part of SpotMe works as follows:

- with probability p, the mobile phone chooses the location k uniformly at random, and
- with probability $1 - p$, it chooses the true location.

Thus, this is

- $|\mathcal{X}| = k$,
- The definition of $\mathsf{Rand_0}$ is $p_0 = 1 - p$.

– The definition of Rand$_1$ is

$$p_{\text{tA},l} = \frac{1}{k},$$

in the simplified randomized response mechanism.

By applying Lemma 1, we conclude that SpotMe [26] satisfies the ϵ-differential privacy, where

$$\epsilon = \ln \frac{k - (k-1)p}{p}.$$

For example, if $p = 0.5$ and $k = 100$ [26], SpotMe satisfies the $\ln 101$-differential privacy.

Private Collaborative Recommendation [25]: Polat and Du proposed a private collaborative recommendation mechanism. In this mechanism, a user poses a rating $\text{tA} \in \{0, 1\}$ of some item and submits it to the system privately. To do so,

– with probability $1 - q$, the user submits a randomly chosen rating in $\{0, 1\}$, and
– with probability q, it chooses the true rating.

Since this is exactly the same as Warner's mechanism, we can conclude that private collaborative recommendation proposed by Polat and Du [25] satisfies the ϵ-differential privacy, where

$$\epsilon = \max \left\{ \ln \frac{1-q}{q}, \ln \frac{q}{1-q} \right\}.$$

5 Concluding Remarks

In this paper, we evaluate the randomized response techniques by using differential privacy. First, we presented a useful lemma (Lemma 1) evaluating the simplified randomized response technique. By using this lemma, we evaluated several well studied randomized response techniques and their applications. We obtained results in which neither ϵ nor sampling variance could be improved. We hope that our results help application designers to choose suitable parameters in the randomized response.

References

1. Abul-Ela, A.L.A., Greenberg, G.G., Horvitz, D.G.: A multi-proportions randomized response model. J. Am. Stat. Assoc. **62**(319), 990–1008 (1967)
2. Andrés, M.E., Bordenabe, N.E., Chatzikokolakis, K., Palamidessi, C.: Geo-indistinguishability: differential privacy for location-based systems. In: Proceedings of the 2013 ACM SIGSAC Conference on Computer and Communications Security, pp. 901–914. ACM (2013)

3. Aoki, S., Iwai, M., Sezaki, K.: Limited negative surveys: privacy-preserving participatory sensing. In: 2012 IEEE 1st International Conference on Cloud Networking (CLOUDNET), pp. 158–160. IEEE (2012)
4. Aoki, S., Sezaki, K.: Negative surveys with randomized response techniques for privacy-aware participatory sensing. IEICE Trans. Commun. **97**(4), 721–729 (2014)
5. Dankar, F.K., El Emam, K.: The application of differential privacy to health data. In: Proceedings of the 2012 Joint EDBT/ICDT Workshops, pp. 158–166. ACM (2012)
6. Dietz, P., Striegel, H., Franke, A.G., Lieb, K., Simon, P., Ulrich, R.: Randomized response estimates for the 12-month prevalence of cognitive-enhancing drug use in university students. Pharmacother. J. Hum. Pharmacol. Drug Ther. **33**(1), 44–50 (2013)
7. Du, W., Zhan, Z.: Using randomized response techniques for privacy-preserving data mining. In: Proceedings of the ninth ACM SIGKDD International Conference on Knowledge Discovery and Data Mining, pp. 505–510. ACM (2003)
8. Dwork, C.: Differential privacy. In: Bugliesi, M., Preneel, B., Sassone, V., Wegener, I. (eds.) ICALP 2006. LNCS, vol. 4052, pp. 1–12. Springer, Heidelberg (2006)
9. Dwork, C., Roth, A.: The algorithmic foundations of differential privacy. Found. Trends Theor. Comput. Sci. **9**(3–4), 211–407 (2014)
10. Eichhorn, B.H., Hayre, L.S.: Scrambled randomized response methods for obtaining sensitive quantitative data. J. Stat. Plan. Infer. **7**(4), 307–316 (1983)
11. Erlingsson, Ú., Pihur, V., Korolova, A.: RAPPOR: randomized aggregatable privacy-preserving ordinal response. In: Proceedings of the 2014 ACM SIGSAC Conference on Computer and Communications Security, pp. 1054–1067. ACM (2014)
12. Esponda, F.: Negative surveys. arXiv preprint math/0608176 (2006)
13. Fidler, D.S., Kleinknecht, R.E.: Randomized response versus direct questioning: two data-collection methods for sensitive information. Psychol. Bull. **84**(5), 1045 (1977)
14. Friedman, A., Schuster, A.: Data mining with differential privacy. In: Proceedings of the 16th ACM SIGKDD International Conference on Knowledge Discovery and Data Mining, pp. 493–502. ACM (2010)
15. Greenberg, B.G., Abul-Ela, A.L.A., Simmons, W.R., Horvitz, D.G.: The unrelated question randomized response model: theoretical framework. J. Am. Stat. Assoc. **64**(326), 520–539 (1969)
16. Groat, M.M., Edwards, B., Horey, J., He, W., Forrest, S.: Enhancing privacy in participatory sensing applications with multidimensional data. In: 2012 IEEE International Conference on Pervasive Computing and Communications (PerCom), pp. 144–152. IEEE (2012)
17. Horey, J., Forrest, S., Groat, M.: Reconstructing spatial distributions from anonymized locations. In: 2012 IEEE 28th International Conference on Data Engineering Workshops (ICDEW), pp. 243–250. IEEE (2012)
18. Horey, J., Groat, M.M., Forrest, S., Esponda, F.: Anonymous data collection in sensor networks. In: Fourth Annual International Conference on Mobile and Ubiquitous Systems: Networking and Services, MobiQuitous 2007, pp. 1–8. IEEE (2007)
19. Huang, Z., Du, W.: OptRR: optimizing randomized response schemes for privacy-preserving data mining. In: IEEE 24th International Conference on Data Engineering, ICDE 2008, pp. 705–714. IEEE (2008)
20. Inan, A., Kantarcioglu, M., Ghinita, G., Bertino, E.: Private record matching using differential privacy. In: Proceedings of the 13th International Conference on Extending Database Technology, pp. 123–134. ACM (2010)

21. Kuk, A.Y.: Asking sensitive questions indirectly. Biometrika **77**(2), 436–438 (1990)
22. Lara, D., García, S.G., Ellertson, C., Camlin, C., Suárez, J.: The measure of induced abortion levels in Mexico using random response technique. Sociol. Meth. Res. **35**(2), 279–301 (2006)
23. Mangat, N.S.: An improved randomized response strategy. J. R. Stat. Soc. Ser. B (Methodol.) **56**(1), 93–95 (1994)
24. Mangat, N., Singh, R.: An alternative randomized response procedure. Biometrika **77**(2), 439–442 (1990)
25. Polat, H., Du, W.: Achieving private recommendations using randomized response techniques. In: Ng, W.-K., Kitsuregawa, M., Li, J., Chang, K. (eds.) PAKDD 2006. LNCS (LNAI), vol. 3918, pp. 637–646. Springer, Heidelberg (2006)
26. Quercia, D., Leontiadis, I., McNamara, L., Mascolo, C., Crowcroft, J.: SpotME if you can: randomized responses for location obfuscation on mobile phones. In: 2011 31st International Conference on Distributed Computing Systems (ICDCS), pp. 363–372. IEEE (2011)
27. Warner, S.L.: Randomized response: a survey technique for eliminating evasive answer bias. J. Am. Stat. Assoc. **60**(309), 63–69 (1965)

Models and Algorithms for Graph Watermarking

David Eppstein[1], Michael T. Goodrich[1], Jenny Lam[2(✉)], Nil Mamano[1],
Michael Mitzenmacher[3], and Manuel Torres[1]

[1] Department of Computer Science, University of California, Irvine, CA, USA
[2] Department of Computer Science, San José State University, San José, CA, USA
jenny.lam@sjsu.edu
[3] Department of Computer Science, Harvard University, Cambridge, MA, USA

Abstract. We introduce models and algorithmic foundations for graph watermarking. Our approach is based on characterizing the feasibility of graph watermarking in terms of keygen, marking, and identification functions defined over graph families with known distributions. We demonstrate the strength of this approach with exemplary watermarking schemes for two random graph models, the classic Erdős-Rényi model and a random power-law graph model, both of which are used to model real-world networks.

1 Introduction

In the classic media watermarking problem, we are given a digital representation, R, for some media object, O, such as a piece of music, a video, or an image, such that there is a rich space, $\mathcal{R}$, of possible representations for O besides R that are all more-or-less equivalent. Informally, a *digital watermarking* scheme for O is a function that maps R and a reasonably short random message, m, to an alternative representation, R', for O in $\mathcal{R}$. The verification of such a marking scheme takes R and a presumably-marked representation, R'' (which was possibly altered by an adversary), along with the set of messages previously used for marking, and it either identifies the message from this set that was assigned to R'' or it indicates a failure. Ideally, it should be difficult for an adversary to transform a representation, R' (which he was given), into another representation R'' in $\mathcal{R}$, that causes the identification function to fail. Some example applications of such digital watermarking schemes include steganographic communication and marking digital works for copyright protection.

With respect to digital representations of media objects that are intended to be rendered for human performances, such as music, videos, and images, there is a well-established literature on digital watermarking schemes and even well-developed models for such schemes (e.g., see Hopper *et al.* [8]). Typically, such watermarking schemes take advantage of the fact that rendered works have many possible representations with almost imperceptibly different renderings from the perspective of a human viewer or listener.

In this paper, we are inspired by recent systems work on *graph watermarking* by Zhao *et al.* [18], who propose a digital watermarking scheme for graphs, such

© Springer International Publishing Switzerland 2016
M. Bishop and A.C.A. Nascimento (Eds.): ISC 2016, LNCS 9866, pp. 283–301, 2016.
DOI: 10.1007/978-3-319-45871-7_18

as social networks, protein-interaction graphs, etc., which are to be used for commercial, entertainment, or scientific purposes. This work by Zhao *et al.* presents a system and experimental results for their particular method for performing graph watermarking, but it is lacking in formal security and algorithmic foundations. For example, Zhao *et al.* do not provide formal proofs for circumstances under which graph watermarking is undetectable or when it is computationally feasible. Thus, as complementary work to the systems results of Zhao *et al.*, we are interested in the present paper in providing models and algorithms for graph watermarking, in the spirit of the watermarking model provided by Hopper *et al.* [8] for media files. In particular, we are interested in providing a framework for identifying when graph watermarking is secure and computationally feasible.

Additional Related Work. Under the term "graph watermarking," there is some additional work, although it is not actually for the problem of graph watermarking as we are defining it. For instance, there is a line of research involving software watermarking using graph-theoretic concepts and encodings. In this case, the object being marked is a piece of software and the goal of a "graph watermarking" scheme is to create a graph, G, from a message, m, and then embed G into the control flow of a piece of software, S, to mark S. Examples of such work include pioneering work by Collberg and Thomborson [6], as well as subsequent work by Venkatesan, Vazirani, and Sinha [16] and Collberg *et al.* [5]. This work on software watermarking differs from the graph watermarking problem we study in the present paper, however, because in the graph watermarking problem we study an input graph is provided and we want to alter it to add a mark. In the graph-based software watermarking problem, a graph is instead created from a message to have a specific, known structure, such as being a permutation graph, and then that graph is embedded into the control flow of the piece of software.

A line of research that is more related to the graph watermarking problem we study is anonymization and de-anonymization for social networks. One of the closest examples of such prior work is by Backstrom, Dwork, and Kleinberg [1], who show how to introduce a small set of "rogue" vertices into a social network and connect them to each other and to other vertices so that if that same network is approximately replicated in another setting it is easy to match the two copies. Such work differs from graph watermarking, however, because the set of rogue vertices are designed to "stand out" from the rest of the graph rather than "blend in," and it may in some cases be relatively easy for an adversary to identify and remove such rogue vertices. In addition to this work, also of note is work by Narayanan and Shmatikov [13], who study the problem of approximately matching two social networks without marking, as well as the work on Khanna and Zane [9] for watermarking road networks by perterbing vertex positions (which is a marking method outside the scope of our approach).

Our Results. In this paper, we introduce a general graph watermarking framework that is based on the use of key generation, marking, and identification functions, as well as a hypothetical watermarking security experiment (which would be performed by an adversary). We define these functions in terms of graphs taken over random families of graphs, which allows us to quantify situations in which graph watermarking is provably feasible.

We also provide some graph watermarking schemes as examples of our framework, defined in terms of the classic Erdős-Rényi random-graph model and a random power-law graph model. Our schemes extend and build upon previous results on graph isomorphism for these graph families, which may be of independent interest. In particular, we design simple marking schemes for these random graph families based on simple edge-flipping strategies involving high- and medium-degree vertices. Analyzing the correctness of our schemes is quite nontrivial, however, and our analysis and proofs involve intricate probabilistic arguments, some of which we include in the ePrint version of this paper [7]. We provide an analysis of our scheme against adversaries that can themselves flip edges in order to defeat our mark identification algorithms. In addition, we provide experimental validation of our algorithms, showing that our edge-flipping scheme can succeed for a graph without specific knowledge of the parameters of its deriving graph family.

2   Our Watermarking Framework

Suppose we are given an undirected graph, $G = (V, E)$, that we wish to mark. To define the security of a watermarking scheme for G, G must come from a family of graphs with some degree of entropy [19]. We formalize this by assuming a probability distribution $\mathcal{D}$ over the family $\mathcal{G}$ of graphs from which G is taken.

Definition 1. *A graph watermarking scheme is a tuple* (keygen, mark, identify) *over a set, $\mathcal{G}$, of graphs where*

- keygen : $\mathbb{N} \times \mathbb{N} \rightarrow$ Aux *is a private key generation function, such that* keygen(ℓ, n) *is a list of ℓ (pseudo-)random graph elements, such as vertices and/or vertex pairs, defined over a graph of n vertices. These candidate locations for marking are defined independent of a specific graph; that is, vertices in Aux are identified simply by the numbering from 1 to n. For example,* keygen(ℓ, n) *could be a small random graph, H, and some random edges to connect H to a larger input graph [19], or* keygen(ℓ, n) *could be a set of vertex pairs in an input graph that form candidate locations for marking.*
- mark : Aux $\times \mathcal{G} \rightarrow \mathbb{N} \times \mathcal{G}$ *takes a private key z generated by* keygen, *and a specific graph G from $\mathcal{G}$, and returns a pair, $S = (\text{id}, H)$, such that* id *is a unique identifier for H and H is the graph obtained by adding the mark determined by* id *to G in the location determined by the private key z.* mark *is called every time a different marked copy needs to be produced, with the i-th copy being denoted by $S_i = (\text{id}_i, H_i)$. Therefore, the unique identifiers should be thought of as being generated randomly. To associate a marked graph H_i with the user who receives it, the watermarking scheme can be augmented with a table storing user name and unique identifiers. Alternatively, the identifiers*

can be generated pseudo-randomly as a hash of a private key provided by the user.

- identify : $\mathsf{Aux} \times \mathcal{G} \times \mathbb{N}^k \times \mathcal{G} \to \mathbb{N} \cup \{\bot\}$ *takes a private key from* Aux, *the original graph,* G, k *identifiers of previously-marked copies of* G, *and a test graph,* G', *and it returns the identifier,* id_i, *of the watermarked graph that it is identifying as a match for* G'. *It may also return* $\bot$, *as an indication of failure, if it does not identify any of the graphs* H_i *as a match for* G'.

In addition, in order for a watermarking scheme to be effective, we require that with high probability[1] over the graphs from $\mathcal{G}$ *and* k *output pairs,* $S_1, \ldots, S_k$ *of* $\mathsf{mark}(z, G)$, *for any* $(\mathsf{id}, G') = S_i$, *we have* $\mathsf{identify}(z, G, \mathsf{id}_1, \ldots, \mathsf{id}_k, G') = \mathsf{id}$.

Algorithm 1 shows a hypothetical security experiment for a watermarking scheme with respect to an adversary, $A : \mathcal{G} \to \mathcal{G}$, who is trying to defeat the scheme. Intuitively, in the hypothetical experiment, we generate a key z, choose a graph G, from family $\mathcal{G}$ according to distribution $\mathcal{D}$ (as discussed above), and then generate k marked graphs according to our scheme (for some set of k messages). Next, we randomly choose one of the marked graphs, G', and communicate it to an adversary. The adversary then outputs a graph G_A that is similar to G' where his goal is to cause our identification algorithm to fail on G_A.

Algorithm 1. Hypothetical Watermarking Security Experiment

experiment(A, k, ℓ, n):
1. $z \leftarrow \mathsf{keygen}(\ell, n)$
2. $G \leftarrow_{\mathcal{D}} \mathcal{G}$
3. $S_i \leftarrow \mathsf{mark}(z, G)$, for $i = 1, \ldots, k$
4. randomly choose $S_i = (\mathsf{id}, G')$ from $\{S_1, \ldots, S_k\}$
5. $G_A \leftarrow A(G')$

In order to characterize differences between graphs, we assume a similarity measure $\mathsf{dist} : \mathcal{G} \times \mathcal{G} \to \mathbb{R}$, defining the distance between graphs in family $\mathcal{G}$. We also include a similarity threshold θ, that defines the advantage of an adversary performing the experiment in Algorithm 1. Specifically, the *advantage* of an adversary, $A : \mathcal{G} \to \mathcal{G}$ who is trying to defeat our watermarking scheme is

$$\mathbb{P}\left[\mathsf{dist}(G, G_A) < \theta \text{ and } \mathsf{identify}(z, G, \mathsf{id}_1, \ldots, \mathsf{id}_k, G_A) \neq \mathsf{id}\right].$$

The watermarking scheme is $(\mathcal{D}, \mathsf{dist}, \theta, k, \ell)$-secure against adversary A if the similarity threshold is θ and A's advantage is *polynomially negligible* (i.e., is $O(n^{-a})$ for some $a > 0$).

Examples of adversaries could include the following:

- *Arbitrary edge-flipping adversary*: a malicious adversary who can arbitrarily flip edges in the graph. That is, the adversary adds an edge if it is not already there, and removes it otherwise.

---

[1] Or "*whp*," that is, with probability at least $1 - O(n^{-a})$, for some $a > 0$.

- *Random edge-flipping adversary*: an adversary who independently flips each edge with a given probability.
- *Arbitrary adversary*: a malicious adversary who can arbitrarily add and/or remove vertices and flip edges in the graph.
- *Random adversary*: an adversary who independently adds and/or removes vertices with a given probability and independently flips each edge with a given probability.

Random Graph Models. As defined above, a graph watermarking scheme requires that graphs to be marked come from some distribution. In this paper, we consider two families of random graphs—the classic Erdős-Rényi model and a random power-law graph model—which should capture large classes of applications where graph watermarking would be of interest.

Definition 2 (The Erdős-Rényi model). *A random graph $G(n,p)$ is a graph with n vertices, where each of the $\binom{n}{2}$ possible edges appears in the graph independently with probability p.*

Definition 3 (The random power-law graph model, Sect. 5.3 of [4]). *Given a sequence $\mathbf{w} = (w_1, w_2, \ldots, w_n)$, such that $\max_i w_i^2 < \sum_k w_k$, the general random graph $G(\mathbf{w})$ is defined by labeling the vertices 1 through n and choosing each edge (i,j) independently from the others with probability $p[i,j] = \rho w_i w_j$, where $\rho = 1/\sum_j w_j$.*

We define a random power-law graph $G(\mathbf{w}^\gamma)$ parameterized by the maximum degree m and average degree w. Let $w_i = ci^{-1/(\gamma-1)}$ for values of i in the range between i_0 and $i_0 + n$, where

$$c = \frac{\gamma - 2}{\gamma - 1} w n^{\frac{1}{\gamma-1}}, \qquad i_0 = n\left(\frac{w(\gamma-2)}{m(\gamma-1)}\right)^{\gamma-1}. \tag{1}$$

This definition implies that each edge (i,j) appears with probability

$$P[i,j] = K_0 \left(n^{\gamma-3}ij\right)^{-\frac{1}{\gamma-1}}, \qquad where \ K_0 \stackrel{def}{=} \left(\frac{\gamma-2}{\gamma-1}\right)^2 w. \tag{2}$$

Graph Watermarking Algorithms. We discuss some instantiations of the graph watermarking framework defined above. Unlike previous watermarking or de-anonymization schemes that add vertices [1,19], we describe an effective and efficient scheme based solely on edge flipping. Such an approach would be especially useful for applications where it could be infeasible to add vertices as part of a watermark.

Our scheme does not require adding labels to the vertices or additional objects stored in the graph for identification purposes. Instead, we simply rely on the structural properties of graphs for the purposes of marking. In particular, we focus on the use of vertex degrees, that is, the number of edges incident on each vertex. We identify high and medium degree vertices as candidates for finding edges that can be flipped in the course of marking. The specific degree thresholds for what we mean by "high-degree" and "medium-degree" depend on the graph family, however, so we postpone defining these notions precisely until our analysis sections.

Algorithms providing an example implementation of our graph watermarking scheme are shown in Algorithm 2. The keygen algorithm randomly selects a set of candidate vertex pairs for flipping, from among the high- and medium-degree vertices, with no vertex being incident to more than a parameter t of candidate pairs. We introduce a procedure, label(G), which labels high-degree vertices by their degree ranks and each medium-degree vertex, w, by a bit vector identifying its high-degree adjacencies. This bit vector has a bit for each high-degree vertex, which is 1 for neighbors of w and 0 for non-neighbors. The algorithm mark(z, G), takes a random set of candidate edges and a graph, G, and it flips the corresponding edges in G according to a resampling of the edges using the distribution $\mathcal{D}$. The algorithm, approximate-isomorphism(G, H), returns a mapping of the high- and medium-degree vertices in G to matching high- and medium-degree vertices in H, if possible. The algorithm, identify$(z, G, \mathrm{id}_1, \ldots, \mathrm{id}_k, H)$, uses the approximate isomorphism algorithm to match up high- and medium-degree vertices in G and H, and then it extracts the bit-vector from this matching using z.

As mentioned above, we also need a notion of distance for graphs. We use two different such notions. The first is the graph edit distance, which is the minimum number of edges needed to flip to go from one graph to another. The second is vertex distance, which intuitively is an edge-flipping metric localized to vertices.

Definition 4 (Graph distances). *Let $\mathcal{G}$ be the set of graphs on n vertices. If $G, H \in \mathcal{G}$, define Π as the set of bijections between the vertex sets $V(G)$ and $V(H)$. Define the* graph edit distance $\mathrm{dist}_e : \mathcal{G} \times \mathcal{G} \to \mathbb{N}$ *as*

$$\mathrm{dist}_e(G, H) = \min_{\pi \in \Pi} |E(G) \oplus_\pi E(H)|,$$

where $\oplus_\pi$ is the symmetric difference of the two edge sets under correspondence π. Define the vertex distance $\mathrm{dist}_v : \mathcal{G} \times \mathcal{G} \to \mathbb{N}$ *as*

$$\mathrm{dist}_v(G, H) = \min_{\pi \in \Pi} \max_{v \in V(G)} |E(v) \oplus_\pi E(\pi(v))|,$$

where $E(v)$ is the set of edges incident to v.

3   Identifying High- and Medium-Degree Vertices

We begin analyzing our proposed graph watermarking scheme by showing how high- and medium-degree vertices can be identified under our two random graph distributions. We ignore low-degree vertices: their information content and distinguishability are low, and they are not used by our example scheme.

We first find a threshold number k such that the k vertices with highest degree are likely to have distinct and well-separated degree values. We call these k

---

Algorithm 2. Watermarking scheme for random graphs.

---

t: the maximum number of flipped edges that can be adjacent to the same vertex. keygen(ℓ, n):

1. Let x denote the total number of high- and medium-degree vertices
2. $X = \{(u, v) \mid 1 \leq u < v \leq x\}$
3. Let z be a list of ℓ pairs randomly sampled (without replacement) from X such that no end vertex appears more than t times
4. return z

label(G):

1. sort the vertices in decreasing order by degree and identify the high- and medium-degree vertices
2. if the degrees of high-degree vertices are not unique, return failure
3. label each high-degree vertex with its position in the vertex sequence
4. label each medium-degree vertex with a bit vector encoding its high-degree adjacencies
5. if the bit vectors are not unique, return failure
6. otherwise, return the labelings

mark(z, G):

1. $S = \varnothing$
2. V is the set of high- and medium-degree vertices of G, sorted lexicographically by their labels given by $L = \text{label}(G)$
3. generate an ℓ-bit string id where each bit i is independently set to 1 with probability $p_{z[i]}$, where $p_{z[i]}$ is the probability of the edge $z[i]$ in $\mathcal{D}$
4. let H be a copy of G
5. for j from 1 to ℓ:
6. $\quad (u, v) = z[j]$
7. $\quad$ if id[j] is 1:
8. $\quad\quad$ insert edge $(V[u], V[v])$ in H
9. $\quad$ else:
10. $\quad\quad$ remove edge $(V[u], V[v])$ from H
11. return (id, H)

approximate-isomorphism(G, H):

1. call label(G) and label(H), returning failure if either of these fail.
2. match each of G's high-degree vertices with the vertex in H with the same label.
3. match each of G's medium-degree vertices with the vertex in H whose label is closest in Hamming distance.
4. if H has a vertex that is matched more than once, return failure.
5. otherwise, return the (partial) vertex assignments between G and H.

identify($z, G, \text{id}_1, \ldots, \text{id}_k, H$):

1. find an approximate-isomorphism(G, H), returning $\perp$ if failure occurred at any step.
2. V is the set of high- and medium-degree vertices of G, sorted lexicographically by their labels given by $L = \text{label}(G)$
3. V' is the set of vertices of H identified as corresponding to those in V, in that same order.
4. id is an empty bit string
5. for (u, v) in z (from left to right):
6. $\quad b = 1$ iff there is an edge between $V'[u]$ and $V'[v]$ in H.
7. $\quad$ append b to id
8. return among the id_i's the one closest to id

---

vertices the *high-degree* vertices. Next, we look among the remaining vertices for those that are well-separated in terms of their high-degree neighbors. Specifically, the (high-degree) *neighborhood distance* between two vertices is the number of high-degree vertices which are connected to exactly one of the two vertices. Note that we will omit the term "high-degree" in "high-degree neighborhood distance" from now on, as it will always be implied.

In the Erdős-Rényi model, we show that all vertices that are not high-degree nevertheless have well-separated high-degree neighborhoods whp. In the random power-law graph model, however, there will be many lower-degree vertices whose high-degree neighborhoods cannot be separated. Those that have well-separated high-degree neighborhoods with high probability form the medium-degree vertices, and the rest are the low-degree vertices.

For completeness, we include the following well-known Chernoff concentration bound, which we will refer to time and again.

Lemma 5 (Chernoff inequality [4]). *Let $X_1, \ldots, X_n$ be independent random variables with*
$$\mathbb{P}[X_i = 1] = p_i, \qquad \mathbb{P}[X_i = 0] = 1 - p_i.$$
We consider the sum $X = \sum_{i=1}^n X_i$, with expectation $\mathbb{E}[X] = \sum_{i=1}^n p_i$. Then

$$\mathbb{P}[X \le \mathbb{E}[X] - \lambda] \le e^{-\frac{\lambda^2}{2\mathbb{E}[X]}},$$

$$\mathbb{P}[X \ge \mathbb{E}[X] + \lambda] \le e^{-\frac{\lambda^2}{2\mathbb{E}[X]+\lambda/3}}.$$

Vertex Separation in the Erdős-Rényi Model. Let us next consider vertex separation results for the classic Erdős-Rényi random-graph model. Recall that in this model, each edge is chosen independently with probability p.

Definition 6. *Index vertices in non-increasing order by degree. Let d_i represent the i-th highest degree in the graph. Given $h = O(n)$, we say that a vertex is high-degree with respect to d_h if it has degree at least d_h. Otherwise, we say that the vertex is medium-degree.*

Note that in this random-graph model, there are no low-degree vertices.

Definition 7. *A graph is (d, d')-separated if all high-degree vertices differ in their degree by at least d and all medium-degree vertices are neighborhood distance d' apart.*

Note: this definition depends on how high-degree or medium-degree vertices are defined and will therefore be different for the random power-law graph model.

Lemma 8 (Extension of Theorem 3.15 in [2]). *Suppose $m = o(pqn/\log n)^{1/4}$, $m \to \infty$, and $\alpha(n) \to 0$. Then with probability*

$$1 - m\alpha(n) - 1/\left[m\left(\log(n/m)\right)^2\right],$$

$G(n,p)$ is such that

$$d_i - d_{i+1} \geq \frac{\alpha(n)}{m^2} \left(\frac{pqn}{\log n} \right)^{1/2} \quad \text{for every } i < m,$$

where $q = 1 - p$.

Proof.   See the ePrint version [7].   □

Lemma 9 (Vertex separation in the Erdős-Rényi model). *Let* $0 < \varepsilon < 1/9$, $d \geq 3$, $C \geq 3$, $h = n^{(1-\varepsilon)/8}$. *Suppose* $0 < p = p(n) \leq \frac{1}{2}$ *is such that* $p = \omega(n^{-\varepsilon} \log n)$. *Then* $G(n,p)$ *is* $(d, C \log n)$-*separated with probability* $1 - O(n^{-(1-\varepsilon)/8})$.

Proof.   See the ePrint version [7].   □

Thus, high-degree vertices are well-separated with high probability in the Erdős-Rényi model, and the medium-degree vertices are distinguished with high probability by their high-degree neighborhoods.

Vertex Separation in the Random Power-Law Graph Model. We next study vertex separation for a random power-law graph model, which can match the degree distributions of many graphs that naturally occur in social networking and science. For more information about power-law graphs and their applications, see e.g. [3,12,14].

In the random power-law graph model, vertex indices are used to define edge weights and therefore do not necessarily start at 1. The lowest index that corresponds to an actual vertex is denoted i_0. So vertex indices range from i_0 to $i_0 + n$. Additionally, there are two other special indices i_H and i_M, which we define in this section, that separate the three classes of vertices.

Definition 10. *The vertices ranging from* i_0 *to* i_H *are the* high-degree *vertices, those that range from* $i_H + 1$ *to* i_M *are the* medium-degree *vertices, and those beyond* i_M *are the* low-degree *vertices.*

In this model, the value of i_0 is constrained by the requirement that $P[i_0, i_0] < 1$. When $\gamma \geq 3$, this constraint is not actually restrictive. However, when $\gamma < 3$, i_0 must be asymptotically greater than $n^{-(\gamma-3)/2}$. The constraints on i_0 also constrain the value of the maximal and average degree of the graph.

We define i_H and i_M to be independent of i_0, but dependent on parameters that control the amount and probability of separation at each level. The constraints that $i_0 < i_H$ and $i_H < i_M$ translate into corresponding restrictions on the valid values of γ, namely that $\gamma > 5/2$ and $\gamma < 3$. We define i_H in the following lemma.

Lemma 11 (Separation of high-degree vertices). *In the* $G(\mathbf{w}^\gamma)$ *model, let* $\delta_i = |w_{i+1} - w_i| / 2$. *Then,*

$$\frac{c}{2(\gamma - 1)} (i+1)^{-\frac{\gamma}{\gamma-1}} \leq \delta_i \leq \frac{c}{2(\gamma-1)} i^{-\frac{\gamma}{\gamma-1}}. \tag{3}$$

Moreover, for all ε_1 satisfying $0 < \varepsilon_1 \leq 1$ and $C_1 > 0$, the probability that

$$|\deg(i) - w_i| < \varepsilon_1 \delta_i \quad \text{for all } i \leq i_H \stackrel{\text{def}}{=} \left(\frac{c\varepsilon_1^2}{16(\gamma-1)^2 C_1 \log n} \right)^{\frac{\gamma-1}{2\gamma-1}}$$

is at least $1 - n^{-C_1}$.

Proof. The first statement follows from the fact that w_i is a convex function of i and from taking its derivative at i and $i + 1$.

For the second statement, let $C > 0$ and let $i'_H \stackrel{\text{def}}{=} \left(\frac{c\varepsilon_1^2}{8(\gamma-1)^2 C \log n} \right)^{\frac{\gamma-1}{2\gamma-1}}$. We will show that if $i \leq i'_H$, then

$$\mathbb{P}\left[|\deg(i) - w_i| \geq \varepsilon_1 \delta_i\right] < n^{-C}. \tag{4}$$

Now we choose C such that $C_1 + \log i_H / \log n < C \leq 2C_1$. The inequality $C \leq 2C_1$ implies that $i_H \leq i'_H$ and (4) holds for all $i \leq i_H$. By the union bound applied to (4)

$$\mathbb{P}\left[\exists i \leq i_H, |\deg(i) - w_i| \geq \varepsilon_1 \delta_i\right] \leq i_H n^{-C}.$$

Since $C_1 + \log i_H / \log n < C$, the right hand side is bounded above by n^{-C_1}. This proves the result.

Now, we prove (4). Clearly, since $\delta_i = (w_i - w_{i+1})/2$, we have that $w_i \geq \delta_i$. So if $\varepsilon_1 \leq 1$ and $\lambda_i = \varepsilon_1 \delta_i$, then $w_i \geq \lambda_i/3$. This implies that

$$\frac{\lambda_i^2}{w_i + \lambda_i/3} \geq \frac{\lambda_i^2}{2w_i} \geq \frac{c\varepsilon_1^2}{8(\gamma-1)^2} i^{-\frac{2\gamma-1}{\gamma-1}},$$

where the second inequality follows from (3) and the definition of w_i given in Definition 3. If $i \leq i'_H$, the right hand side is lower-bounded by $C \log n$. The result follows by applying a Chernoff bound (Lemma 5).   □

For simplicity, we often use the following observation.

Observation 12. *Rewriting i_H to show its dependence on n, we have*

$$i_H(\varepsilon_1, C_1) = K_1(\varepsilon_1, C_1) \, n^{\frac{1}{2\gamma-1}} (\log n)^{-\frac{\gamma-1}{2\gamma-1}}, \quad K_1(\varepsilon_1, C_1) \stackrel{\text{def}}{=} \left(\frac{\gamma-2}{(\gamma-1)^3} \frac{w\varepsilon_1^2}{16C_1} \right)^{\frac{\gamma-1}{2\gamma-1}}. \tag{5}$$

For the graph model to make sense, the high-degree threshold must be asymptotically greater than the lowest index. In other words, we must have that $i_0 = o(i_H)$. Since $i_0 = \Omega(n^{-(\gamma-3)/2})$, this implies that $\gamma > 5/2$.

We next define i_M, the degree threshold for medium-degree vertices, in the following lemma.

Lemma 13 (Separation of medium-degree vertices). *Let K_0 be defined as in Definition 3, $K_1(\varepsilon_1, C_1)$ be defined as in (5), and*

$$K_2(\varepsilon_1, C_1, \varepsilon_2, C_2) \stackrel{\text{def}}{=} \frac{K_0^{\gamma-1} K_1^{\gamma-2}(\varepsilon_1, C_1)}{(C_2 + 2\Gamma + 2\log(K_0^{\gamma-1} K_1^{\gamma-2}(\varepsilon_1, C_1)) + 2\varepsilon_2)^{\gamma-1}}. \tag{6}$$

*Let X_{ij} denote the neighborhood distance between two vertices i and j in $G(\mathbf{w}^\gamma)$.
If $5/2 < \gamma < 3$, for every $\varepsilon_2 > 0$ and $C_2 > 0$, the probability that*

$$X_{ij} > \varepsilon_2 \log n, \quad \text{for all } i_H \leq i, j \leq i_M$$

where

$$i_M(\varepsilon_1, C_1, \varepsilon_2, C_2) \overset{def}{=} K_2(\varepsilon_1, C_1, \varepsilon_2, C_2)\, n^\Gamma (\log n)^{-\frac{3(\gamma-1)^2}{2\gamma-1}}, \quad \Gamma \overset{def}{=} -\frac{2\gamma^2 - 8\gamma + 5}{2\gamma - 1}, \quad (7)$$

is at least $1 - n^{-C_2}$ for sufficiently large n.

Proof. Let $C > 0$ and let

$$i'_M \overset{def}{=} \left(\frac{C_2 + 2\Gamma + 2\log(K_0^{\gamma-1} K_1^{\gamma-2}) + 2\varepsilon_2}{C + 2\varepsilon_2} \right)^{\gamma-1} i_M.$$

We claim that if $i_H \leq i, j \leq i'_M$, then

$$\mathbb{P}\left[X_{ij} \leq \varepsilon_2 \log n\right] \leq n^{-C}. \quad (8)$$

If we choose $C = C_2 + 2\Gamma + 2\log K_0^{\gamma-1} K_1^{\gamma-2}$, we have that $i_M = i'_M$, so that (8)
applies to all i, j such that $i, j \leq i_M$. Moreover, since

$$i_M \leq K_0^{\gamma-1} K_1^{\gamma-2} n^\Gamma \leq n^{\log(K_0^{\gamma-1} K_1^{\gamma-2})} n^\Gamma,$$

our choice of C implies that $i_M^2\, n^{-C} \leq n^{-C_2}$. By applying the union bound to
(8), we have

$$\mathbb{P}\left[\exists i, j \text{ s.t. } i_H \leq i, j \leq i_M,\ X_{ij} \leq \varepsilon_2 \log n\right] \leq i_M^2 n^{-C} \leq n^{-C_2},$$

which establishes the lemma.

Let us now prove the claim. Observe that X_{ij} is the sum over the high-degree
vertices k, of indicator variables X_{ij}^k for the event that vertex k is connected to
exactly one of the vertices i and j. It i For fixed i and j, these are independent
random variables. Therefore, we can apply a Chernoff bound. The probability
that $X_{ij}^k = 1$ is

$$P[i,k](1 - P[j,k]) + P[j,k](1 - P[i,k]) \geq 2P[i_M, i_H](1 - P[i_0, i_H]).$$

Since $P[i_0, i_H] \to 0$, for sufficiently large n, this expression is bounded below by
$P[i_M, i_H]$, and

$$\mathbb{E}\left[X_{ij}\right] \geq i_H P[i_M, i_H] \geq (C + 2\varepsilon_2) \log n,$$

by (2), (5) and (7), as can be shown by a straightforward but lengthy computation. Let $d = \varepsilon_2 \log n$. This implies that

$$\frac{(\mathbb{E}\left[X_{ij}\right] - d)^2}{\mathbb{E}\left[X_{ij}\right]} \geq \mathbb{E}\left[X_{ij}\right] - 2d \geq C \log n.$$

Therefore, applying the Chernoff bound (Lemma 5) to the X_{ij}^k for fixed i and j
and all high-degree vertices k proves the claim. $\qquad\square$

Observation 14. *We would have the undesirable situation that $i_M = o(1)$ whenever $\frac{2\gamma^2 - 8\gamma + 5}{2\gamma - 1} > 0$, or equivalently when $\gamma > 2 + \sqrt{3/2} > 3$. In fact, in order for $i_H = o(i_M)$, we must have $\gamma < 3$.*

We illustrate the breakpoints for high-, medium-, and low-degree vertices in Fig. 1.

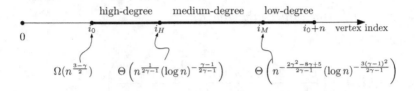

Fig. 1. Degree breakpoints for the random power-law graph model.

The next lemma summarizes the above discussion and provides the forms of i_H and i_M that we use in our analysis.

Lemma 15 (Vertex separation in the power-law model). *Let $5/2 < \gamma < 3$. Fix $\varepsilon > 0, C_1 > 0, C_2 > 0$. Let $i_H = i_H(\varepsilon_1, C_1)$ and $i_M = i_M(\varepsilon_1, C_1, \varepsilon_2, C_2)$ where $\varepsilon_1 = 1$ and $\varepsilon_2 = \varepsilon$. Let*

$$d = n^{\frac{1}{2\gamma - 1}} \quad and \quad d' = \log n.$$

For sufficiently large n, the probability that a graph $G(\mathbf{w}^\gamma)$ is not $(\varepsilon d, \varepsilon d')$-separated is at most $n^{-C_1} + n^{-C_2}$.

Proof. Let δ_i be defined as in Lemma 11. A straightforward computation using (1), (3), and (5) shows that

$$\delta_{i_H} \geq \text{constant} \cdot n^{\frac{1}{2\gamma - 1}} (\log n)^{\frac{\gamma}{2\gamma - 1}}.$$

So for sufficiently large n, we have $\delta_{i_H} \geq 3\varepsilon d/2$. For all $i \leq i_H$, the average degrees w_i of consecutive vertices are at least $3\varepsilon d/2$ apart. So for two high-degree vertices to be within εd of each other, at least one of the two must have degree at least $(3\varepsilon/2 - \varepsilon/2)d$ away from its expected degree. By Lemma 11, the probability that some high-degree vertex i satisfies $|\deg(i) - w_i| > \delta_{i_H}$ is at most n^{-C_1}.

By Lemma 13, the probability that there are two medium-degree vertices with neighborhood distance less than $\varepsilon d'$ is at most n^{-C_2}.   □

Thus, our marking scheme for the random power-law graph model is effective.

4   Adversary Tolerance

In this section, we study the degree to which our exemplary graph watermarking scheme can tolerate an arbitrary edge-flipping adversary.

Theorem 16 (Security against an arbitrary edge-flipping adversary in the Erdős-Rényi model). *Let $0 < \varepsilon < 1/9$, $d \geq 3$, $h = n^{(1-\varepsilon)/8}$ and $p \leq 1/2$ such that $p = \omega(n^{-\varepsilon} \log n)$. Let d be sufficiently large so that*

$$\varepsilon \, \frac{d+1}{d-1} < 1. \tag{9}$$

Suppose the similarity measure is the vertex distance dist_v, the similarity threshold is $\theta = d$, we have a number $k = n^C$ of watermarked copies, and their identifiers are generated using $\ell = 8(2C + C')n^\varepsilon$ bits. Suppose also that the identifiers map to sets of edges of a graph constrained by the fact that no more than $t = d$ edges can be incident to any vertex. The watermarking scheme defined in Algorithm 2 is $(G(n,p), \mathsf{dist}_v, \theta, k, \ell)$-secure against any deterministic adversary.

The proof of this theorem relies on two lemmas. Lemma 17 identifies conditions under which a set of bit vectors with bits independently set to 1 is unlikely to have two close bit vectors. Lemma 18 states that a deterministic adversary's ability to guess the location of the watermark is limited. Informally, this is because the watermarked graph was obtained through a random process, so that there are many likely original graphs that could have produced it.

Lemma 17 (Separation of IDs). *Consider $k = n^C$ random bit strings of length ℓ, where each bit is independently set to 1, and the i-th bit is 1 with probability q_i satisfying $p \leq q_i \leq 1/2$ for a fixed value p. The probability that at least two of these strings are within Hamming distance $D = 4(2C + C') \log n$ of each other is at most $n^{-C'}$ if $\ell p \geq 2D$.*

Proof.   See the ePrint version [7].   □

Lemma 18 (Guessing power of adversary). *Consider a complete graph on N vertices, and let r of its edges be red. Let s be a sample of ℓ edges chosen uniformly at random among those that satisfy the constraint that no more than t edges of the sample can be incident to any one vertex. Suppose also that ℓ, N and t are non-decreasing functions of n such that*

$$\frac{\ell^{t+1}}{N^{t-1}} \to 0 \text{ as } n \to \infty. \tag{10}$$

For sufficiently large N, the probability that s contains at least $R = 8\ell r/N^2$ red edges is bounded by $4\exp\left(-12\ell r/(7N^2)\right)$. Moreover, if $\ell r/N^2 \to 0$, then the probability that s contains at least $R = 1$ red edge is bounded by $4\exp\left(-cN^2/(\ell r)\right)$, for some $c > 0$ and for sufficiently large N.

Proof.   See the ePrint version [7].   □

Proof (Theorem 16). An upper bound on the advantage of any deterministic adversary $A : \mathcal{G} \to \mathcal{G}$ on graphs on n vertices is given by the conditional probability

$$\mathbb{P}\left[\text{identify}(z, G, \text{id}_1, \ldots, \text{id}_k, G_A) \neq \text{id} \mid \text{dist}_v(G, G_A) < \theta\right],$$

where the parameters passed to identify are defined according to the experiment in Algorithm 1. We show that this quantity is polynomially negligible.

For G_A to be successfully identified, it is sufficient for the following three conditions to hold:

1. the original graph $G = G(n, p)$ is $(4d, 4d)$-separated;
2. the Hamming distance between any two id and id$'$ involved in a pair in S is at least $D = 4(2C + C') \log n$;
3. A changes no edges of the watermark.

These are sufficient conditions because we only test graphs whose vertices had at most d incident edges modified by the adversary, and another d incident edges modified by the watermarking. So for original graphs that are $(4d, 4d)$-separated, the labeling of the vertices can be successfully recovered. Finally, if the adversary does not modify any potential edge that is part of the watermark, the id of the graph is intact and can be recovered from the labeling.

Now, by Lemma 9, the probability that $G(n, p)$ is not $(4d, 4d)$-separated is less than $O(n^{-(1-\varepsilon)/8})$. Moreover, since $\ell p \geq 2D$, by Lemma 17, the probability that there are two identifiers in S that are within D of each other is at most $n^{-C'}$.

Finally, for graphs in which an adversary makes fewer than d modifications per vertex, the total number of edges the adversary can modify is $r \leq dn/2$. Since all vertices are high- and medium-degree vertices in this model, $N = n$. Therefore, $\ell r / N^2 = O(1/n^{(1-\varepsilon)}) \to 0$. Eq. (9) guarantees that the hypothesis given by (10) of Lemma 18 is satisfied. Consequently, the probability that A changes one or more adversary edges is $O(\exp[cn^{1-\varepsilon}])$ for some constant c.

This proves that each of the three conditions listed above fails with polynomially negligible probability, which implies that the conditional probability is also polynomially negligible.   □

Theorem 19 (Security against an arbitrary edge-flipping adversary in the random power-law graph model). *Let $5/2 < \gamma < 3$, $C > 0$, $i_H = i_H(\varepsilon_1, C_1)$ and $i_M = i_M(\varepsilon_1, C_1, \varepsilon_2, C_2)$ where $\varepsilon_1 = 1$, $\varepsilon_2 = 8(C + 1)$ and $C_1 = C_2 = C$.*

Let $p = P[i_M, i_M]$. Suppose the similarity measure is a vector of distances $\text{dist} = (\text{dist}_e, \text{dist}_v)$, that the corresponding similarity threshold is the vector $\theta = (r, \log n)$ where $r = p(i_M)^2/32$ is the maximum number of edges the adversary can flip in total, and $\log n$ the maximum number of edges it can flip per vertex. Suppose that we have $k = n^{C''}$ watermarked copies of the graph, that we use $\ell = 8(2C'' + C')(\log n)/p$ to watermark a graph.

Suppose also that the identifiers map to sets of edges of a graph constrained by the fact that no more than $t = \log n$ edges can be incident to any vertex. Then the watermarking scheme defined in Algorithm 2 is $(G(\mathbf{w}^\gamma), \text{dist} = (\text{dist}_e, \text{dist}_v), \theta = (r, \log n), k, \ell)$-secure against any deterministic adversary.

Proof.   See the ePrint version [7].   □

Discussion. It is interesting to note how the differences in the two random graph models translate into differences in their watermarking schemes. The Erdős-Rényi model, with its uniform edge probability, allows for constant separation of high-degree vertices, at best. But all the vertices tend to be well-separated. On the other hand, the skewed edge distribution that is characteristic of the random power-law model allows high-degree vertices to be very well-separated, but a significant number of vertices—the low-degree ones, will not be easily distinguished.

These differences lead to the intuition that virtually all edges in the Erdős-Rényi model are candidates for use in a watermark, as long as only a constant number of selected edges are incident to any single vertex. Therefore, both our watermarking function and the adversary are allowed an approximately linear number of changes to the graph. Theorem 16 confirms this intuition with a scheme that proposes $O(n^\varepsilon)$ bits for the watermark, and a nearly linear number $O(n)$ bits that the adversary may modify.

In contrast, the number of edges that can be used as part of a watermark in the random power-law graph model is limited by the number of distinguishable vertices, which is on the order of i_M or $O(n^\varepsilon)$, where $\varepsilon = -\frac{2\gamma^2 - 8\gamma + 5}{2\gamma - 1}$.

5   Experiments

Although our paper is a foundational complement to the systems work of Zhao *et al.* [18], we nevertheless provide in this section the results of a small set of empirical tests of our methods, so as to experimentally reproduce the hypothetical watermarking security experiment from Algorithm 1. Our experiments are performed on two large social network graphs, Youtube [17] from the SNAP library [10], and Flickr [11], as well as a randomly generated graph drawn from the random power-law graph model distribution. Table 1 illustrates the basic properties of the networks. To generate the random power-law graph, we set the number of nodes to $n = 10000$, the maximum degree to $m = 1000$, the average degree to $w = 20$, and $\gamma = 2.75$.

To adapt our theoretical framework to the rough-and-tumble world of empirical realities, we made three modifications to our framework for the sake of our empirical tests.

Table 1. Network statistics

Network	# nodes	# edges	Max. degree	Avg. degree	Unique degree	Estimated γ
Power-law	10, 000	94, 431	960	18.89	14	—
Youtube	1, 134, 890	2, 987, 624	28, 754	5.27	29	1.48
Flickr	1, 715, 256	15, 554, 181	27, 203	18.14	130	1.62

Table 2. Experiment parameters

Network	# high-degree	# medium-degree	Key size	Marking dK-2 deviation
Power law	64	374	219	0.065
Youtube	256	113	184	0.033
Flickr	300	5901	3250	0.002

First, instead of using the high-degree and medium-degree thresholds derived from Lemmas 11 and 13, for the power-law distribution, to define the cutoffs for high-degree and medium-degree vertices, we used these and the other lemmas given above as justifications for the existence of such distinguishing sets of vertices and we then optimized the number of high- and medium-degree vertices to be values that work best in practice. The column, "Unique degree," from Table 1 shows, for each network, the number of consecutive nodes with unique degree when considering the nodes in descending order of degree. Since this value is too small in most cases, we applied the principles of Lemmas 9 and 11 again, in a second-order fashion, to distinguish and order the high-degree nodes. In particular, in addition to the degree of each high-degree vertex, we also label each vertex with the list of degrees of its neighbors, sorted in decreasing order. With this change, we are not restricted in our choice of number of high-degree nodes as required by applying these lemmas only in a first-order fashion. Table 2 shows the values used in our experiments based on this second-order application. As medium-degree vertices, we picked the maximum number such that there are no collisions among their bit vectors of high-degree node adjacencies.

Second, instead of returning failure if (a) two high-degree nodes have the same degree and list of degrees of their neighbors, (b) two medium-degree nodes have the same bit vector, or (c) the approximate isomorphism is not injective, we instead proceed with the algorithm. Despite the existence of collisions, the remaining nodes often provide enough information to conclude successfully.

Finally, we simplified how we resampled (and flipped) edges in order to create a graph watermark, using our approach for the Erdős-Rényi model even for power-law graphs, since resampling uniformly among our small set of marked edges is likely not to cause major deviations in the graph's distribution and, in any case, it is empirically difficult to determine the value of γ for real-world social networks. Therefore, we set the resampling probability to 0.5 so that it is consistent with the Erdős-Rényi model and so that each bit in the message is represented uniformly and independently.

Experiment Parameters. For the experiment parameters other than the original network and the number of high- and medium-degree nodes, we set the following values.

Maximum Flips Adjacent to Any Given Node During Marking: 1.
Key size: We set this to the maximum possible value (i.e., the number of high- and medium-degree vertices divided by two, as shown in Table 2), because the

numbers of high- and medium- degree nodes are not large. This effectively means that every high- and medium-degree node has exactly one edge added or removed.

Number of Marked Graphs: 10.

Adversary: We used a time-efficient variation of the *arbitrary edge-flipping adversary*. This adversary selects a set of pairs of nodes randomly, and flips the potential edge among each pair.

Results. We evaluated how much distortion the adversary can introduce before our method fails to identify the leaked network correctly. For this purpose, we compared the identification success rate to the amount of distortion under different fractions of modified edges by the adversary. To estimate the success rate, we ran the experiment 10 times and reported the fraction of times that the leaked network was identified correctly. As a measure of distortion, we used the dK-2 deviation [18] between the original network and the version modified by

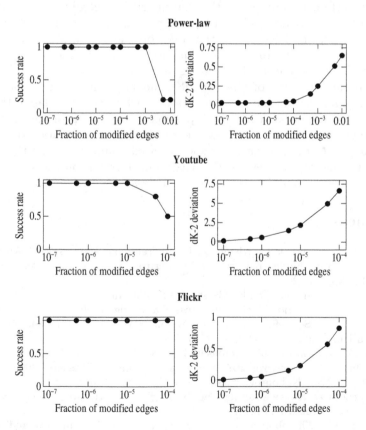

Fig. 2. Success rate and dK-2 deviation under different fractions of modified potential edges by the adversary, for the Power law, Youtube, and Flickr networks.

the adversary. The dK-2 deviation is the euclidean distance between the dK-2 series [15] of the two graphs, normalized by the number of tuples in the dK-2 series. The dK-2 deviation captures the differences between the joint degree distributions of the networks, that is, the probability that a randomly selected edge has as endpoints nodes with certain degrees. We average the dK-2 deviation among the 10 runs. Figure 2 shows the outcome of our experiments. Moreover, Table 2 shows the dK-2 deviation introduced by the marking alone.

Based on our experiments, the success rate of our scheme is high but it drops after a certain threshold. This demonstrates that there is a distinct range of adversarial edge flips that can be tolerated by our scheme. Specifically, our scheme worked well when the fraction of potential edges flipped by the adversary is up to 10^{-3} and 10^{-5} for the random power-law and Youtube networks, respectively. For these graphs, this number of flipped potential edges corresponds to 52.9% and 215.6% of the number of edges in the original graphs, respectively. For the Flickr network, the runtime of the adversary modification became excessive before the success rate could decrease, at a fraction of 10^{-4} of potential edges flipped.

The distortion introduced by the watermark is negligible compared to the distortion caused by the number of flips that the scheme can tolerate. On average, the marking modifies half of the edges on the key, which corresponds to $1.1 \cdot 10^{-3}, 3 \cdot 10^{-5}$, and 10^{-4} of the number of edges in the original random power-law, Youtube, and Flickr networks, respectively.

For the same number of flips, the dK-2 deviation in the Youtube network was much larger than in the Flickr network, which in turn was larger than that of the random power-law network. A possible explanation for this is that any set of uniform edge flips has a bigger effect on the dk2-deviation of a skewed graph than on the dK-2 deviation of a less skewed graph. Note that the Youtube network has the largest skew, as the maximum degree is on the same order as the Flickr network, but the average degree is less.

References

1. Backstrom, L., Dwork, C., Kleinberg, J.: Wherefore art thou r3579x?: Anonymized social networks, hidden patterns, and structural steganography. Commun. ACM **54**(12), 133–141 (2011)
2. Bollobás, B.: Random Graphs, Cambridge Studies in Advanced Mathematics, vol. 73, 2nd edn. Cambridge University Press, Cambridge (2001)
3. Caldarelli, G.: Scale-Free Networks: Complex Webs in Nature and Technology. Oxford University Press, Oxford (2013)
4. Chung, F., Lu, L.: Complex graphs and networks. In: CBMS Regional Conference Series in Mathematics, vol. 107. American Mathematical Society (2006)
5. Collberg, C.S., Kobourov, S.G., Carter, E., Thomborson, C.: Graph-based approaches to software watermarking. In: Bodlaender, H.L. (ed.) WG 2003. LNCS, vol. 2880, pp. 156–167. Springer, Heidelberg (2003)
6. Collberg, C., Thomborson, C.: Software watermarking: models and dynamic embeddings. In: ACM Symposium on Principles of Programming Language (POPL), pp. 311–324 (1999)

7. Eppstein, D., Goodrich, M.T., Lam, J., Mamano, N., Mitzenmacher, M., Torres, M.: Models and algorithms for graph watermarking. ArXiv ePrint abs/1605.09425 (2016). http://arxiv.org/abs/1605.09425
8. Hopper, N.J., Molnar, D., Wagner, D.: From weak to strong watermarking. In: Vadhan, S.P. (ed.) TCC 2007. LNCS, vol. 4392, pp. 362–382. Springer, Heidelberg (2007)
9. Khanna, S., Zane, F.: Watermarking maps: hiding information in structured data. In: 11th ACM-SIAM Symposium on Discrete Algorithms (SODA), pp. 596–605 (2000). http://dl.acm.org/citation.cfm?id=338219.338612
10. Leskovec, J., Sosič, R.: SNAP: a general purpose network analysis and graph mining library in C++. http://snap.stanford.edu/snap
11. Mislove, A., Marcon, M., Gummadi, K.P., Druschel, P., Bhattacharjee, B.: Measurement and analysis of online social networks. In: 5th ACM/Usenix Internet Measurement Conference (IMC) (2007)
12. Mitzenmacher, M.: A brief history of generative models for power law and lognormal distributions. Internet Math. **1**(2), 226–251 (2004)
13. Narayanan, A., Shmatikov, V.: De-anonymizing social networks. In: IEEE Symposium on Security and Privacy (SP), pp. 173–187 (2009)
14. Newman, M., Barabasi, A.L., Watts, D.J.: The Structure and Dynamics of Networks. Princeton Studies in Complexity. Princeton University Press, Princeton (2006)
15. Sala, A., Cao, L., Wilson, C., Zablit, R., Zheng, H., Zhao, B.Y.: Measurement-calibrated graph models for social network experiments. In: 19th International Conference on the World Wide Web (WWW), pp. 861–870 (2010)
16. Venkatesan, R., Vazirani, V.V., Sinha, S.: A graph theoretic approach to software watermarking. In: Moskowitz, I.S. (ed.) IH 2001. LNCS, vol. 2137, pp. 157–168. Springer, Heidelberg (2001)
17. Yang, J., Leskovec, J.: Defining and evaluating network communities based on ground-truth. CoRR abs/1205.6233 (2012). http://arxiv.org/abs/1205.6233
18. Zhao, X., Liu, Q., Zheng, H., Zhao, B.Y.: Towards graph watermarks. In: 2015 ACM Conference on Online Social Networks (COSN), pp. 101–112 (2015)
19. Zhao, X., Liu, Q., Zhou, L., Zheng, H., Zhao, B.Y.: Graph watermarks. ArXiv ePrint abs/1506.00022 (2015). http://arxiv.org/abs/1506.00022

Software Security

Policy-Based Implicit Attestation for Microkernel-Based Virtualized Systems

Steffen Wagner[1(✉)] and Claudia Eckert[2]

[1] Fraunhofer Institute AISEC, Munich, Germany
steffen.wagner@aisec.fraunhofer.de
[2] Technische Universität München, Munich, Germany
eckert@sec.in.tum.de

Abstract. We present an attestation mechanism that enables a remote verifier to implicitly evaluate the trustworthiness of the prover's system through policies. Those policies are verified and enforced by a TPM 2.0, when the attestor interacts with a virtualized hardware component of the prover's system. For instance, when the verifier reads a virtualized sensor device and requests integrity-protected sensor data, such as the average temperature, a heartbeat value, or an anomaly detection score, the prover's TPM, which acts as a trust anchor, checks and enforces the policies specified by the verifier. The prover, in turn, is also able to define policies, which can limit access to certain hardware components and are also enforced by the TPM. As a result, both parties have to cooperate for a successful attestation, which implicitly creates verifiable proof of the prover's trustworthiness using mainly symmetric instead of expensive asymmetric cryptographic operations like digital signatures.

Keywords: Remote attestation · Trusted platform module · Policy · Data integrity · Microkernel

1 Introduction

With hardware-based virtualization technologies, such as Intel VT [12] or ARM's Virtualization Extensions [3,5], isolating rich operating systems like Linux from each other and the rest of the system is a very effective way to ensure overall system security. This level of security can be even further increased if a micro-kernel, such as *L4/Fiasco.OC* [11,18], serves as the basis for an unprivileged hypervisor. Since microkernels implement all non-essential system components as user-space tasks, strictly separate those tasks, and have a very small code size, a microkernel-based hypervisor in user space reduces the system's attack surface significantly. As a result, such microkernel-based virtualized systems are suited even for the most security critical applications, e.g., in mobile, industrial, automotive, or avionic systems. However, since the virtualized rich operating systems usually still require some degree of access to physical hardware, they cannot be completely isolated and a virtual machine monitor (VMM) must provide

M. Bishop and A.C.A. Nascimento (Eds.): ISC 2016, LNCS 9866, pp. 305–322, 2016.
DOI: 10.1007/978-3-319-45871-7_19

mechanisms to make selected hardware components available to the virtualized systems.

One mechanism to give a virtualized system access to a physical hardware component, such as a display or camera, directly maps the component's physical memory address to the address space of a virtualized system, which is then able to exclusively use this component. However, not all components can be directly assigned to a specific virtualized system, because some hardware components, such as the physical network interface, a mobile broadband modem, or sensors, are shared. In addition, this simple and naive mechanism does not allow for inspecting, dynamically restricting, and multiplexing the access to a component. That is why hardware components are usually virtualized, which means that access requests, i.e., read and write operations, have to go through the virtual machine monitor and to device drivers, which support virtualization.

On the other hand, virtualizing a hardware component, such as a sensor, also presents a number of challenges. For example, a virtualized system cannot be sure that the access to a component was handled as requested, because it is not able to directly access that component and make the request itself. Using *device emulation* techniques, a hypervisor could simulate a hardware component, particularly a sensor, and modify, for example, the result of a read operation before it is returned to the virtualized system. That is why a (remote) user or system, which interacts with the virtualized rich operating system, needs to be able to verify the integrity of the underlying system to be able to trust the data from a hardware component. Unfortunately, most existing mechanisms to (remotely) verify the integrity of a system, such as IBM's Integrity Measurement Architecture (IMA) [14] in combination with a Trusted Platform Module (TPM), are not able to attest a microkernel-based system acting as hypervisor.

To overcome these challenges, we first present a microkernel-based system architecture, which uses ARM's Virtualization Extensions to run multiple rich operating systems and ARM TrustZone together with a TPM 2.0 [17] to securely handle critical operations and store sensitive information, like keys. Our main contribution is a security protocol, which leverages our system design to protect the integrity of data while implicitly verifying the trustworthiness of the system. The proposed protocol uses efficient symmetric cryptographic operations, such as hashing and hash-based message authentication codes (HMACs), and only optionally utilizes asymmetric cryptographic operations during the setup phase. Our third contribution is a prototype implementation of our microkernel-based system architecture and our proposed attestation protocol. Due to the lack of a dedicated hardware TPM 2.0, however, it features a fully functional simulator, which we have extracted from the public PDF version of the TPM Library Specification [17] using a Python script [19] that we have made open source.

The rest of the paper is structured as follows. In Sect. 2, we discuss related work with a focus on existing remote attestation mechanisms. Section 3 outlines our scenario and attacker model, which is the basis for the design of our overall system architecture described in Sect. 4. We then present our main contribution, the integrity protection and policy-based attestation protocol, in Sect. 5, while reserving details about the prototype implementation for Sect. 6. Finally, we discuss the security of our protocol in Sect. 7 and conclude with Sect. 8.

2 Related Work

A remote attestation is a cryptographic process for creating verifiable proof that enables a remote verifier to detect modifications to the prover's system, thus allowing the attestor to determine the prover's trustworthiness.

In a *hash-based remote attestation* as specified by the Trusted Computing Group (TCG) [16,17], the prover's system calculates *static load-time integrity measurements* for all relevant software components, which are securely stored inside so-called platform configuration registers (PCRs) of the TPM and can be used to prove the system's integrity to a remote verifier. More precisely, each boot component hashes the next software component during *authenticated boot* starting from an immutable Core Root of Trust for Measurement (CRTM). After the boot process has been completed, the operating system continues to measure software binaries through integrity verification mechanisms such as IMA. For a hash-based remote attestation, the integrity measurements inside the PCRs are signed by the TPM and sent to the remote verifier. With the corresponding public key and a so-called *stored measurement log* (SML), the remote party is able to verify the signature and check the entries of the SML against expected measurements provided the prior signature verification was successful. To address privacy concerns related to the public key, the TCG alternatively also adopted a remote attestation primitive called Direct Anonymous Attestation (DAA) [6], which aims to preserve the prover's privacy using zero-knowledge proofs.

However, since both primitives specified by the TCG, traditional remote attestation and DAA, primarily focuses on hash-based load-time integrity measurements for software binaries only, other schemes, such as *property-based* [13], *semantic* [8], *group-based* [1], or *logical attestation* [15], have tried to extend and generalize the attestation mechanism. For example, the idea behind property-based attestation is to prove certain security characteristics and qualities rather than to verify the hash-based integrity of certain software components. Similarly, logical attestation is based on verifiable statements about software properties, which are expressed in a logic. Group-based attestation, in turn, uses Chameleon signatures [9] to enhance privacy and the ability to manage software integrity.

Unfortunately, those characteristics, qualities, and logical properties are not enforced by the TPM, but the operating system. Our attestation mechanism, on the other hand, is based on hash-based cryptographic policies, which are enforced by a TPM 2.0. In addition, our approach does not rely on expensive cryptographic operations, such as digital signatures, to create verifiable proof for the integrity of the prover's system. Furthermore, our remote attestation mechanism, which is designed for, but not limited to microkernel-based virtualized systems, enables a verifier to protect the integrity of data, e.g., from a virtualized device, while implicitly verifying the trustworthiness of the prover's system.

3   Scenario and Attacker Model

In this section, we describe the scenario, which outlines the settings for our integrity protection and attestation protocol. We also specify the attacker model.

3.1   Data Integrity Protection and Attestation Scenario

For our data integrity protection, which enables the detection of unauthorized modifications and also implicitly attests the integrity of the underlying system, we define a prover ($\mathcal{P}$) and a verifier ($\mathcal{V}$). The prover is a microkernel-based system, such as an industrial control system, a smartphone, or a vehicle. $\mathcal{P}$ is equipped with a TPM 2.0 and able to virtualize rich operating systems, such as Linux or Android, though hardware-based virtualization technologies like ARM's Virtualization Extensions. Since we assume the prover also executes safety- or security-critical applications as native microkernel tasks, which must be strictly isolated from the rich operating systems (and sometimes even the VMM), the prover's system additionally provides hardware-enforced separation mechanism like ARM's Security Extension, also known as TrustZone. $\mathcal{V}$, on the other hand, is a remote verifier, which is considered honest and trustworthy in the context of our security protocol. Like the prover, the verifier $\mathcal{V}$ is also equipped with a TPM 2.0 to store sensitive information, such as cryptographic keys.

Without loss of generality, we assume that the prover is an *industrial control system* with at least one rich operating system and a *set of sensors* monitoring its state, environmental conditions, and a fixed number of attached components. In our scenario, the verifier is allowed to log into the rich operating system with credentials provided by the prover's administrator and, hence, is able to interact with certain sensors. However, since the rich operating system is virtualized and device access has to go through the hypervisor, which is controlled by the prover, the verifier can only access said sensors in a very controlled and restricted way. As a result, the verifier cannot be sure that the data from a virtualized device has not been modified. For example, the data of a heartbeat sensor might have been modified by the prover to reflect a system working without any interruptions or anomalies, while the system was, in fact, not available for some time. Obviously, we also have to assume that an attacker might try to modify the data when sent to the verifier's system for further evaluation if that is part of the scenario.

3.2   Attacker Model

In our data integrity and remote attestation scenario, an adversary ($\mathcal{A}$) can read messages sent between the prover $\mathcal{P}$ and the verifier $\mathcal{V}$ as long as those messages are not encrypted with a scheme that is still considered secure. An adversary, which can include $\mathcal{P}$ if the prover acts dishonestly, is also able to manipulate data if its integrity is not protected, e.g., by a message authentication code (MAC). As a result, $\mathcal{A}$ can only decrypt an encrypted message or forge a correct MAC for a modified message if the attacker has access to the correct key. Furthermore, an attacker is not able to invert cryptographic hash functions.

In addition, we assume that hardware attacks are not feasible, as specified for most remote attestation protocols. In particular, security mechanisms, which are integrated into the chip, like ARM TrustZone, or provided by a TPM cannot be compromised by an attacker. That means we assume that the implementation of hardware-based security features, e.g., cryptographic engines or security extensions, and any firmware components implemented in software is correct.

4   Microkernel-Based System Architecture with TPM 2.0

In this section, we describe the design of our proposed microkernel-based system architecture, which includes a TPM 2.0 and makes use of hardware-based virtualization and security mechanisms of modern ARM system-on-chips (SoCs).

As the name suggests, microkernels only have a fraction of the code size of regular monolithic kernels like Linux. In addition, microkernel-based systems implement all non-essential system components, such as drivers, as user-space tasks, strictly separate those tasks, and provide only a small number of system calls. That is why they are ideally suited even for the most safety- and security-critical systems. Hence, we use a microkernel as the basis for our system architecture.

As shown in Fig. 1, the design of our system architecture separates a secure execution environment (right side) from a non-secure environment (left side). Both execution environments accommodate a microkernel-based system, which consists of at least a kernel component *core* in privileged levels PL1+ and a user-space component *init* in the unprivileged level PL0. The separation into two isolated execution environments can be realized, for example, through the ARM TrustZone mechanism, which basically assigns system components, devices, and memory to either *Secure World* or *Non-secure World*. In the *Secure World*, the microkernel-based system or, more precisely, its TrustZone VMM (*tzvmm*), handles all request to security-critical tasks and devices, such as the TPM 2.0. In the *Non-secure World*, a second virtual machine monitor (*vmm*) similarly handles calls (and traps) to the hypervisor, in this case however, with the goal to virtualize a rich operating system like Linux.

The virtualized rich operating system, which can be an conventional Linux or Android, provides the usual services for a user to interact with the system. That means a user can log into the rich operating system and, for example, read data from a hardware device, such as a sensor. If the device is virtualized, access is trapped to the virtual machine monitor. The VMM, in turn, either forwards the request to a device driver, which also resides in the *Non-secure World*, or it uses a hardware-based interface, a so-called *Secure Monitor Call* (SMC), to access a device driver implemented in *Secure World* as depicted in Fig. 1. If the request is handled by the virtual machine monitor in the TrustZone, *tzvmm* forwards it to the corresponding device driver. That way, a security-critical device like a TPM 2.0 can be accessed and shared between multiple rich operating systems, while the VMM is able to monitor, restrict, and deny access if necessary.

At this point it is very important to note that the minimal hardware-based TrustZone interface, the SMC, which is even less complex than the small set of

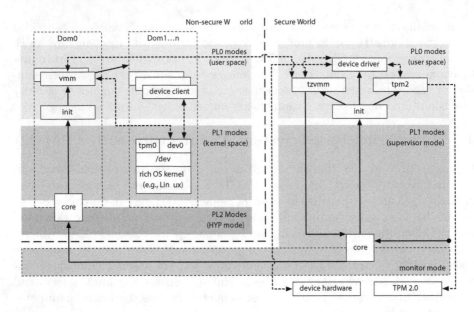

Fig. 1. Microkernel-based system architecture with TPM 2.0 on an ARM SoC

system calls used by microkernels, is also the reason for having two microkernels in our system architecture. That way, there is a strong possibility that the system can still function and actively recover, even if the *Non-secure World* was compromised. Assuming the system architecture was implemented correctly, an adversary needs to successfully attack the *core* component in the *Secure World* (while using mainly SMC calls or traps) in order to fully compromise the system.

However, even with a separation of resources, such as memory or devices, though hardware-based virtualization and security mechanisms like TrustZone, the design of our system architecture also includes a TPM 2.0, which is used to securely create and store sensitive information, particularly cryptographic keys. In contrast to the ARM's TrustZone, a TPM is a dedicated non-programmable hardware security module, which not only implements cryptographic engines in hardware, but also establishes trust, precisely because it provides assurances that its firmware cannot be easily modified by any user or remote attacker.

In addition to acting as a trust anchor, a TPM 2.0 provides mechanisms to store cryptographic integrity measurements. Those measurements are usually collected during *authenticated boot*, where the current component measures the next component in the boot chain and extends the hash into one of the PCRs. In a remote attestation as specified by the TCG, those PCRs are digitally signed with an asymmetric key to create proof about the trustworthiness of the system. Since the PCRs are only reset when the system is reset and can only be updated or, more precisely, extended with new measurements, an attacker is not able to modify a boot component without detection. On top of that, PCRs can also be used to cryptographically bind a key to specific values, which means that the key can only be used if the current values in the PCRs match the specified ones. With the TPM 2.0, the TCG generalized this idea and developed a concept called *Extended Authorization* (EA). With EA, the TPM allows the use of a cryptographic key if the user can satisfy a policy. A TPM 2.0 policy is represented by a cryptographic hash, which needs to match a specific value, and can include, for example, a hashed secret, the value of NV memory inside the TPM, or PCR values. That is why we use TPM 2.0 policies as the basis for our attestation.

5   Data Integrity Protection with Implicit Attestation

In this section, we present our data integrity protection and attestation protocol. In contrast to a traditional remote attestation as specified by the TCG, our protocol uses efficient symmetric operation instead of relying on expensive asymmetric cryptographic operations to create verifiable proof of the system's integrity. As the main contribution, however, our implicit attestation protocol makes use of the *Extended Authorization* mechanism provided by the TPM 2.0, which allows for a flexible definition of authorization and attestation policies.

The main idea of our proposed protocol is based on the fact that both parties, the prover $\mathcal{P}$ and the verifier $\mathcal{V}$, each control a cryptographic key with a policy. For a successful policy-based attestation, $\mathcal{P}$ has to satisfy the verifier's policy in order to be able to load $\mathcal{V}$'s key, when $\mathcal{V}$ requests integrity-protected data. $\mathcal{V}$'s policy, thus, may include trusted PCR values for the prover's system, which means the key can only be loaded if $\mathcal{P}$'s system is still in a trustworthy state. In turn, $\mathcal{V}$ must act according to the prover's policy, which convinces $\mathcal{P}$ to load the integrity key on behalf of $\mathcal{V}$ and create integrity-protected data. Hence, $\mathcal{P}$'s policy could, for example, specify that $\mathcal{V}$ may only access a virtualized hardware resource if the device is enabled and the access pattern meets certain criteria. As a result, each policy must be satisfied by the other party for a successful implicit attestation when requesting integrity-protected data from a virtualized device.

In the following sections, we first define the notations and the cryptographic keys in Sects. 5.1 and 5.2. After that, we specify the setup phase in Sect. 5.3 and then focus on the integrity protection and attestation mechanism in Sect. 5.4.

5.1   Notation

In general, a hash function H can compress arbitrary-length input to an output with length l, which depends on a specific algorithm, that is $H : \{0,1\}^* \rightarrow \{0,1\}^l$. A cryptographic hash function is such a one-way hash function with collision and pre-image resistance, which both describe additional security properties.

A *message authentication code (MAC)* is a cryptographic value, which can be calculated based on a cryptographic hash function and a shared symmetric key, and used to verify the authenticity and integrity of a message. Formally, a MAC algorithm is a function that generates a message digest d with fixed length l for a secret key K and a given input m with virtually arbitrary size as $MAC(K, m) = d = \{0,1\}^l$. As an example for a hash-based MAC, a *HMAC* calculates a message authentication digest for data m based on a key symmetric key K as $HMAC(K, m) = H((K \oplus opad) \,||\, H((K \oplus ipad) \,||\, m))$, where $||$ denotes a concatenation, $\oplus$ the exclusive or, *opad* the outer and *ipad* the inner padding.

Cryptographic hash functions are also used to measure the integrity of software components before they are loaded. We assume that the load-time integrity of a microkernel-based system can be adequately described by a set of measurement values, which are securely stored in the PCRs of a TPM. Such a set of PCR values are referred to as *platform configuration* $PC := (PCR[i_1], \ldots, PCR[i_k])$, where $i \in \{0 \ldots r-1\}$, $k \leq r$, and r is the number of available PCRs.

To store an integrity measurement value μ in a PCR with index i, the current value inside the TPM is combined with the new measurement value using $PCR\_Extend(PCR[i], \mu)$, which is specified as $PCR[i] \leftarrow H(PCR[i] \,||\, \mu)$. For the sake of simplicity, we assume that the PCR values of our microkernel-based system are public and, hence, known to a verifier in advance.

Similar to integrity measurements, a policy P is represented as a cryptographic hash, which can be used to authorize TPM operations. For example, to use a key K for signing, the initial empty hash P_0 is extended with a command code for signing, i.e., $P_{sign} \leftarrow H(P_0 \,||\, \texttt{TPM\_CC\_PolicyCommandCode} \,||\, \texttt{TPM2\_Sign})$. If the TPM can verify that the resulting policy hash P_{sign} matches the policy hash P' assigned to a key K, this key can be used according to that specific policy. Since multiple policies can also be combined, e.g., with $TPM2\_PolicyOR$, it is possible to create larger, more complex, yet flexible policies.

5.2   Cryptographic Keys

For our data integrity protection and attestation protocol, we assume that a primary storage key K_{PSK}, which is created from a TPM primary seed PS, and a storage key K_{SK} already exist in the prover's TPM. As shown in Fig. 2, we only require for the sake of simplicity that K_{SK} can be loaded with a policy, i.e., $P_{K_{SK}} \leftarrow H(P_0 \,||\, \texttt{TPM\_CC\_PolicyCommandCode} \,||\, \texttt{TPM2\_Load})$.

In addition to the storage key K_{SK}, the prover needs a second key K_p, which acts as a intermediate parent key for the integrity key K_{int}. The key allows for a policy-based import of child keys, which is encoded by specifying the command code $\texttt{TPM2\_Import}$ as shown in Fig. 2. Like K_{SK}, this key is bound to the prover's TPM as well as its dedicated parent, which is enforced by the TPM though the key attributes $\texttt{fixedTPM}$ and $\texttt{fixedParent}$. The second policy attached to K_p is used in our protocol to control access to hardware devices, such as sensors, by binding the content of NV memory areas to the load command. More precisely, if a bit in *NV1* is set, the hardware component assigned to this bit is enabled, which is a requirement to access it. In *NV2*, a minimum threshold, e.g., for the number of sensor values used in a average function or a anomaly detection score, can be stored. That way, the administrator of the prover's system can restrict access to a device if the number of access requests stored in *NV2* do not match the specified threshold of the policy or the pattern indicates malicious behavior.

Finally, we specify a keyed hash key K_{int}, which is used to symmetrically sign data with a *HMAC* to protect the integrity on behalf of the verifier.

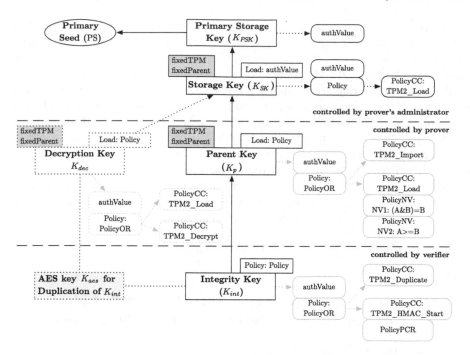

Fig. 2. Cryptographic key hierarchy for our policy-based implicit attestation protocol

This key is initially created by the verifier's TPM and migrated to prover's system during the setup phase of our protocol. Consequently, the verifier alone is able to define the policies that have to be satisfied in order to be able to use the key as indicated in Fig. 2. For our protocol, we at least require that the key allows for the calculation of a *HMAC* if the current PCRs values, which must include the microkernel-based system, i.e., *core*, *init*, *tzvmm*, etc., match the specified ones of a trusted platform configuration PC_t. For the sake of simplicity, we also specify a policy-based authorization for a *duplication*, which is the TPM 2.0 term for migration, from the verifier's to the prover's TPM. Since we encrypt K_{int} during the duplication, we optionally define a decryption key K_{dec}, which can be used to decrypt the AES encryption key K_{aes}. However, please note that this encryption and decryption step is only required once during the setup phase.

5.3 Phase 1: Setup

In the setup phase of our protocol, the administrator of the prover's system first creates two non-volatile memory areas, *NV1* and *NV2*, inside the TPM using TPM2_NV_DefineSpace. The first NV location is used to enable access to a device, whereas the second one allows for a more fine grained access control. Both NV areas can be read with a policy P_{NV_Read} and used with TPM_CC_PolicyNV.

In *NV1*, the administrator can set certain bits to enable the corresponding hardware components. For instance, if bit 0 of *NV1* is assigned to a device with an index 0, the administrator can set this bit to 1 in order to enable access to the device. Additionally, the administrator can use *NV2* in a policy to specify a minimum granularity or threshold. For example, if we assume the device is a sensor and the verifier should only be able to read the average value of at least n sensor values, the current value in *NV2* could be compared to the reference value n specified in the policy. *NV2* could also be used to store the result of an anomaly detection algorithm, which needs to be below a certain threshold t defined in a policy to be able to load and use the key K_{int}.

After setting up the non-volatile memory locations, which are used as part of the policies for K_p, the storage key K_{SK} is loaded using the policy of K_{PSK}. Then, K_p is created with the policy described in the previous section, which is

$$P_{K_p} = H(P_{base} \,\|\, \texttt{TPM\_CC\_PolicyOR} \,\|\, P_{Import} \,\|\, P_{Load\_NVs}), \tag{1}$$

where

$$P_{Import} = H(P_0 \,\|\, \texttt{TPM\_CC\_PolicyCommandCode} \,\|\, \texttt{TPM\_CC\_Import}),$$
$$P_{Load\_NVs} = H(P_{NVs} \,\|\, \texttt{TPM\_CC\_PolicyCommandCode} \,\|\, \texttt{TPM\_CC\_Load}), \tag{2}$$

and P_{base} is either P_{Import} or P_{Load_NVs}.

The value P_{NVs}, in turn, is calculated based on the following equations, which use a cryptographic hash value generated during the initialization of the NV locations as the respective name of *NV1* and *NV2*:

$$P_{NV1} = H(P_0 \,\|\, \texttt{TPM\_CC\_PolicyNV} \,\|\, args \,\|\, \text{nvIndex} \to \text{Name}) \tag{3}$$
$$P_{NVs} = H(P_{NV1} \,\|\, \texttt{TPM\_CC\_PolicyNV} \,\|\, args \,\|\, \text{nvIndex} \to \text{Name}) \tag{4}$$

with

$$args = H(\text{operandB.buffer} \,\|\, \text{offset} \,\|\, \text{operation})$$

where *operandB* is the value used for the comparison, *offset* is the start value of the NV data, and *operation* is the type of comparison. For *NV1*, the operation is $(A\&B) = B$, which checks that all bits in B are set in A, while the operation for *NV2* is $A \geq B$, which enables the prover to specify a minimum value.

Once the policy P_{K_p} is successfully generated, the key K_p is created with **TPM2_Create**, which calculates a new (ordinary) key. For this command, a public template specifies the properties of the key to be generated by the TPM, e.g., the type of key and the associated policy. The command returns the public and encrypted private key as well as data about the creation, which can be certified.

When the intermediate parent key K_p was created, an asymmetric decryption key pair K_{dec} can be optionally generated using **TPM2_Create** while the storage key K_{SK} is still loaded. Like K_p, this key is also attached with a combined policy, which allows for loading and decryption:

$$P_{K_{dec}} = H(P_{base} \,\|\, \texttt{TPM\_CC\_PolicyOR} \,\|\, P_{Load} \,\|\, P_{Decrypt}), \tag{5}$$

where

$$P_{Load} = H(P_0 \,||\, \texttt{TPM\_CC\_PolicyCommandCode} \,||\, \texttt{TPM\_CC\_Load}),$$
$$P_{Decrypt} = H(P_0 \,||\, \texttt{TPM\_CC\_PolicyCommandCode} \,||\, \texttt{TPM\_CC\_Decrypt}),$$

and P_{base} is either P_{Load} or $P_{Decrypt}$.

As described in the previous section, this asymmetric key is only used to securely transfer an AES key, which is used to encrypt $\mathcal{V}$'s integrity key K_{int}, from the verifier to the prover. This optional step only executed once and, hence, has no significant impact on our protocol.

On the verifier's system, $\mathcal{V}$ generates the keyed hash key K_{int} as part of the setup process. This key is a symmetric signing key, which can be migrated to a new TPM (fixedTPM is CLEAR) and is cryptographically bound to the integrity measurements of the prover's microkernel-based system. As a result, it can only be loaded if the current values of the PCRs match the ones specified by the verifier. The policy for this key, which—for the sake of simplicity—allows for a duplication without a strong authentication, is calculated as

$$P_{K_{int}} = H(P_{base} \,||\, \texttt{TPM\_CC\_PolicyOR} \,||\, P_{Dup} \,||\, P_{PCR\_HMAC}), \tag{6}$$

with

$$P_{Dup} = H(P_0 \,||\, \texttt{TPM\_CC\_PolicyCommandCode} \,||\, \texttt{TPM\_CC\_Duplicate})$$

and P_{base} is either P_{Dup} or P_{PCR_HMAC}. The value P_{PCR_HMAC}, in turn, is calculated by the TPM as

$$P_{PCR} = H(P_0 \,||\, \texttt{TPM\_CC\_PolicyPCR} \,||\, pcrs \,||\, digestTPM)$$
$$P_{PCR\_HMAC} = H(P_{PCR} \,||\, \texttt{TPM\_CC\_PolicyCC} \,||\, \texttt{TPM\_CC\_HMAC\_Start})$$

where $pcrs$ is a structure specifying the bits corresponding to the PCRs and $digestTPM$ is the digest of the selected PCRs provided by the verifier using a so-called *trial session*. This type of session allows for specifying the expected PCR values, whereas in a *non-trial session* the TPM would use the internal PCR values to calculate the digest.

Once the policy $P_{K_{int}}$ is successfully generated, the key K_{int} can be created using the command TPM2_Create. Again, the policy and key type (keyed hash key) can be specified in the public template, which is used by the TPM to create a key accordingly. To duplicate or migrate the keyed hash key K_{int} to the prover, the TPM cryptographically binds the key to its new parent key K_p, whose integrity and authenticity can be verified using the certified creation data produced by the prover's TPM, when K_p was created. For the actual duplication of K_{int} to the prover's system, the verifier runs TPM2_Duplicate with an optional AES encryption key and the public portion of the storage key K_p as input. Note that this implicitly also restricts the duplication to the prover's TPM, since the attribute fixedTPM is SET for K_p. The result of TPM2_Duplicate is an AES-encrypted key structure, which includes all necessary information to import the key into the target TPM. To complete the migration of K_{int}, the AES key K_{aes} is encrypted with the public portion of K_{dec} and sent to the prover together with the encrypted K_{int}.

On the prover's system the AES encryption key is decrypted using the private portion of K_{dec} and, in turn, used to decrypt K_{int} while it is imported to its new parent K_p. Please note that since all of those commands, TPM2_Duplicate, TPM2_Decrypt, and TPM2_Import, are part of the policy of their respective keys, this process does not require any interactive authorization by an administrator.

5.4   Phase 2: Data Integrity Protection with Implicit Attestation

In this section, we describe our data integrity protection and attestation protocol, which implicitly creates verifiable proof that $\mathcal{P}$'s system is still in a trustworthy state while protecting the integrity of data from a virtualized device for $\mathcal{V}$.

To read data from a device, such as a sensor, which is virtualized by the microkernel-based system, the verifier $\mathcal{V}$ first configures the device through a mechanism provided by the rich operating system, e.g., ioctl. This configuration includes the setting of the *granularity* or *threshold* n and a *nonce*$_\mathcal{V}$. However, since the prover $\mathcal{P}$ does not allow the kernel or device drivers of the rich operating system to configure the device directly, the operation is trapped to the hypervisor in the *Non-secure World* as shown in the top half of Fig. 3 above the dashed line. The hypervisor, in turn, evaluates the configuration—in particular, the value of the granularity n—which is used, for example, in an average function or simply to limit access to the hardware device. If the configuration is valid and matches the criteria set by the prover's administrator, the hypervisor forwards the request to the secure device driver implemented in the *Secure World*, which is able to configure the physical hardware device. In parallel, the VMM in TrustZone stores the granularity n in *NV2*, which is a critical step for using the key K_p and part of a correct behavior of *tzvmm* that is assumed to be reflected in the PCRs.

After the configuration, the client application in the rich operating system can read data from the device, which is, again, trapped and forwarded to the secure device driver in the *Secure World*. For a sensor, the device driver in the *Secure World* then reads the necessary sensor values based on the granularity n and, for example, calculates average. To protect the integrity of the result, the TrustZone VMM then requests the TPM to calculate a HMAC *mac* as

$$mac = HMAC\big(K_{int}, (nonce_\mathcal{V} \, \| \, data)\big). \tag{7}$$

This is only possible if the keyed hash key K_{int} is loaded and the PCRs of the microkernel-based systems, i.e., *core*, *init*, *tzvmm*, *vmm*, etc., match the specified values. K_{int}, however, can only be loaded under the parent key K_p, if the device is enabled in *NV1* and the granularity n stored in *NV2* is above the threshold specified by the prover's administrator in the policy P_{NV1}, P_{NVs}, and P_{K_p}.

Consequently, the prover needs to re-create the policy P_{K_p} (cf. Eq. 1) in a *policy session* inside the TPM to be able to load the key K_{int} on behalf of the verifier. More precisely, $\mathcal{P}$ has to calculate P_{Load_NVs} (cf. Eq. 2), which is only

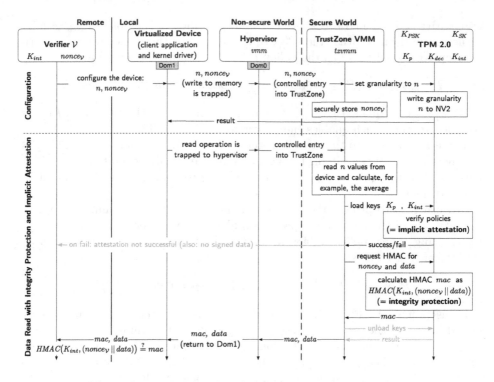

Fig. 3. Data access with policy-based implicit attestation

possible if the values in the NV indices satisfy the respective policies that are calculated as follows:

$$P'_{NV1} = H(P_0 \,\|\, \mathtt{TPM\_CC\_PolicyNV} \,\|\, args \,\|\, \text{nvIndex} \to \text{Name})$$

with: $\quad args = H(\text{operandB.buffer} \,\|\, \text{offset} \,\|\, \text{operation}),$

where *operandB* is the value in *NV1*, *offset* is the start value (0), and *operation* is the type of comparison, i.e., $(A\&B) = B$ for *NV1* in our example. This policy verifies that the device is enabled. If the comparison returns true, P'_{NV1} equals P_{NV1}, which means the device is enabled. The policy P'_{NVs} can then be calculated as

$$P'_{NVs} = H(P'_{NV1} \,\|\, \mathtt{TPM\_CC\_PolicyNV} \,\|\, args \,\|\, \text{nvIndex} \to \text{Name})$$

with: $\quad args = H(\text{operandB.buffer} \,\|\, \text{offset} \,\|\, \text{operation}),$

where *operandB* is the value n in *NV2*, *offset* is the start value (0), and *operation* is a comparison of $A \geq B$, that is $n \geq n_{P_{NVs}}$. This policy checks that the granularity or threshold n is above the value specified in the policy P_{NVs}. Again, if the comparison returns true, P'_{NVs} equals P_{NVs} and the device access pattern is accepted. Based on P'_{NVs}, the policy P'_{Load_NVs} can be calculated as

$$P'_{Load\_NVs} = H(P'_{NVs} \,||\, \texttt{TPM\_CC\_PolicyCommandCode} \,||\, \texttt{TPM\_CC\_Load}).$$

If P'_{Load_NVs} equals P_{Load_NVs}, this policy can satisfy the OR-policy P_{K_p} as specified in (1), which enables the prover to load K_{int}. The prover only has to combine P'_{Load_NVs} with the pre-calculated value of P_{Import} using $\texttt{TPM2\_PolicyOR}$ to generate P'_{K_p}. By specifying the session with the freshly generated policy P'_{K_p}, which should be equal to P_{K_p}, the prover is able to load the key K_{int} on behalf of the verifier.

To use K_{int} to protect the integrity of the device data and implicitly verify the integrity of the system, the prover simply has to re-create the policy $P_{K_{int}}$. By creating a new policy session inside the TPM and using $\texttt{TPM2\_PolicyPCR}$, the prover first creates P_{PCR} as

$$P'_{PCR} = H(P_0 \,||\, \texttt{TPM\_CC\_PolicyPCR} \,||\, pcrs \,||\, digest TPM).$$

For this policy, the PCRs of the microkernel-based system, which are usually stored in one of the lower PCRs, must be specified. We assume that the prover and the verifier agree on the selection of PCRs, since both aim for a successful attestation. The policy P'_{PCR} is then used in $\texttt{TPM2\_PolicyCommandCode}$ to calculate P'_{PCR_HMAC}, which is combined with the pre-calculated value P_{Dup} to generate $P'_{K_{int}}$:

$$P'_{PCR\_HMAC} = H(P'_{PCR} \,||\, \texttt{TPM\_CC\_PolicyCC} \,||\, \texttt{TPM\_CC\_HMAC\_Start})$$
$$P'_{K_{int}} = H(P'_{PCR\_HMAC} \,||\, \texttt{TPM\_CC\_PolicyOR} \,||\, P_{Dup} \,||\, P'_{PCR\_HMAC}).$$

If $P'_{K_{int}}$ equals $P_{K_{int}}$, the policy can finally be used to create the HMAC mac over $data$ and $nonce_V$, which is used to prove freshness, as described in (7).

The $data$ and the HMAC mac are then returned to the rich operating system. The HMAC-protected $data$ can then be transferred to the verifier's system, where a fresh HMAC mac' can be generated with the verifier's K_{int} as

$$mac' = HMAC(K_{int}, (nonce_V \,||\, data)). \tag{8}$$

If the freshly generated HMAC mac' matches the HMAC mac from the prover and $nonce_V$ is the expected nonce, the verifier does not only know that the data has not been modified, but also that the prover's system is still trustworthy.

6   Implementation

In this section, we present details about our proof-of-concept implementation, which we realized on an *Arndale* board. This development board, which is a so-called single-board computer, features an *Exynos 5250* SoC with a *Cortex-A15 MPCore* [4] that includes both ARM's Virtualization and Security Extensions.

The main components of our prototype comprise a microkernel-based system in TrustZone, which includes a TPM 2.0 simulator (*tpm2sim*) as native microkernel tasks, and a similar system in the *Non-secure World*, which acts as a hypervisor and virtualizes a rich operating system.

As our microkernel-based system, we employ the bare-metal kernel of the *Genode* [10] *base-hw* project, which combines *Genode's core* component with a small kernel library, only has a size of about 17 thousand of lines of code (KLOC) and supports TrustZone as well as virtualization. On top of the microkernel, we use regular *Genode* user-space components, such as *init*, which is about 3 KLOC, and the virtual machine monitors, *vmm* in the *Non-secure World* and *tzvmm* in the *Secure World*. For our protocol, we mainly adapted the trapping and the *World Switch* mechanism to be able to transfer data to and from the virtual machine monitors in *Non-secure World* and in TrustZone.

In the *Secure World*, we have extended the virtual machine monitor *tzvmm* to handle requests from the *Non-secure World* and also be able to execute the appropriate TPM commands. Since we did not have a hardware TPM 2.0 when we started the implementation, we created a Python script to extract a working simulator from the public PDF version of the TPM 2.0 Library Specification, which we have made open source. However, since the code of the specification is primarily written for Windows, we had to port the code to *Genode* to be able to run the simulator as a native microkernel task. This included modifications to the random number generation, the NV memory subsystem, and the communication, which was socket-based and uses inter-process communication (IPC) with a shared memory area for the commands and responses in our *Genode* port.

For our rich operating system, which is a conventional unmodified Linux 4.0 kernel with a *BusyBox* [2], we created a device client application and a kernel module to implement the device driver for a hardware component. In our prototype, this device driver, which would normally configure and access the hardware component directly, is trapped to the hypervisor of the *Non-secure World*. To be able to transfer data from the rich operating system to the hypervisor, we added

Table 1. Code size of relevant native components (calculated with cloc [7])

Component	Original Size	Difference		Total
core (Secure World)	17572	+ 215	+ 1,2%	17787
tzvmm	651	+ 956	+ 146,9%	1607
tpm2sim	0	+ 40469	+ 100,0%	40469
↳ *tpm2sim_server*	0	+ 305		305
↳ *tpm2sim_libplatform.lib.so*	0	+ 448		448
↳ *tpm2sim_libCryptoEngine.lib.so*	0	+ 5501		5501
↳ *tpm2sim_libTPM.lib.so*	0	+ 22258		22258
↳ *include* and *tpm/include*	0	+ 11957		11957
core (Non-Secure World)	17572	+ 154	+ 0,9%	17726
vmm	1132	+ 698	+ 61,7%	1830

a memory trap to the hypervisor configuration and additionally implemented an smc-based *World Switch*, which uses shared memory locations to transfer data to and from the TrustZone.

To put our prototype implementation in perspective, the modifications to the existing *Genode* components, such as *core*, *tzvmm*, or *vmm*, only amount to a few hundred lines of code per component as shown in Table 1. The reason for that is the fact that most of the protocol is handled by the TPM 2.0 simulator, which we extracted from the specification document and has about 40 KLOC, and the virtual machine monitors in the *Secure* and *Non-secure World*. The rest of the system uses mechanisms, such as IPC and shared memory, which are already part of the microkernel-based system provided by *Genode*.

7 Security Discussion

In this security discussion, we analyze the key security aspects of our protocol. Since our proposed protocol combines data access and integrity protection with an implicit attestation, we first focus on the integrity of the data, which is transmitted from the prover to the verifier. After that, we discuss the security of our policy-based implicit attestation mechanism in detail.

To protect the integrity, the prover's TPM calculates a message authentication code over the data, e.g., from a sensor, using the shared HMAC key K_{int}. This key, which is created and controlled by the verifier, is encrypted and migrated from the verifier's TPM to the prover's TPM during the setup phase and can only be used inside those respective TPMs. That way, an attacker is not able to easily forge a correct HMAC for data with unauthorized modification, because it is not able to intercept the HMAC key, decrypt it, and use it in an arbitrary TPM. In addition, please note that the identity of the prover is implicitly included in the HMAC if the verifier creates a distinct key for each prover. More precisely, since the HMAC key, which has been created for a particular prover, is duplicated specifying the public key of that prover's K_p, the HMAC key is cryptographically bound to the identity of that prover and its TPM. Similarly, the AES encryption key, which is used during duplication, is also cryptographically bound to the prover's K_{dec}, which is fixed to the prover's TPM and, thus, cannot be migrated to an arbitrary TPM.

Furthermore, our protocol includes a nonce for freshness, which has to be checked by the verifier to make sure that the HMAC has been generated for most recent request. For an adversary, this eliminates the possibility to replay old data, which has been protected with a correct HMAC, but for data that is potentially no longer valid. This is particularly relevant for devices, such as a heartbeat sensor, where the verifier must be able to detect a replay attack, where an attacker (or even the prover) might try to convince the verifier that the system still functions without any downtime or anomalies.

In order to create an correct HMAC on behalf of the verifier, the prover needs to load the HMAC key K_{int}, which is implicitly used for an attestation. In our attestation mechanism, the verifier creates and controls the key K_{int}, which means the verifier is able to define the policies, which have to be satisfied by the prover. That allows the verifier to specify, for example, the exact PCR values, which we assume reflect a known and trusted platform configuration, particularly of the microkernel-based systems in *Secure* and *Non-secure World*. Since the authentication value (*authValue*) for the key is only known to the verifier, the prover cannot change this policy later. However, the prover is able to define policies for the parent key K_p, which enables the prover to restrict access to certain data sources.

During the attestation, the prover's policies are evaluated first. If the policies cannot be satisfied, the verifier does not get access to a device, which effectively allows the prover to limit access to devices. However, if the prover's policies can be met, the policies defined by the verifier, which include at least a trusted set of PCR values, are evaluated before the HMAC for the device data is calculated. At this point, it is important to note that the policies are not verified by the operating system, as it is usually the case in policy-based authorization schemes. In our protocol, the policies are instead verified by the TPM, which also moves the point of enforcement inside the TPM. Consequently, a successful attestation is only possible if the TPM ensure that policies are satisfied, which means, for example, that the prover's system is in a trustworthy state as reflected by the PCRs. If the prover or any attacker has modified the system, the final policy of the verifier cannot be met and the prover is not able to load the HMAC key K_{int} on behalf of the verifier. As a result, the prover cannot protect the integrity of the data and the attestation eventually fails, because the verifier does not receive a fresh and valid HMAC.

8   Conclusion

In this paper, we presented a policy-based implicit attestation protocol, which does not rely on expensive asymmetric cryptographic operations traditionally required in a remote attestation. Instead, our attestation mechanism mainly uses hash-based message authentication codes and policies that are enforced by a TPM. In particular, our policy-based approach enables the verifier to create a key, which is used for integrity protection, and cryptographically bind a policy, which specifies the characteristics of a trustworthy system, to that key. For a successful attestation, the prover is expected to use that key to protect the integrity of the requested data from a virtualized hardware component, such as a sensor. Consequently, the verifier can implicitly evaluate the trustworthiness of the prover's system, whenever it accesses a virtualized device and the requested data is protected with the key, which has been bound to an attestation policy. As a result, our approach enables the verifier to not only ensure that the data requested from a virtualized device has not been modified, but also to implicitly verify the integrity and trustworthiness of the prover's system.

Acknowledgments. Parts of this work were funded by the *Industrial Data Space* project (GN: 01IS15054) of the German Federal Ministry of Education and Research. We also like to thank Sergej Proskurin and Tamas Bakos for contributing to our prototype and to the *TPM 2.0 Simulator Extraction Script*.

References

1. Alsouri, S., Dagdelen, Ö., Katzenbeisser, S.: Group-based attestation: enhancing privacy and management in remote attestation. In: Acquisti, A., Smith, S.W., Sadeghi, A.-R. (eds.) TRUST 2010. LNCS, vol. 6101, pp. 63–77. Springer, Heidelberg (2010)
2. Andersen, E., Landley, R., Vlasenko, D., et al.: Busybox. https://busybox.net
3. ARM Ltd.: Virtualization extensions architecture specification (2010). http://infocenter.arm.com
4. ARM Ltd.: ARM Cortex-A15 technical reference manual. ARM DDI 0438C, September 2011
5. ARM Ltd.: ARM architecture reference manual. ARMv7-A and ARMv7-R edition. ARM DDI 0406C.b, July 2012
6. Brickell, E., Camenisch, J., Chen, L.: Direct anonymous attestation. In: Proceedings of the 11th ACM Conference on Computer and Communications Security, CCS 2004, pp. 132–145. ACM, New York (2004). http://doi.acm.org/10.1145/1030083.1030103
7. Danial, A.: CLOC - Count Lines of Code. Version 1.67. https://github.com/AlDanial/cloc
8. Haldar, V., Chandra, D., Franz, M.: Semantic remote attestation: a virtual machine directed approach to trusted computing. In: Proceedings of the 3rd Conference on Virtual Machine Research and Technology Symposium, Berkeley, CA, USA (2004)
9. Krawczyk, H., Rabin, T.: Chameleon hashing and signatures. IACR Cryptology ePrint Archive (1998)
10. Genode Labs. http://www.genode.org
11. Liedtke, J.: Microkernels must and can be small. In: Proceedings of the 5th IEEE International Workshop on Object-Orientation in Operating Systems (IWOOOS). Seattle, WA, October 1996. http://l4ka.org/publications/
12. Neiger, G., Santoni, A., Leung, F., Rodgers, D., Uhlig, R.: Intel virtualization technology: hardware support for efficient processor virtualization. Intel Technol. J. **10**(3), 167–177 (2006)
13. Sadeghi, A.R., Stüble, C.: Property-based attestation for computing platforms: caring about properties, not mechanisms. In: Proceedings of the 2004 Workshop on New Security Paradigms, NSPW 2004, pp. 67–77. ACM, New York (2004)
14. Sailer, R., Zhang, X., Jaeger, T., van Doorn, L.: Design and implementation of a TCG-based integrity measurement architecture. In: Proceedings of the 13th Conference on USENIX Security Symposium, vol. 13, Berkeley, CA, USA (2004)
15. Sirer, E.G., de Bruijn, W., Reynolds, P., Shieh, A., Walsh, K., Williams, D., Schneider, F.B.: Logical attestation: an authorization architecture for trustworthy computing. In: Proceedings of the Twenty-Third ACM Symposium on Operating Systems Principles, SOSP 2011, pp. 249–264. ACM, New York (2011)
16. Trusted Computing Group (TCG): TPM Main Specification Version 1.2 rev. 116. http://www.trustedcomputinggroup.org/resources/tpm_main_specification
17. Trusted Computing Group (TCG): Trusted Platform Module Library Specification. Family "2.0". Level 00, Revision 01.16. http://www.trustedcomputinggroup.org/resources/tpm_library_specification
18. TU Dresden OS Group: L4/Fiasco.OC. http://os.inf.tu-dresden.de/fiasco/
19. Wagner, S., Proskurin, S., Bakos, T.: TPM 2.0 Simulator Extraction Script (2016). https://github.com/stwagnr/tpm2simulator

Generalized Dynamic Opaque Predicates: A New Control Flow Obfuscation Method

Dongpeng Xu, Jiang Ming, and Dinghao Wu[(⊠)]

College of Information Sciences and Technology,
The Pennsylvania State University, University Park, PA 16802, USA
{dux103,jum310,dwu}@ist.psu.edu

Abstract. Opaque predicate obfuscation, a low-cost and stealthy control flow obfuscation method to introduce superfluous branches, has been demonstrated to be effective to impede reverse engineering efforts and broadly used in various areas of software security. Conventional opaque predicates typically rely on the invariant property of well-known number theoretic theorems, making them easy to be detected by the dynamic testing and formal semantics techniques. To address this limitation, previous work has introduced the idea of dynamic opaque predicates, whose values may vary in different runs. However, the systematical design and evaluation of dynamic opaque predicates are far from mature. In this paper, we generalize the concept and systematically develop a new control flow obfuscation scheme called *generalized dynamic opaque predicates*. Compared to the previous work, our approach has two distinct advantages: (1) We extend the application scope by automatically transforming more common program structures (e.g., straight-line code, branch, and loop) into dynamic opaque predicates; (2) Our system design does not require that dynamic opaque predicates to be strictly adjacent, which is more resilient to the deobfuscation techniques. We have developed a prototype tool based on LLVM IR and evaluated it by obfuscating the GNU core utilities. Our experimental results show the efficacy and generality of our method. In addition, the comparative evaluation demonstrates that our method is resilient to the latest formal program semantics-based opaque predicate detection method.

Keywords: Software protection · Obfuscation · Opaque predicate · Control flow obfuscation

1 Introduction

Predicates are conditional expressions that evaluate to true or false. An opaque predicate means its value are known to the obfuscator at obfuscation time, but it is difficult for an attacker to figure it out afterwards. Used together with junk code, the effect of opaque predicates results in a heavily cluttered control flow graph with redundant infeasible paths. Therefore, any further analysis based on the control flow graph will turn into arduous work. Compared with

© Springer International Publishing Switzerland 2016
M. Bishop and A.C.A. Nascimento (Eds.): ISC 2016, LNCS 9866, pp. 323–342, 2016.
DOI: 10.1007/978-3-319-45871-7_20

other control flow graph obfuscation methods such as control flow flattening [26] and call stack tampering [24], opaque predicates are more stealthy because it is difficult to differentiate opaque predicates from original path conditions in binary code [4,5]. Also, another benefit of opaque predicates is they have a small impact on the runtime performance and code size. First proposed by Collberg et al. [6], opaque predicates have been applied widely in various ways, such as software diversification [10,15], metamorphic malware mutation [2,3], software watermarking [1,21], and Android Apps obfuscation [14]. Due to the low-cost and stealthy properties, most real-world obfuscation toolkits have supported inserting opaque predicates into a program, through link-time program rewriting or binary rewriting [8,13,18].

On the other hand, opaque predicate detection has attracted many security researchers' attention. Plenty of approaches have been proposed to identify opaque predicates inside programs. For instance, Preda et al. [23], Madou [17] and Udupa et al. [25] did research on opaque predicate detection based on the fact that the value of opaque predicate doesn't change during multiple executions. The invariant property of those "static" opaque predicates leads to the fact that they are likely to be detected by program analysis tools. Furthermore, recent research work [19] shows that even dynamic opaque predicate, which is more complicated and advanced than traditional static opaque predicates, can also be detected by their deobfuscation tool. Dynamic opaque predicates overcome the invariant weakness of static opaque predicates by using a set of correlated predicates. The authors claims that they can detect static and dynamic opaque predicates inside an execution binary trace.

Essentially, existing opaque predicates detection techniques utilize several weaknesses of opaque predicates. First, as mentioned above, the invariant property of traditional algebraic based opaque predicates reveals their existence. Second, the design of dynamic opaque predicate is far from mature. Existing technique can only insert dynamic opaque predicates into a piece of straight-line code. It cannot spread dynamic opaque predicates across branch conditions. This limitation leads to the consequence that all predicates constituting a dynamic opaque predicate are adjacent, which is utilized by advanced opaque predicates detection tools such as LOOP [19].

In order to overcome the limitations of current opaque predicates, we present a systematic design of a novel control flow obfuscation method, *Generalized Dynamic Opaque Predicates*, which is able to inject diversified dynamic opaque predicates into complicated program structures such as branch and loop. Being compared with the previous technique which can only insert dynamic opaque predicates into straight-line program, our new method is more resilient to program analysis tools. We have implemented a prototype tool based on the LLVM compiler infrastructure [16]. The tool first performs fine-grained data flow analysis to search possible insertion locations. After that it automatically transforms common program structures to construct dynamic opaque predicates. We have tested and evaluated the tool by obfuscating several hot functions of GNU core utilities with different obfuscation levels. The experimental results show that our method is effective and general in control flow obfuscation. Besides, we demonstrate that our obfuscation can defeat the commercial binary difference analysis tools and the state-of-the-art formal program semantics-based deobfuscation methods. The performance data indicate that our proposed obfuscation only introduces negligible overhead.

In summary, we make the following contributions.

- First, we propose an effective and generalized opaque predicate obfuscation method. Our method outperforms existing work by automatically inserting opaque predicates into more general program structures like branches and loops, whereas previous work can only work on straight-line code.
- Second, we demonstrate our obfuscation is very resilient to the state-of-art opaque predicate detection tool.
- Third, we have implemented our method on top of LLVM and the source code is available.

The rest of the paper is organized as follows. Section 2 introduces the related work on opaque predicates and state-of-the-art opaque predicate detection methods. Section 3 presents our new obfuscation method, generalized dynamic opaque predicates in detail. Section 4 presents our implementation details. We evaluate our method in Sect. 5 and conclude the paper in Sect. 6.

2   Related Work

In this section, we first introduce the related work on static and dynamic opaque predicates. Then we discuss the drawbacks of current opaque predicate detection methods, which also inspires us to propose the generalized dynamic opaque predicates.

2.1   Static Opaque Predicates

Static opaque predicates indicates the opaque predicates whose value is fixed during runtime. Basically, there are two categories of static opaque predicates: invariant opaque predicates and contextual opaque predicates. According to previous research work [19], invariant opaque predicates refer to those predicates whose value always evaluates to true or false for all possible inputs. The predicate is opaque since it is difficult to know the value in advance except the obfuscator. Usually invariant opaque predicates are constructed by utilizing some algebraic theorems [21] or quadratic residues [1] as follows.

$$\forall x \in Z. \ (4x^2 + 4) \bmod 19 \not\equiv 0$$

As a result of its simplicity, there are large numbers of invariant opaque predicates candidates. On the other hand, the invariant feature also leads to the shortage of this category of opaque predicates. One possible way to detect invariant opaque predicates is to observe the branches that never change at run time with fuzzing testing [17].

The other kind of static opaque predicates is contextual opaque predicate. It is proposed by Drape [11] to avoid an opaque predicate always produces the fixed value for all inputs. Contextual opaque predicate only evaluates to always true or false under a given precondition. Typically this kind of opaque predicate is an implication relation between two predicates, which is elaborately constructed in a particular program context. An example of contextual opaque predicates is presented as follows.

$$\forall x \in Z. \ (7x - 5) \bmod 3 \equiv 0 \Rightarrow (28x^2 - 13x - 5) \bmod 9 \equiv 0$$

In this example, the predicate $(28x^2 - 13x - 5) \bmod 9 \equiv 0$ is always true given $(7x - 5) \bmod 3 \equiv 0$ and x is an integer. In addition, the constant value in contextual opaque predicates can be further obfuscated so as to hide the implication relation [20].

2.2   Conventional Dynamic Opaque Predicates

Palsberg et al. [22] first introduce the concept of dynamic opaque predicates, which consist of a family of *correlated* predicates that all present the same value in one given execution, but the value may be changed in another execution. Thus the values of the dynamic opaque predicates switch dynamically at run time. Here we use the term "conventional dynamic opaque predicates" to distinguish it from the generalized dynamic opaque predicates we proposed in this paper. Particularly, since its design is still immature although the concept is noval, conventional dynamic opaque predicate can only be injected into straight-line programs, which results in that all predicates are set adjacently. We provide a conventional dynamic opaque predicate example as follows.

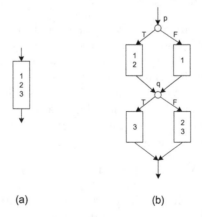

(a)   (b)

Fig. 1. An example of conventional dynamic opaque predicates.

Figure 1(a) shows the original straight-line code and Fig. 1(b) shows the obfuscated version using conventional dynamic opaque predicates. In this paper, we use a rectangle to represent a basic block and the numbers inside to indicate instructions. The small circles represent predicates. Section 3.2 provides more detailed description of those symbols. In Fig. 1(b), p and q are two correlated predicates. They are evaluated to both true or false in any given run. In the original program as shown in Fig. 1(a), three instructions are executed one by one: [1 2 3]. In the obfuscated version, each execution either follows all left branches ($p \wedge q$ holds) or all right branches ($\neg p \wedge \neg q$ holds). The same instruction sequence is executed in both cases: [1 2]->[3] when taking the left branches and [1]->[2 3] vice versa. Since the predicate q split the two paths into different segments, p and q have to be adjacent to maintain the semantic equivalence.

2.3 Opaque Predicate Detection

Collberg et al. [6] first propose the idea of opaque predicates to prevent malicious reverse engineering attempts. In addition, the authors also provide some ad-hoc detection methods, such as "statistical analysis". This approach utilize the assumption that, if a predicate that always produces the same result over a larger number of test cases, it is likely to be an opaque predicate. Due to the low coverage of inputs, statistical analysis could lead to high false positive rates.

Preda et al. [23] propose to detect opaque predicates by another method called abstract interpretation. However, their approach can only handle a specific type of known invariant opaque predicates. Madou [17] first identifies candidate branches that never changes at run time, and then verifies such predicates by fuzz testing with a considerably high error rate. Furthermore, Udupa et al. [25] utilize static path feasibility analysis to determine whether an execution path is feasible. Note that their approaches are still based on detection of invariant features such as infeasible branches, so they cannot detect the dynamic opaque predicates.

Currently, the state-of-the-art work on opaque predicate detection is LOOP [19], a logic oriented opaque predicate detection tool for obfuscated binary code. The authors propose an approach based on symbolic execution and theorem proving techniques to automatically detect static and dynamic opaque predicates. When detecting invariant opaque predicates, LOOP perform symbolic execution on an execution trace and check whether one branch condition is always true or false. Furthermore, it runs an logic implication check to decide whether one predicate is a contextual opaque predicate.

Particularly, LOOP is also able to detect the conventional dynamic opaque predicate. The detection is based on the fact that each predicate in a dynamic opaque predicates is semantically equivalent, or in another word, they implies each other logically, such as p and q in Fig. 1(b). Therefore, LOOP performs two implication check on two adjacent predicates. One is on the execution trace and the other is on the execution with the inverted path condition. Taking the example in Fig. 1, LOOP decides it is a dynamic opaque predicate when

$p \Rightarrow q$ and $\neg p \Rightarrow \neg q$ both hold. By this approach, LOOP is able to check the conventional dynamic opaque predicate.

However, to our knowledge, LOOP still utilizes the limitation of conventional dynamic opaque predicates that all predicates should be adjacently injected into a straight-line code. When testing $\neg p \Rightarrow \neg q$, LOOP first generates a new trace by negating the path condition p. Then it tests whether the next predicate in the new trace is equivalent to $\neg q$. If so, LOOP further checks whether the new trace is semantically equivalent to the original trace. This procedure is very time consuming, which limits LOOP's searching capacity. Therefore, LOOP's heuristic is only checking two adjacent predicates such as p and q in Fig. 1. In the following sections, we present that our method overcome the limitation of existing conventional dynamic opaque predicate and lead to LOOP's poor detection ratio on our generalized dynamic opaque predicates.

3   Generalized Dynamic Opaque Predicates

In this section, we present the details of the generalized dynamic opaque predicates method. First, we introduce the concept of correlated predicate. After that, we explain how to insert generalized dynamic opaque predicates into straight-line programs, branches and loops.

3.1   Correlated Predicates

Correlated predicate, as briefly discussed in Sect. 2.2, is a basic concept in dynamic opaque predicate. In this section, we present the formal definition of correlated predicate. First we need to define *correlated variables. Correlated variables* is a set of variables that are always evaluated to the same value in any program execution. One common example of correlated variables is the aliases of the same variable, like the pointers in C or the references in C++ or Java.

Correlated predicates are a set of predicates that are composed of correlated variables and have a fixed relation of their true value. The fixed relation means that, given a set of correlated predicates, if the true value of one of them is given, all other predicates' true value are known. Usually, it is intuitive to construct correlated predicates using correlated variables. Table 1 shows some examples of correlated predicates. The integer variables x, y and z in the first column is the correlated variables (CV). The CP_1, CP_2 and CP_3 columns show three sets of different correlated predicates.

Here we take the CP_2 column as an example to show how correlated predicates work. First, since x, y and z are correlated integer variables, they are always equivalent. There are three predicates in CP_2, x%2 == 1, y%2 == 0 and z%2 == 1. Note that x, y and z are integer variables, so they are either even or odd. Therefore, given the true value of any one of these predicates, we can immediately get the others' true values. Furthermore, it is not necessary that correlated predicates have similar syntax form. We can use semantically equivalent operations to create correlated predicates. CP_3 shows such an example.

Table 1. Examples of correlates predicates.

CV	CP_1	CP_2	CP_3
x	x > 0	x % 2 == 1	x + x > 0
y	y > 0	y % 2 == 0	2 * y <= 0
z	z <= 0	z % 2 == 1	z << 1 > 0

Although the syntax of each predicate is different from others, they still meet the definition of correlated predicates.

One problem we need to pay attention to is that the value of the correlated variables should not be changed during the dynamic opaque predicates, which ensures that every correlated variable are evaluated to the same value in all dynamic opaque predicates in one execution. Therefore, we compute the def-use chain inside a function and choose the section between two definitions of a variable as the candidate to be obfuscated. Note that pointer access operations could still cause the variable's value changes. Our solution is performing a simple alias analysis to decide whether the pointer is an alias of the variable. If not, we can include the pointer access instructions inside the dynamic opaque predicates; otherwise not. Since alias analysis is complicated and difficult, we only run a light-weighted address-taken algorithm [12] in our implementation. It is flow-insensitive and context-insensitive. If the analysis cannot tell whether the pointer is an alias of the correlated variable, we will conservatively consider that it could point to the variable and exclude it from the dynamic opaque predicates candidates.

3.2 Straight-Line Code

In this section, we present how to insert dynamic opaque predicate into a straight-line code. Before digging into the details, we first explain the symbols in the figures as follows.

1. A *rectangle* is a basic block.
2. A *number* in a rectangle represents one instruction.
3. A *circle* indicates a correlated predicate.
4. An *arrow* between two basic blocks indicates the control flow transfer. Typically, it is a conditional or unconditional jump. If there is only one arrow between two blocks, it is an unconditional jump; otherwise, it is a conditional jump.

Given the definition above, Fig. 2(a) shows a straight-line code which contains only one basic block, in which there are five sequential instructions. If a straight-line code comprises multiple basic blocks which are connected by unconditional jumps, it can be merged into one basic block. So for the ease of understanding, we use the single basic block example to present straight-line code.

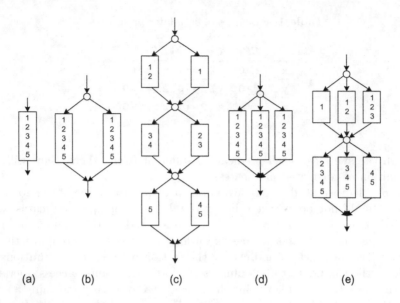

Fig. 2. Dynamic opaque predicate insertion in straight-line code.

When inserting dynamic opaque predicates into straight-line code, we have two strategies, depth-first and breadth-first, whose obfuscation result is shown in Fig. 2(c) and (e). Here we introduce the depth-first style first and briefly discuss the breadth-first later since they are similar. When inserting dynamic opaque predicates in depth-first style, we select the first correlated predicate and then make a copy of the original basic block, as shown in Fig. 2(b). After that, the two basic blocks are split at different locations so as to create two chains of basic blocks in which each basic block are different with each other. At last, we insert other correlated predicates to ensure that the control flow takes either all left branches or all right branches.

Furthermore, when inserting depth-first dynamic opaque predicates, we could insert as many correlated predicates as we can by splitting the basic blocks at different locations. As shown in Fig. 2(c), those basic blocks constitute two chains, in which the execution flow will either take every left branch or right branch. We call the basic block sequence that consists of all left or right branches an *opaque trace*. In this paper, the multiple execution traces caused by the effect of opaque predicates are called *opaque trace*. As shown in Fig. 2(c), if the execution flow takes all the left branches, the opaque trace is `[1 2]->[3 4]->[5]`. Similarly, when taking all right branches, the opaque trace is `[1]->[2 3]->[4 5]`. Therefore, Fig. 2(c) contains two opaque traces.

Generally speaking, the steps to insert depth-first dynamic opaque predicates to a single basic block BB are described as follows.

1. Select a correlated variable and creating the first correlated predicate accordingly.
2. Clone a new basic block BB' from BB.

3. Split BB and BB' at different locations to create two sequences of basic blocks, or say, two opaque traces T_1 and T_2:

$$T_1 = BB_1 \rightarrow BB_2 \rightarrow \cdots \rightarrow BB_n$$
$$T_2 = BB'_1 \rightarrow BB'_2 \rightarrow \cdots \rightarrow BB'_n$$

4. Create and insert the remaining $n - 1$ correlated predicates.
5. Insert conditional or unconditional jumps into the end of each basic block to create the correct control flow.

The other strategy is breadth-first inserting dynamic opaque predicates. It create more opaque traces via correlated predicates that have multiple branches. The inserting process is similar as depth-first. Assuming each predicate has three branches, the first step is to select and insert the first correlated predicate and create two copies of the original basic block as shown in Fig. 2(d). Then split the three basic blocks at different offsets so as to create three opaque traces. At last, insert the other correlated predicates and other jump instructions to adjust the CFG. The result is shown in Fig. 2(e).

Furthermore, we can easily create more complicated generalized dynamic opaque predicates by iteratively applying depth first and breadth first injection. For example, the basic block [1,2,3] can also be split to create a depth first generalized dynamic opaque predicate. Note that it naturally breaks the adjacency of the two predicates in Fig. 2(e). Being compared with the conventional dynamic opaque predicate shown in Sect. 2.2 which only has two adjacent predicates p and q, our method can insert more generalized and non-adjacent dynamic opaque predicates in straight-line code.

3.3   Branches

In the previous section, we present the approach to inserting dynamic opaque predicates into straight-line code. However, real world programs also consist of other structures such as branches and loops. When considering inserting dynamic opaque predicates into branches or loops, one straight forward idea is only inserting dynamic opaque predicates into basic blocks independently by treating them as straight-line code. However, this idea has one obvious problem: it doesn't spread the dynamic opaque predicates across the branch or loop condition, so essentially it is still the same as what we have done in Sect. 3.2.

In this section, we describe the process to insert dynamic opaque predicates into a branch program, which improves the program obfuscation level. For the ease of presenting our approach, we consider the branch program which contains three basic blocks as shown in Fig. 3(a). Our solution can also be applied to more complicated cases such as each branch contains multiple basic blocks. As shown in Fig. 3, Cond is the branch condition. BB_1 is located before the branch condition. BB_2 is the true branch and BB_3 is the false branch.

As the first step of inserting branch dynamic opaque predicate, we backwards search for an instruction that is independent from all instructions until

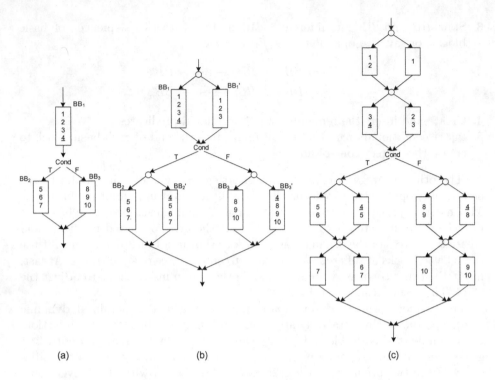

Fig. 3. Dynamic opaque predicate insertion in a branch program.

the branch condition, and also independent from the branch instruction. In this paper, this instruction is called a *branch independent instruction*. Essentially, it can be moved across the branch condition so as to create the offset in different opaque traces. In Fig. 3(a), the underlined instruction 4 is a branch independent instruction. Based on our observation, there are plenty of branch independent instructions. For example, the Coreutils program ls contains 289 branch conditions, in each of which we find at least one branch independent instruction. Typically, these instructions prepare data which are used both in the true and false branch.

After identifying the branch independent instruction, we select and insert the correlated variables, then make a copy of each basic blocks. Moreover, we move the instruction 4 along the right opaque trace across the branch condition and Fig. 3(b) shows the result. Note that due to instruction 4 is branch independent, so moving it to the head of basic blocks in the branches will not change the original program's semantics. At last, we create straight-line code dynamic opaque predicates for BB_1, BB_2 and BB_3. The final result of the obfuscated CFG is shown in Fig. 3(c). We briefly summarize the steps of inserting dynamic opaque predicates into a branch program as follows.

1. Find the branch independent instruction in BB_1.
2. Select and insert the correlated predicates.

3. Clone BB_1, BB_2 and BB_3 as BB_1', BB_2' and BB_3'.
4. Move the branch independent instruction from BB_1' to BB_2' and BB_3'.
5. Split basic blocks and create dynamic opaque predicates as in straight-line code.

3.4   Loops

Previous sections present the details about how to insert dynamic opaque predicates into straight-line code and branch programs. In this section, we consider inserting dynamic opaque predicates into a loop. In this paper, a loop refers to a program which contains a backward control flow, such as Fig. 4(a). BB_1 is the first basic block of the loop body and BB_2 is the last one. The dashed line indicates other instructions in the basic block. The dashed arrow means other instructions in the loop body, which could be a basic block, branch or even another loop. Particularly, if there is only one basic block in the loop body, BB_1 and BB_2 refer to the same basic block.

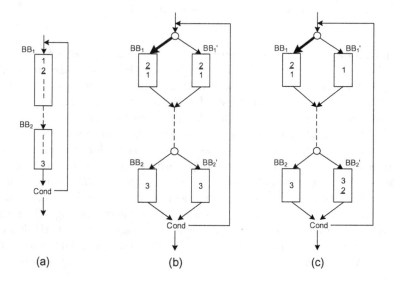

Fig. 4. Dynamic opaque predicate insertion in a loop.

The key idea in inserting dynamic opaque predicates to a loop program is finding a *loop independent instruction* and moving it across the loop condition in the same opaque trace. We define *loop independent instruction* as an instruction in a loop whose operands are all loop invariants. Loop invariant is a classical concept in compiler optimization. A variable is called loop invariant if its value never changes no matter how many times the loop is executed. For instance, Fig. 5 shows a loop invariant. The variable m is defined outside the loop and is never changed inside the loop. Each iteration of the loop accesses the same array element A[m] and assigns it to the variable x. Therefore, m, A[m] and x are loop invariant. Note that here we use the C source code to present the idea. Actually we are working on the compiler IR level, where every instruction is close to a machine instruction. As a result, in the IR level, all instructions that only operate the loop invariants are loop independent instructions. For instance, the instruction that load the value of A[m] from memory to x is an loop independent instruction. Based on our observation, there are plenty of loop independent instructions inside a loop body, such as the instructions to compute a variable's offset address. In the experiment, we find at least loop independent instruction for each of the 61 loops in the Coreutils program ls.

```
1  for (i = 0; i < 10; i++) {
2      x = A[m];     /* loop invariant */
3      B[i] = x * i;
4  }
```

Fig. 5. An example of loop invariants.

In traditional compiler optimization, the loop independent instructions are extracted out of the loop body so as to reduce the loop body size and further improve the runtime performance. All compiler frameworks implement a data flow analysis to analyze and identify the loop invariants. In this paper, we take advantage of the loop independent instructions to create the offset between opaque traces. Consider the example shown in Fig. 4(a). First, we search and identify that instruction 2 is a loop independent instruction. Second, we lift the instruction 2 to the beginning of the loop body, since other instructions might need the output of instruction 2. Then we make copies of BB_1 and BB_2 as BB_1' and BB_2'. After that we select the correlated predicates and initialize the first one to ensure that it takes the left branch. The bold arrow in Fig. 4(b) indicates the initialized predicates. We will soon discuss the reason. At last, the loop independent instruction 2 is moved from BB_1' to BB_2' and the final result is shown in Fig. 4(c). We summarize the steps of creating loop dynamic opaque predicates as follows.

1. Find the loop independent instruction I_i.
2. Lift I_i to the beginning of the loop body in BB_1.
3. Select the correlated predicates and initialize the first one correctly.
4. Clone BB_1 and BB_2 as BB_1' and BB_2'.
5. Remove I_i' from BB_1' and add it to the end of BB_2'.
6. Add dynamic opaque predicates as separate basic blocks and according jumps to build correct control flow.

Note that at the third step, we initialize the correlated variables so as to ensure the control flow goes to the left branch at the first iteration. The reason is that we have to make the loop invariant instructions executed at least once at the first iteration of the loop in order to assure all loop invariants loaded, computed and stored correctly. The value of correlated variables may change during the dashed part of the loop body so as to divert the execution flow to each opaque trace. Particularly, when the execution reaching the last iteration of the loop, there is a redundant instruction 2 if the execution follows the right branch. Since instruction 2 is loop independent, it doesn't affect the semantic of the program execution.

4   Implementation

Our implementation is based on Obfuscator-LLVM [13], an open source fork of the LLVM compilation suite that aims to improve the software security via code obfuscation and tamper-proofing. The architecture of our system is shown in Fig. 6. The generalized dynamic opaque predicate obfuscator (GDOP obfuscator) is surrounded with dashed lines. Basically, our automatic GDOP obfuscator works as a pass in LLVM framework. The workflow contains three steps. First, the LLVM frontend Clang read the source code and translate it into LLVM IR. Second, GDOP obfuscator reads the IR and inserts generalized dynamic opaque predicates to the appropriate location. At last, the LLVM backend outputs the executable program based on the obfuscated IR files.

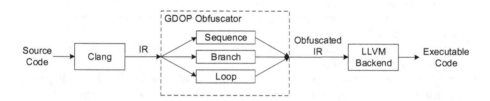

Fig. 6. The architecture of dynamic opaque predicate obfuscator.

Particularly, we implement the procedure of inserting generalized dynamic opaque predicates to a straight-line, branch and loop program as three separate passes, which includes 1251 lines of C++ code in total. We also write a driver program to invoke the three passes so as to insert all kinds of generalized dynamic opaque predicates. In addition, we implement a junk code generator to insert useless code into functions, such as redundancy branches and extra dependencies. Moreover, we provide a compiler option for users to configure the probability for inserting generalized dynamic opaque predicates. For each basic block, our obfuscator generates a random number between zero and one. If the number is smaller than the given probability, it tries to insert generalized dynamic opaque predicates into the basic block; otherwise it skips the basic block.

5   Evaluation

We conduct our experiments with several objectives. First, we want to evaluate whether our approach is effective to obfuscate control flow graph. To this end, we measure control flow complexity of GNU Coreutils with three metrics. We also test our tool with a commercial binary diffing tool which is based on control flow graph comparison. Last but not least, we want to prove our approach can defeat the state-of-the-art deobfuscation tool. Our testbed consists of an Intel Core i7-3770 processor (Quad Core with 3.40 GHz) and 8 GB memory, running Ubuntu Linux 12.04 LTS. We turn off other compiler optimization options by using -g option.

5.1   Obfuscation Metrics with Coreutils

This section shows our evaluation result of inserting generalized dynamic opaque predicates into the GNU Coreutils 8.23. Since the generalized dynamic opaque predicate is an intra-procedural obfuscation [22], we evaluate it by comparing the control flow complexity of the modified function before and after the generalized dynamic opaque predicate obfuscation. In this experiments, we choose five hot functions in the Coreutils program set by profiling. At the same time, we make sure all the functions containing at least ten basic blocks[1]. After profiling, the five hot functions we select are as follows.

1. get_next: This function is defined in tr.c. It returns the next single character of the expansion of a list.
2. make_format: This function is defined in stat.c. It removes unportable flags as needed for particular specifiers.
3. length_of_file_name_and_frills: This function is defined in ls.c for counting the length of file names.
4. print_file_name_and_frills: This function is also defined in ls.c. It prints the file name with appropriate quoting with file size and some other information as requested by switches.
5. eval6: This function is defined in eval6.c to handle sub-string, index, quoting and so on.

The metrics that we choose to show the CFG complexity are the number of CFG edges, the number of basic blocks and the cyclomatic number. The cyclomatic number is calculated as $e - n + 2$ where e is the number of CFG edges and n is the number of basic blocks. The cyclomatic number is considered as the amount of decision points in a program [9] and has been used as the metrics for evaluating obfuscation effects [7]. We first insert generalized dynamic opaque predicates into the hot functions with two different probability level: 50 % and 100 %. After that, we perform functionality testing to make sure our obfuscation is semantics-preserving. Table 2 shows the obfuscation metrics of the original clean version and the obfuscated version. The data shows that our dynamic opaque predicate obfuscation can significantly increase the program complexity.

---

[1] We do not consider dynamic link library functions because our approach takes the target program source code as input.

Table 2. Obfuscation metrics and BinDiff scores of hot functions in Coreutils.

Function	# of Basic blocks			# of CFG edges			Cyclomatic number			Bindiff score	
	Orig	50%	100%	Orig	50%	100%	Orig	50%	100%	50%	100%
1	43	171	229	62	258	338	21	89	111	0.05	0.02
2	20	75	105	30	114	158	12	41	55	0.02	0.01
3	30	94	120	49	141	177	21	49	59	0.02	0.02
4	46	138	208	80	220	320	36	84	114	0.04	0.01
5	76	272	376	117	425	573	43	155	199	0.05	0.02

To test the control flow graph after our obfuscation is heavily cluttered, we also evaluate our approach with BinDiff[2], which is a commercial binary diffing tool by measuring the similarity of two control flow graphs. We run BinDiff to compare the 50% and 100% obfuscated versions with the original five programs and the similarity score is presented in the fifth column in Table 2. The low scores indicate that the obfuscated program is very different from the original version.

5.2   Resilience

In this experiment, we evaluate the resilience to deobfuscation by applying LOOP [19], the latest formal program semantics-based opaque predicate detection tool. The authors present a program logic-based and obfuscation resilient approach to the opaque predicate detection in binary code. Their approach represents the characteristics of various opaque predicates with logical formulas and verifies them with a constraint solver. According to the authors, LOOP is able to detect various opaque predicates, including not only simple invariant opaque predicates, but also advanced contextual and dynamic opaque predicates.

In our evaluation, we run two round of 100% obfuscation on the five Coreutils functions and use LOOP to check them. The results are presented in Table 3.

Table 3. The result of LOOP detection.

Function	Straight line DOP			Branch DOP			Loop DOP		
	Total	Detected	Ratio	Total	Detected	Ratio	Total	Detected	Ratio
1	52	3	5.77%	21	0	0.00%	8	0	0.00%
2	28	2	7.14%	15	0	0.00%	6	0	0.00%
3	27	2	7.41%	23	0	0.00%	6	0	0.00%
4	54	5	9.26%	26	0	0.00%	8	0	0.00%
5	82	8	9.76%	52	0	0.00%	14	0	0.00%

---

[2] http://www.zynamics.com/bindiff.html.

As shown in Table 3, LOOP can detect very few number of the generalized dynamic opaque predicates inserted in straight-line code but fails to detect all those in branches and loops. We look into every generalized dynamic opaque predicate that is detected by LOOP and find that they are all conventional adjacent dynamic opaque predicates. We also check verify that LOOP fails to detect the remaining generalized dynamic opaque predicates.

We carefully analyze LOOP's report and find several reasons that lead to LOOP's poor detection ratio on generalized dynamic opaque predicates. First, iterative injection causes LOOP fails to detect majority of the generalized dynamic opaque predicates in straight-line code. Our obfuscation method can be iteratively executed on a candidate function, which means we are able to insert generalized dynamic opaque predicates into the same function several times. Note that each time we choose different correlated variables and different correlated predicates. Therefore, the generalized dynamic opaque predicates that are inserted by the later pass will break the adjacency of those inserted by the previous pass. In addition, junk code injection is another reason that prevents LOOP's detection.

Second, generalized dynamic opaque predicates spread across the branch or loop structure so they naturally break the adjacency property, which causes LOOP detects none of the generalized dynamic opaque predicates in branches and loops. For example, when we execute the loop shown in Fig. 4, there are two correlated but not adjacent predicates. They are separated by the instructions in the dashed line and the loop condition. Therefore, the detection method in the LOOP paper fails to detect the generalized dynamic opaque predicates.

5.3   Cost

This section presents the cost evaluation of our generalized dynamic opaque predicate obfuscation. We evaluate the cost from two aspects: binary code size and execution time. For binary code size, we measure and compare the number of bytes of the compiled programs that contain the five hot functions. For instance, we compare the size of tr's binary code when inserting generalized dynamic opaque predicates to function get_next with different probabilities such as 50 % and 100 %. For the evaluation of execution time, we record and compare the

Table 4. Cost evaluation of the dynamic opaque predicate obfuscation.

Function	Program	Binary size (bytes)				Execution time (ms)		
		Orig	50 %	100 %	Ratio	Orig	50 %	100 %
1	tr	132,084	132,826	133,491	0.53 %	2.2	2.2	2.4
2	stat	210,864	211,355	211,710	0.20 %	4.0	4.0	4.1
3	ls	350,076	350,916	351,527	0.21 %	23.2	23.4	23.7
4	ls	350,076	351,083	351,742	0.24 %	23.2	23.3	23.8
5	expr	129,696	130,836	131,409	0.66 %	0.6	0.6	0.6

Table 5. Obfuscation metrics of `sort_files`.

Function	# of basic blocks			# of CFG edges			Cyclomatic number		
	Orig	Round 1	Round 2	Orig	Round 1	Round 2	Orig	Round 1	Round 2
`sort_files`	19	160	405	25	255	539	8	97	136

execution time of clean version and the obfuscated program. We configure the switches and input files so as to ensure the control flow touches the obfuscated function.

Table 4 shows the evaluation result. We can observe that our approach slightly increases the binary code size, which is less than 0.7 %. Moreover, according to our experiments, the generalized dynamic opaque predicates have a small impact

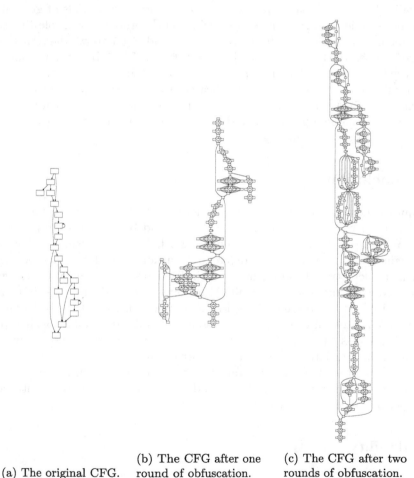

(a) The original CFG.

(b) The CFG after one round of obfuscation.

(c) The CFG after two rounds of obfuscation.

Fig. 7. Comparison between CFGs after different rounds of dynamic opaque predicate obfuscation.

on program performance. The execution time of most programs stays the same when inserting generalized dynamic opaque predicates with 50 % probability and increases a little when inserting with 100 % probability.

5.4   Case Study

As mentioned in Sect. 5.2, our generalized dynamic opaque predicates can be iteratively apply to a candidate program so as to create more obfuscated result. In this section, we provide a case study to show the result of generalized dynamic opaque predicate obfuscation iteration.

The target function is the `sort_files` function in the `ls` program. We choose this function since its CFG size is appropriate and it contains straight-line codes, branches and loops, which are suited for inserting all three categories of generalized dynamic opaque predicates. We perform two rounds of generalized dynamic opaque predicate obfuscation with 100 % probability. Table 5 presents the same measures as shown in the last section and Fig. 7 shows the result CFG. Figure 7(a) shows the original CFG of `sort_files` Fig. 7(b) presents the CFG after the first round of dynamic opaque predicate obfuscation. Next, we perform another round of dynamic opaque predicate obfuscation on (b) and the result is shown in Fig. 7(c). The comparison of the three CFGs clearly indicates that our generalized dynamic opaque predicate obfuscation can significantly modify the intra-procedural control flow graph.

6   Conclusion

Opaque predicate obfuscation is a prevalent control flow obfuscation method and has been widely applied both in malware and benign software protection. Dynamic opaque predicate obfuscation is regarded as a promising method since the predicate values may vary in different executions and thus make them more resilient to detection. However, little work discusses the systematical design of dynamic opaque predicates in detail. Also, some recent advanced deobfuscation tools utilize certain specific properties as ad hoc heuristics to detect dynamic opaque predicates. In this paper, we present generalized dynamic opaque predicates to address these limitations. Our method automatically inserts dynamic opaque predicates into common program structures and is hard to be detected by the state-of-the-art formal program semantics-based deobfuscation tools. The experimental results show the efficacy and resilience of our method with negligible performance overhead.

Availability

To better facilitate future research, we have released the source code of our dynamic opaque predicate obfuscation tool at https://github.com/s3team/gdop.

Acknowledgements. We thank the anonymous reviewers for their valuable feedback. This research was supported in part by the National Science Foundation (NSF) grants CNS-1223710 and CCF-1320605, and the Office of Naval Research (ONR) grants N00014-13-1-0175 and N00014-16-1-2265.

References

1. Arboit, G.: A method for watermarking Java programs via opaque predicates. In: Proceedings of 5th International Conference on Electronic Commerce Research (ICECR-5) (2002)
2. Bruschi, D., Martignoni, L., Monga, M.: Detecting self-mutating malware usingcontrol-flow graph matching. In: Proceedings of Detection of Intrusions and Malware and Vulnerability Assessment (DIMVA 2006) (2006)
3. Bruschi, D., Martignoni, L., Monga, M.: Code normalization for self-mutating malware. IEEE Secur. Priv. **5**(2), 46–54 (2007)
4. Cappaert, J., Preneel, B.: A general model for hiding control flow. In: Proceedings of the 10th Annual ACM Workshop on Digital Rights Management (DRM 2010) (2010)
5. Chen, H., Yuan, L., Wu, X., Zang, B., Huang, B., Yew, P.C.: Control flow obfuscation with information flow tracking. In: Proceedings of the 42nd Annual IEEE/ACM International Symposium on Microarchitecture (MICRO 42) (2009)
6. Collberg, C., Thomborson, C., Low, D.: A taxonomy of obfuscating transformations. The University of Auckland, Technical report (1997)
7. Collberg, C., Thomborson, C., Low, D.: Manufacturing cheap, resilient, and stealthy opaque constructs. In: Proceedings of the 25th ACM SIGPLAN-SIGACT Symposium on Principles of Programming Languages (POPL 1998) (1998)
8. Collberg, C., Myles, G., Huntwork, A.: Sandmark-a tool for software protection research. IEEE Secur. Priv. **1**(4), 40–49 (2003)
9. Conte, S.D., Dunsmore, H.E., Shen, V.Y.: Software Engineering Metrics and Models. Benjamin-Cummings Publishing Co. Inc., REdwood City (1986)
10. Coppens, B., De Sutter, B., Maebe, J.: Feedback-driven binary code diversification. ACM Trans. Architect. Code Optim. (TACO) **9**(4), 24:1–24:26 (2013)
11. Drape, S.: Intellectual property protection using obfuscation. Technical report, RR-10-02, Oxford University Computing Laboratory (2010)
12. Hind, M., Pioli, A.: Which pointer analysis should i use?. In: Proceedings of the ACM SIGSOFT International Symposium on Software Testing and Analysis (ISSTA 2000), pp. 113–123. ACM (2000)
13. Junod, P., Rinaldini, J., Wehrli, J., Michielin, J.: Obfuscator-LLVM - software protection for the masses. In: Proceedings of the 1st International Workshop on Software Protection (SPRO 2015) (2015)
14. Kovacheva, A.: Efficient code obfuscation for Android. Master's thesis, University of Luxembourg (2013)
15. Larsen, P., Homescu, A., Brunthaler, S., Franz, M.: SoK: automated software diversity. In: Proceedings of the 2014 IEEE Symposium on Security and Privacy (SP 2014) (2014)
16. Lattner, C., Adve, V.: LLVM: a compilation framework for lifelong program analysis and transformation. In: Proceedings of the International Symposium on Code Generation and Optimization (CGO 2004) (2004)
17. Madou, M.: Application security through program obfuscation. Ph.D. thesis, Ghent University (2007)

18. Madou, M., Van Put, L., De Bosschere, K.: LOCO: an interactive code (de)obfuscation tool. In: Proceedings of the 2006 ACM SIGPLAN Symposium on Partial Evaluation and Semantics-Based Program Manipulation (PEPM 2006) (2006)

19. Ming, J., Xu, D., Wang, L., Wu, D.: LOOP: logic-oriented opaque predicate detection in obfuscated binary code. In: Proceedings of the 22nd ACM SIGSAC Conference on Computer and Communications Security (CCS 2015) (2015)

20. Moser, A., Kruegel, C., Kirda, E.: Limits of static analysis for malware detection. In: Proceedings of the 23th Annual Computer Security Applications Conference (ACSAC 2007), December 2007

21. Myles, G., Collberg, C.: Software watermarking via opaque predicates: implementation, analysis, and attacks. Electron. Commer. Res. **6**(2), 155–171 (2006)

22. Palsberg, J., Krishnaswamy, S., Kwon, M., Ma, D., Shao, Q., Zhang, Y.: Experience with software watermarking. In: Proceedings of the 16th Annual Computer Security Applications Conference (ACSAC 2000) (2000)

23. Preda, M.D., Madou, M., Bosschere, K.D., Giacobazzi, R.: Opaque predicate detection by abstract interpretation. In: Proceedings of 11th International Conference on Algebriac Methodology and Software Technology (AMAST 2006) (2006)

24. Roundy, K.A., Miller, B.P.: Binary-code obfuscations in prevalent packer tools. ACM J. Name **1**, 21 (2012)

25. Udupa, S.K., Debray, S.K., Madou, M.: Deobfuscation: Reverse engineering obfuscated code. In: Proceedings of the 12th Working Conference on Reverse Engineering (WCRE 2005) (2005)

26. Wang, C., Hill, J., Knight, J.C., Davidson, J.W.: Protection of software-based survivability mechanisms. In: Proceedings of the 2001 International Conference on Dependable Systems and Networks (DSN 2001) (2001)

A Bayesian Cogntive Approach to Quantifying Software Exploitability Based on Reachability Testing

Guanhua Yan[1](✉), Yunus Kucuk[1,2], Max Slocum[1], and David C. Last[3]

[1] Department of Computer Science,
Binghamton University, State University of New York, Binghamton, USA
{ghyan,ykucuk1,mslocum1}@binghamton.edu
[2] Defense Sciences Institute, Turkish Military Academy, Ankara, Turkey
ykucuk@kho.edu.tr
[3] Resilient Synchronized Systems Branch,
Air Force Research Laboratory, Rome, USA
david.last.1@us.af.mil

Abstract. Computer hackers or their malware surrogates constantly look for software vulnerabilities in the cyberspace to perform various online crimes, such as identity theft, cyber espionage, and denial of service attacks. It is thus crucial to assess accurately the likelihood that a software can be exploited before it is put into practical use. In this work, we propose a cognitive framework that uses Bayesian reasoning as its first principle to quantify software exploitability. Using the Bayes' rule, our framework combines in an organic manner the evaluator's prior beliefs with her empirical observations from software tests that check if the security-critical components of a software are reachable from its attack surface. We rigorously analyze this framework as a system of non-linear equations, and henceforth perform extensive numerical simulations to gain insights into issues such as convergence of parameter estimation and the effects of the evaluator's cognitive characteristics.

1 Introduction

Software flaws are difficult, if not impossible, to avoid, either due to the limited cognitive capacities of the programmers to test all corner cases, or the fundamental weaknesses of the programming languages used. Software defects enable cybercriminals or their malware surrogates to perform a wide spectrum of malicious online activities, such as identity theft, cyber espionage, and denial of service attacks. As evidenced by numerous hacks that have occurred in the past, vulnerable software can result in significant economic losses and reputation damages. For instance, it was estimated that the revelation of the Shellshock vulnerability had led to one billion attacks [2], and an announced software vulnerability costs a firm an average loss of 0.5 % value in stock price [33].

When a software system is put into practical use, its operator is concerned with the likelihood that it can be exploited maliciously. For security-critical

© Springer International Publishing Switzerland 2016
M. Bishop and A.C.A. Nascimento (Eds.): ISC 2016, LNCS 9866, pp. 343–365, 2016.
DOI: 10.1007/978-3-319-45871-7_21

applications, a software system can only be trusted for operational use if its operator's confidence level in its unexploitability exceeds a certain threshold, say, 99 %. The challenge, then, is: *how can we derive such confidence levels to assist human operators with decision-making?* This problem is largely unexplored in the literature. There are some publicly available sources to find known software vulnerabilities, such as National Vulnerability Database [4], Exploit Database [5], and OSVDB (Open Sourced Vulnerability Database) [6]. However, these sources contain only known vulnerabilities in typically popular software, and thus cannot be solely relied upon to evaluate the security of a software system. Moreover, containing vulnerabilities does not necessarily mean that the software is exploitable in a certain running environment, as a successful software exploitation requires the existence of a realizable execution path from the attack surface of the program to its vulnerable software components [24,27,36].

Quantifiable measures of software exploitability can guide human operators in deciding whether it is sufficiently secure to put a software into operation. The Common Vulnerability Scoring System (CVSS) [28] is widely used in the industry, but its design is more of an *art* rather than a *science*. For example, it assesses the security of a vulnerable software with an overly simplistic equation: $BaseScore = 1.176 \times (3I/5 + 2E/5 - 3/2)$, where impact factor I and exploitability factor E take circumstance-specific values. Although this equation has surely been thoroughly meditated, there lack rigorous scientific arguments on why its parameters are so chosen.

In this work, we model the evaluation of software exploitability as a dynamic process done by an evaluator, who has her prior belief in software exploitability based upon some of its static features (e.g., its size, type, or some other metrics). Henceforth, she uses reachability testing tools to check whether there exists an injection vector from its attack surface that enables reachability of its security-critical components, such as a system call capable of privilege escalation or a potential buffer overflow vulnerability. The exploitability of the software is then characterized as the evaluator's subjective belief dynamically adjusted with the reachability testing results presented to her. During this process, the evaluator also continuously updates her perceptions about the performances of the tools used.

To model human cognition, we adopt a first-principled approach that integrates an evaluator's prior belief in software exploitability with her empirical observations from the reachability tests in a *Bayesian* manner. Bayesian reasoning is performed in a probabilistic paradigm, where given a hypothesis H and the evidence E, the posterior probability, or the probability of hypothesis H after seeing evidence E is calculated based upon the Bayes' rule: $\mathbb{P}\{H|E\} = \frac{\mathbb{P}\{E|H\} \cdot \mathbb{P}\{H\}}{\mathbb{P}\{E\}}$. Although there lacks evidence that humans reason in a Bayesian way at the neural level, psychological experiments show that humans behave consistently with the model at a functional level in a number of scenarios [17,20,29].

In a nutshell, our contributions can be summarized as follows:

– We propose a Bayesian cognitive framework that quantifies software exploitability as the evaluator's belief in whether an injection vector can be found from the attack surface of a software to enable the execution of a sensitive code block (e.g., one invoking a system call that leads to privilege escalation). The evaluator's belief is dynamically updated with the Bayes' rule, which uses the past performances of the reachability testing tools to calculate the likelihood functions for each hypothesis.

- We represent the Bayesian cognitive framework for quantifying software exploitability with a system of nonlinear equations, and rigorously analyze its time and space complexity, its sensitivity to the order of reachability tests, and the conditions under which the evaluator's belief in software exploitability improves or deteriorates.

- We use numerical simulations to analyze the Bayesian cognitive framework, including the convergence of the evaluator's beliefs, convergence of estimated parameters, effects of the evaluator's prior beliefs, effects of the ordering of software reachability tests, effects of dependency among different reachability testing tools, effects of short memory in parameter estimation, and effects of lazy evaluation. Our analysis shows that the nature of nonlinear equations leads to interesting observations that are not so intuitive.

From a high level, our work suggests a continuous and adaptive methodology for quantifiable cybersecurity, which is hard for an environment like the Internet that is open, dynamic and adversarial [34]. Although put in the context of software exploitability evaluation, the proposed Bayesian cognitive framework can be applied to various cybersecurity problems, such as malware detection and anomaly detection. Moreover, such cognitive frameworks allow us to further design autonomous systems that mimic the decision-making process of human defenders, thus preventing human errors.

2   Related Work

A large body of research has been dedicated to identifying security-sensitive software bugs in an efficient manner. One of the most widely used methods for finding software bugs in practice is black-box fuzzing, which generates malformed inputs in a brute-force manner to force crashes. The key challenge facing black-box fuzzing is lack of efficiency when dealing with large software systems, and there have been some recent works aimed at improving its performance [16,30]. In contrast to black-box fuzzing, white-box fuzzing takes advantage of knowledge of the internal structures of the program to find software bugs. The key enabling technology behind effective white-box fuzzing is the so-called concolic execution or dynamic symbolic execution [13], which allows systematic exploration of program branches for whole-program security testing. Notable white-box fuzzing tools include EXE [12], KLEE [11] and SAGE [18,19]. One step further, a few tools have been developed to automate the process of finding software exploits, such as APEG [10], AEG [8] and MAYHEM [15]. Many aforementioned tools can be used, directly or indirectly, for software reachability testing. Black-box fuzzing tools, for instance, can be used to test software reachability in an opportunistic manner. Symbolic or concolic execution tools can be adapted to find satisfiable paths reaching security-critical code blocks of interest.

Our work on quantifying software exploitability intersects with existing efforts on security metrics, which are valuable to strategic support, quality assurance, and tactical oversight in cyber security operations [22]. Although security metrics are important for cyber security to progress as a scientific field [25], it is hard to develop practically useful security metrics due to the dynamic and adversarial nature of the cyberspace [9,22,34]. As desirable properties of security metrics include objectivity and repeatability, software exploitability quantified by our proposed scheme does not qualify as a security metric. However, useful metrics indicative of software exploitability can be incorporated into our cognitive framework as the evaluator's prior belief. As the landscape of software exploitation is changing over time [26], these metrics may gradually lose their predictive power. Our cognitive framework allows the evaluator to adjust her beliefs with observations from new exploitation tests.

Our work finds inspirations from recent advances in modeling human cognition. A number of psychological experiments have shown that humans tend to behave consistently with the Bayesian cognitive model at the functional level [17,20,29]. Cognition-inspired methods have found a few applications in cyber security, such as malware family identification [23] and cyber-attack analysis [37]. Such cognition-based methods can be used in autonomous cyber defense systems to mimic the decision-making process of human operators and prevent human mistakes or their intrinsic cognitive biases [32].

3   Software Exploitation Based on Reachability Testing

An experienced hacker would narrow down the attack target to a few security-sensitive code blocks, a technique called *red pointing* [21]. Successful software exploitation requires both the existence of a software defect and the ability of the attacker to exploit it to achieve his attack goal [8]. With a software bug as the target, if there exists an execution path from the attack surface (which is controllable by the attacker) to invoke the software bug, the bug is deemed as *exploitable*. Note that our definition of software exploitability is different from that in [8], where a software bug is considered to be exploitable only if it is reachable from the attack surface of the program *and* the runtime environment satisfies the user-defined exploitation predicate after the control flow is hijacked (e.g., the shellcode is well-formed in memory and will be eventually executed). Consider the following C program with a buffer overflow bug:

```
#include <stdio.h>
#include <fcntl.h>
void innocent() { return; }
void vulnerable() { char buf[8]; gets(buf); }
int main(int argc, char** argv) {
    if (argc != 2) { return -1; }
    int fn = open(argv[1], O_RDONLY);
    char c, d = 0;
    int i;
```

```
for (i = 0; i < 10; i++) {
    if(read(fn, &c, 1) == 1) d = d^c; else break;
}
if (d == 0) vulnerable(); else innocent();
close(fn);
}
```

To reach `vulnerable()` with a buffer overflow bug, we need to find an input file the XOR of whose first 10 bytes is 0. We tried the following on a commodity PC:

Black-box Fuzzing: A black-box fuzzer randomly generates input files to force program crashes. We add `assert(0);` at the beginning of function `vulnerable()` to cause a crash when it is called, and then use BFF [14] to fuzz against the program. Using a single seed file of size 1,805 bytes, BFF can find the first crash within a second.

Symbolic Execution: Symbolic execution does not need to execute the program concretely. Rather, it relies upon symbolic evaluation to find an input that causes a part of the program to be executed. We use the Z3 tool developed by Microsoft Research [31] to find a satisfiable condition that enables the execution of function `vulnerable()`. As Z3 does not support the `char` type explicitly, we use bit-vectors of size 8 (in Z3 parlance, they are defined with: `Z3_sort bv_sort = Z3_mk_bv_sort(ctx, 8)` where `ctx` is a Z3 context) to perform bit-wise XOR operations. With 10 symbolic variables of type bv_sort defined, Z3 can find within a few milliseconds their assignments such that the condition for entering function `vulnerable()` is satisfied.

Concolic Execution: Concolic execution combines symbolic execution with concrete execution to speed up code exploration. We first try the CREST tool [1] to find solutions to the 10 symbolic variables of type `CREST_char`, each corresponding to a byte read from a file. However, as CREST uses Yices 1 as its SMT solver for satisfiability of formulas [7], which does not support bit-vector operations, it does not find a condition that leads to the execution of function `vulnerable()`. Another popular concolic execution tool is KLEE [11], which works on object files in the LLVM bitcode format and uses the STP solver supporting bit-vectors and arrays [3]. Similarly, by defining 10 symbolic variables using `klee_make_symbolic`, each corresponding to a byte read from the input file, we are able to use KLEE to find quickly their proper assignments that enable the execution of function `vulnerable()`.

For a large and complex software, some of the tools may not find exploits enabling reachability of its security-critical components. A security evaluator may need multiple tools for a software exploitation task, and intuitively, her memory of the past performances of these tools affects her evaluation of software exploitability.

4   A Bayesian Cognitive Framework

Motivated by the example in the previous section, we model software exploitation as a process of finding a proper injection vector in the attack surface of a software that enables its execution to reach one of its security-critical code blocks, using some reachability testing tools. Our goal is to quantify software exploitability as the likelihood that, given a security-critical target in the software, there exists such an injection vector that successfully leads to its execution. We assume that the evaluation of software exploitability is performed by an evaluator. Intuitively, if she has already found such an injection vector, her perception of the exploitability of this software is certain. Otherwise, she is *uncertain* about the exploitability of the software: there may exist an execution path that reaches the target from the attack surface but she just cannot find it at the moment. The evaluator may proceed to use some other tools to check the existence of such an injection vector, and with more failed attempts, she should be increasingly confident in the notion that the security-critical target of the software system is not exploitable.

Some notations are needed to describe the probabilistic model characterizing this cognitive process. We define the software-target pair (s, x) as an *exploitation task*, whose goal is to find whether target x is reachable in software s from its attack surface. We consider a null hypothesis $H_0(s, x)$, which simply states that target x is *unreachable* in software s from its attack surface. Hence, the unexploitability of target x in software s is quantified by the probability with which the null hypothesis is true, i.e., $\mathbb{P}\{H_0(s, x) \text{ is true}\}$, or simply $\mathbb{P}\{H_0(s, x)\}$. For ease of presentation, we let the null hypothesis $H_0(s, x)$ be the evaluator's *belief* in the unreachability of target x in software s and $\mathbb{P}\{H_0(s, x)\}$ her *belief level*.

Available to the evaluator is also a list of reachability testing tools, as discussed in Sect. 3, for finding an injection vector from a software's attack surface to reach a security-critical target of interest. Let Z denote such a list of tools, each of which works as follows: given target x in software s, tool $z \in Z$ either outputs that x is not reachable in s from its attack surface, or an injection vector that it detects to be able to reach target x. Given an injection vector v by a tool, the evaluator can execute the software with the injection vector v to validate whether target x can be reached. Like any other security detector, a reachability testing tool may wrongly report that target x is unreachable in software s, or misdetect a wrong injection vector as being able to reach target x.

Table 1. Tool parameters (IV: injection vector)

Truth/Result	Unreachable	Reachable, correct IV	Reachable, wrong IV
Unreachable	α	0	$1 - \alpha$
Reachable	β	γ	$1 - \beta - \gamma$

We thus model the performance of a reachability testing tool as probabilities in Table 1. Each tool has three performance parameters, α, β, and γ: (1) *The truth is that the target is unreachable.* A binomial process is used to characterize the output of the tool, which returns a result of being unreachable with probability α, and a result of being reachable with probability $1 - \alpha$. (2) *The truth is that the target is reachable.* The tool behaves as a multinomial process: it classifies the target as being unreachable with probability β, as being reachable with a correct injection vector with probability γ, and as being reachable with a wrong injection vector with probability $1 - \beta - \gamma$.

The rationale behind choosing the binomial and multinomial processes in our model is two-fold: they not only lead to a parsimonious model of human recognition of tool performances (by simple counting), but also provide algebraic convenience as their conjugate priors are well known. In a more fine-grained model, for the same tool z, the evaluator may associate different parameter values with some properties of the software (e.g., its size, type, or some other metrics). To deal with such subtleties, for each tool z, the evaluator can associate different values of parameters $\alpha^{(z,k)}$, $\beta^{(z,k)}$, and $\gamma^{(z,k)}$ when it is applied on software of type k. Moreover, to reflect the dynamics of these parameters, we use subscript t to indicate their values at time t. For example, $\alpha_t^{(z,k[s])}$ gives the value of parameter α at time t when tool z is used on the type of software $k[s]$.

Next we discuss how the evaluator, after using tool z for a new reachability test, updates her posterior belief in the reachability of target x in software s. Let the new observation made at time t be O_t, which falls into one of the following types:

- **Type E_0**: The tool detects target x to be unreachable in software s.
- **Type E_1**: The tool detects that target x is reachable in software s, and also returns an injection vector v, which is verified to be *true* by the evaluator.
- **Type E_2**: The tool detects that the target x is reachable in software s, and also returns an injection vector v, which is verified to be *false* by the evaluator.

After performing a reachability test with tool z and observing O_t from the test at time t, her belief level in the unreachability of target x in software s is updated to be the posterior probability $\mathbb{P}\{H_0(s,x)|O_t\}$ according to Eqs. (1–3) in Fig. 1.

$$\mathbb{P}\{H_0(s,x)|O_t = E_0\} = \frac{\mathbb{P}\{H_0(s,x)\} \cdot \alpha_t^{(z,k[s])}}{\mathbb{P}\{H_0(s,x)\} \cdot \alpha_t^{(z,k[s])} + (1 - \mathbb{P}\{H_0(s,x)\}) \cdot \beta_t^{(z,k[s])}} \quad (1)$$

$$\mathbb{P}\{H_0(s,x)|O_t = E_1\} = 0 \quad (2)$$

$$\mathbb{P}\{H_0(s,x)|O_t = E_2\} = \frac{\mathbb{P}\{H_0(s,x)\} \cdot (1 - \alpha_t^{(z,k[s])})}{\mathbb{P}\{H_0(s,x)\} \cdot (1 - \alpha_t^{(z,k[s])}) + (1 - \mathbb{P}\{H_0(s,x)\}) \cdot (1 - \beta_t^{(z,k[s])} - \gamma_t^{(z,k[s])})} \quad (3)$$

Fig. 1. Calculation of posterior probability after seeing the result from a reachability test

The calculation of Eqs. (1–3) is based upon the Bayes' rule and the performance of the reachability testing tool in Table 1. In Eq. (1), the observation is that the tool detects the target to be unreachable. As the hypothesis $H_0(s, x)$ states that the target is unreachable, the probability that the observation results from the hypothesis being true is $\alpha_t^{(z,k[s])}$. If the opposite hypothesis holds (the target is reachable), the observation occurs with probability $\beta_t^{(z,k[s])}$. Hence, Eq. (1) naturally follows based on the Bayes' rule. Moreover, when it is observed that the tool classifies the target to be reachable with a correct injection vector, it is certain that hypothesis $H_0(s, x)$ must not hold any more. This can be confirmed by Eq. (2) as $\mathbb{P}\{E_1|H_0(s, x)\}$ equals 0. Similarly, we can reason about the case when the tool classifies the target as being reachable but provides a wrong injection vector, and derive Eq. (3).

5   Parameter Updating

In this section, we discuss how the evaluator dynamically updates the values of the performance parameters (i.e., α, β, and γ) associated with each reachability testing tool based on the Bayes' rule. To evaluate the performance of a reachability testing tool, it would help if the ground truth is known to the evaluator. For example, if it is known that target x is surely reachable from the attack surface of software s, any tool that reports it being unreachable has a false negative error. One important observation, however, is that *if it is true that target x is unreachable in software s, it may never be verifiable by the evaluator for a large software, although the opposite is not true: as long as a single injection vector is found to reach target x, it is certain that the target must be reachable.* Hence, when no verifiable injection vector has been found yet to reach target x from the attack surface of software s, a *"relative fact"* reflecting whether a target has been found reachable is used to replace the truth in Table 1. Therefore, for each reachability testing tool $z \in Z$ used on software of type k, the evaluator keeps a *performance counting table*, or $PCT^{(z,k)}$, which contains five performance counters $c_0^{(z,k)}$, ..., $c_4^{(z,k)}$ as in Table 2. When the context is clear, we drop the superscript (z, k).

Table 2. The performance counting table for tool z used on software of type k, i.e., $PCT^{(z,k)}$.

"Relative fact"/Result	Unreachable	Reachable, correct IV	Reachable, wrong IV
Unreachable	$c_0^{(z,k)}$	N/A	$c_1^{(z,k)}$
Reachable	$c_2^{(z,k)}$	$c_3^{(z,k)}$	$c_4^{(z,k)}$

The evaluator performs a sequence of software reachability tests, $Q = \{q_0, q_1, ..., q_t, ...\}$, where in $q_t = (s_t, x_t, z_t, o_t)$, tool z_t is used to test the reachability of x_t in software s_t at time step t with observed test result o_t. For ease of explanation, we further define subsequences of software exploitation tests, each corresponding to a specific software exploitation task (s, x):

$$Q_{s,x} = \{q_t \mid s_t = s \wedge x_t = x\}, \tag{4}$$

and the first element in $Q_{s,x}$ is given as $Q_{s,x}[0]$.

For exploitation task (s, x), parameters are updated based upon its mode $m(s, x)$: *pre-exploitation* and *post-exploitation*. In the pre-exploitation mode, the evaluator has not found any injection vector that enables reachability of target x in software s, and by contrast, in the post-exploitation mode, such an injection vector has already been found. Initially, for every software exploitation task (s, x) its mode $m(s, x)$ is set to be *pre-exploitation*.

Consider the software reachability tests in Q sequentially. Given a new test (s, x, z, o) in Q, which corresponds to the i-th one in $Q_{s,x}$ (i.e., $Q_{s,x}[i] = (s, x, z, o)$), the evaluator uses the following rules to update the performance counters in table $PCT^{(z,k[s])}$, where $k[s]$ is the type of software s:

- **Rule I** applies to the case when $o = E_0$. If $m(s, x)$ is *pre-exploitation*, $c_0^{(z,k[s])}$ increases by 1; otherwise, $c_2^{(z,k[s])}$ increases by 1.
- **Rule II** applies to the case when $o = E_1$. If $m(s, x)$ is *post-exploitation*, $c_3^{(z,k[s])}$ increases by 1. Otherwise, if $m(s, x)$ is *pre-exploitation*, the evaluator has just found an injection vector to reach target x in software s. After increasing $c_3^{(z,k[s])}$ by 1, mode $m(s, x)$ is changed from *pre-exploitation* to *post-exploitation*. *During this change of mode, the evaluator also needs to update the performance counters for those tools that have been previously used to test the software, as the "relative fact" that has been used to update these counters previously turns out to be false.* Hence, for every j with $0 \leq j < i$, supposing that $Q_{s,x}[j] = (s, x, z', o')$, the following *revision steps* are applied: (1) if $o' = E_0$, then decrease $c_0^{(z',k[s])}$ by 1 and increase $c_2^{(z',k[s])}$ by 1; (2) if $o' = E_2$, then decrease $c_1^{(z',k[s])}$ by 1 and increase $c_4^{(z',k[s])}$ by 1. Note that it is impossible to have $o' = E_1$ (otherwise, the mode must have already been changed to *post-exploitation* after o' is seen). Hence, the evaluator needs to revise the performance counts based on the newly found truth that target x is reachable from the attack surface of software s.
- **Rule III** applies to the case when $o = E_2$. If $m(s, x)$ is *pre-exploitation*, $c_1^{(z,k[s])}$ increases by 1; otherwise, $c_4^{(z,k[s])}$ increases by 1.

The performance counters in table $PCT^{(z,k)}$ can be used to estimate the parameters $\alpha_t^{(z,k)}$, $\beta_t^{(z,k)}$, and $\gamma_t^{(z,k)}$ at the current time t. We let the values of the performance counters in table $PCT^{(z,k)}$ at time t be $c_i^{(z,k)}(t)$, for $i = 0, ..., 4$. Using a frequentist's view, parameters $\alpha_t^{(z,k)}$, $\beta_t^{(z,k)}$, and $\gamma_t^{(z,k)}$ could be estimated as their relative frequencies. When few tests have been done, however, the estimated values of $\alpha_t^{(z,k)}$, $\beta_t^{(z,k)}$, and $\gamma_t^{(z,k)}$ as derived may not be sufficiently reliable to characterize the performance of the reachability testing tool. This resembles the scenario that a person, whose prior belief is that any coin is fair, would not believe that the coin will always produce **head** even after seeing three **heads** in a row.

Our model, again, takes the evaluator's prior belief into account when estimating these parameters. After tool z is used to test whether target x is reachable in software s, which is of type k, if $m(s, x)$ is still *pre-exploitation*, the truth may not be known to the evaluator. Without knowing the truth, the evaluator relies on the "relative fact" that target x is not reachable from the attack surface of software s. Therefore, depending on the current mode of exploitation task (s, x), she updates the parameters as follows:

– If $m(s, x)$ is *pre-exploitation*, tool z works as a Binomial process where it returns a result of being unreachable with probability $\alpha^{(z,k)}$. As the conjugate prior for a Binomial process is a Beta distribution, we assume that the prior for parameter $\alpha^{(z,k)}$ takes a $Beta(d_0^{(z,k)} + 1, d_1^{(z,k)} + 1)$ distribution. We use the MAP (Maximum A Posteriori) estimate to update $\alpha^{(z,k)}$:

$$\alpha_t^{(z,k)} = \frac{d_0^{(z,k)} + c_0^{(z,k)}(t)}{d_0^{(z,k)} + c_0^{(z,k)}(t) + d_1^{(z,k)} + c_1^{(z,k)}(t)} \tag{5}$$

– If $m(s, x)$ is *post-exploitation*, tool z behaves as a multinomial process where it returns being unreachable with probability $\beta^{(z,k)}$, being reachable with a correct injection vector $\gamma^{(z,k)}$, and being reachable with a wrong injection vector $1 - \beta^{(z,k)} - \gamma^{(z,k)}$. Similarly, as the conjugate prior for a multinomial process is the Dirichlet distribution, we assume that the prior for parameter $(\beta^{(z,k)}, \gamma^{(z,k)})$ follows a Dirichlet distribution $Dir(d_2^{(z,k)} + 1, d_3^{(z,k)} + 1, d_4^{(z,k)} + 1)$. We again use the MAP estimate to update $\beta^{(z,k)}$ and $\gamma^{(z,k)}$:

$$\beta_t^{(z,k)} = \frac{d_2^{(z,k)} + c_2^{(z,k)}(t)}{\sum_{i=2}^{4} d_i^{(z,k)} + \sum_{i=2}^{4} c_i^{(z,k)}(t)} \tag{6}$$

$$\gamma_t^{(z,k)} = \frac{d_3^{(z,k)} + c_3^{(z,k)}(t)}{\sum_{i=2}^{4} d_i^{(z,k)} + \sum_{i=2}^{4} c_i^{(z,k)}(t)} \tag{7}$$

The evaluator assumes target x to be unreachable from the attack surface of software s if mode $m(s, x)$ is pre-exploitation, and this assumption is used as the relative fact to update the performance counters in related PCTs. However, when a later test finds an exploitation for the task (s, x), which invalidates the assumption, the parameters of those tools whose values have been previously estimated based upon this relative fact should be updated to reflect this change of mode. Mechanically, however, the evaluator can simply maintain PCTs like Table 2, and whenever it is necessary to use parameters α, β, and γ in Eq. (1–3), the tables are used to calculate their latest values based on Eq. (5–7).

6   Model Analysis

Space Complexity. The space used in the cognitive model includes those PCTs that the evaluator uses to keep the aggregate results from previous software reachability tests. It is noted that the prior information for parameter updating (i.e., d_0-d_4) can be put in the tables as initial values; hence, each entry in the table represents $c_i^{(z,k)}(t) + d_i^{(z,k)}$ where $0 \leq i \leq 4$. Supposing that there are $|Z|$ reachability testing tools and $|K|$ software types, as each PCT contains 5 entries (see Table 2), it requires $5|Z||K|$ to store the tables. Clearly, as the space is linear with $|K|$, more fine-grained categorization of software would bring more cognitive burden to the evaluator unless auxiliary methods are used to help remember these tables.

For every exploitation task (s, x), it is necessary to remember the evaluator's belief level $\mathbb{P}\{H_0(s, x)\}$ and its current mode $m(s, x)$. When an exploitation task is in the *pre-exploitation* mode, the evaluator also needs to remember the tools that have been previously used for the task, so if later an exploit is found, the evaluator can take the revision steps to correct the performance counters associated with these tools (Rule II of parameter updating). Therefore, if no specific ordering scheme on the exploitation tools is used, the amount of tests that the evaluator has to remember may be large, and in the worse case, it is $|Q|$.

To alleviate her cognitive burden, the evaluator may use auxiliary devices (e.g., papers) for remembering the information needed in the model, or simplify the model. For example, all the tools are numbered, and for every exploitation task, these tools are always used in an increasing order. Rules can be used to check if a tool is applicable for an exploitation task. Hence, when the mode of an exploitation task changes from *pre-exploitation* to *post-exploitation*, the evaluator can simply revise the PCTs of those applicable tools that are numbered lower than the one finding the exploitation.

Time Complexity. Given the input Q, it is assumed that executing each of Eqs. (1–3) takes a constant amount of time. For an exploitation task (s, x), changing its mode from *pre-exploitation* to *post-exploitation* requires updating the performance counters of those tools that have previously been used on them. However, for each reachability test in Q, revision of its result occurs at most once. Therefore, the time complexity of the model is $O(|Q|)$.

We can thus establish the following theorem regarding the complexity of the model:

Theorem 1. *The space and time complexity of the cognitive model is $O(|Z||K|+ n + |Q|)$ and $O(|Q|)$, respectively, where $|Z|$ is the number of reachability testing tools, $|K|$ is the number of software types, n is the number of exploitation tasks, and $|Q|$ is the total number of reachability tests done by the evaluator.*

6.1 Order Sensitivity

Equations (1–3) and (5–7) form a complex nonlinear system, whose input is comprised of sequence Q, the initial states of the PCTs for all tools in Z, and the prior values of $\mathbb{P}\{s, x\}$ for every exploitation task (s, x). We say that the cognitive model is *order insensitive* if no matter how we change the order of tests in Q, the following conditions are satisfied after all tests: (1) the evaluator's final belief level for every exploitation task is the same, and (2) the states of all the PCTs are the same. It is noted that the mode of each exploitation task must not change with the order of tests in Q: For any exploitation task (s, x), if its mode is *post-exploitation* before tests in Q, its mode remains the same after all tests in Q; otherwise, if there exists any test in Q for this task that leads to observation E_1, regardless of its order in Q, the mode of the task must be changed to *post-exploitation*, or otherwise if no such test exists, its mode should be *pre-exploitation*.

To understand under what circumstances the cognitive model is order insensitive, we first start with a simple case where there are only two reachability tests in Q. We can establish the following lemma (proof in [35]):

Lemma 1. *For any* $Q = [(s_0, x_0, z_0, o_0), (s_1, x_1, z_1, o_1)]$ *and* $Q' = [(s_1, x_1, z_1, o_1), (s_0, x_0, z_0, o_0)]$, *if* $(s_0, x_0) = (s_1, x_1)$ *or* $z_0 \neq z_1$, *the cognitive model is order insensitive.*

Now we consider the general case of array Q which may have more than two tests. According to Lemma 1, for any two consecutive reachability tests in a sequence, as long as they do not use the same reachability testing tool on two different exploitation tasks, we can swap their order. We call such a swapping of consecutive reachability tests a *safe swapping*. Given a sequence of reachability tests in Q, we can freely perform safe swappings on two consecutive tests without affecting the evaluator's final beliefs. We can thus establish the following theorem (proof in [35]):

Theorem 2. *For any sequence Q of software exploitation tests and Q' one of its permutations, assume that for every reachability testing tool, the relative order of reachability tests using this tool is the same in Q and Q'. Then the evaluator's final belief in every exploitation task must be the same after finishing Q and Q'.*

6.2   Exploitability Analysis

We now consider under what conditions a new reachability test, (s, x, z, o), improves the posterior probability $\mathbb{P}\{H_0(s, x)\}$. We consider the following cases. Without loss of generality, we drop the subscripts of the parameters.

Observation $o = E_0$: Given Eq. (1), in order to have $\mathbb{P}\{H_0(s, x) \mid O_t = E_0\} > \mathbb{P}\{H_0(s, x)\}$, we must have both $\alpha > \beta$ and $0 < \mathbb{P}\{H_0(s, x)\} < 1$. If $\mathbb{P}\{H_0(s, x)\} = 1$, the evaluator is certain that the target is not reachable a priori and thus any new evidence does not improve the posterior probability. On the other hand, if $\mathbb{P}\{H_0(s, x)\} = 0$, the Bayes' rule tells us that the posterior probability is also 0. With $\alpha > \beta$, it means that an unreachable target is detected to be unreachable with a higher probability than a reachable target being mistakenly classified as unreachable. Therefore, when a new test shows that the target is unreachable, it is better to use the former as the explanation than the latter, which suggests that the posterior probability $\mathbb{P}\{H_0(s, x)|E_0\}$ becomes higher after the test.

Observation $o = E_1$: Given Eq. (2), if the mode is still *pre-exploitation*, then seeing the test result lowers the evaluator's belief; otherwise, her belief level remains to be 0.

Observation $o = E_2$: Given Eq. (1), in order to have $\mathbb{P}\{H_0(s,x) \mid O_t = E_2\} > \mathbb{P}\{H_0(s,x)\}$, we must have: $\alpha < \beta + \gamma$ and $0 < \mathbb{P}\{H_0(s,x)\} < 1$. The same argument holds when $\mathbb{P}\{H_0(s,x)\} = 0$ or 1 as in the case when $o = E_0$. With $\alpha < \beta + \gamma$ or equivalently $1 - \alpha > 1 - (\beta + \gamma)$, it is more likely that an unreachable target is detected by the tool to be reachable with a wrong injection vector than a reachable target being detected as reachable but with a wrong input vector; hence, given the same observation E_2, it is better to use the former than the latter to explain the observation.

The above analysis leads to the following theorem:

Theorem 3. *For an exploitation task in a pre-exploitation mode, with a reachability testing tool of parameters α, β, and γ for the type of software in the task, the test result by this tool boosts the evaluator's belief level if and only if the evaluator's prior belief is in $(0,1)$ and we have $\alpha > \beta$ if E_0 is observed or $\alpha < \beta + \gamma$ if E_2 is observed.*

7   Numerical Results

We perform experiments that simulate the Bayesian cognitive model, a system of non-linear equations. The baseline configuration of an experiment is shown in Table 3. The reachability testing tools are those discussed in Sect. 3. As a reachability testing software may behave differently under different configurations, they are treated as different tools in our experiments. Parameter ϕ denotes the true probability that an exploitation task is achievable. For the test ordering, the tests are first ordered by the software to be exploited and then for each software, it is tested with the tools in the same order. The experiments mentioned in this section use parameter settings in Table 3 unless stated otherwise. We assume that the tests performed by all the tools are independent. For each tool, as the initial counts in its PCTs are all 1's, the evaluator's prior estimations of α, β, and γ are 1/2, 1/3, and 1/3, respectively.

Convergence of Estimated Parameters α, β, γ. In this set of experiments, we study how the estimated parameters converge over time. We consider 10 reachability testing tools, which are used to test 10,000 software. For each tool, its parameters α, β, and γ have true values, 0.75, 0.1, and 0.5, respectively. The others are the same as in Table 3.

Table 3. Parameter settings in baseline cases

Parameter	Value	Parameter	Value
Number of tools	100	Initial counts in PCTs	All 1's
Number of software	100	Parameter α	[0.2, 0.4, 0.6, 0.8]
Prior belief level	0.5	Parameters β, γ	[0.1, 0.2, 0.3, 0.4, 0.5]
Parameter ϕ	0.3	Test ordering	Order by software then tools

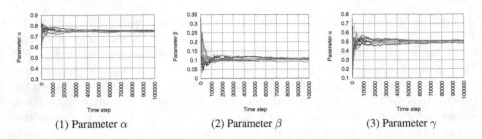

Fig. 2. Convergence of parameters α, β, and γ. The true values of these parameters are 0.75, 0.1, and 0.5, respectively. In each time step, a reachability test is performed. The ranges of these estimated parameters among the 10 tools in the last time step are 0.0152, 0.0248, and 0.0292, which are 2.0%, 24.8%, and 5.8% of their true values, respectively.

Figure 2 shows the convergence of the parameters estimated by the evaluator. *We observe that the estimation of each parameter eventually converges towards its true value, but the convergence occurs slowly.* For instance, even after performing reachability tests for 1000 software (i.e., after time step 10000 as each software uses 10 time steps, one by each tool), the estimated value of each parameter is still not very stable. Also, although the 10 tools have the same true values for their parameters, there is significant variation among these tools after 10000 reachability tests.

Convergence of Belief Levels. In this set of experiments, we study the convergence of the evaluator's belief levels. Figure 3 presents, for each combination of parameter settings, the average number of tests the evaluator needs to reach a belief level of 99% for a truly unexploitable software (left), along with the average number of tests to find an exploit for a truly exploitable software (right).

We first examine the results for truly unexploitable software. From Fig. 3(1), we observe that for a truly unexploitable software, the average number of tests required to reach a belief level of 99% ranges from 3.6 to 27.2, showing a wide variation across different combinations of parameter settings. We also observe that given the same parameters α and β, increasing γ reduces the number of tests needed. This is because that the evaluator's belief level is affected by γ only through Eq. (3), where a higher γ boosts her belief level. The observation also agrees well with Theorem 3,

The effects of parameter β, however, are not as straightforward with the same α and γ. We observe that when α is small, a higher β reduces the number of tests required, but when α is large, increasing β would also increase the number of tests. This can be explained as follows. Note that both observations E_0 and E_2 allow β to affect the evaluator's belief. When α is higher, the number of observations of type E_0 increases, and the importance of Eq. (1) becomes higher, where a higher β decreases the evaluator's belief level; by contrast, when α is smaller, the number of observations of type E_2 increases, which increases the importance of Eq. (3), where a higher β increases the evaluator's belief level.

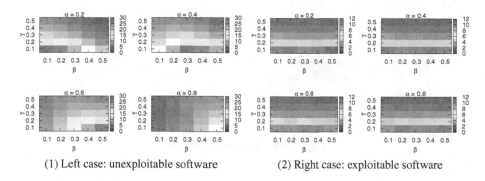

(1) Left case: unexploitable software   (2) Right case: exploitable software

Fig. 3. Convergence of belief levels. The left case gives the average number of tests before the evaluator's belief level reaches 99 % for a truly unexplotable software, and the right one the average number of tests before the evaluator's belief level reaches 0 % for a truly exploitable software.

We next study the results for truly exploitable software. From Fig. 3(2), we observe that the range of tests required for the subject to find a successful exploit is from 1.9 to 11.0. The dominating factor is γ, where a higher γ reduces the number of tests needed. This agrees well with our intuition that with tools that are more capable of finding exploits, the evaluator needs fewer tests to find exploits.

Effects of Prior Beliefs. In this set of experiments, we vary the evaluator's prior belief levels to study their effects. Figure 4 presents the average number of tests for the evaluator to reach a belief level of 99 % for a truly unexploitable software and the average number of tests to find an exploit for a truly exploitable software. For the former, it is observed that a higher prior belief reduces the number of tests to reach a certain belief level. This is because regardless of the observation types (E_0 or E_2), the posterior belief increases monotonically with the prior belief as seen in both Eqs. (1) and (3). At one extreme, if the evaluator holds her prior belief firmly that the target must be reachable, any observation that no exploitation has been found against the software does not change that belief at all. That is to say, the number of tests for her to reach a belief of 99 % would be infinity. At the other extreme, if the evaluator is certain that the software is not exploitable, obviously it does not need any test for her to reach a belief level of at least 99 %.

Furthermore, as reachability tests are performed independently, the average number of tests to find an exploit for a truly exploitable software is always $1/\gamma$, irrespective of the evaluator's prior belief level. This is confirmed by Fig. 4(2), where the evaluator's belief level does not change with the average number of tests needed to find an exploit.

Effects of Test Ordering on Belief Convergence. We now study how changing the order of software reachability tests affects the evaluator's belief convergence. We perform three groups of experiments: In the first group

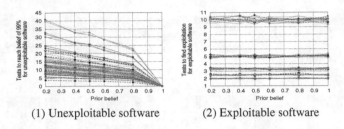

(1) Unexploitable software   (2) Exploitable software

Fig. 4. Effects of prior beliefs

(*order-by-software-then-tools*), the tests are first ordered by the software to be exploited and then for each software, we test it using 100 tools in the same order. In the second group (*order-by-tools-then-software*), the tools are first ordered, and then for each tool, it is used to exploit the 100 software consecutively in the same order. In the third group *order-randomly*, the reachability tests are ordered randomly. Figure 5 again shows the average number of tests needed to reach a belief level of 99 % for a truly unexploitable software (left) and the average number of tests to find an exploit for a truly exploitable software (right).

Interestingly, we observe that given a truly unexploitable software, on average it takes more tests to reach a certain belief level in the group of order-by-software-then-tools than those in the group of order-by-tools-then-software. The key difference is illustrated by a simple example shown in Fig. 5(3), where three tools, 1, 2, and 3, are used to test software A, B, and C. The test results of applying tools 1, 2, and 3 on software A are E_0, E_2, and E_1, respectively. For ease of explanation, we assume that before the tests, the performance counters c_0, c_1, c_2, c_3, and c_4 of all the tools are all initialized to be 1. If the tests are first ordered by software and then tools (the upper row), then the first three tests are performed with the three tools on software A. After these three tests, because tool 3 finds an exploitable path, the performance counters of the three tools are: $(1,1,2,1,1)$, $(1,1,1,1,2)$, and $(1,1,1,2,1)$. These counts will be used to update the posterior belief levels of software B and C later. By contrast, if the tests are first

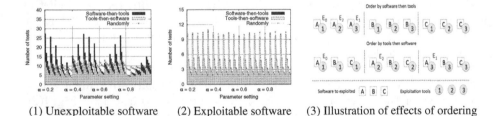

(1) Unexploitable software   (2) Exploitable software   (3) Illustration of effects of ordering

Fig. 5. Effects of test ordering on belief convergence. The left case gives the average number of tests needed to reach a belief level of 99 % for a truly unexploitable software, and the right one the average number of tests to find an exploit for a truly exploitable software. For each α setting, the tests are sorted by the increasing order of tuple (β, γ).

ordered by tools and then software (the bottom row), after the first test (tool 1 used on software A), the performance counters of tool 1 becomes (2,1,1,1,1) and these counts are used to update the posterior belief levels on software B and C in the second and third tests. Similarly, after the fourth test (tool 2 used on software A), the performance counters of tool 2 becomes (1,2,1,1,1), which will be used to update the posterior beliefs on software B and C next.

Hence, when the tests are first ordered by software and then tools, if any tool can find an exploitable path of software, this fact can change the mode of the software from *pre-exploitation* to *post-exploitation* and the performance counters of the tools previously used to test this software are updated to reflect this fact before they are used for testing other software. In contrast, if the tests are first ordered by the tools and then software, when the mode of the software is changed from *pre-exploitation* to *post-exploitation*, the performance counters of the tools previously used to test this software were updated assuming that the software is unexploitable, and then used to update the posterior beliefs of those software that were tested with these tools before the mode change.

How does such a difference affect the evaluator's posterior belief levels? For the same observation E_0, the performance counter c_0 increases by 1 if the mode is *pre-exploitation*, or c_2 increases 1 if the mode is *post-exploitation*. As we have:

$$\frac{\beta}{\alpha} = \frac{c_2/(c_2+c_3+c_4)}{c_0/(c_0+c_1)} = \frac{1+c_1/c_0}{1+(c_3+c_4)/c_2}, \tag{8}$$

post-exploitation updating increases $\frac{\beta}{\alpha}$ compared to pre-exploitation updating, which further decreases the evaluator's belief level after she sees E_0 according to Eq. (1).

Similarly, for the same observation E_2, the performance counter c_1 increases by 1 in the mode of *pre-exploitation*, or c_4 increases 1 in the mode of *post-exploitation*. Since

$$\frac{1-\beta-\gamma}{1-\alpha} = \frac{c_4/(c_2+c_3+c_4)}{c_1/(c_0+c_1)} = \frac{1+c_0/c_1}{1+(c_2+c_3)/c_4}, \tag{9}$$

post-exploitation updating increases $\frac{1-\beta-\gamma}{1-\alpha}$ compared to pre-exploitation updating, which further decreases the evaluator's belief level after E_2 is observed according to Eq. (3).

In summary, post-exploitation updating always reduces the evaluator's belief level for the software exploitation task at hand. This explains why more tests are needed for the evaluator to reach a certain belief level when tests are first ordered by software and then tools than when they are first ordered by the tools and then software, because the former case has more post-exploitation updatings than the latter, as seen in Fig. 5(1). To confirm this, we did the experiments without any observations of type E_1 and then the differences in Fig. 5(1) between order-by-software-then-tools and order-by-tools-then-software disappeared. Hence, there seems to be an irony: *postponing knowing that some software are exploitable helps improve the evaluator's belief level in the unexploitability of the others!*

In Fig. 5(2), we present the average number of tests for the evaluator to find a successful exploit for a truly exploitable software. It is seen that the effect of the order of the reachability tests is little. This is because the test results by different reachability testing tools are assumed to be independent. With a probability of γ for any tool to find the proper injection vector for an exploitable software, the average number of tests needed is thus $1/\gamma$.

Effects of Short Memory. Recall that in the basic cognitive model, the evaluator has to remember the test results for each software exploitation task that is still in the pre-exploitation mode. According to Theorem 1, this may cause high cognitive burden to the evaluator. Hence, in a new set of experiments, we study the effects of short memory, with which the evaluator omits the revision steps in Rule II of parameter updating.

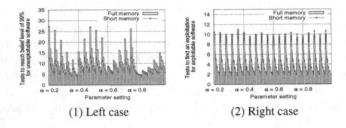

(1) Left case   (2) Right case

Fig. 6. Effects of short memory. The left case shows the average number of tests needed to reach a belief level of 99 % for an unexploitable software, and the right case the average number of tests needed to find an exploit for an exploitable software.

Figure 6 shows the effects of having a short memory in parameter updating on the evaluator's belief convergence. We observe that *due to a shorter memory, the evaluator needs fewer tests for her to reach a belief level of 99 % for a truly unexploitable software, but the average number of tests for her to find a proper injection vector for a truly exploitable software changes little.* Equations (8) and (9) can be used again to explain the smaller number of tests needed to reach a certain belief level for a truly unexploitable software. When the mode of an exploitation task changes from pre-exploitation to post-exploitation, having a short memory has the following effect for any tool that is previously used for this task:

- If the observation in that test was E_0, having a short memory omits moving 1 from c_0 to c_2. This makes β/α smaller based on Eq. (8), which increases the evaluator's belief level with a new observation E_0 according to Eq. (1), but makes $(1-\beta-\gamma)/(1-\alpha)$ larger due to Eq. (9), which decreases the evaluator's belief level with a new observation E_2 according to Eq. (3).
- If E_2 was observed in that test, having a short memory omits moving 1 from c_1 to c_4. This makes β/α larger based on Eq. (8), which decreases the evaluator's belief level with a new observation E_0 due to Eq. (1), but makes $(1 - \beta - \gamma)/(1 - \alpha)$ smaller due to Eq. (9), which improves the evaluator's belief level with observation E_2 due to Eq. (3).

At first glance, having a short memory has mixed effects on a latter observation, be it E_0 or E_2. However, the key observation here is that *the impact of having a short memory on improving the evaluator's belief level is positive if the same type of observation is made later, and is negative otherwise.* Hence, if the distribution of observations is stationary over time as assumed in the experiments, the positive impact outweighs the negative one. This resembles the positive externality in economics. Therefore, having a short memory helps improve the convergence of the evaluator's belief level when the software is truly unexploitable. On the other hand, as having a short memory does not affect the estimation of parameter γ, the average number of tests to find a proper injection vector for a truly exploitable software, which is $1/\gamma$, is not affected by a short memory in parameter updating.

Effects of Dependency. In another set of experiments, we evaluate effects of dependency on the evaluator's belief convergence. To model the dependency among the test results, we use the first tool to test a software independently. For any other tool, with probability p the test result is exactly the same as that done by the first one, and with probability $1 - p$ the result is independent of those from the other tests. We vary dependence parameter p among 0.0, 0.2, and 0.4. Figure 7 gives how the average number of tests needed to reach a belief level of 99 % for a truly unexploitable software (left) and the average number of tests needed to find an exploit for a truly exploitable software (right) change with parameter p.

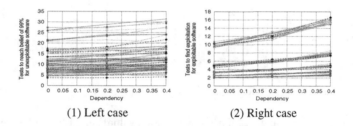

(1) Left case	(2) Right case

Fig. 7. Effects of dependency. The left case shows the average number of tests needed to reach a belief level of 99 % for a truly unexploitable software, and the right case the average number of tests needed to find an exploit for a truly exploitable software.

Clearly, *when the test results by the tools become more similar, the evaluator needs to perform more tests to reach the same belief level for a truly unexploitable software, and also more tests to find an exploit for a truly exploitable software.* To explain this phenomenon, we examine the distribution of observations per software when $\alpha = 0.4$, $\beta = 0.2$, and $\gamma = 0.2$. As the parameter setting is the same for all the tools, we find that the total number of observations of each type (E_0, E_1, or E_2) over all software is similar. However, when $p = 0.4$, the distribution of these observations per software is more bursty than that when

$p = 0.0$. That is to say, when $p = 0.4$, the variation of the numbers of the same type of observations is higher across different software than that when $p = 0.0$.

Different types of observations increases (or decreases) the evaluator's posterior belief to different degrees. For example, when $\beta/\alpha > (1 - \beta - \gamma)/(1 - \alpha)$ or equivalently, $\beta > \alpha(1 - \gamma)$, the evaluator's posterior belief after seeing E_0 is lower than that after seeing E_2. As the rule of updating posterior beliefs is *nonlinear*, the average number of tests required to reach a certain belief level on a truly unexploitable software, or to find an exploit for a truly exploitable software, is not the same if we skew the distribution of different types of observations among different software even though the total numbers of observations for the same types of observations remain the same among all software.

Effects of Lazy Evaluation. In this set of experiments, the reachability tests are first ordered by software and then by tools. There are 100 tools and 100 software to be exploited. We model a "lazy" evaluator who, after observing the software is exploitable (i.e., seeing E_1), stops using the remaining tools to test it.

Figure 8(1,2) shows the average number of tests needed for the evaluator to reach a belief level of 99 % for a truly unexploitable software and the average number of tests to find an exploit for a truly exploitable software. The parameters in the plots are ordered first by α, then β, and lastly γ. According to Fig. 8(2), lazy evaluation does not affect the number of tests to find an exploitation, which is obvious as reachability tests are omitted only after the first exploit has been found for each software.

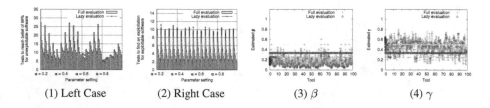

(1) Left Case   (2) Right Case   (3) β (4) γ

Fig. 8. Comparison of lazy evaluation with full evaluation (1,2) and estimation of parameters β and γ (3,4). In lazy evaluation, the evaluator stops testing a software after an exploit has been found. In contrast, full evaluation tests a software with all the tools.

The effects of lazy evaluation on the number of tests for the evaluator to reach a belief level of 99 % for a truly unexploitable software are mixed: in some cases, more tests are needed, and in others fewer are necessary. We examine the estimated values of parameters α, β and γ when their true values are 0.2, 0.1, and 0.5, respectively. Lazy evaluation does not affect much the estimation of parameter α, but it only estimates the values of parameters β and γ for a few tools, as seen in Fig. 8(3,4)! That is to say, for the majority of the tools, parameters β and γ remain to be their initial values, which are $1/3$ and $1/3$, respectively.

The differences between lazy evaluation and full evaluation as seen in Fig. 8 boil down to the differences in the estimated values of parameters β and γ. If an observation of type E_0 is seen, a larger β reduces the evaluator's posterior belief level (see Eq. (1)). On the other hand, if the new observation is of type E_2, then a larger β or γ helps improve the evaluator's posterior belief level (see Eq. (3)). With these observations, we can explain some cases where lazy evaluation requires more tests for belief convergence than full evaluation in Fig. 8(1). First, when α is small, there are more observations of type E_2; as the majority of the tools in lazy evaluation have parameters β and γ set to be both $1/3$, if their true values are higher than $1/3$, lazy evaluation tends to underestimate their true values and thus reduces the evaluator's posterior belief level, which leads to more tests needed compared to full evaluation. The effect of parameter γ is more prominent than that of β as the latter is mixed in Eqs. (1) and (2). On the other hand, when α is large, there are more observations of type E_0. If the true value of β is smaller than $1/3$, lazy evaluation always overestimates it and thus reduces the evaluator's posterior belief level according to Eq. (1), which leads to more tests needed for belief convergence than full evaluation.

8 Concluding Remarks

In this work, we propose a new cognitive framework using Bayesian reasoning as its first principle to quantify software exploitability. We rigorously analyze this framework, and also use intensive numerical simulations to study the convergence of parameter estimation and the effects of the evaluator's cognitive characteristics. In our future work, we plan to extend this work by integrating into this framework some real-world tools (e.g., software fuzzers and concolic execution tools) that can be used to exploit vulnerable software. We also plan to enrich the cognitive model used in this work.

Acknowledgment. We acknowledge the support of the Air Force Research Laboratory Visiting Faculty Research Program for this work.

References

1. Crest: Concolic test generation tool for c. https://jburnim.github.io/crest/
2. http://www.securityweek.com/shellshock-attacks-could-already-top-1-billion-report
3. Stp constraint solver. http://stp.github.io/
4. https://nvd.nist.gov/
5. https://www.exploit-db.com/
6. http://www.osvdb.org/
7. The Yices SMT Solver. http://yices.csl.sri.com
8. Avgerinos, T., Cha, S.K., Hao, B.L.T., Brumley, D.: AEG: automatic exploit generation. NDSS **11**, 59–66 (2011)
9. Bellovin, S.M.: On the brittleness of software and the infeasibility of security metrics. IEEE Secur. Priv. **4**(4), 96 (2006)

10. Brumley, D., Poosankam, P., Song, D., Zheng, J.: Automatic patch-based exploit generation is possible: techniques and implications. In: IEEE Symposium on Security and Privacy (2008)
11. Cadar, C., Dunbar, D., Engler, D.R.: KLEE: unassisted and automatic generation of high-coverage tests for complex systems programs. OSDI **8**, 209–224 (2008)
12. Cadar, C., Ganesh, V., Pawlowski, P.M., Dill, D.L., Engler, D.R.: EXE: automatically generating inputs of death. ACM Trans. Inf. Syst. Secur. (TISSEC) **12**(2), 10 (2008)
13. Cadar, C., Sen, K.: Symbolic execution for software testing: three decades later. Commun. ACM **56**(2), 82–90 (2013)
14. CERT. Basic fuzzing framework (bff). https://www.cert.org/vulnerability-analysis/tools/bff.cfm?
15. Cha, S.K., Avgerinos, T., Rebert, A., Brumley, D.: Unleashing mayhem on binary code. In: IEEE Symposium on Security and Privacy (SP), pp. 380–394. IEEE (2012)
16. Cha, S.K., Woo, M., Brumley, D.: Program-adaptive mutational fuzzing. In: Proceedings of the IEEE Symposium on Security and Privacy (2015)
17. Cooper, G.F.: The computational complexity of probabilistic inference using Bayesian belief networks. Artif. Intell. **42**(2), 393–405 (1990)
18. Godefroid, P., Levin, M.Y., Molnar, D.: SAGE: whitebox fuzzing for security testing. Queue **10**(1), 20 (2012)
19. Godefroid, P., Levin, M.Y., Molnar, D.A.: Automated whitebox fuzz testing. In: Proceedings of Network and Distributed System Security Symposium (NDSS) (2008)
20. Griffiths, T.L., Kemp, C., Tenenbaum, J.B.: Bayesian models of cognition (2008)
21. Hoglund, G., McGraw, G.: Exploiting Software: How to Break Code. Addison-Wesley, Boston (2004)
22. Jansen, W.: Directions in Security Metrics Research. Diane Publishing, Collingdale (2010)
23. Lebiere, C., Bennati, S., Thomson, R., Shakarian, P., Nunes, E.: Functional cognitive models of malware identification. In: Proceedings of International Conference on Cognitive Modeling (2015)
24. Manadhata, P.K., Wing, J.M.: An attack surface metric. IEEE Trans. Soft. Eng. **37**(3), 371–386 (2011)
25. McMorrow, D.: Science of cyber-security. Technical report, JASON Program Office (2010)
26. Nagaraju, S., Craioveanu, C., Florio, E., Miller, M.: Software vulnerability exploitation trends (2013)
27. Nayak, K., Marino, D., Efstathopoulos, P., Dumitraş, T.: Some vulnerabilities are different than others. In: Stavrou, A., Bos, H., Portokalidis, G. (eds.) RAID 2014. LNCS, vol. 8688, pp. 426–446. Springer, Heidelberg (2014)
28. Forum of Incident Response and Security Teams (FIRST). Common vulnerabilities scoring system (cvss). http://www.first.org/cvss/
29. Perfors, A., Tenenbaum, J.B., Griffiths, T.L., Xu, F.: A tutorial introduction to bayesian models of cognitive development. Cognition **120**(3), 302–321 (2011)
30. Rebert, A., Cha, S.K., Avgerinos, T., Foote, J., Warren, D., Grieco, G., Brumley, D.: Optimizing seed selection for fuzzing. In: Proceedings of the USENIX Security Symposium (2014)
31. Microsoft Research. Z3. https://github.com/Z3Prover/z3
32. Smith, S.W.: Security and cognitive bias: exploring the role of the mind. IEEE Secur. Priv. **5**, 75–78 (2012)

33. Telang, R., Wattal, S.: An empirical analysis of the impact of software vulnerability announcements on firm stock price. IEEE Trans. Soft. Eng. **33**(8), 544–557 (2007)
34. Verendel, V.: Quantified security is a weak hypothesis: a critical survey of results and assumptions. In: Proceedings of the 2009 Workshop on New Security Paradigms Workshop. ACM (2009)
35. Yan, G., Kucuk, Y., Slocum, M., Last, D.C.: A Bayesian cogntive approach to quantifying software exploitability based on reachability testing (extended version). http://www.cs.binghamton.edu/~ghyan/papers/extended-isc16.pdf
36. Younis, A., Malaiya, Y.K., Ray, I.: Assessing vulnerability exploitability risk using software properties. Soft. Qual. J **24**(1), 1–44 (2016)
37. Zhong, C., Yen, J., Liu, P., Erbacher, R., Etoty, R., Garneau, C.: An integrated computer-aided cognitive task analysis method for tracing cyber-attack analysis processes. In: Proceedings of the 2015 Symposium and Bootcamp on the Science of Security. ACM (2015)

Control Flow Integrity Enforcement with Dynamic Code Optimization

Yan Lin[1(✉)], Xiaoxiao Tang[1], Debin Gao[1], and Jianming Fu[2]

[1] School of Information Systems, Singapore Management University, `
Singapore, Singapore
yanlin0816@gmail.com
[2] Computer School, Wuhan University, Wuhan, China

Abstract. Control Flow Integrity (CFI) is an attractive security property with which most injected and code reuse attacks can be defeated, including advanced attacking techniques like Return-Oriented Programming (ROP). However, comprehensive enforcement of CFI is expensive due to additional supports needed (e.g., compiler support and presence of relocation or debug information) and performance overhead. Recent research has been trying to strike the balance among reasonable approximation of the CFI properties, minimal additional supports needed, and acceptable performance. We investigate existing dynamic code optimization techniques and find that they provide an architecture on which CFI can be enforced effectively and efficiently. In this paper, we propose and implement *DynCFI* that enforces security policies on a well established dynamic optimizer and show that it provides comparable CFI properties with existing CFI implementations while lowering the overall performance overhead from 28.6 % to 14.8 %. We further perform comprehensive evaluations and shed light on the exact amount of savings contributed by the various components of the dynamic optimizer including basic block cache, trace cache, branch prediction, and indirect branch lookup.

Keywords: Control Flow Integrity · Return-oriented programming · Dynamic code optimization

1 Introduction

Control Flow Integrity (CFI) introduced by Abadi et al. [2] provides attractive security features because of its effectiveness in defending against most injected and code reuse attacks, including the recent and advanced attacking techniques like Return-Oriented Programming (ROP) [22]. Its basic idea is to enforce a control-flow graph (usually built from static analysis) so that the program only makes control transfers to intended target locations.

However, having an accurate and practical enforcement of CFI is known to be hard [2, 13, 18, 23, 27]. First, it is generally difficult to accurately identify the target locations for all control transfers. Existing solutions typically apply

© Springer International Publishing Switzerland 2016
M. Bishop and A.C.A. Nascimento (Eds.): ISC 2016, LNCS 9866, pp. 366–385, 2016.
DOI: 10.1007/978-3-319-45871-7_22

a coarse-grained policy (e.g., to allow indirect calls to any functions [25]) or require compiler support or presence of relocation or debug information [4,19,27], which may not be applicable to Commercial Off-The-Shelf (COTS) software. Second, intercepting control transfers and doing the necessary checking typically result in large performance overhead [7,10]. Many have proposed ways of striking the balance among reasonable approximation of the CFI properties, minimizing additional supports needed, and acceptable performance [20,28]. Therefore, any noticeable reduction in the performance overhead would likely lead to more practical implementation and potentially better security properties.

An interesting observation is that prior to the introduction of CFI in 2005, there have already been a lot of research on dynamic code optimization to improve performance of dynamic program interpreters, e.g., Wiggins/ Redstone [11], Dynamo [3], Mojo [8], and DynamoRIO [5]. Dynamo and DynamoRIO are among the more popular and mature ones. Dynamo targets a PA-RISC machine and uses a speculative scheme MRET (Most Recently Executed Tail) to pick hot traces without doing any path or branch profiling. DynamoRIO uses the same scheme to pick hot traces, except that it targets the x86-64 system. Although most of these were not proposed by the security community, there is at least one noticeable work called *program shepherding* [15] which makes use of a general purpose dynamic optimizer RIO [5] to enforce security policies. DynamoRIO and program shepherding provide nice interfaces for enforcing security policies on control transfers, which makes us believe that they can be good candidate architectures for CFI enforcement. Since these well established and mature dynamic code optimizers are proven to introduce minimal overhead, we believe that they could result in a system that outperforms existing CFI implementations.

In this paper, we propose *DynCFI* that enforces a set of security policies on top of DynamoRIO for CFI properties. We show that *DynCFI* achieves similar security properties when compared to a number of existing CFI implementations while experiencing a much lower performance overhead of 14.8 % as opposed to 28.6 % of *BinCFI*. We stress that *DynCFI* is not necessarily a CFI enforcement implementation that has the lowest performance overhead. Instead, our contribution lies on the utilization of the dynamic code optimization system which is a matured system proposed and well studied before CFI was even introduced. To the best of our knowledge, *DynCFI* is the first implementation of CFI enforcement on top of a dynamic code optimizer.

In the second half of this paper, we further investigate the exact contribution to this performance improvement. We propose a three-dimensional design space and perform comprehensive experiments to evaluate the contribution of each axis in the design space in terms of performance overhead. Among many interesting findings, we show that traces in the dynamic optimizer, which consist of cached basic blocks stitched together, had contributed the most performance improvement. Results show that traces have decreased the performance overhead from 22.7 % to 14.8 %. We also evaluate how branch prediction and indirect branch lookup have changed the performance. To the best of our knowledge, this is the first comprehensive evaluation on the performance overhead contributed by various components of the system, and we believe that this detailed understanding would aid future research and development of efficient CFI enforcement systems.

The remainder of this paper is structured as follows. Section 2 summarizes related work and outlines our motivation of using a dynamic optimizer. Section 3 introduces the security policies of *DynCFI* we enforce on top of DynamoRIO and compares them with a number of existing CFI enforcement implementations. In Sect. 4, we propose a three-dimensional design space for *DynCFI* and present a set of experiments to evaluate the contributing factors of various components of the dynamic optimizer. We present our security evaluation and some discussion in Sect. 5. In the end, we conclude in Sect. 6.

2   Related Work and Motivation

In this section, we first cover some important related work on CFI and dynamic code optimization, and then motivate our idea of enforcing CFI on top of one of the most well-established dynamic optimizers.

2.1   Control Flow Integrity

Control-flow Integrity (CFI) was first introduced by Abadi et al. [2]. The basic idea of CFI is to mark the valid targets of indirect branches with unique identifiers and then insert ID-checks into the program before each indirect branch transfer. Since its introduction in 2005, there have been a large body of CFI variants introduced [4,10,12,20,25,27,28].

Some of these proposals focus on extracting accurate targets of indirect transfers. For example, CFL [4] requires recompilation of the target application to obtain such target information, and performs a "lock" operation before each indirect control flow transfer with a corresponding "unlock" operation at valid destinations only. ROPdefender [10] makes use of the dynamic binary instrumentation tool Pin [16] to implement a shadow stack where the return addresses are recorded and later compared with the return target address executed. It suffers from performance issues due to its checking for every return instruction executed. CFIMon [25] makes use of BTS [14] supported by hardware to collect in-flight branch transfers. Once the BTS buffer is full, a monitor process will start to detect whether these branch transfers are valid. However, BTS is a debugging mechanism that records all branches in a user-defined memory area, and there will be high overhead because of the large number of memory accesses. In BinCFI [28], potential candidates of indirect branch targets are recorded and all indirect branches are instrumented to be a jump to a CFI validation routine. BinCFI will cause high performance overhead as it has to translate all indirect branch targets executed, especially for programs which have a large percentage of indirect branches. CFIGuard [26] uses Last Branch Record and Performance Monitor Unit supported by hardware to record source and target addresses for indirect branches, and then compare them with valid targets obtained ahead of time, but it relies on source code to obtain valid targets for indirect calls. *Lockdown* [17] is implemented in a dynamic binary translation platform called *libdetox*, which also uses shadow stack similar to ROPdefender to restrict the targets of return branches. However, its security policy for indirect jumps is relatively weak in allowing the target of a jump instruction to be any function entry points or any addresses inside the current function. This gives a lot of flexibility to attackers in using various gadgets.

Others focus on efficient ways of enforcing the CFI property for lower performance overhead. For example, in CCFIR [27], all control flow targets for indirect branches are allocated on a so-called springboard section, and indirect branches are only allowed to use control flow targets contained in the springboard section.The main restriction is that it requires relocation information to be included in the binaries. kBouncer [20] uses LBR [14] on Intel to record branch transfers. It checks whether the target of a return instruction is call-proceeded when a system call is invoked. It can be bypassed because the LBR mechanism only records limited number of branch transfers. ROPGuard [12] also performs CFI validation on Windows API calls. Like kBouncer, it requires that return addresses are call-preceded and the memory word before each return address is the start address of the API function.

In general, all existing proposals of CFI implementation enforce an approximation of the original and strict security policies due to the lack of accurate indirect transfer target information and performance considerations. Many have to trade security for better performance of the resulting system. Research has shown that some of these approximated CFI implementation are vulnerable to various attacks [6,13,21]. Therefore, any noticeable reduction in the performance overhead not only would lead to better user acceptance, but might translate into a better approximation of the CFI security policy.

2.2 Dynamic Code Optimization

We notice that another body of work called dynamic code optimization, mostly done by the software engineering community, could potentially be useful for improving the performance overhead. Most of them build hot traces for blocks frequently executed to boost execution. Dynamo [3], a dynamic optimizer for a PA-RISC machine, acts as a native interpreter which allows it to observe runtime behavior without instrumentation. Wiggins/Redstone [11] uses performance counters on the Alpha to build traces. Mojo [8] uses the same mechanism in Dynamo to pick hot traces and targets Windows NT running on IA-32. DynamoRIO [5] is an x86 system based on Dynamo. Some of these platforms provide nice interfaces of intercepting control flow transfers of the target program with very low overhead, e.g., DynamoRIO [5], to the extent that the overhead could be negative (performance improvement) for some situations. Such platforms could be perfect candidates on top of which CFI properties are enforced.

We are not the first to make use of such systems for security purposes. Program Shepherding [15] successfully makes use of DynamoRIO to restrict code origins and control transfers. DynamoRIO provides a suitable platform for security enforcement because the sandboxing checks added cannot be bypassed [15]. Due to this reason and the fact that it provides efficient interfaces of intercepting control flow transfers, we choose it for our CFI enforcement, too.

2.3 DynamoRIO

Figure 1 shows an overview of *DynamoRIO* [5], with darker shading indicating the application code to be monitored.

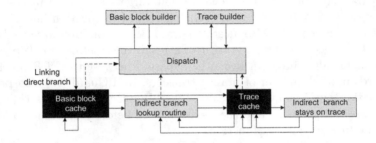

Fig. 1. Overview of *DynamoRIO*

DynamoRIO first copies basic blocks into the basic block cache. If a target basic block is present in the code cache and is targeted via a direct branch, *DynamoRIO* links the two blocks together with a direct jump. If the basic block is targeted via an indirect branch, *DynamoRIO* goes to the indirect branch lookup routine to translate its target address to the code cache address. Basic blocks that are frequently executed in a sequence are stitched together into the trace cache. When connecting beyond a basic block that ends in an indirect branch, a check is inserted to ensure that the actual target of the branch will stay on the trace. If the check fails, it will go to the indirect branch lookup routine to find the translated address.

To make itself a secure platform on which programs are executed, *DynamoRIO* splits the user-space address into two modes: the untrusted application mode and the trusted and protected RIO mode. This design protects *DynamoRIO* against memory corruption attacks. Meanwhile, the beauty of *DynamoRIO* (and the corresponding good performance) come mainly from the indirect branch lookup which is very efficient in determining control transfer targets with a hashtable. This hashtable maps the original target addresses with addresses in the basic block cache and trace cache so that most control transfers require minimal processing. We delay further details of *DynamoRIO* to Sects. 3 and 4 when we explain policies to be enforced on top of it and when we evaluate the improved performance achieved by individual components of *DynamoRIO*.

3 Design, Implementation, and Security Comparison

As discussed in Sect. 1, our motivation is to use *DynamoRIO* to enforce CFI properties in anticipation for improved performance. Our objective is to design a practical and efficient CFI enforcement without the extra requirement of re-compilation or dependency on debug information. In this section, we first present the design of *DynCFI* that can be effectively enforced on *DynamoRIO* and the implementation of it, and then compare the security property it achieves with some existing CFI (and related defense) approaches.

3.1 Returns

The most frequently executed indirect control transfer instructions are returns. *DynCFI* maintains a shadow call stack for each thread to remember caller information and the corresponding return address. The whole process is shown in Fig. 2. For a call instruction, we store the return address on our shadow stack. For a return instruction, we check whether the address on the shadow stack equals to the address stored at the stack memory specified by %esp. Such a shadow stack enables *DynCFI* to apply a strict policy that only returning to the caller is allowed, although a relaxed version could also be applied to reduce overhead (see Sect. 3.4 for more discussion). *DynCFI* also takes care of the following exceptions in special cases.

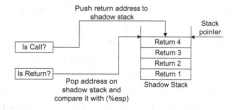

Fig. 2. Shadow stack operations

- **Signals:** A signal comes with a return address but not a call instruction. Fortunately, *DynamoRIO* records all necessary signal information for us to maintain a correct shadow stack.
- **Lazy binding:** The procedure `dl_runtime_resolve()` in lazy binding uses `ret` (without a corresponding `call`) to perform a `jmp` operation. The pattern of the code is fairly easy to identify though.
- **setjmp and longjmp:** `setjmp` and `longjmp` allow bypassing of multiple stack frames. We `pop` out return addresses continuously until a match is found or when the shadow stack is empty.
- **C++ exception handling:** We use the second argument of `Unwind_SetIP` as the return address for proper enforcement of our policy.

3.2   Indirect Jumps and Indirect Calls

We further classify indirect jumps into normal indirect jumps and PLT jumps, such as `jmp offset (base_register)`, which are used to call functions in other modules, target of which can only be exported from other modules. To obtain target information for every indirect branch, we use the static analysis engine provided by another well-known CFI enforcement *BinCFI* [28], which combines linear and recursive disassembling techniques and uses static analysis results to ensure correct disassembling. Targets of indirect calls are function entry points and targets of indirect jumps are function entry points and targets of returns. Meanwhile, targets of PLT jumps are exported symbol addresses. These valid jump and call targets are organized into three different hashtables to improve performance—one for indirect jumps, one for indirect calls, and one for PLT jumps. Most importantly, the shadow stack and hashtables readable only in the untrusted application mode, so attackers cannot modify them.

3.3   Implementation

As discussed in Sect. 2, *DynamoRIO* maintains a hashtable that maps original control transfer target addresses with addresses of code caches. The hashtable has to be built when the control transfer occurs the first time though. This process, together with the dispatcher which is invoked when matches are not found in the hashtable (see Fig. 1), become the natural place of our CFI enforcement, since CFI mainly concerns control transfer targets.

We obtained the source code of *DynamoRIO* version 5.0.0 from the developer's website [1], and added more than 700 lines of code (in C) to implement *DynCFI*. Most of the additional code is added to the dispatcher where checks of control flow transfers are performed. Some code is also added to basic block cache building to implement our shadow call stack and to initialize *DynamoRIO* to load the valid jump/call target addresses into our own hashtables.

DynCFI does not implement the full sets of CFI properties originally proposed by Abadi et al. [2]. We only perform checks on indirect control transfers at the first time when the target of an indirect branch occurs. However, it does not really impact security, and it is exactly the reason why *DynamoRIO* is widely accepted as an efficient dynamic optimizer—original code is cached in short sequences and security policies, if any, need only be checked the first time the code cache is executed [5]. Subsequent executions of the same code cache will be allowed (without checking) as long as the control transfer targets remain unchanged. Any violations to our policy will miss the (very efficient) indirect branch hashtable lookup and go back to the dynamic interpreter which will consider the control transfer a first timer and perform all the checks (inefficient).

3.4   Security Comparison

Table 1 shows the security policy of *DynCFI* when compared with some existing CFI implementations and ROP defense solutions. A caveat here is that we make

Table 1. Security comparison with other CFI and ROP defenses

Approach	Policy			
	Return	Indirect jump	Indirect call	PLT jump
BinCFI [28]	Call-preceded	Function entry, return address	Function entry	Exported symbol address
CCFIR [27]	Corresponding springboard section			Nil
CFIMon [25]	Call-preceded	Any address in the training set	Any function entry	Nil
ROPdefender [10]	Caller	Nil	Nil	Nil
kBouncer [20]	Call-preceded	Nil	Nil	Nil
LockDown [17]	Caller	Function entry, instruction in the current function	Function entry	Nil
DynCFI	First execution: Caller, Others: Call-preceded	Function entry, return address	Function entry	Exported symbol address

use of the shadow call stack information only when a new target is added to the hashtable. This will make the policy effectively call-proceeded only. Since call-proceeded policy is widely considered as adequate by many other approaches, we apply this performance improvement in our subsequent evaluation. This relaxed policy also enables a fair comparison between *DynCFI* and other CFI enforcement schemes since many others also use a call-proceeded policy.

DynCFI achieves similar security when compared with these existing approaches. In particular, *DynCFI* is mostly comparable to *BinCFI* in that both maintain a list of valid target addresses to be checked at runtime, with one noticeable difference in the enforcement mechanism: *BinCFI* enforces the policies with static instrumentation to translate indirect target address while *DynCFI* uses *DynamoRIO* as the interpreter platform. This makes *BinCFI* the perfect candidate for performance overhead comparison with *DynCFI*, which is the topic of our next Section.

4   Detailed Performance Profiling

In this section, we conduct a comprehensive set of experiments on the performance overhead of *DynCFI*. Besides the overall performance overhead, we run some detailed performance profiling to find out the contribution to such overhead by various components of the dynamic optimizer. We wish that such a detailed profiling could shed light on the part that contributes most to the performance overhead, and give guidance to future research in further improvement.

To better understand our evaluation strategy, we present our first attempt in the profiling, show the results, and explain the limitation of this attempt. We then choose an existing CFI implementation for the detailed comparison with *DynCFI*. We analyze the design space of CFI enforcement implementation and organize it along three axes on which the two systems under comparison could be clearly identified. Lastly, we perform a sequence of experiments by modifying individual components of *DynCFI* so that the contribution of each to performance overhead can be evaluated.

4.1   Target Applications

To evaluate the performance overhead, we need to subject *DynCFI* (and another CFI implementation for comparison purposes) to some applications. To enable fair comparison with existing work, we used twelve pure C/C++ programs we can find in SPEC CPU2006, which are also used in the evaluation of the original work of *BinCFI* [28], as our benchmarking suite.

Experiments were executed on a desktop computer with an i7 4510u CPU and 8 GB of memory running x86 version of Ubuntu 12.04. Each individual experiment was conducted 10 times, average of which is reported in this paper.

4.2   First Attempt in Performance Profiling

As an initial attempt to understanding the performance overhead contributed by various components of *DynCFI*, we use program counter sampling to record the amount of time spent in various components of *DynCFI*. We use the ITIMER_VIRTUAL timer which counts down only when the process is executing and delivers a signal when it expires. The handler used for this signal records the program counter of the process at the time the signal is delivered. We sample the program counter every ten milliseconds.

Table 2 shows the percentage of time each application spends in various steps in *DynCFI*. It suggests that more than 90 % of the time is spent on the application's code on average. Other non-negligible processes include Indirect Branch Lookup (IBL) inlined with the application's code and that not inlined, basic block and trace cache building, as well as the dispatcher.

In an attempt to explain why some applications, e.g., gcc, omnetpp, soplex, and povray, incur larger overhead, we count the number of different control transfers in each application (runtime) and present statistics in Table 3. The correlation between the two tables suggests that larger number of control transfers could lead to the higher overhead.

Although it sounds like we have obtained detailed understanding of the performance overhead, there is one important factor that we have overlooked so far—the overhead contribution of the dynamic optimizer on executing the application's code (second column of Table 2). In other words, Table 2 does not tell us if the dynamic optimizer had sped up or slowed down the execution of the application's code, and what had contributed to that speedup or slowdown. Our further comparison verifies this suspicion, see Table 4, as there is noticeable difference in the amount of time spent.

Therefore, we want to further investigate the contribution of various components of the dynamic optimizer in speeding up or slowing down the application's code. We present our second attempt in the rest of this section. With the objective of finding out contributions to the performance overhead by individual components of the dynamic optimizer, our strategy is to

Table 2. Percentage of time spent on various components

Application	Application code	IBL inlined	IBL not inlined	Basic block building	Trace building	Dispatch	Others
bzip2	97.99	0.60	0.00	0.20	1.20	0.00	0.00
gcc	86.78	7.46	0.26	0.91	3.42	1.10	0.07
mcf	97.48	0.42	1.26	0.14	0.07	0.14	0.49
gobmk	80.00	1.08	0.00	2.70	11.35	4.86	0.00
sjeng	94.10	5.67	0.11	0.02	0.09	0.02	0.00
libquantum	99.51	0.49	0.00	0.00	0.00	0.00	0.00
omnetpp	84.88	14.50	0.38	0.06	0.15	0.03	0.01
astar	94.36	4.79	0.78	0.00	0.01	0.04	0.01
namd	99.89	0.69	0.00	0.00	0.02	0.00	0.00
soplex	74.21	25.42	0.03	0.10	0.10	0.10	0.02
povray	89.71	6.88	0.82	0.76	1.01	0.76	0.06
lbm	99.99	0.00	0.00	0.00	0.01	0.00	0.00
Average	91.57	5.62	0.30	0.41	1.45	0.59	0.06

Table 3. Statistics of different types of control transfers

Application	%Indirect call	%Indirect jump	%Return	%Direct branch	Total
bzip2	0.002	0.002	0.774	99.222	2813437750
gcc	0.434	1.958	7.767	89.841	40789466606
mcf	0.001	0.029	5.402	94.568	5000155956
gobmk	0.001	0.027	4.811	95.161	687830197
sjeng	1.072	2.289	4.718	91.921	122978889385
libquantum	0.000	0.000	0.242	99.758	706839248554
omnetpp	1.609	1.763	33.998	62.630	87535408451
astar	1.698	0.049	19.738	78.515	30621019276
namd	0.000	0.008	3.292	96.700	115933566091
soplex	0.002	0.018	23.239	76.741	73160950993
povray	2.776	0.154	26.279	70.791	8195937460
lbm	0.000	0.017	0.035	99.948	15270883768

1. Find an existing CFI implementation X for comparison.
2. Continuously disable or modify individual components of *DynCFI* so that the modified system eventually becomes similar to the implementation of X.
3. In every step of disabling or modifying the components, perform experiments to find the corresponding (difference in) performance overhead.

4.3   Picking BinCFI for Detailed Comparison

With this strategy, it is important that we choose an X that

Table 4. Time spent in application code

Application	in *DynCFI* (sec)	Natively (sec)	Overhead (%)
bzip2	4.88	4.86	0.41
gcc	60.73	56.25	7.96
mcf	13.91	14.19	−1.97
gobmk	1.48	1.35	9.62
sjeng	158.93	150.01	5.95
libquantum	813.12	821.63	−1.04
omnetpp	138.54	122.23	13.34
astar	76.16	75.44	0.95
namd	735.51	733.73	0.24
soplex	64.81	61.15	5.98
povray	14.21	14.12	0.64
lbm	375.45	388.14	−3.27

– Is an independent, state-of-the-art implementation of CFI enforcement;
– Shares the same high-level idea with *DynCFI* while validating control transfers with a different approach (e.g., by binary instrumentation) from that of the dynamic optimizer/interpreter as in *DynCFI*.

so that our evaluation could attribute the difference in performance overhead to the dynamic optimizer.

As discussed at the end of Sect. 3, *BinCFI* and *DynCFI* are similar in that both maintain a set of valid control transfer targets and use a centralized validation routine for CFI enforcement. In both cases, the validation routine maintains a hashtable for the valid control transfer targets. Figure 3 shows the work-flow of *BinCFI*.

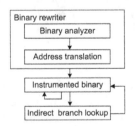

Fig. 3. Overview of *BinCFI*

The difference between *BinCFI* and *DynCFI* is that *BinCFI* obtains the valid target addresses of indirect branches statically and records their corresponding instrumented target addresses into the hashtable, and then replaces the indirect

instructions with a direct jump to the CFI validation routine. *BinCFI* satisfies
our requirements for the performance comparison, and is therefore chosen for
our subsequent detailed evaluation.

4.4   Overall Comparison and the Design Space

The overall performance overhead of executing the benchmarking applications
under (original, unmodified) *DynamoRIO*, *DynCFI*, and (original, unmodified)
BinCFI is shown in Fig. 4. Results are shown in terms of percentage overhead
beyond natively executing the applications on an unmodified Linux Ubuntu
system. We obtained the source code implementation of *BinCFI* [28] from its
authors.

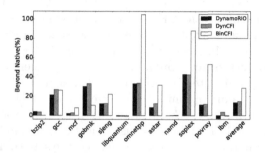

Fig. 4. Overall performance overhead

An interesting observation is that the original *DynamoRIO* and *DynCFI* do
not differ much in terms of overhead (a relatively small 1.3 % difference). This
shows that the interfaces provided by *DynamoRIO* are convenient and effective
for CFI enforcement, which confirms our intuition since *DynamoRIO* intercepts
all control transfers and no additional intercepting is needed in our modification
to *DynamoRIO*.

DynCFI experiences a significantly smaller overhead of 14.8 % compared to
BinCFI at 28.6 %. This suggests that the dynamic optimizer provides a more efficient platform for CFI enforcement compared to existing approaches like binary
instrumentation as in *BinCFI*. That said, the two systems differ in other aspects
and therefore this overall evaluation result is insufficient in attributing the majority of the performance gain to mechanisms of the dynamic optimizer.

As discussed in Sect. 4.3, our strategy to this difficulty is to continuously disable or modify individual components of *DynCFI* so that eventually it becomes
similar to *BinCFI*, in terms of their operating mechanism as well as the performance overhead. By doing so, we would likely observe degradation of performance (increase in overhead) of the modified system which is definitely due to the
corresponding feature disabled or modified. The question is – which individual
component or feature to disable or modified?

To answer this question, we analyze the internal validation mechanisms of the two approaches and identify three main factors that could significantly contribute to the different performance overhead.

1. **Trace.** Trace is the most important mechanism in *DynamoRIO* to speed up indirect transfers. Traces are formed by stitching together basic blocks that are frequently executed in a sequence. Benefits include avoiding indirect branch lookups by inlining a popular target of an indirect branch into a trace (with a check to ensure that the target stays on the trace and otherwise fall back to the full security check), eliminating inter-block branches, and helping branch prediction. Trace is unique in *DynamoRIO* and is not in *BinCFI*.
2. **Branch prediction.** Modern processors maintain buffers for branch prediction, e.g., Branch Target Buffer (BTB) and Return Stack Buffer (RSB). The effectiveness of these predictors could get seriously affected due to the modifications to the control transfers. For example, turning a return instruction into a indirect jump would make RSB useless in the branch prediction, potentially leading to an increase in the performance overhead.
3. **Indirect branch lookup routine.** Besides implementation details that are not necessarily due to the architectural design (to be discussed more in Sect. 4.5), a dynamic optimizer could use a single lookup routine for the entire application including the dynamically loaded libraries, while systems that apply static analysis and binary instrumentation would likely have to use a dedicated lookup routine for each module because some dynamically loaded libraries might not have been statically analyzed or instrumented. This could contribute to noticeable differences in performance overhead.

We want to explore details into these three axes to see how each of them affects the performance overhead. Other factors that might contribute to the overhead in *DynCFI* which we do not further investigate include

- Building basic block caches;
- Building trace caches;
- Inserting new entries into hashtables;
- Context switches between DynamoRIO and code caches.

4.5   Profiling Along the Three Axes

With identification of the three axes, we make our second attempt in detailed understanding of the performance overhead of the two systems. Since executing on *DynCFI* and executing on the original unmodified *DynamoRIO* experience about the same overhead (see Fig. 4), our subsequent experiments will only focus on comparing *DynCFI* and *BinCFI*. Also recall that our strategy is to disable or modify one component of *DynCFI* at a time and observe the corresponding change in performance overhead.

4.5.1   Traces. Traces are unique in dynamic optimizers like *DynamoRIO* and *DynCFI*. There are potentially two ways in which traces impact the performance overhead. First, the stitching of basic blocks together eliminates some inter-block branches. Second, each trace has inlined code to check if the control transfer target is still on the trace (we call this InT). If the target is still on the trace, execution will just carry on without further checking; otherwise, a second inlined code (we call this InH) is executed to perform hashtable lookup without collisions. If collision happens, execution will go to the full indirect branch lookup routine (denoted as R). We examine contribution of InT and InH by disabling them individually. We also examine the effect of traces overall and present the results in Fig. 5.

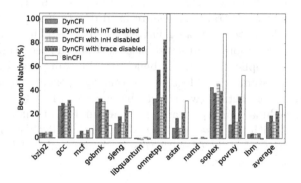

Fig. 5. Impact of trace on overhead

Figure 5 shows that the contribution due to InT is big, averaging to 5.5 %. Exceptions go to `bzip2` and `soplex` which do not gain much with InT mainly because the fall-back of InH is very effective on them (which can be verified from the next-to-zero time spent in IBL not inlined in Table 2).

Although performance overhead increases when disabling InT (see Fig. 5), *DynCFI* is still better than *BinCFI*. When disabling traces altogether, the overhead of *DynCFI* increases from 14.8 % to 22.7 % on average, with some going over the overhead in *BinCFI*. This shows that traces are contributing significantly in the low overhead of *DynCFI*. For applications with a large percentage of indirect branches (see Table 3), *DynCFI* with traces disabled still outperforms *BinCFI*. This suggests that there are other contributing factors in *DynCFI* which we have not evaluated.

4.5.2   Branch Prediction. The way in which *DynCFI* and *BinCFI* intercept and deliver control flow transfers has an implicit effect on branch prediction. Branch prediction is typically achieved by remembering a history of control transfer targets by the same instruction. Both *DynCFI* and *BinCFI* could weaken branch prediction due to R using the same instruction (an indirect jump) to execute control transfers originally executed by different instructions in the application [5,28]. Table 5 summaries how indirect control transfers in an application are executed in *DynCFI* and *BinCFI*.

Table 5. Execution of indirect control transfers

Original transfer		Return	Indirect call/jump
DynCFI	Basic block cache	Jump to R, indirect jump to target	
	Trace cache	InT or InH or jump to R, indirect jump to target	
BinCFI		Return	jump to R, indirect jump to target

In summary, *DynCFI* leads *BinCFI* in retaining branch prediction for indirect calls and jumps when trace caches are used due to InT and InH; however, *BinCFI* would perform better than *DynCFI* for returns. That said, note that there are typically far more return instructions than indirect calls and jumps executed for all the applications in our benchmarking suite, see Table 3.

To better understand the effect of various components of *DynCFI* and *BinCFI* on branch prediction, we count the number of mispredictions when executing the benchmarking applications on a number of different settings – *DynCFI*, *DynCFI* with InT disabled, *DynCFI* with InH disabled, *DynCFI* with traces disabled, *BinCFI*, *BinCFI* with returns being replaced by jumps to R, and present the results in Fig. 6.

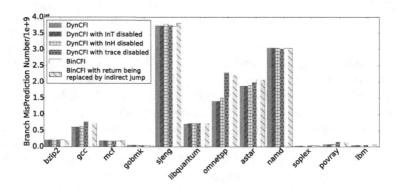

Fig. 6. Impact of traces on the number of branch mispredictions

We observe that disabling InH has a larger impact on branch prediction than disabling InT in general. This shows that the inlined hashtable lookup has its fair share of its contribution on lower overhead. It also indirectly shows that the hashtable implementation in *DynCFI* is good in that collisions do not happen often (since R not inlined is not executed often as shown in Table 2). Another interesting finding is that replacing returns with indirect jumps on *BinCFI* adds a large number of mispredictions for some programs. In terms of overhead, this translates to about 2 % more in the overhead as shown in Fig. 7.

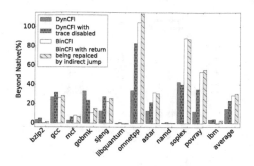

Fig. 7. Impact of branch prediction on overhead

4.5.3 Indirect Branch Lookup Routine R. The indirect branch lookup routine in *DynCFI* and *BinCFI* very much shares the same strategy. Both use an efficient implementation of a hashtable to record valid control transfer targets. One noticeable difference, though, is that *BinCFI* requires an extra step to check if the target resides within the same software module before directing control to the corresponding R. Each software module has to implement its own copy of R because some dynamically loaded libraries might not have been statically analyzed or instrumented and *BinCFI* cannot use a centralized R for all modules.

On the other hand, *DynCFI* executes the application on top of a dynamic interpreter without static analysis or binary instrumentation, and therefore has three centralized R (one for returns, one for indirect jumps, and one for indirect calls) for all software modules. This architectural difference contributes to some additional performance overhead to *BinCFI*.

Besides the difference due to the architectural design, there are also lower level differences in implementing R between *DynCFI* (inhcriting the same R from *DynamoRIO*) and *BinCFI*. In particular, they differ in the indirect jump instructions used (*DynCFI* uses a register to specify the target while *BinCFI* uses a memory), the number of registers used throughout the algorithm (and as a result the number of registers to be saved and restored), and efficiency of the hashtable lookup algorithm.

To evaluate the contribution of R in the overall performance overhead, we replace R in both *DynCFI* (with traces disabled) and *BinCFI* (with returned replaced with indirect jumps) with R', our (supposedly more efficient) implementation of the algorithm, and show the resulting performance overhead in Fig. 8.

Comparing these results with those shown in Fig. 7, we find that such low-level details in the implementation of R translates to significant differences in the overhead. In particular, the difference between *DynCFI* and *BinCFI* shrinks with R' replacing R, indicating that the original R used in *DynCFI* is more efficient than that in *BinCFI*.

4.5.4 Summary. Recall that our strategy in the second attempt of detailed profiling of *DynCFI* is to continuously disable or modify various components to

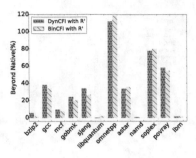

Fig. 8. Performance overhead with unified R′

find the contribution of them in terms of performance overhead. Figure 8 shows the comparison between *DynCFI* with traces disabled (bringing both systems to the same configuration on the first axis) and *BinCFI* with returns replaced by indirect jumps (bringing both systems to the same configuration on the second axis) while they use the same R′ (bringing both systems to the same configuration on the third axis). They are fairly close to each other in their performance, confirming that we manage to attribute their originally large difference being successfully attributed to the three axes.

We therefore believe that this second attempt provides a successful and accurate detailed profiling for *DynCFI* and *BinCFI*. With the detailed understanding of the contribution of each components on the three axes, we hope that future research could improve the performance further by, e.g., designing an indirect branch lookup routine that results in better branch prediction.

5   Security Evaluation and Discussions

5.1   Real World Exploits

We use a publicly available intrusion prevention evaluator RIPE [24] to verify that *DynCFI* offers comparable security properties with existing CFI proposals (as analysis presented in Sect. 3.4). In particular, we check if *DynCFI* can detect exploits that employ the advanced Return-Oriented Programming (ROP) techniques.

RIPE contains 140 return-to-libc exploits out of which 60 exploit return instructions and 80 exploit indirect call instructions. For the 60 exploits on return instructions, our experiments confirm that *DynCFI* manages to detect all of them because they violate the call-preceded policy we enforced on return instructions. RIPE also contains 10 ROP attacks using return instructions, which are all successfully detected by *DynCFI* as the targets of these gadgets are not call-preceded.

DynCFI and *BinCFI* share the weakness in detecting exploits that change the value of a function pointer to a valid entry point of a function. Such attacks cannot be detected by most other CFI implementations either [25].

5.2   Average Indirect Target Reduction

Zhang and Sekar [28] propose a metric for measuring the strength of CFI called Average Indirect target Reduction (AIR). As *DynCFI* uses different policy on return branches, we apply the same metric to test *DynCFI* when applied to the SPEC benchmarking suite. We compares the AIR metrics for *DynCFI* and *BinCFI*. We can find that average AIR for *DynCFI* is 98.80 %, which is comparable to 98.86 % for the case of *BinCFI*.

5.3   Shadow Stack in Full Enforcement

As described in Sect. 3, in order to improve the performance, we do not check the shadow call stack if the target address is found in our hashtable (in which all addresses have already been fully checked when they were first added to the hashtable).

We understand that a full enforcement of the shadow call stack is more secure as it ensures that every return jumps to its caller; however, its high performance overhead is also well documented in previous research [9, 10]. To verify such high performance overhead, we modify *DynCFI* to check the shadow call stack for every return instruction, and show the results in Fig. 9.

Figure 9 shows that *DynCFI* with full enforcement of the shadow stack runs with an average performance overhead of 29.8 %, a big jump from our optimized implementation at 14.8 %. Although such a full enforcement of the shadow stack takes away the performance advantage of *DynCFI* compared to *BinCFI*, *DynCFI* now offers much better security. We check the AIR metric and find that AIR for *DynCFI* with full enforcement of the shadow stack increases from 98.80 % to 99.66 % for SPEC CPU2006, which is better than that of *BinCFI* at 98.86 %. Our experiments also show that *DynCFI* can now detect some more advanced ROP attacks, e.g., the ROP attack constructed by Goktas et al. [13] using callpreceded gadgets. A call-proceeded-only policy, e.g., that used in *BinCFI*, would miss such advanced attacks.

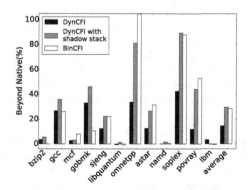

Fig. 9. Performance overhead with shadow stack

6   Conclusion

In this paper, we propose *DynCFI*, a new implementation of CFI properties on top of a well-studied dynamic code optimization platform. We show that *DynCFI* achieves comparable CFI security properties with many existing CFI proposals while enjoying much lower performance overhead of 14.8 % on average compared to that of a state-of-the-art CFI implementation *BinCFI* at 28.6 %. Our detailed profiling of *DynCFI* shows that traces, a mechanism in the dynamic code optimization platform, contribute the most to such performance improvement.

Acknowledgment. This work was supported by No. 61373168 and No. 2012014 1110002.

References

1. DynamoRIO. http://www.dynamorio.org/
2. Abadi, M., Budiu, M., Erlingsson, U., Ligatti, J.: Control-flow integrity. In: Proceedings of the 12th ACM Conference on Computer and Communications Security, pp. 340–353. ACM (2005)
3. Bala, V., Duesterwald, E., Banerjia, S.: Dynamo: a transparent dynamic optimization system. In: ACM SIGPLAN Notices, vol. 35, pp. 1–12. ACM (2000)
4. Bletsch, T., Jiang, X., Freeh, V.: Mitigating code-reuse attacks with control-flow locking. In: Proceedings of the 27th Annual Computer Security Applications Conference, pp. 353–362. ACM (2011)
5. Bruening, D.: Efficient, transparent, and comprehensive runtime code manipulation. Ph.D. thesis. Massachusetts Institute of Technology (2004)
6. Carlini, N., Wagner, D.: Rop is still dangerous: breaking modern defenses. In: USENIX Security Symposium (2014)
7. Chen, P., Xing, X., Han, H., Mao, B., Xie, L.: Efficient detection of the return-oriented programming malicious code. In: Jha, S., Mathuria, A. (eds.) ICISS 2010. LNCS, vol. 6503, pp. 140–155. Springer, Heidelberg (2010)
8. Chen, W.-K., Lerner, S., Chaiken, R., Gillies, D.M.: Mojo: a dynamic optimization system. In: 3rd ACM Workshop on Feedback-Directed and Dynamic Optimization (FDDO-3), pp. 81–90 (2000)
9. Dang, T.H., Maniatis, P., Wagner, D.: The performance cost of shadow stacks and stack canaries. In: ACM Symposium on Information, Computer and Communications Security, ASIACCS, vol. 15 (2015)
10. Davi, L., Sadeghi, A.-R., Winandy, M.: ROPdefender: a detection tool to defend against return-oriented programming attacks. In: Proceedings of the 6th ACM Symposium on Information, Computer and Communications Security, pp. 40–51. ACM (2011)
11. Deaver, D., Gorton, R., Rubin, N., Wiggins, R.: An on-line program specializer. In: Proceedings of the IEEE Hot Chips XI Conference (1999)
12. Fratric, I.: Runtime Prevention of Return-Oriented Programming Attacks. University of Zagreb (2012)
13. Goktas, E., Athanasopoulos, E., Bos, H., Portokalidis, G.: Out of control: overcoming control-flow integrity. In: 2014 IEEE Symposium on Security and Privacy (SP), pp. 575–589. IEEE (2014)

14. Intel Corporation. Intel®64 and IA-32 Architectures Software Developer's Manual (2015)
15. Kiriansky, V., Bruening, D., Amarasinghe, S.P.: Secure execution via program shepherding. In: USENIX Security Symposium, vol. 92 (2002)
16. Luk, C.-K., Cohn, R., Muth, R., Patil, H., Klauser, A., Lowney, G., Wallace, S., Reddi, V.J., Hazelwood, K.: Pin: building customized program analysis tools with dynamic instrumentation. In: ACM Sigplan Notices, vol. 40, pp. 190–200. ACM (2005)
17. Mathias, P., Antonio, B., Thomas, R.: Fine-grained control-flow integrity through binary hardening. In: Almgren, M., Gulisano, V., Maggi, F. (eds.) Detection of Intrusions and Malware, and Vulnerability Assessment. LNCS, vol. 9148, pp. 144–164. Springer, Cham (2015)
18. Mohan, V., Larsen, P., Brunthaler, S., Hamlen, K., Franz, M.: Opaque control-flow integrity. In: Symposium on Network and Distributed System Security (NDSS) (2015)
19. Niu, B., Tan, G.: Modular control-flow integrity. In: Proceedings of the 35th ACM SIGPLAN Conference on Programming Language Design and Implementation, p. 58. ACM (2014)
20. Pappas, V., Polychronakis, M., Keromytis, A.D.: Transparent ROP exploit mitigation using indirect branch tracing. In: USENIX Security, pp. 447–462 (2013)
21. Schuster, F., Tendyck, T., Pewny, J., Maaß, A., Steegmanns, M., Contag, M., Holz, T.: Evaluating the effectiveness of current anti-ROP defenses. In: Stavrou, A., Bos, H., Portokalidis, G. (eds.) RAID 2014. LNCS, vol. 8688, pp. 88–108. Springer, Heidelberg (2014)
22. Shacham, H.: The geometry of innocent flesh on the bone: return-into-libc without function calls (on the x86). In: Proceedings of the 14th ACM Conference on Computer and Communications Security, pp. 552–561. ACM (2007)
23. van der Veen, V., Göktas, E., Contag, M., Pawlowski, A., Chen, X., Rawat, S., Bos, H., Holz, T., Athanasopoulos, E., Giuffrida, C.: A tough call: mitigating advanced code-reuse attacks at the binary level. In: IEEE Symposium on Security and Privacy (S&P) (2016)
24. Wilander, J., Nikiforakis, N., Younan, Y., Kamkar, M., Joosen, W.: RIPE: Runtime Intrusion Prevention Evaluator. In: Proceedings of the 27th Annual Computer Security Applications Conference, pp. 41–50. ACM (2011)
25. Xia, Y., Liu, Y., Chen, H., Zang, B.: CFIMon: detecting violation of control flow integrity using performance counters. In: 2012 42nd Annual IEEE/IFIP International Conference on Dependable Systems and Networks (DSN), pp. 1–12. IEEE (2012)
26. Yuan, P., Zeng, Q., Ding, X.: Hardware-assisted fine-grained code-reuse attack detection. In: Bos, H., et al. (eds.) Raid 2015. LNCS, vol. 9404, pp. 66–85. Springer, Heidelberg (2015). doi:10.1007/978-3-319-26362-5_4
27. Zhang, C., Wei, T., Chen, Z., Duan, L., Szekeres, L., McCamant, S., Song, D., Zou, W.: Practical control flow integrity and randomization for binary executables. In: 2013 IEEE Symposium on Security and Privacy (SP), pp. 559–573 (2013)
28. Zhang, M., Sekar, L.: Control flow integrity for COTS binaries. In: Proceedings of the 22th USENIX Security Symposium, pp. 337–352 (2013)

Encryption, Signatures and Fundamentals

Impossibility on the Provable Security of the Fiat-Shamir-Type Signatures in the Non-programmable Random Oracle Model

Masayuki Fukumitsu[1(✉)] and Shingo Hasegawa[2]

[1] Faculty of Information Media, Hokkaido Information University,
Nishi-Nopporo 59-2, Ebetsu, Hokkaido 069-8585, Japan
fukumitsu@do-johodai.ac.jp
[2] Graduate School of Information Sciences, Tohoku University,
41 Kawauchi, Aoba-ku, Sendai, Miyagi 980-8576, Japan
hasegawa@cite.tohoku.ac.jp

Abstract. On the security of Fiat-Shamir (FS) type signatures, some negative circumstantial evidences were given in the non-programmable random oracle model (NPROM). Fischlin and Fleischhacker first showed an impossibility for specific FS-type signatures via a single-instance reduction. In ISC 2015, Fukumitsu and Hasegawa found another conditions to prove such an impossibility, however their result requires a strong condition on a reduction, i.e. a key-preserving reduction. In this paper, we focus on a *non-key-preserving reduction*, and then we show that an FS-type signature cannot be proven to be secure in the NPROM via a *sequentially multi-instance reduction* from the security of the underlying ID scheme. Our result can be interpreted as a generalization of the two impossibility results introduced above.

By applying our impossibility result, the security incompatibility between the DL assumption and the security of the Schnorr signature in the NPROM via a sequentially multi-instance reduction can be shown. Our incompatibility result means that the security of the Schnorr signature is not likely to be proven in the NPROM.

Keywords: Fiat-Shamir transformation · Schnorr signature · Non-programmable random oracle model · Meta-reduction · Static message attack

1 Introduction

The Fiat-Shamir (FS) transformation is known as a general way to yield an efficient signature from a canonical identification (ID) scheme. By using this method, several famous signature schemes can be constructed such as the Schnorr signature [28] and the Guillou-Quisquater (GQ) signature [20].

© Springer International Publishing Switzerland 2016
M. Bishop and A.C.A. Nascimento (Eds.): ISC 2016, LNCS 9866, pp. 389–407, 2016.
DOI: 10.1007/978-3-319-45871-7_23

The security of *FS-type signatures*, the signatures yielded via the FS transformation, is discussed in several literatures. Pointcheval and Stern [27] showed that an FS-type signature is strongly existentially unforgeable against the chosen-message attack (sEUF-CMA) in the random oracle model (ROM) if the underlying ID scheme is a honest-verifier zero-knowledge proof of knowledge. This implies that the Schnorr signature can be proven to be sEUF-CMA in the ROM from the discrete logarithm (DL) assumption. Subsequently, Abdalla, An, Bellare and Namprempre [1] relaxed the condition on the underlying ID scheme. They showed the equivalence between the sEUF-CMA security of an FS-type signature in the ROM and the security of the underlying ID scheme against an impersonation under the passive attack (imp-pa security).

On the other hand, Paillier and Vergnaud [25] gave a negative circumstantial evidence on the provable security of FS-type signatures in the *standard model*. More precisely, they showed the impossibility of proving the security of the Schnorr signature in the standard model via an algebraic reduction from the DL assumption as long as the one-more (OM) DL assumption holds. In a similar manner, such an impossibility was also proven for the GQ signature in the standard model.

One can observe that the security of the FS-type signatures can be proven in the ROM, whereas it may not be proven in the standard model. The main reason of this difference is whether or not the *programming technique* of a hash function can be utilized. The programming of a hash function means that a reduction $\mathcal{R}$, which aims to prove the security of a cryptographic scheme, is allowed to program a value of a hash function for an input in the security proof. By using this technique, the security of several cryptographic schemes including FS-type signatures was proven in the ROM. Although the programming property is valuable to prove the security of cryptographic schemes, this is known to be strong [32]. On the difference between the ROM and the standard model, one of the interests of the theoretical cryptography is how one can constrain the programmability on the security proof of cryptographic schemes [13,14,32]. In order to discuss this topic, several variants of the ROM were proposed. One of these is the *non-programmable random oracle model (NPROM)* [23]. In the NPROM, a hash value is obtained from the random oracle as well as in the ROM, but the random oracle is dealt with the independent party in the security proof. Namely, the reduction in the security proof cannot program hash values in the NPROM. Fischlin, Lehmann, Ristenpart, Shrimpton, Stam and Tessaro [14] discussed the provable security of several cryptographic schemes such as the FDH signature and the Shoup's trapdoor-permutation-based key encapsulation scheme [30] in the NPROM.

Fischlin and Fleischhacker [13] first gave the impossibility of proving the security of some specific FS-type signatures in the NPROM. More specifically, they showed that the Schnorr signature cannot be proven to be EUF-CMA in the NPROM via a single-instance reduction from the DL assumption as long as the OM-DL assumption holds. The *single-instance* reduction is a reduction which can invoke a forger $\mathcal{F}$ of the designated signature in the security proof only once, but it is allowed to rewind $\mathcal{F}$ many times. They also mentioned that such an impossibility result is applicable to other FS-type signatures which satisfy the following two conditions: (1) the "one-more" assumption related to the assumption from which the security of the signature schemes is proven in the ROM holds, (2) for any public key pk of the signature schemes, there exists the unique secret key corresponding to pk. On the other hand, it is not known whether or not a similar impossibility holds for FS-type signatures which do not satisfy above conditions such as [21,24]. In ISC 2015, Fukumitsu and Hasegawa [16] found another abstract conditions by which the impossibility of proving the security of many FS-type signatures in the NPROM holds. Their conditions are to restrict the type of reduction to being *key-preserving* and to require the underlying ID scheme to being secure against an impersonation under the *active attack* (imp-aa security). The key-preserving reduction means that a reduction invokes a forger $\mathcal{F}$ with the public key which is the same one given to the reduction. Namely, they showed that an FS-type signature cannot be proven to be secure in the NPROM via a key-preserving reduction from the imp-pa security of the underlying ID scheme as long as the ID scheme is imp-aa secure. Although their result likely covers many FS-type signatures than the result in [13], the condition of the key-preserving reduction is a bit strong as noted in [32]. In fact, the public key with which the key-preserving reduction invokes $\mathcal{F}$ is wholly preserved to that given to $\mathcal{R}$, whereas the public key of the single-instance reduction $\mathcal{R}$ is just partially limited in a sense that the public key is required to contain the same group description as that given to $\mathcal{R}$. Therefore, the result by [16] may not be regarded as a generalization of the result by [13]. It eventually remains open whether or not the impossibility of the provable security of FS-type signatures in the NPROM via a *non-key-preserving* reduction can be proven.

1.1 Our Result

In this paper, we give an impossibility of FS-type signatures in the NPROM via a non-key-preserving reduction by employing the technique introduced by [4]. Namely, we show that FS-type signatures cannot be proven to have the new security, i.e. *security against the static message attack (SMA security)* which is weaker than EUF-CMA, by the following theorem.

Theorem 6 (Informal). *If the underlying ID scheme is imp-pa secure, then an FS-type signature cannot be proven to be SMA secure in the NPROM via a sequentially multi-instance reduction from the imp-pa security of the ID scheme.*

We prove this theorem by the *meta-reduction* technique [8]. This technique is frequently used to show the impossibility of proving the security of cryptographic schemes such as [2–5, 10, 11, 13, 15, 16, 18, 22, 25, 26, 29, 32], and the relationships among cryptographic assumptions [8, 9, 17, 31].

We now explain our conditions. The first is to restrict an underlying ID scheme to being imp-pa secure. We consider that this condition is natural. This is because an ID scheme is in general proven to be secure against an impersonation

Table 1. Comparison of restriction of impossibility results

	Type of a reduction	Assumptions on the theorem
[13]	Single-instance	The signature is unique key,
		The related OM assumption holds
[16]	Key-preserving	The underlying ID scheme is imp-aa secure
[Ours]	Sequentially multi-instance	The underlying ID scheme is imp-pa secure

Table 2. Type of reductions concerned in the impossibility results

	Type of a reduction	# of invocation of $\mathcal{F}$	Public key given to $\mathcal{F}$
[13]	Single-instance	Once	Any
[16]	Key-preserving	Many times	Preserved to pk given to $\mathcal{R}$
[Ours]	Sequentially multi-instance	Many times	Any

under the concurrent attack which is stronger than the imp-pa security. Moreover, our condition is weaker than that of [16], namely [16] applied the imp-aa security as a condition. Table 1 summarizes conditions where the impossibility results on FS-type signatures requires.

The second is to focus only on a *sequentially multi-instance reduction*. The sequentially multi-instance reduction $\mathcal{R}$ is a reduction which can invoke a forger $\mathcal{F}$ polynomially many times, but it is prohibited to invoke the clones of $\mathcal{F}$ during an invocation of $\mathcal{F}$. Table 2 shows the comparison of the three types of reductions. Namely, the restriction of the sequentially multi-instance reduction is weaker than that of the single-instance reduction and that of the key-preserving reduction. Note that reductions concerned in the general security proofs e.g. [1,6,7,10,13,25,30] belong to this type.

In Theorem 6, we give the impossibility of proving the SMA security of FS-type signatures. A signature scheme is said to be SMA secure if there exists no probabilistic polynomial-time (PPT) forger $\mathcal{F}$ such that on given any polynomially many messages $m_1, m_2, \ldots, m_q$, $\mathcal{F}$ can forge one m_{j*} of the massages by utilizing a tuple of signatures σ_i of the messages m_i other than m_{j*}. For the relationship among other security notions, Bader, Jager and Li, Schäge [4] showed that the SMA security is weaker than the EUF-CMA security. Therefore, the impossibility of proving that a signature scheme is SMA secure means the impossibility of proving that the signature is EUF-CMA.

One can interpret Theorem 6 as a generalized result of [13,16]. Our result indicates that the security of FS-type signatures may not be proven in the NPROM by employing general proof techniques only. However, this does not exclude the possibility that the security of FS-type signatures is proven in the NPROM via a reduction which is allowed to invoke multiple clones of the forger $\mathcal{F}$ concurrently.

By applying Theorem 6 to the Schnorr signature, one can prove the *security incompatibility* between the DL assumption and the EUF-CMA security of the

Table 3. Type of reductions concerned in the incompatibility results

	Type of a reduction	# of invocation of $\mathcal{F}$	Public key given to $\mathcal{F}$
[16]	Single-instance and key-preserving	Once	Preserved to pk given to $\mathcal{R}$
[Ours]	Sequentially multi-instance	Many times	Any

Schnorr signature in the NPROM via a sequentially multi-instance reduction. This security incompatibility means that the EUF-CMA security of the Schnorr signature in the NPROM is not compatible with the DL assumption. By employing Theorem 6, we specifically show that the Schnorr signature cannot be proven to be EUF-CMA in the NPROM via a sequentially multi-instance reduction from the DL assumption as long as the DL assumption holds. Such an incompatibility was first proven in the previous work [16]. They showed the incompatibility via a strongly restricted reduction, namely single-instance key-preserving reduction. Table 3 shows the comparison of the type of reduction concerned in our incompatibility result and theirs. The advantage of our incompatibility is that the restriction on a reduction is weaker than that of theirs. Since it is believed that the DL assumption holds, our incompatibility result means that the EUF-CMA security of the Schnorr signature is not likely to be proven in the NPROM.

It should be noted that Fischlin and Fleischhacker [13] showed that such an incompatibility cannot be proven from the DL assumption. We now explain that our incompatibility result does not contradict to theirs by comparing the *meta-reduction* concerned in ours with that in theirs. By employing the meta-reduction technique, one shows an impossibility as follows: assume that there exists a PPT reduction $\mathcal{R}$ which proves the security of a designated cryptographic scheme. Then, one aims to construct a PPT meta-reduction algorithm $\mathcal{M}$ which breaks some cryptographic assumption. In [13], they formally showed that if there exists a PPT meta-reduction $\mathcal{M}$ which proves the impossibility of the provable security of the Schnorr signature in the NPROM from the DL assumption as long as the DL assumption holds, then the Schnorr signature is not sEUF-CMA. In their result, they only consider the meta-reduction which does not execute another clone of an assumed reduction $\mathcal{R}$ during the execution of $\mathcal{R}$, nevertheless it can execute $\mathcal{R}$ polynomially many times. On the other hand, we circumvent their result by constructing the meta-reduction which executes the polynomially many clones of $\mathcal{R}$ concurrently. Thus our impossibility result does not contradict to theirs.

2   Preliminaries

For any natural number n, let $\mathbb{Z}_n$ be the residue ring $\mathbb{Z}/n\mathbb{Z}$. The notation $[n]$ denotes the set of all natural numbers $1 \leq i \leq n$. We write $x \in_U D$ to denote that an element x is chosen uniformly at random from a finite set D. We say that the family $\{D_\lambda\}_\lambda$ of finite sets is *polynomial-time samplable* if there exists an probabilistic polynomial-time (PPT) algorithm that outputs an element y which is uniformly distributed over D_λ on input λ. By $x := y$, we mean that an element x is defined or substituted by y. For any algorithm $\mathcal{A}$, $y \leftarrow \mathcal{A}(x)$ means that $\mathcal{A}$ outputs y on input x. When $\mathcal{A}$ is a probabilistic algorithm, $y \leftarrow \mathcal{A}(x; r)$ denotes that $\mathcal{A}$ outputs y on input x with random coins r, and $\mathcal{A}(x)$ is the random variable for the output of $\mathcal{A}$ on input x where the randomness is taken over the internal coin flips r of $\mathcal{A}$. For an algorithms $\mathcal{A}$ and $\mathcal{O}$, $\mathcal{A}^{\mathcal{O}}$ means that $\mathcal{A}$ has $\mathcal{O}$ as an oracle. A function ν is said to be *negligible* in λ if for any polynomial f, there exists a natural number λ_0 such that $\nu(\lambda) < 1/f(\lambda)$ for any $\lambda \geq \lambda_0$.

2.1   Digital Signature Scheme

A *signature scheme* Sig consists of three algorithms (KGen, Sign, Ver). KGen is a PPT key generation algorithm which on input 1^{λ}, outputs a key pair (sk, pk) of a secret key sk and a corresponding public key pk. Sign is a PPT signing algorithm which on input (sk, pk, m) of a key pair (sk, pk) and a message m, outputs a signature σ on the message m under the public key pk. Ver is a deterministic polynomial-time verifying algorithm which on input (pk, m, σ) of a public key pk, a message m and a signature σ, outputs 1 if σ is a signature on the message σ under the public key pk.

Let Sig = (KGen, Sign, Ver) be a signature scheme. We define the existential unforgeability against the chosen-message attack (EUF-CMA) [19]. The EUF-CMA is defined by the existentially forging game of Sig against the chosen-message attack (the EF-CMA game). This game is played by two parties (algorithms), a challenger $\mathcal{C}$ and a forger $\mathcal{F}$, in the following way: $\mathcal{C}$ first generates a key pair $(sk, pk) \leftarrow$ KGen(1^{λ}), and then invokes $\mathcal{F}$ with the public key pk. Then $\mathcal{F}$ aims to output a pair (m^*, σ^*) of a message m^* and its signature σ^* under pk. Here, $\mathcal{F}$ can make polynomially many queries m_i to $\mathcal{C}$ adaptively in order to obtain its signature σ_i. When $\mathcal{F}$ eventually outputs a pair (m^*, σ^*), $\mathcal{C}$ outputs 1 if m^* has not been queried to $\mathcal{C}$ and Ver$(pk, m^*, \sigma^*) = 1$. $\mathcal{F}$ is said to *win the EF-CMA game of* Sig if $\mathcal{C}$ outputs 1 in the EUF-CMA game of Sig between $\mathcal{C}$ and $\mathcal{F}$. Then, Sig is *EUF-CMA* if there exists no PPT forger $\mathcal{F}$ that wins EF-CMA game of Sig with non-negligible probability.

We also define the security notion, the security against the static message attack (SMA security) given by [4]. This security is defined by the SMA game between a challenger $\mathcal{C}$ and a forger $\mathcal{F}$ in a similar manner to the EF-CMA game. In this game, the behavior of $\mathcal{F}$ is divided into two sub-algorithms $\mathcal{F}_1$ and $\mathcal{F}_2$. This game is proceeded as depicted in Fig. 1, where $q := q(\lambda)$ is a polynomial in λ. $\mathcal{F}$ is said to *win the SMA game of* Sig if $\mathcal{C}$ outputs 1 in the SMA game of Sig between $\mathcal{C}$ and $\mathcal{F} = (\mathcal{F}_1, \mathcal{F}_2)$. Then, Sig is *SMA secure* if there exists no PPT forger $\mathcal{F} = (\mathcal{F}_1, \mathcal{F}_2)$ that wins the SMA game of Sig with non-negligible probability.

For these two security notions, the following relation is known.

Proposition 1 ([4]). *Let* Sig *be a signature scheme. If* Sig *is EUF-CMA, then* Sig *is SMA secure.*

SMA Game of Sig between a challenger $\mathcal{C}$ and a forger $\mathcal{F} = (\mathcal{F}_1, \mathcal{F}_2)$

On input 1^λ, $\mathcal{C}$ proceeds as follows:

Init. $\mathcal{C}$ invokes $\mathcal{F}_1\Big(pk, (m_i)_{i\in[q]}; r_\mathcal{F}\Big)$, where it chooses pk, $(m_i)_{i\in[q]}$ and $r_\mathcal{F}$ in the following way:

    1. Generate a key pair $(sk, pk) \leftarrow \mathsf{KGen}(1^\lambda)$ and random coins $r_\mathcal{F}$.

    2. Choose q messages $m_1, m_2, \ldots, m_q$ randomly and disjointly.

Select. When $\mathcal{F}_1$ outputs a pair $(j^*, \mathsf{st}_\mathcal{F})$ of a challenge index j^* and a state $\mathsf{st}_\mathcal{F}$ of $\mathcal{F}$,

    $\mathcal{C}$ invokes $\mathcal{F}_2\Big((\sigma_i)_{i\in[q]\setminus\{j^*\}}, \mathsf{st}_\mathcal{F}\Big)$, where for each index $i \in [q] \setminus \{j^*\}$, the signature σ_i is generated by $\sigma_i \leftarrow \mathsf{Sign}(sk, pk, m_i)$.

Challenge When $\mathcal{F}_2$ outputs σ^*, $\mathcal{C}$ outputs 1 if $\mathsf{Ver}(pk, m_{j^*}, \sigma^*) = 1$.

Fig. 1. SMA game of Sig

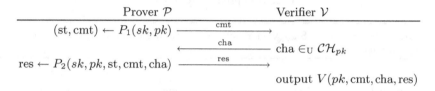

Fig. 2. Canonical ID scheme

2.2 Canonical ID Scheme

A *canonical identification (ID) scheme* consists of a tuple $(K, P_1, P_2, \mathcal{CH}, V)$. K is a PPT key generation algorithm which on input 1^λ, outputs a key pair (sk, pk). $\mathcal{CH} := \{\mathcal{CH}_{pk}\}_{pk \in \mathsf{PK}_\lambda}$ is a family of polynomial-time samplable sets $\mathcal{CH}_{pk}$ of all challenges cha indexed by a public $pk \in \mathsf{PK}_\lambda$, where PK_λ is a set of all public keys which can be generated by $K(1^\lambda)$. Both P_1 and P_2 are algorithms for a prover $\mathcal{P}$. Namely, P_1 outputs a pair $(\mathsf{st}, \mathsf{cmt})$ of a state st and a commitment cmt on input (sk, pk). P_2 outputs a response res on input $(sk, pk, \mathsf{st}, \mathsf{cmt}, \mathsf{cha})$. V is a verifying algorithm for a verifier $\mathcal{V}$. On input $(pk, \mathsf{cmt}, \mathsf{cha}, \mathsf{res})$, V outputs 1 if it accepts $\mathcal{P}$. The communication between the prover $\mathcal{P}$ and the verifier $\mathcal{V}$ is depicted as in Fig. 2.

Let $\mathsf{ID} = (K, P_1, P_2, \mathcal{CH}, V)$ be an ID scheme. We define the security of ID against an impersonation under the passive attack (imp-pa) [1]. In a similar manner to [16], this security is defined by an imp-pa game of ID between a challenger $\mathcal{C}$ and an impersonator $\mathcal{I}$ depicted in Fig. 3. An impersonator $\mathcal{I}$ is said to *win the imp-pa game of* ID if $\mathcal{C}$ finally outputs 1 in the imp-pa game of ID between $\mathcal{C}$ and $\mathcal{I}$. Then, ID is *imp-pa secure* if there exists no PPT impersonator $\mathcal{I}$ that wins the imp-pa game of ID with non-negligible probability.

imp-pa game of ID between a challenger $\mathcal{C}$ and an impersonator $\mathcal{I}$

Given 1^λ, $\mathcal{C}$ proceeds in the following way:

Init. $\mathcal{C}$ executes $\mathcal{I}(pk)$, where the public key pk is generated by $(sk, pk) \leftarrow K(1^\lambda)$.

Transcript. When $\mathcal{I}$ makes a t-th query, $\mathcal{C}$ replies a transcript $(\mathrm{cmt}_t, \mathrm{cha}_t, \mathrm{res}_t) \leftarrow \mathsf{Tr}^{\mathsf{ID}}_{sk,pk}$ to $\mathcal{I}$.

Impersonate. When $\mathcal{I}$ finally outputs a commitment $\hat{\mathrm{cmt}}$, $\mathcal{C}$ proceeds as follows:
  1. send a challenge $\hat{\mathrm{cha}} \in_U \mathcal{CH}_{pk}$ to $\mathcal{I}$;
  2. after receiving a response $\hat{\mathrm{res}}$ from $\mathcal{I}$, output $V(pk, \hat{\mathrm{cmt}}, \hat{\mathrm{cha}}, \hat{\mathrm{res}})$.

Transcript Oracle $\mathsf{Tr}^{\mathsf{ID}}_{sk,pk}$

  1. $(\mathrm{st}_t, \mathrm{cmt}_t) \leftarrow P_1(sk, pk)$;
  2. $\mathrm{cha}_t \in_U \mathcal{CH}_{pk}$;
  3. $\mathrm{res}_t \leftarrow P_2(sk, pk, \mathrm{st}_t, \mathrm{cmt}_t, \mathrm{cha}_t)$;
  4. return $(\mathrm{cmt}_t, \mathrm{cha}_t, \mathrm{res}_t)$.

Fig. 3. imp-pa game of ID

Signature FS-Sig $=$ (KGen, Sign, Ver) Yielded by Applying Fiat-Shamir Transformation to ID Scheme ID

KGen coincides with K.
Sign on input (sk, pk, m), issues a signature $\sigma := (\mathrm{cmt}, \mathrm{res})$ in the following way:
  1. $(\mathrm{st}, \mathrm{cmt}) \leftarrow P_1(sk, pk)$;
  2. $\mathrm{cha} := H_{pk}(\mathrm{cmt}, m)$;
  3. $\mathrm{res} \leftarrow P_2(sk, pk, \mathrm{st}, \mathrm{cmt}, \mathrm{cha})$.
Ver on input (pk, m, σ), sets $c := H_{pk}(\mathrm{cmt}, m)$, and then outputs $V(pk, \mathrm{cmt}, c, \mathrm{res})$.

Fig. 4. Fiat-Shamir transformation

2.3 Fiat-Shamir Transformation

Let ID $= (K, P_1, P_2, \mathcal{CH}, V)$ be an ID scheme, and let $\{H_{pk} : \{0,1\}^* \to \mathcal{CH}_{pk}\}_{\lambda, pk \in \mathsf{PK}_\lambda}$ be a family of hash functions indexed by security parameters λ and public keys $pk \in \mathsf{PK}_\lambda$ generated by $K(1^\lambda)$. Then, the signature FS-Sig is yielded by the Fiat-Shamir (FS) transformation [12] as depicted in Fig. 4. The signature FS-Sig is referred to as the *FS-type signature*.

3   Impossibility of Proving the SMA Security of FS-Type Signature

In this section, we prove the impossibility of proving the SMA security of an FS-type signature in the NPROM from the imp-pa security of the underlying ID scheme as long as the ID scheme is imp-pa secure. Let ID $= (K, P_1, P_2, \mathcal{CH}, V)$ be an ID scheme, and let $\{H_{pk} : \{0,1\}^* \to \mathcal{CH}_{pk}\}_{\lambda, pk \in \mathsf{PK}_\lambda}$ be a family of hash functions. Then $\mathsf{FS\text{-}Sig} = (\mathsf{KGen}, \mathsf{Sign}, \mathsf{Ver})$ denotes the FS-type signature yielded by applying the FS transformation to ID. Before explaining the statement of the impossibility, we formally describe the situation where $\mathsf{FS\text{-}Sig}$ is proven to be SMA secure in the NPROM from the imp-pa security of the underlying ID scheme ID. The definition of this situation is given by black-box reduction such as [4,13]. Namely, this situation holds if there exist a black-box reduction $\mathcal{R}$ and a non-negligible function $\epsilon_{\mathcal{R}}$ such that $\mathcal{R}$ wins the imp-pa game of ID with the probability $\epsilon_{\mathcal{R}}$ by black-box accessing to a forger $\mathcal{F} = (\mathcal{F}_1, \mathcal{F}_2)$ which wins the SMA game of $\mathsf{FS\text{-}Sig}$ with the non-negligible probability. Through the black-box access, $\mathcal{R}$ would play the SMA game of $\mathsf{FS\text{-}Sig}$ with $\mathcal{F}$ in which $\mathcal{R}$ is placed at the challenger's position. Here $\mathcal{R}$ is supposed to invoke $\mathcal{F}$ at most I times.

3.1 Case: Simple Reduction

For ease of the explanation of the main theorem, we first consider the simple case, namely $I = 1$. We say that such a reduction $\mathcal{R}$ is *simple*. In this case, the behavior of $\mathcal{R}$ is separated into a tuple $(\mathcal{R}_1, \mathcal{R}_2, \mathcal{R}_3, \mathcal{R}_4)$ of four sub-algorithms as follows:

$\mathcal{R}_1$ on a public key pk given from the imp-pa challenger $\mathcal{C}$ with random coins $r_{\mathcal{R}}$, outputs a pair $\left(\left(\overline{pk}, (m_i)_{i \in [q]}, r_{\mathcal{F}}\right), \mathsf{st}_{\mathcal{R}_2}\right)$ of an input $\left(\overline{pk}, (m_i)_{i \in [q]}, r_{\mathcal{F}}\right)$ to $\mathcal{F}_1$ and a state $\mathsf{st}_{\mathcal{R}_2}$. Then, $\mathcal{F}_1(\overline{pk}, (m_i)_{i \in [q]}; r_{\mathcal{F}})$ is invoked.

$\mathcal{R}_2$ on a challenge index $j^* \in [q]$ output by $\mathcal{F}_1(\overline{pk}, (m_i)_{i \in [q]}; r_{\mathcal{F}})$ and the state $\mathsf{st}_{\mathcal{R}_2}$, outputs a pair $\left((\sigma_i)_{i \in [q] \setminus \{j^*\}}, \mathsf{st}_{\mathcal{R}_3}\right)$ of a sequence $(\sigma_i)_{i \in [q] \setminus \{j^*\}}$ of signatures $\sigma_i = (\mathsf{cmt}_i, \mathsf{res}_i)$ replied to $\mathcal{F}_2$ and a state $\mathsf{st}_{\mathcal{R}_3}$. Then, by using a state $\mathsf{st}_{\mathcal{F}}$ output by $\mathcal{F}_1$, $\mathcal{F}_2((\sigma_i)_{i \in [q] \setminus \{j^*\}}, \mathsf{st}_{\mathcal{F}})$ is invoked.

$\mathcal{R}_3$ on a challenge signature $\sigma^* = (\mathsf{cmt}^*, \mathsf{res}^*)$ output by $\mathcal{F}_2((\sigma_i)_{i \in [q] \setminus \{j^*\}}, \mathsf{st}_{\mathcal{F}})$ and the state $\mathsf{st}_{\mathcal{R}_3}$, moves onto **Challenge** phase of the imp-pa game with sending a commitment $\hat{\mathsf{cmt}}$ to $\mathcal{C}$.

$\mathcal{R}_4$ on a challenge $\hat{\mathsf{cha}}$ received from $\mathcal{C}$ and the state $\mathsf{st}_{\mathcal{R}_4}$, replies a response $\hat{\mathsf{res}}$ to $\mathcal{C}$.

The configuration of $\mathcal{R}$ is described in Fig. 5. Here $\mathcal{R}$ plays a role of the challenger of the SMA game. Therefore, it needs to reply for $\mathcal{F}_1$'s queries. However, $\mathcal{R}$, namely $\mathcal{R}_2$, may fail to reply since $\mathcal{R}$ is a PPT algorithm. Namely, $\mathcal{R}_2$ may output a sequence $(\sigma_i)_{i \in [q] \setminus \{j^*\}}$ such that there exists an index $i_0 \in [q] \setminus \{j^*\}$ such that $\mathsf{Ver}(\overline{pk}, m_{i_0}, \sigma_{i_0}) \neq 1$. In this case, $\mathcal{F}_2$ is allowed to output any symbol.

Moreover, we suppose that the states $\mathsf{st}_{\mathcal{R}_2}$, $\mathsf{st}_{\mathcal{R}_3}$ and $\mathsf{st}_{\mathcal{R}_4}$ contains the random coins $r_{\mathcal{R}}$ which are given to $\mathcal{R}_1$, and $\mathcal{R}_2$, $\mathcal{R}_3$ and $\mathcal{R}_4$ are invoked with $r_{\mathcal{R}}$.

An imp-pa impersonator $\mathcal{R}$ can query to the transcript oracle $\mathsf{Tr}^{\mathsf{ID}}_{sk,pk}$ in **Transcript** phase before moving onto **Challenge** phase. In other words, $\mathcal{R}_1$, $\mathcal{R}_2$ and $\mathcal{R}_3$ are allowed to query to $\mathsf{Tr}^{\mathsf{ID}}_{sk,pk}$ adaptively. In the non-programmable random oracle model (NPROM), all of the parties $\mathcal{R}_1$, $\mathcal{R}_2$, $\mathcal{R}_3$, $\mathcal{R}_4$, $\mathcal{F}_1$ and $\mathcal{F}_2$ obtain hash values from the random oracle. Here $\mathcal{R} = (\mathcal{R}_1, \mathcal{R}_2, \mathcal{R}_3, \mathcal{R}_4)$ can

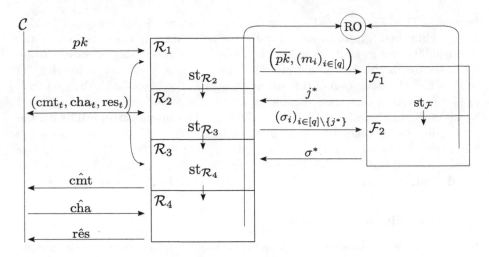

Fig. 5. Configuration of $\mathcal{R}$

observe all of random oracle queries by $\mathcal{F} = (\mathcal{F}_1, \mathcal{F}_2)$. However, $\mathcal{R}$ is prohibited to program hash values.

We have described the behavior of the reduction $\mathcal{R}$: it proceeds to (I) execute $\mathcal{F}_1$, (II) execute $\mathcal{F}_2$, and then (III) move on to **Challenge** phase of the imp-pa game. It should be noted that it has the possibility that the processes of $\mathcal{R}$ are ordered by (I), (III) and (II), and by (III), (I) and (II). One can discuss the impossibility in a similar manner even in such cases. Therefore, we only consider the case where the processes of $\mathcal{R}$ are ordered by (I), (II) and (III) in this paper.

Theorem 2. *Assume that* FS-Sig *is proven to be SMA secure in the NPROM via a simple reduction from the imp-pa security of the underlying ID scheme* ID. *Then* ID *is not imp-pa secure.*

Proof (Sketch). Assume that FS-Sig is proven to be SMA secure in the NPROM via a simple reduction from the imp-pa security of ID. Then there exist a PPT simple reduction $\mathcal{R} = (\mathcal{R}_1, \mathcal{R}_2, \mathcal{R}_3, \mathcal{R}_4)$ and a non-negligible function $\epsilon_{\mathcal{R}}$ such that $\mathcal{R}$ wins the imp-pa game of ID with the probability $\epsilon_{\mathcal{R}}$ by accessing an SMA forger $\mathcal{F} = (\mathcal{F}_1, \mathcal{F}_2)$ which wins the SMA game of FS-Sig with non-negligible probability. Here, the simple reduction $\mathcal{R}$ can invoke $\mathcal{F}$ only once without any rewind. In this proof, we aim to construct a PPT meta-reduction $\mathcal{M}$ which wins the imp-pa game of ID with non-negligible probability by utilizing $\mathcal{R}$. Note that $\mathcal{R}$ can win the imp-pa game with non-negligible probability if a winning SMA forger $\mathcal{F}$ is provided. We first define a hypothetical computationally unbounded forger $\tilde{\mathcal{F}} = \left(\tilde{\mathcal{F}}_1, \tilde{\mathcal{F}}_2 \right)$. It is noted that $\mathcal{R}$ should win the imp-pa game with non-negligible probability even if such a forger $\tilde{\mathcal{F}}$ is provided. We next construct the meta-reduction $\mathcal{M}$ so that $\mathcal{M}$ executes $\mathcal{R}$ with the simulation of the forger $\tilde{\mathcal{F}} = \left(\tilde{\mathcal{F}}_1, \tilde{\mathcal{F}}_2 \right)$ in polynomial time.

Hypothetical Forger $\tilde{\mathcal{F}} = \left(\tilde{\mathcal{F}}_1, \tilde{\mathcal{F}}_2 \right)$

$\tilde{\mathcal{F}}_1(\overline{pk}, (m_i)_{i \in [q]}; r_{\mathcal{F}})$ chooses a challenge index $j^* \in_U [q]$, sets a state $\mathrm{st}_{\mathcal{F}} :=$ $\left(j^*, \overline{pk}, (m_i)_{i \in [q]}, r_{\mathcal{F}} \right)$, and then outputs a pair $(j^*, \mathrm{st}_{\mathcal{F}})$.

$\tilde{\mathcal{F}}_2((\sigma_i)_{i \in [q] \setminus \{j^*\}}, \mathrm{st}_{\mathcal{F}})$ outputs σ^* as chosen in the following way:

    if for any index $i \in [q] \setminus \{j^*\}$, $\mathsf{Ver}(\overline{pk}, m_i, \sigma_i) = 1$; then finds $\sigma^* = (\mathrm{cmt}^*, \mathrm{res}^*)$ satisfying $\mathsf{Ver}(\overline{pk}, m_{j^*}, \sigma^*) = 1$. Here, it makes a sequence Q of queries, where Q is a set of all of queries (cmt_i, m_i) for each $i \in [q] \setminus \{j^*\}$ and $(\mathrm{cmt}^*, m_{j^*})$, to obtain the corresponding hash values.

    otherwise; sets $\sigma^* := \bot$.

Fig. 6. Hypothetical forger $\tilde{\mathcal{F}}$

Hypothetical Forger $\tilde{\mathcal{F}}$. We depict the hypothetical forger $\tilde{\mathcal{F}} = \left(\tilde{\mathcal{F}}_1, \tilde{\mathcal{F}}_2 \right)$ in Fig. 6. Let $(m_i)_{i \in [q]}$ be a sequence of q messages m_i given to $\tilde{\mathcal{F}}_1$. For a challenge index j^* output by $\tilde{\mathcal{F}}_1$, $\tilde{\mathcal{F}}_2$ correctly outputs a signature $\sigma^* = (\mathrm{cmt}^*, \mathrm{res}^*)$ on the message m_{j^*}, if $\tilde{\mathcal{F}}_2$ is given all of correct signatures $\sigma_i = (\mathrm{cmt}_i, \mathrm{res}_i)$ on the messages m_i other than the challenge message m_{j^*}. Otherwise, $\tilde{\mathcal{F}}_2$ is allowed to output any symbol as mentioned in the situation of $\mathcal{R}$. Therefore, $\tilde{\mathcal{F}} = \left(\tilde{\mathcal{F}}_1, \tilde{\mathcal{F}}_2 \right)$ is a winning SMA forger with non-negligible probability. In the NPROM, $\tilde{\mathcal{F}}_2$ needs to obtain any hash value from the random oracle. Here, $\tilde{\mathcal{F}}_2$ obtains the hash values including the sequence Q of all of the pairs (cmt_i, m_i) and the pair $(\mathrm{cmt}^*, m_{j^*})$ from the random oracle.

    Note that the running time of $\tilde{\mathcal{F}}_2$ would not be bounded in polynomial. We will show that $\mathcal{M}$ can be constructed in a way that it can simulate $\tilde{\mathcal{F}}_2$ in polynomial time sooner.

Meta-reduction $\mathcal{M}$. Recall that $\mathcal{R}$ can win the imp-pa game of ID with the probability $\epsilon_{\mathcal{R}}$ if the hypothetical forger $\tilde{\mathcal{F}}$ is provided. Here, we construct a meta-reduction $\mathcal{M}$ depicted in Fig. 7 which wins the imp-pa game of ID by executing the reduction $\mathcal{R}$ with the simulation of $\tilde{\mathcal{F}}$.

Correctness and Success Probability of $\mathcal{M}$. We show that $\mathcal{M}$ wins the imp-pa game of ID with non-negligible probability. As in Fig. 7, $\mathcal{M}$ attempts to win its game by executing the reduction $\mathcal{R}$ with the simulation of the forger $\tilde{\mathcal{F}}$. In the execution of $\mathcal{M}$, when $\mathcal{R}_1$, $\mathcal{R}_2$ and $\mathcal{R}_3$ make a query, $\mathcal{M}$ forwards a transcript $(\mathrm{cmt}, \mathrm{cha}, \mathrm{res})$ by querying to $\mathsf{Tr}^{\mathsf{ID}}_{sk, pk}$ provided by the imp-pa challenger $\mathcal{C}$. In **Challenge** phase of the imp-pa game between $\mathcal{C}$ and $\mathcal{M}$, $\mathcal{M}$ just intermediates between $\mathcal{C}$ and $\mathcal{R}$ as described in (M-5) and (M-6). It follows that $\mathcal{M}$ can win the imp-pa game of ID if $\mathcal{R}$ wins its game. Recall that $\mathcal{R}$ can win the imp-pa game of ID with the non-negligible probability $\epsilon_{\mathcal{R}}$ if a winning SMA forger $\tilde{\mathcal{F}}$ of FS-Sig is provided. In a nutshell, $\mathcal{M}$ can win the imp-pa game of ID with

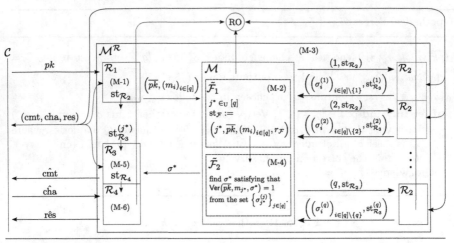

Meta-Reduction $\mathcal{M}$

On a public key pk given from the imp-pa challenger $\mathcal{C}$, $\mathcal{M}$ proceeds as follows:

(M-1) choose random coins $r_{\mathcal{R}}$, and then execute $\left(\overline{pk}, (m_i)_{i \in [q]}, r_{\mathcal{F}}, \mathrm{st}_{\mathcal{R}_2}\right) \leftarrow$
$\mathcal{R}_1^{\mathrm{Tr}_{sk,pk}^{\mathrm{ID}}}(pk; r_{\mathcal{R}})$.

(M-2) simulate $\tilde{\mathcal{F}}_1$ by choosing a challenge index $j^* \in_U [q]$, and then setting $\mathrm{st}_{\mathcal{F}} :=$
$\left(j^*, \overline{pk}, (m_i)_{i \in [q]}, r_{\mathcal{F}}\right)$.

(M-3) for each $j \in [q]$, execute $\left(\left(\sigma_i^{(j)}\right)_{i \in [q] \setminus \{j\}}, \mathrm{st}_{\mathcal{R}_3}^{(j)}\right) \leftarrow \mathcal{R}_2^{\mathrm{Tr}_{sk,pk}^{\mathrm{ID}}}(j, \mathrm{st}_{\mathcal{R}_2})$.

(M-4) simulate $\tilde{\mathcal{F}}_2$ in the following way:

if for any $i \in [q] \setminus \{j^*\}$, $\mathsf{Ver}\left(\overline{pk}, m_i, \sigma_i^{(j^*)}\right) = 1$; **then**

if there exists an index $j_0 \in [q] \setminus \{j^*\}$ **such that** $\mathsf{Ver}\left(\overline{pk}, m_{j^*}, \sigma_{j^*}^{(j_0)}\right) = 1$;

then set $\sigma^* := \sigma_{j^*}^{(j_0)}$. Here, it makes a sequence Q of queries, where Q is a set of all of queries (cmt_i, m_i) for each $i \in [q] \setminus \{j^*\}$ and $(\mathrm{cmt}^*, m_{j^*})$, to obtain the corresponding hash values.

otherwise; abort.

otherwise; set $\sigma^* := \perp$.

(M-5) execute $\left(\hat{\mathrm{cmt}}, \mathrm{st}_{\mathcal{R}_4}\right) \leftarrow \mathcal{R}_3^{\mathrm{Tr}_{sk,pk}^{\mathrm{ID}}}\left(\sigma^*, \mathrm{st}_{\mathcal{R}_3}^{(j^*)}\right)$, and move onto **Challenge** phase of the imp-pa game with sending $\hat{\mathrm{cmt}}$ to $\mathcal{C}$.

(M-6) given $\hat{\mathrm{cha}}$ from $\mathcal{C}$, execute $\hat{\mathrm{res}} \leftarrow \mathcal{R}_4\left(\hat{\mathrm{cha}}, \mathrm{st}_{\mathcal{R}_4}\right)$, and then output $\hat{\mathrm{res}}$.

Fig. 7. Meta-reduction $\mathcal{M}$

non-negligible probability if it succeeds in the simulation of $\tilde{\mathcal{F}}$. We show that $\mathcal{M}$ indeed simulates $\tilde{\mathcal{F}}_1$ and $\tilde{\mathcal{F}}_2$ correctly.

Lemma 3. $\mathcal{M}$ simulates $\tilde{\mathcal{F}}_1$ perfectly in the viewpoint of $\mathcal{R}$.

Proof. Let $\left(\overline{pk}, (m_i)_{i \in [q]} \right)$ and $r_{\mathcal{F}}$ be the input and the random coins of $\tilde{\mathcal{F}}_1$ output by $\mathcal{R}_1$ in (M-1), respectively. In (M-2), $\mathcal{M}$ chooses a challenge index $j^* \in_U [q]$, and then sets $\mathrm{st}_{\mathcal{F}} := \left(j^*, \overline{pk}, (m_i)_{i \in [q]}, r_{\mathcal{F}} \right)$ in the same way as $\tilde{\mathcal{F}}_1$. Therefore, $\mathcal{M}$ perfectly simulates $\tilde{\mathcal{F}}_1$ in the viewpoint of $\mathcal{R}$.   □

Lemma 4. *$\mathcal{M}$ simulates $\tilde{\mathcal{F}}_2$ perfectly in the viewpoint of $\mathcal{R}$ if it does not abort in (M-4).*

Proof. Let $\left(\left(\sigma_i^{(j^*)} \right)_{i \in [q] \setminus \{j^*\}}, \mathrm{st}_{\mathcal{F}} \right)$ be the input to $\tilde{\mathcal{F}}_2$, where $\left(\sigma_i^{(j^*)} \right)_{i \in [q] \setminus \{j^*\}}$ is a sequence of signatures output by the j^*-th execution of $\mathcal{R}_2$ in (M-3) and $\mathrm{st}_{\mathcal{F}}$ is the state set in (M-2). In (M-4), $\mathcal{M}$ first checks if for any $i \in [q] \setminus \{j^*\}$, $\mathsf{Ver}(\overline{pk}, m_i, \sigma_i^{(j^*)}) = 1$.

If its check passes, $\mathcal{M}$ is required to find a signature $\sigma^* = (\mathrm{cmt}^*, \mathrm{res}^*)$ satisfying $\mathsf{Ver}(\overline{pk}, m_{j^*}, \sigma^*) = 1$ as the behavior of $\tilde{\mathcal{F}}_2$. $\mathcal{M}$ does it by utilizing the sub-algorithm $\mathcal{R}_2$ as in (M-3). Note that on each index $j \in [q]$, $\mathcal{R}_2$ would output a sequence $\left(\sigma_i^{(j)} \right)_{i \in [q] \setminus \{j\}}$ of correct signatures σ_i on the messages m_i other than the j-th message m_j. In particular, there is the possibility that $\sigma_{j^*}^{(j_0)}$ which is output by $\mathcal{R}$ on some index $j_0 \neq j^*$ is a correct signature on the challenge message m_{j^*}. Now, we assume that $\mathcal{M}$ does not abort in (M-4). Namely, it is guaranteed that there exists a signature on m_{j^*} in the set $\left\{ \sigma_{j^*}^{(j)} \right\}_{j \in [q] \setminus \{j^*\}}$ of the signatures obtained in (M-3). Therefore, $\mathcal{M}$ finds a signature σ^* such that $\mathsf{Ver}(\overline{pk}, m_{j^*}, \sigma^*) = 1$ from $\left\{ \sigma_{j^*}^{(j)} \right\}_{j \in [q] \setminus \{j^*\}}$. As the behavior of $\tilde{\mathcal{F}}_2$, $\mathcal{M}$ also makes the sequence Q of the queries to the random oracle. Therefore, $\mathcal{M}$ indeed simulates $\tilde{\mathcal{F}}_2$ in such a case. Otherwise, $\mathcal{M}$ sets $\sigma^* := \bot$ in the same way as $\tilde{\mathcal{F}}_2$. Thus, $\mathcal{M}$ simulates $\tilde{\mathcal{F}}_2$ perfectly in the viewpoint of $\mathcal{R}$ if it does not abort.   □

We now evaluate the probability that $\mathcal{M}$ wins the imp-pa game of ID. Recall that $\mathcal{M}$ can win the imp-pa game of ID if it succeeds in the simulation of $\tilde{\mathcal{F}}$. It follows from Lemmas 3 and 4 that $\mathcal{M}$ succeeds in the simulation if it does not abort in (M-4). Therefore, we show that $\mathcal{M}$ does not abort with non-negligible probability. $\mathcal{M}$ does not abort if there exists an index $j_0 \in [q] \setminus \{j^*\}$ such that $\mathsf{Ver}(\overline{pk}, m_{j^*}, \sigma_{j^*}^{(j_0)}) = 1$. It suffices that there exists an index $j_1 \in [q] \setminus \{j^*\}$ such that for any index $i \in [q] \setminus \{j_1\}$, $\mathsf{Ver}(\overline{pk}, m_i, \sigma_i^{(j_1)}) = 1$ for the j_1-th sequence $\left(\sigma_i^{(j_1)} \right)_{i \in [q] \setminus \{j_1\}}$ issued in (M-3). In other words, $\mathcal{M}$ does not abort if on some index $j_1 \neq j^*$, $\mathcal{R}_2$ correctly replies the sequence $\left(\sigma_i^{(j_1)} \right)_{i \in [q] \setminus \{j_1\}}$ in (M-3). On the other hand, one can show the following lemma on the success probability of $\mathcal{R}_2$.

Lemma 5. *If the probability that $\mathcal{R}_2$ correctly replies a sequence $(\sigma_i)_{i \in [q] \setminus \{j\}}$ on input $j \in [q]$ is negligible, then ID is no longer imp-pa secure.*

Proof. Assume that the probability that $\mathcal{R}_2$ correctly replies $(\sigma_i)_{i \in [q] \setminus \{j\}}$ on input $j \in [q]$ is negligible. This means that the probability that for any $i \in [q] \setminus \{j\}$, $\mathsf{Ver}(\overline{pk}, m_i, \sigma_i) = 1$ for signatures σ_i output by $\mathcal{R}_2$ is negligible. Then, we now show that ID is not imp-pa secure by constructing a PPT reduction $\mathcal{R}'$ that wins the imp-pa game of ID without any black-box access. On a public key pk given by the imp-pa challenger $\mathcal{C}$, $\mathcal{R}'$ proceeds as follows:

(R'-1) choose random coins $r_\mathcal{R}$, and then execute $\left(\overline{pk}, (m_i)_{i \in [q]}, r_\mathcal{F}, \mathsf{st}_{\mathcal{R}_2} \right) \leftarrow \mathcal{R}_1^{\mathsf{Tr}_{sk,pk}^{\mathsf{ID}}}(pk; r_\mathcal{R})$.

(R'-2) simulate $\mathcal{F}_1$ by choosing a challenge index $j^* \in_U [q]$, and then setting $\mathsf{st}_\mathcal{F} := \left(j^*, \overline{pk}, (m_i)_{i \in [q]}, r_\mathcal{F} \right)$.

(R'-3) execute $\left((\sigma_i)_{i \in [q] \setminus \{j^*\}}, \mathsf{st}_{\mathcal{R}_3} \right) \leftarrow \mathcal{R}_2^{\mathsf{Tr}_{sk,pk}^{\mathsf{ID}}}(j^*, \mathsf{st}_{\mathcal{R}_2})$.

(R'-4) simulate $\mathcal{F}_2$ by aborting if for any $i \in [q] \setminus \{j^*\}$, $\mathsf{Ver}(\overline{pk}, m_i, \sigma_i) = 1$, or setting $\sigma^* := \bot$ otherwise.

(R'-5) execute $(\hat{\mathsf{cmt}}, \mathsf{st}_{\mathcal{R}_4}) \leftarrow \mathcal{R}_3^{\mathsf{Tr}_{sk,pk}^{\mathsf{ID}}}(\sigma^*, \mathsf{st}_{\mathcal{R}_3})$, and move onto **Challenge** phase of the imp-pa game with sending $\hat{\mathsf{cmt}}$ to $\mathcal{C}$.

(R'-6) given $\hat{\mathsf{cha}}$ from $\mathcal{C}$, execute $\hat{\mathsf{res}} \leftarrow \mathcal{R}_4 \left(\hat{\mathsf{cha}}, \mathsf{st}_{\mathcal{R}_4} \right)$, and then output $\hat{\mathsf{res}}$.

Since $\mathcal{F}_2$ is allowed to reply any symbol if $\mathcal{R}_2$ fails to reply, namely there exists an index $i_0 \in [q] \setminus \{j^*\}$ such that $\mathsf{Ver}(\overline{pk}, m_{i_0}, \sigma_{i_0}) \neq 1$, the reduction $\mathcal{R}$, hence $\mathcal{R}'$ would win the imp-pa game of ID with non-negligible probability $\epsilon_\mathcal{R}$ if it does not abort. On the other hand, $\mathcal{R}'$ aborts if for any $i \in [q] \setminus \{j^*\}$, $\mathsf{Ver}(\overline{pk}, m_i, \sigma_{j^*, i}) = 1$. Namely, $\mathcal{R}'$ aborts if $\mathcal{R}_2$ correctly replies the sequence $(\sigma_i)_{i \in [q] \setminus \{j^*\}}$ in (R'-3). By the assumption, such a probability is negligible. Therefore, $\mathcal{R}'$ can win the imp-pa game of ID with non-negligible probability without any black-box access. Thus ID is not imp-pa secure. $\square$

It follows from Lemma 5 that Theorem 2 holds if the sub-algorithm $\mathcal{R}_2$ of the assumed reduction $\mathcal{R}$ correctly replies with negligible probability. Otherwise, it holds that the probability that $\mathcal{R}_2$ correctly replies the sequence $\left(\sigma_i^{(j_1)} \right)_{i \in [q] \setminus \{j_1\}}$ on input $j_1 \in [q]$ is non-negligible. Namely, $\mathcal{M}$ does not abort with non-negligible probability. Let ϵ_{Sim} be the probability of the non-abortion of $\mathcal{M}$. Thus the success probability of $\mathcal{M}$ is evaluated as follows:

$$\Pr[\mathsf{Succ}_\mathcal{M}] \geq \Pr[\mathsf{Succ}_\mathcal{M} \wedge \mathsf{Sim}]$$
$$= \Pr[\mathsf{Sim}] \Pr[\mathsf{Succ}_\mathcal{M} \mid \mathsf{Sim}]$$
$$\geq \Pr[\mathsf{Sim}] \Pr[\mathsf{Succ}_\mathcal{R} \mid \mathsf{Sim}]$$
$$\geq \epsilon_{\mathsf{Sim}} \epsilon_\mathcal{R},$$

where $\mathsf{Succ}_\mathcal{M}$ and $\mathsf{Succ}_\mathcal{R}$ denote the events of winning the imp-pa game of $\mathcal{M}$ and $\mathcal{R}$, respectively. Sim stands for the event of the correct simulation of $\tilde{\mathcal{F}}$.

Running Time of $\mathcal{M}$. In (M-1), (M-5) and (M-6), $\mathcal{M}$ executes the PPT sub-algorithms $\mathcal{R}_1$, $\mathcal{R}_3$ and $\mathcal{R}_4$ of $\mathcal{R}$ once, respectively. Moreover, it also executes the PPT sub-algorithm $\mathcal{R}_2$ $q = q(\lambda)$ times in (M-3). Observe that the processes of (M-2) and (M-4) runs in polynomial time as in (M-2) and (M-4). Therefore, the meta-reduction $\mathcal{M}$ runs in polynomial time.

Thus, ID is not imp-pa secure.   □

3.2   Case: Sequentially Multi-Instance Reduction

We also consider the case where the reduction $\mathcal{R}$ is sequentially multi-instance. The *sequentially multi-instance* reduction means that the reduction invokes a forger $\mathcal{F}$ polynomially many times, namely I is polynomial. In this case, the reduction $\mathcal{R}$ is separated into a tuple $\left(\mathcal{R}_0, (\mathcal{R}_{k,1}, \mathcal{R}_{k,2}, \mathcal{R}_{k,3})_{k \in [I]}, \mathcal{R}_4, \mathcal{R}_5\right)$ of sub-algorithms:

$\mathcal{R}_0$ on a public key pk given from the imp-pa challenger $\mathcal{C}$ with random coins $r_{\mathcal{R}}$, outputs a state $\mathrm{st}_{\mathcal{R}_{1,1}}$.

$\mathcal{R}_{k,1}, \mathcal{R}_{k,2}, \mathcal{R}_{k,3}$ for each $k \in [I]$, the k-th invocation of the SMA forger $\mathcal{F} = (\mathcal{F}_1, \mathcal{F}_2)$ is done in the following way:

$\mathcal{R}_{k,1}$ on the state $\mathrm{st}_{\mathcal{R}_{k,1}}$ output by the previous sub-algorithm, outputs a pair $\left(\left(\overline{pk}_k, (m_{k,i})_{i \in [q]}, r_{k,\mathcal{F}}\right), \mathrm{st}_{\mathcal{R}_{k,2}}\right)$ of a k-th input $\left(\overline{pk}_k, (m_{k,i})_{i \in [q]}, r_{k,\mathcal{F}}\right)$ to $\mathcal{F}_1$ and a state $\mathrm{st}_{\mathcal{R}_{k,2}}$. Then, $\mathcal{F}_\infty(\overline{pk}_k, (m_{k,i})_{i \in [q]}; r_{k,\mathcal{F}})$ is invoked.

$\mathcal{R}_{k,2}$ on   a   challenge   index $j_k^* \in [q]$ which is queried by $\mathcal{F}_1(\overline{pk}_k, (m_{k,i})_{i \in [q]}; r_{k,\mathcal{F}})$ and the state $\mathrm{st}_{\mathcal{R}_{k,2}}$, out-puts a pair $\left((\sigma_{k,i})_{i \in [q] \setminus \{j_k^*\}}, \mathrm{st}_{\mathcal{R}_{k,3}}\right)$ of a sequence $(\sigma_{k,i})_{i \in [q] \setminus \{j_k^*\}}$ of sig-natures replied to $\mathcal{F}_2$ and a state $\mathrm{st}_{\mathcal{R}_{k,3}}$. Then, $\mathcal{F}_2((\sigma_{k,i})_{i \in [q] \setminus \{j_k^*\}}, \mathrm{st}_{k,\mathcal{F}})$ is invoked, where $\mathrm{st}_{k,\mathcal{F}}$ is a state which is output by the k-th invocation of $\mathcal{F}_1$.

$\mathcal{R}_{k,3}$ on a challenge signature σ_k^* output by $\mathcal{F}_2$, $((\sigma_{k,i})_{i \in [q] \setminus \{j_k^*\}}$ and the state $\mathrm{st}_{\mathcal{R}_{k,3}}$, outputs the state $\mathrm{st}_{\mathcal{R}_{k+1,1}}$ which will be used to execute the next sub-algorithm.

$\mathcal{R}_4$ on the state $\mathrm{st}_{\mathcal{R}_{I+1,1}}$ output by $\mathcal{R}_{I,3}$, moves onto **Challenge** phase of the imp-pa game with sending a commitment $\hat{cmt}$ to $\mathcal{C}$. Simultaneously, it outputs a state $\mathrm{st}_{\mathcal{R}_5}$.

$\mathcal{R}_5$ on a challenge $\hat{cha}$ received from $\mathcal{C}$ and the state $\mathrm{st}_{\mathcal{R}_5}$, replies a response $\hat{res}$ to $\mathcal{C}$.

As the simple reduction, we suppose that the states $\left(\mathrm{st}_{\mathcal{R}_{k,1}}, \mathrm{st}_{\mathcal{R}_{k,2}}, \mathrm{st}_{\mathcal{R}_{k,3}}\right)_{k \in [I]}$, $\mathrm{st}_{\mathcal{R}_{I+1,1}}$ and $\mathrm{st}_{\mathcal{R}_5}$ contain the random coins $r_{\mathcal{R}}$ which are given to $\mathcal{R}_0$, and the algorithms $\left((\mathcal{R}_{k,1}, \mathcal{R}_{k,2}, \mathcal{R}_{k,3})_{k \in [I]}, \mathcal{R}_4, \mathcal{R}_5\right)$ use the ran-dom coins $r_{\mathcal{R}}$ in these executions. $\left(\mathcal{R}_0, (\mathcal{R}_{k,1}, \mathcal{R}_{k,2}, \mathcal{R}_{k,3})_{k \in [I]}, \mathcal{R}_4\right)$ can query to the transcript oracle $\mathrm{Tr}_{sk,pk}^{\mathrm{ID}}$ and the random oracle. $\mathcal{R}_5$, $\mathcal{F}_1$ and $\mathcal{F}_2$ are allowed to query to the random oracle.

Note that the reduction which is allowed to rewind $\mathcal{F}$ can be converted into a sequentially multi-instance reduction by regarding the rewind of $\mathcal{F}$ as the new invocation of $\mathcal{F}$ as mentioned in the proof of Theorem 1 on [16]. In a similar manner to Theorem 2, the following theorem can be proven.

Theorem 6. *Assume that* FS-Sig *is proven to be SMA secure in the NPROM via a sequentially multi-instance reduction from the imp-pa security of the underlying ID scheme* ID. *Then* ID *is not imp-pa secure.*

By combining Theorem 6 with Proposition 1, the following corollary follows.

Corollary 7. *Assume that* FS-Sig *is proven to be EUF-CMA in the NPROM via a sequentially multi-instance reduction from the imp-pa security of the underlying ID scheme* ID. *Then* ID *is not imp-pa secure.*

4   Impossibility of Proving the Security of the Schnorr Signature

In this section, we prove that the DL assumption is a necessary and sufficient condition for the impossibility of proving the EUF-CMA security of the Schnorr signature in the NPROM via a sequentially multi-instance reduction.

Let $\mathbb{G}$ be a group of prime order p with a generator g. GGen is a group generation algorithm which on a security parameter 1^λ, outputs a tuple $(\mathbb{G}, p, g)$ of a group description $\mathbb{G}$, the order p of $\mathbb{G}$ and a generator g of $\mathbb{G}$ such that p is of polynomial length in λ. An algorithm $\mathcal{R}$ is said to *solve the discrete logarithm (DL) problem* if on a pair $((\mathbb{G}, p, g), y)$ of a tuple $(\mathbb{G}, p, g)$ and an element $y \in \mathbb{G}$, $\mathcal{R}$ outputs a solution $x \in \mathbb{Z}_p$ such that $y = g^x$. Then the *DL assumption holds* if there exists no PPT algorithm $\mathcal{R}$ which solves the DL problem for a tuple $(\mathbb{G}, p, g) \leftarrow$ GGen(1^λ) and $y = g^x$ with non-negligible probability, where the probability is taken over the choice of $x \in_U \mathbb{Z}_p$ and the internal coin flips of GGen and $\mathcal{R}$.

The Schnorr signature is derived from the Schnorr ID scheme which is depicted in Fig. 8 via the FS transformation. It is known that the imp-pa security of the Schnorr ID is polynomially equivalent to the DL assumption [28]. By employing this fact and by applying Corollary 7 to the Schnorr signature, we have the following theorem.

Theorem 8. *Assume that the Schnorr signature is proven to be EUF-CMA in the NPROM via a sequentially multi-instance reduction from the DL assumption. Then, the DL assumption does not hold.*

Proof. Assume that the Schnorr signature is proven to be EUF-CMA in the NPROM via a sequentially multi-instance reduction from the DL assumption. Since the imp-pa security of the Schnorr ID is polynomially equivalent to the DL assumption [28], the Schnorr signature is proven to be EUF-CMA in the NPROM via a sequentially multi-instance reduction from the imp-pa security of the Schnorr ID. It follows from Corollary 7 that the Schnorr ID is not imp-pa

---

Schnorr ID [28]

K on a security parameter 1^λ, outputs a key pair $(sk, pk) := (x, ((\mathbb{G}, p, g), y))$ by running $(\mathbb{G}, p, g) \leftarrow \mathsf{GGen}(1^\lambda)$, choosing $x \in_U \mathbb{Z}_p$ and then setting $y := g^x$.

$\mathcal{CH}$ for each $pk \in \mathsf{PK}_\lambda$, let $\mathcal{CH}_{pk} := \mathbb{Z}_p$.

P_1 on a key pair (sk, pk), outputs a pair (st, cmt) of a state st $\in_U \mathbb{Z}_p$ and a commitment cmt $:= g^{\mathrm{st}}$.

P_2 on a key pair (sk, pk), a pair (st, cmt) of an output by P_1, and a challenge cha $\in \mathbb{Z}_p$, outputs a response res $:=$ st $+$ cha $\cdot$ sk mod p.

V on a tuple $(pk, (\mathrm{cmt}, \mathrm{cha}, \mathrm{res}))$ of a public key pk and a transcript (cmt, cha, res), outputs 1 if it holds that cmt $= g^{\mathrm{res}} y^{-\mathrm{cha}}$.

Fig. 8. Schnorr ID

secure. This implies that the DL assumption does not hold by the equivalence between the imp-pa security of the Schnorr ID and the DL assumption.   □

On the other hand, if the DL assumption does not hold, the Schnorr signature is trivially proven to be EUF-CMA in the NPROM from the DL assumption. Therefore, this result shows that the DL assumption is not compatible with the EUF-CMA security of the Schnorr signature in the NPROM.

References

1. Abdalla, M., An, J.H., Bellare, M., Namprempre, C.: From identification to signatures via the Fiat-Shamir transform: necessary and sufficient conditions for security and forward-security. IEEE Trans. Inf. Theor. **54**(8), 3631–3646 (2008)
2. Abe, M., Groth, J., Ohkubo, M.: Separating short structure-preserving signatures from non-interactive assumptions. In: Lee, D.H., Wang, X. (eds.) ASIACRYPT 2011. LNCS, vol. 7073, pp. 628–646. Springer, Heidelberg (2011)
3. Abe, M., Haralambiev, K., Ohkubo, M.: Group to group commitments do not shrink. In: Pointcheval, D., Johansson, T. (eds.) EUROCRYPT 2012. LNCS, vol. 7237, pp. 301–317. Springer, Heidelberg (2012)
4. Bader, C., Jager, T., Li, Y., Schäge, S.: On the impossibility of tight cryptographic reductions. In: Fischlin, M., Coron, J.-S. (eds.) EUROCRYPT 2016. LNCS, vol. 9666, pp. 273–304. Springer, Heidelberg (2016). doi:10.1007/978-3-662-49896-5_10
5. Baldimtsi, F., Lysyanskaya, A.: On the security of one-witness blind signature schemes. In: Sako, K., Sarkar, P. (eds.) ASIACRYPT 2013, Part II. LNCS, vol. 8270, pp. 82–99. Springer, Heidelberg (2013)
6. Bellare, M., Rogaway, P.: Random oracles are practical: a paradigm for designing efficient protocols. In: ACM CCS 1993, Fairfax, Virginia, USA, pp. 62–73. ACM Press, New York (1993)
7. Boneh, D., Franklin, M.: Identity-based encryption from the Weil pairing. In: Kilian, J. (ed.) CRYPTO 2001. LNCS, vol. 2139, pp. 213–229. Springer, Heidelberg (2001)

8. Boneh, D., Venkatesan, R.: Breaking RSA may not be equivalent to factoring. In: Nyberg, K. (ed.) EUROCRYPT 1998. LNCS, vol. 1403, pp. 59–71. Springer, Heidelberg (1998)

9. Bresson, E., Monnerat, J., Vergnaud, D.: Separation results on the "One-More" computational problems. In: Malkin, T. (ed.) CT-RSA 2008. LNCS, vol. 4964, pp. 71–87. Springer, Heidelberg (2008)

10. Chen, Y., Huang, Q., Zhang, Z.: Sakai-Ohgishi-Kasahara identity-based non-interactive key exchange scheme, revisited. In: Susilo, W., Mu, Y. (eds.) ACISP 2014. LNCS, vol. 8544, pp. 274–289. Springer, Heidelberg (2014)

11. Coron, J.-S.: Optimal security proofs for PSS and other signature schemes. In: Knudsen, L.R. (ed.) EUROCRYPT 2002. LNCS, vol. 2332, pp. 272–287. Springer, Heidelberg (2002)

12. Fiat, A., Shamir, A.: How to prove yourself: practical solutions to identification and signature problems. In: Odlyzko, A.M. (ed.) CRYPTO 1986. LNCS, vol. 263, pp. 186–194. Springer, Heidelberg (1987)

13. Fischlin, M., Fleischhacker, N.: Limitations of the meta-reduction technique: the case of Schnorr signatures. In: Johansson, T., Nguyen, P.Q. (eds.) EUROCRYPT 2013. LNCS, vol. 7881, pp. 444–460. Springer, Heidelberg (2013)

14. Fischlin, M., Lehmann, A., Ristenpart, T., Shrimpton, T., Stam, M., Tessaro, S.: Random oracles with(out) programmability. In: Abe, M. (ed.) ASIACRYPT 2010. LNCS, vol. 6477, pp. 303–320. Springer, Heidelberg (2010)

15. Fleischhacker, N., Jager, T., Schröder, D.: On tight security proofs for Schnorr signatures. In: Sarkar, P., Iwata, T. (eds.) ASIACRYPT 2014. LNCS, vol. 8873, pp. 512–531. Springer, Heidelberg (2014)

16. Fukumitsu, M., Hasegawa, S.: Black-box separations on Fiat-Shamir-type signatures in the non-programmable random oracle model. In: López, J., Mitchell, C.J. (eds.) ISC 2015. LNCS, vol. 9290, pp. 3–20. Springer, Heidelberg (2015)

17. Fukumitsu, M., Hasegawa, S., Isobe, S., Koizumi, E., Shizuya, H.: Toward separating the strong adaptive pseudo-freeness from the strong RSA assumption. In: Boyd, C., Simpson, L. (eds.) ACISP. LNCS, vol. 7959, pp. 72–87. Springer, Heidelberg (2013)

18. Fukumitsu, M., Hasegawa, S., Isobe, S., Shizuya, H.: On the impossibility of proving security of strong-RSA signatures via the RSA assumption. In: Susilo, W., Mu, Y. (eds.) ACISP 2014. LNCS, vol. 8544, pp. 290–305. Springer, Heidelberg (2014)

19. Goldwasser, S., Micali, S., Rivest, R.L.: A digital signature scheme secure against adaptive chosen-message attacks. SIAM J. Comput. **17**(2), 281–308 (1988)

20. Guillou, L.C., Quisquater, J.-J.: A practical zero-knowledge protocol fitted to security microprocessor minimizing both transmission and memory. In: Günther, C.G. (ed.) EUROCRYPT 1988. LNCS, vol. 330, pp. 123–128. Springer, Heidelberg (1988)

21. Katz, J., Wang, N.: Efficiency improvements for signature schemes with tight security reductions. In: ACM CCS 2003. pp. 155–164. ACM, New York (2003)

22. Kawai, Y., Sakai, Y., Kunihiro, N.: On the (im)possibility results for strong attack models for public key cryptsystems. JISIS **1**(2/3), 125–139 (2011)

23. Nielsen, J.B.: Separating random oracle proofs from complexity theoretic proofs: the non-committing encryption case. In: Yung, M. (ed.) CRYPTO 2002. LNCS, vol. 2442, pp. 111–126. Springer, Heidelberg (2002)

24. Okamoto, T.: Provably secure and practical identification schemes and corresponding signature schemes. In: Brickell, E.F. (ed.) CRYPTO 1992. LNCS, vol. 740, pp. 31–53. Springer, Heidelberg (1993)

25. Paillier, P., Vergnaud, D.: Discrete-log-based signatures may not be equivalent to discrete log. In: Roy, B. (ed.) ASIACRYPT 2005. LNCS, vol. 3788, pp. 1–20. Springer, Heidelberg (2005)
26. Paillier, P., Villar, J.L.: Trading one-wayness against chosen-ciphertext security in factoring-based encryption. In: Lai, X., Chen, K. (eds.) ASIACRYPT 2006. LNCS, vol. 4284, pp. 252–266. Springer, Heidelberg (2006)
27. Pointcheval, D., Stern, J.: Security arguments for digital signatures and blind signatures. J. Cryptology **13**(3), 361–396 (2000)
28. Schnorr, C.: Efficient signature generation by smart cards. J. Cryptology **4**(3), 161–174 (1991)
29. Seurin, Y.: On the exact security of Schnorr-type signatures in the random oracle model. In: Pointcheval, D., Johansson, T. (eds.) EUROCRYPT 2012. LNCS, vol. 7237, pp. 554–571. Springer, Heidelberg (2012)
30. Shoup, V.: A proposal for an iso standard for public key encryption. Cryptology ePrint Archive, Report 2001/112 (2001). http://eprint.iacr.org/
31. Zhang, J., Zhang, Z., Chen, Y., Guo, Y., Zhang, Z.: Black-box separations for one-more (static) CDH and its generalization. In: Sarkar, P., Iwata, T. (eds.) ASIACRYPT 2014, Part II. LNCS, vol. 8874, pp. 366–385. Springer, Heidelberg (2014)
32. Zhang, Z., Chen, Y., Chow, S.S.M., Hanaoka, G., Cao, Z., Zhao, Y.: Black-box separations of hash-and-sign signatures in the non-programmable random oracle model. In: Au, M.-H., et al. (eds.) ProvSec 2015. LNCS, vol. 9451, pp. 435–454. Springer, Heidelberg (2015). doi:10.1007/978-3-319-26059-4_24

Efficient Functional Encryption
for Inner-Product Values
with Full-Hiding Security

Junichi Tomida[1]([✉]), Masayuki Abe[2], and Tatsuaki Okamoto[2]

[1] Kyoto University, Kyoto, Japan
`tomida@ai.soc.i.kyoto-u.ac.jp`
[2] NTT Secure Platform Laboratories, Tokyo, Japan
`{abe.masayuki,okamoto.tatsuaki}@lab.ntt.co.jp`

Abstract. We construct an efficient non-generic private-key functional encryption (FE) for inner-product values with full-hiding security, where confidentiality is assured not only for encrypted data but also for functions associated with secret keys. Recently, Datta et al. presented such a scheme in PKC 2016 and this is the only scheme that achieved full-hiding security. Our scheme has an advantage over their scheme for the following points.

1. *More efficient:* our scheme is two times faster in encryption and decryption, and a master secret key, secret keys and ciphertexts are the half size, compared with their scheme.
2. *Weaker assumption:* our scheme is secure under the decisional linear (DLIN) assumption or its variant, while their scheme is under a stronger assumption, the symmetric external Diffie-Hellman (SXDH) assumption.
3. *More flexible:* we can apply our scheme to any type of bilinear pairing groups, while their scheme is suitable only for type 3 groups.

Keywords: Functional encryption · Inner product · Function privacy

1 Introduction

Functional encryption (FE) is a very useful tool for non-interactive computation on encrypted data [8]. In a FE scheme, an owner of a master secret key msk can create a secret key sk_f for a function f, and it enables users to compute the value of $f(x)$ by decrypting a ciphertext for x without revealing anything else about x. As cloud services are increasing rapidly, users' demand for computation on encrypted data is also increasing because cloud servers are by no means trustful. FE is one solution for this problem, providing a paradigm where users can compute a function f on encrypted data using a secret key sk_f without revealing anything else about the encrypted data to the cloud server.

One of principal interests in FE is what class of of functions $\mathcal{F}$ can be supported and what kind of security can be achieved. It has started from identity-based

© Springer International Publishing Switzerland 2016
M. Bishop and A.C.A. Nascimento (Eds.): ISC 2016, LNCS 9866, pp. 408–425, 2016.
DOI: 10.1007/978-3-319-45871-7_24

encryption (IBE) [7], followed by attribute-based encryption (ABE) [22], inner-product encryption (IPE) [16,17,19,20] and predicate encryption (PE) [16]. Amazingly, recent works realize computation of general polynomial-size circuit [13,14], although they require expensive assumptions like indistinguishability obfuscation and they are far from being practical. Motivated by this unreality, Abdalla et al. [1] introduced a new non-generic FE scheme specialized for computation of inner-product values, which is efficient and constructed from standard assumptions. As Abdalla et al. mentioned in their work, evaluation of inner-product is a very useful tool for statistics because it can provide the *weighted mean*. The scheme of [1] only has a selective security, and following works present adaptively secure schemes [2,5]. Note that FE for inner-product values is different from inner-product encryption (IPE) in the context of predicate encryption, where a secret key for a predicate $\vec{y}$ and a ciphertext for a message m and an attribute $\vec{x}$ yield m iff $\vec{x} \cdot \vec{y} = 0$. On the other hand, FE for inner-product values enables users to compute inner-product values themselves. To avoid confusion, we refer to FE for inner-product values as inner-product value encryption (IPVE) and this is the main topic of our paper.

Function Privacy of Functional Encryption. In terms of security, most of research on FE has concentrated on confidentiality of ciphertexts [8,21]. It is very inconvenient for real applications, however, because one may need to hide a function as well as a plaintext from others. Consider the case where Alice holds encrypted data on an untrusted cloud server and wants to compute a secret algorithm f on the data. Alice may accomplish her purpose by a FE scheme, to create a secret key for the function f, send it to the server and get the result. However, if the FE scheme does not support the privacy of functions associated with secret keys, the secret key for f may reveal sensitive information about f to the server. Consequently, it is important to consider the function privacy of FE, which assure that secret keys do not reveal any information on associated functions. In the public-key setting, we need to put some restrictions on the distribution of the functions to obtain a meaningful security of function privacy [10,11]. It is because users can encrypt any message and examine what the function associated with their secret key is, by decrypting them. On the contrary, we can obtain the strongest notion of indistinguishability-based security in the private-key setting, known as full-hiding security, which guarantees function privacy as well as message privacy [4,9].

Full-Hiding Security. For private-key FE, the notion of full-hiding security considers an adversary that interacts with an encryption oracle Enc_b and a secret key oracle KeyGen_b where b is randomly chosen from $\{0,1\}$. The adversary queries a pair of messages $(x_0^{(\ell)}, x_1^{(\ell)})$ to the encryption oracle and gets a ciphertext for $x_b^{(\ell)}$. It also gets a secret key for $f_b^{(j)}$ from the key generation oracle similarly. In the game for guessing b, the minimum necessary restriction is $f_0^{(j)}(x_0^{(\ell)}) = f_1^{(j)}(x_1^{(\ell)})$ for all ℓ and j, as otherwise the adversary can trivially determine b. After the query phase under the above restriction, if any adversary can guess b only with negligible advantage, then we say that the private-key FE scheme has full-hiding security. This notion implies that any efficient adversary that has secret keys for $f_1, \cdots, f_m$ and encryption of $x_1, \cdots, x_n$, cannot obtain any information other than the values of $\{f_i(x_j)\}_{1 \leq i \leq m, 1 \leq j \leq n}$ [9].

Inner-Product Value Encryption with Full-Hiding Security. For these reasons, we have great interest to construct direct and efficient FE schemes for practical functionalities with full-hiding security, and it is a quite important task for research on FE. Bishop et al. [6] have taken a first step for a private-key IPVE scheme that has function privacy, although their scheme achieved rather weak security. Considering natural application of full-hiding security to private-key IPVE, the restriction of queries that the adversary can make would be $\vec{x}_0^{(\ell)} \cdot \vec{y}_0^{(j)} = \vec{x}_1^{(\ell)} \cdot \vec{y}_1^{(j)}$. However they assumes the following restriction in the game, that is $\vec{x}_0^{(\ell)} \cdot \vec{y}_0^{(j)} = \vec{x}_1^{(\ell)} \cdot \vec{y}_0^{(j)} = \vec{x}_0^{(\ell)} \cdot \vec{y}_1^{(j)} = \vec{x}_1^{(\ell)} \cdot \vec{y}_1^{(j)}$. This unnatural restriction weakens its security guarantee. Very recently, Datta et al. [12] presented a private-key IPVE scheme with natural full-hiding security. Unfortunately, their scheme is wasteful because it needs a $4n$ dimensional vector space to encrypt n dimensional vectors.

1.1 Our Contribution

We construct a more efficient and flexible private-key IPVE scheme with full-hiding security than that of [12], which is the only scheme that achieved full-hiding security. In addition, our scheme is secure under weaker assumptions than their scheme. To ensure correctness, like [6,12], our scheme requires that inner-products are within a polynomial range, where discrete logarithm of $g^{\vec{x} \cdot \vec{y}}$ can be found in polynomial time. This is a reasonable requirement because results of statistical computation, like average, on a polynomial-size database will naturally be in a polynomial range.

Efficiency. Before discussing the efficiency, we recall techniques of private-key IPVE schemes. The schemes of [6,12] and our scheme are all constructed based on dual paring vector spaces (DPVS) introduced by Okamoto et al. [18,19], and they have the same structure. Namely, a master secret key is orthonormal bases of DPVS, secret keys and ciphertexts are vectors of DPVS, both key generation algorithm and encryption algorithms involve scalar multiplications on cyclic groups, and a decryption algorithm involves paring operations on bilinear paring groups. Our scheme is superior to that of [12] with a constant factor 2, in terms of both necessary storage and computational efficiency (Table 1).

Assumption and Flexibility. The schemes of [6,12] are secure under the symmetric external Diffie-Hellman (SXDH) assumption, while our scheme is the decisional linear (DLIN) assumption or its variant (XDLIN). SXDH holds in only type 3 bilinear pairing groups because if there is a isomorphism from one group to the other group, the SXDH problem is trivial. On the other hand, DLIN and XDLIN hold even if there is any isomorphism between two groups. In other

Table 1. Comparison of private-key IPVE schemes, where n is a dimension of a vector to encrypt, msk, sk and ct size are numbers of group elements.

	BJK15 [6]	DDM15 [12]	Our scheme
Security	Not full-hiding	Full-hiding	Full-hiding
msk size	$8n^2 + 8$	$8n^2 + 12n + 28$	$4n^2 + 18n + 20$
sk,ct size	$2n + 2$	$4n + 8$	$2n + 5$
#Scalar multiplications	$2n + 2$	$4n + 8$	$2n + 5$
#Paring operations	$2n + 2$	$4n + 8$	$2n + 5$
Assumptions	SXDH	SXDH	XDLIN or DLIN
Pairing groups	type 3	type 3	type 1,2,3

words, they hold in any type of bilinear pairing groups. In that sense, DLIN and XDLIN are weaker assumptions than SXDH. For the reason, the schemes of [6,12] work in only type 3 groups while we can use our scheme in any type of groups. We often construct cryptographic schemes on type 3 groups in current situation but the future condition is unpredictable, then we believe that the flexibility of our scheme is advantage. See Sect. 2.2 about the types of bilinear paring groups.

Our Technique. From a technical view point, it is generally difficult to achieve more efficient cryptographic schemes from weaker assumptions. In our case, the DLIN assumption indicates that it is hard to guess whether an DLIN instance spans 2 or 3 dimensions, while the SXDH assumption does 1 or 2. Then it seems that we need more dimensions or group elements to construct schemes from DLIN than from SXDH. However, we have overcome this difficulty by developing new two techniques for the security proof. These techniques enable us to change the form of secret keys and ciphertexts in *compact* DPVS over security game transition. Let we give an intuitive explanation. The first technique is, between two hybrid games, to reduce a difference of *one* coefficient in a secret key or a ciphertext, to a DLIN or XDLIN instance. In contrast, Datta et al. have made reductions from a difference of n coefficients to an SXDH instance in their security proof. Our technique can save about n dimensions of vector spaces compared with their technique. The second technique is, in the series of security game, to change directly a secret key including $\vec{y}_0^{(j)}$ into one including $\vec{y}_1^{(j)}$. Datta et al. employed a more complicated but unnecessary technique for this transformation, which requires another n dimensions.

2 Preliminary

2.1 Notation

For a set S, $x \xleftarrow{\mathsf{U}} S$ denotes that x is uniformly chosen from S. For a probability distribution X, $x \xleftarrow{\mathsf{R}} X$ denotes that x is chosen from X according to its distribution. For a prime q, $\mathbb{Z}_q$ denotes a set of integers $\{0, \cdots, q-1\}$, and $\mathbb{Z}_q^\times$ denotes a set of integers $\{1, \cdots, q-1\}$. $\vec{0}$ denotes a zero vector. For a n dimensional vector $\vec{x}$, $x_i (1 \le i \le n)$ denotes the i-th component of $\vec{x}$. For vectors $\vec{x}, \vec{y} \in \mathbb{Z}_q^n$, $\vec{x} \cdot \vec{y}$ denotes inner-product of $\vec{x}$ and $\vec{y}$ over $\mathbb{Z}_q$. For vector components, 0^n denotes a line of n zeros, e.g., $\vec{a} := (0,0,0,1) = (0^3, 1)$. For a security game and an adversary $\mathcal{A}$, $\mathsf{Exp}_{\mathcal{A}}^{\mathsf{Game}}(\lambda) \to b$ denotes an event where $\mathcal{A}$ outputs b in the game. For a function $f : \mathbb{N} \to \mathbb{R}$, $f(\lambda) < \epsilon(\lambda)$ denotes that f is *negligible* in λ, and means that $\forall c > 0, \exists n \in \mathbb{N}, \forall \lambda > n, f(\lambda) < \lambda^{-c}$.

2.2 Bilinear Pairing Groups

Bilinear pairing groups are defined by the tuple $(q, \mathbb{G}_1, \mathbb{G}_2, \mathbb{G}_T, e)$, where q is a prime, $\mathbb{G}_1$, $\mathbb{G}_2$ and $\mathbb{G}_T$ are cyclic groups of order q, and $e : \mathbb{G}_1 \times \mathbb{G}_2 \to \mathbb{G}_T$ is a map that has the following properties:

1. *Bilinear*: $\forall G_1 \in \mathbb{G}_1, \forall G_2 \in \mathbb{G}_2, \forall a, b \in \mathbb{Z}_q, e(aG_1, bG_2) = e(G_1, G_2)^{ab}$
2. *Non-degenerate*: if $\forall G_1 \in \mathbb{G}_1, e(G_1, G_2) = 1$, then $G_2 = 0$.

There are three types of bilinear groups according to whether efficient isomorphisms exist or not between $\mathbb{G}_1$ and $\mathbb{G}_2$ [15]. In the type 1, both the isomorphism $\phi : \mathbb{G}_2 \to \mathbb{G}_1$ and its inverse $\phi^{-1} : \mathbb{G}_1 \to \mathbb{G}_2$ can be computed efficiently, i.e., $\mathbb{G}_1 = \mathbb{G}_2$. In the type 2, the isomorphism $\phi : \mathbb{G}_2 \to \mathbb{G}_1$ is computed efficiently but its inverse is not. Type 3 groups have no efficient isomorphisms between $\mathbb{G}_1$ and $\mathbb{G}_2$. Type1 groups are called symmetric bilinear pairing groups, and type 2 and 3 are called asymmetric bilinear pairing groups. We use type 3 groups for our scheme in this paper, but we can easily apply it to type 1 or 2 groups. Let $\mathcal{G}_{\mathsf{abpg}}$ be an asymmetric bilinear pairing group generator that takes 1^λ and outputs a description of groups $(q, \mathbb{G}_1, \mathbb{G}_2, \mathbb{G}_T, e)$ and generators of groups $G_1 \ne 0 \in \mathbb{G}_1, G_2 \ne 0 \in \mathbb{G}_2$. We denote the tuple $(q, \mathbb{G}_1, \mathbb{G}_2, \mathbb{G}_T, G_1, G_2, e)$ by $\mathsf{param}_\mathbb{G}$.

2.3 Dual Pairing Vector Spaces [18,19]

We will construct our scheme based on dual pairing vector spaces (DPVS). There are two types of DPVS, one is using symmetric bilinear pairing groups and the other is asymmetric bilinear pairing groups [19]. In this paper, we use the asymmetric version of DPVS but we can also use the symmetric version similarly. We briefly explain the notion of DPVS here, and see [19] for more details.

Definition 1 (Dual Pairing Vector Spaces: DPVS). *DPVS are defined by the tuple $(q, \mathbb{V}, \mathbb{V}^*, \mathbb{G}_T, \mathbb{A}, \mathbb{A}^*, \tilde{e})$, which is directly constructed from $\mathsf{param}_\mathbb{G} \xleftarrow{\mathsf{R}} \mathcal{G}_{\mathsf{abpg}}(1^\lambda)$. $\mathbb{V} := \mathbb{G}_1^n$ and $\mathbb{V}^* := \mathbb{G}_2^n$ are n dimensional vector spaces, $\mathbb{A} := (\boldsymbol{a}_1, \cdots, \boldsymbol{a}_n)$ and $\mathbb{A}^* := (\boldsymbol{a}_1^*, \cdots, \boldsymbol{a}_n^*)$ are canonical bases, where $\boldsymbol{a}_i := (0^{i-1}, G_1, 0^{n-i})$, $\boldsymbol{a}_i^* := (0^{i-1}, G_2, 0^{n-i})$, and $\tilde{e} : \mathbb{V} \times \mathbb{V}^* \to \mathbb{G}_T$ is pairing defined below. A prime q and a group $\mathbb{G}_T$ are the same entities as the instance of pairing groups. The pairing is defined by $\tilde{e}(\boldsymbol{x}, \boldsymbol{y}) := \prod_{i=1}^n e(X_i, Y_i) \in \mathbb{G}_T$, where $\boldsymbol{x} := (X_1, \cdots, X_n) \in \mathbb{V}, \boldsymbol{y} := (Y_1, \cdots, Y_n) \in \mathbb{V}^*$.*

Here we consider random dual orthonormal bases:

$$\psi \xleftarrow{\mathsf{U}} \mathbb{Z}_q^\times, \quad \boldsymbol{B} := (b_{i,j})_{1\le i,j\le n} \xleftarrow{\mathsf{U}} GL(n, \mathbb{Z}_q), \quad (b_{i,j}^*)_{1\le i,j\le n} := \psi(\boldsymbol{B}^{\mathrm{T}})^{-1}$$

$$\boldsymbol{b}_i := \sum_{j=1}^n b_{i,j} \boldsymbol{a}_j, \quad \boldsymbol{b}_i^* := \sum_{j=1}^n b_{i,j}^* \boldsymbol{a}_j^* \text{ for } i = 1, \cdots, n,$$

$$\mathbb{B} := (\boldsymbol{b}_1, \cdots, \boldsymbol{b}_n), \quad \mathbb{B}^* := (\boldsymbol{b}_1^*, \cdots, \boldsymbol{b}_n^*), \quad g_T := e(G_1, G_2)^\psi.$$

Let $\mathcal{G}_{\mathsf{ob}}$ be random dual orthonormal basis generator that takes 1^λ and a dimension of bases n and outputs $(\mathsf{param}_\mathbb{G}, \mathbb{B}, \mathbb{B}^*, g_T)$, where $\mathbb{B}, \mathbb{B}^*, g_T$ are computed as demonstrated above. We denote the combination $(\mathsf{param}_\mathbb{G}, g_T)$ by $\mathsf{param}_\mathbb{V}$. For a vector $\vec{x} := (x_1, \cdots, x_n)^{\mathrm{T}} \in \mathbb{Z}_q^n$ and a basis $\mathbb{B} := (\boldsymbol{b}_1, \cdots, \boldsymbol{b}_n)$, we denote $\sum_{i=1}^n x_i \boldsymbol{b}_i$ by $(\vec{x})_\mathbb{B}$. Then it can be seen that

$$\tilde{e}((\vec{x})_\mathbb{A}, (\vec{y})_{\mathbb{A}^*}) = \prod_{i=1}^n e(x_i G_1, y_i G_2) = e(G_1, G_2)^{\sum_{i=1}^n x_i y_i} = e(G_1, G_2)^{\vec{x} \cdot \vec{y}},$$

$$\therefore \; \tilde{e}((\vec{x})_\mathbb{B}, (\vec{y})_{\mathbb{B}^*}) = \tilde{e}\left((\boldsymbol{B}\vec{x})_\mathbb{A}, (\psi(\boldsymbol{B}^{\mathrm{T}})^{-1}\vec{y})_{\mathbb{A}^*}\right) = e(G_1, G_2)^{\psi \boldsymbol{B}\vec{x} \cdot (\boldsymbol{B}^{\mathrm{T}})^{-1}\vec{y}} = g_T^{\vec{x} \cdot \vec{y}}.$$

2.4 External Decisional Linear Assumption

When we construct our scheme on type 3 groups, we assume the following property.

Definition 2 (External Decisional Linear Assumption: XDLIN [3]). *We choose an arbitrary number $x \in \{1, 2\}$. The XDLIN problem is to guess a bit b, given P_b, where*

$$\mathsf{param}_\mathbb{G} \xleftarrow{\mathsf{R}} \mathcal{G}_{\mathsf{abpg}}(1^\lambda), \quad \xi, \kappa, \delta, \sigma, \rho \xleftarrow{\mathsf{U}} \mathbb{Z}_q,$$

$$Y_0 := (\delta + \sigma)G_x, \quad Y_1 := (\delta + \sigma + \rho)G_x,$$

$$P_b := (\mathsf{param}_\mathbb{G}, \xi G_1, \kappa G_1, \delta\xi G_1, \sigma\kappa G_1, \xi G_2, \kappa G_2, \delta\xi G_2, \sigma\kappa G_2, Y_b).$$

For any probabilistic polynomial time (PPT) adversary $\mathcal{A}$, if the advantage of $\mathcal{A}$ for the XDLIN problem is negligible in λ, then we say that the XDLIN assumption holds. Namely,

$$\mathsf{Adv}_\mathcal{A}^{\mathsf{XDLIN}}(\lambda) := \left| \Pr[\mathcal{A}(1^\lambda, P_0) \to 1] - \Pr[\mathcal{A}(1^\lambda, P_1) \to 1] \right| < \epsilon(\lambda).$$

Remark 1. We can also construct our scheme on type 1 or 2 groups, and in that case we use the standard decisional linear (DLIN) assumption in $\mathbb{G}_2$. See Definition 11 in [19] about the DLIN assumption. Roughly speaking, the DLIN assumption in $\mathbb{G}_2$ is sufficient for the security proof of our scheme because we can obtain $\mathbb{G}_1$ elements that have the same coefficients as $\mathbb{G}_2$ elements of a DLIN instance, using a efficient isomorphism $\phi : \mathbb{G}_2 \to \mathbb{G}_1$.

2.5   The Notion of Private-Key Inner-Product Value Encryption

We adopt the general notion of function-private FE in the private-key setting, introduced in [4], to the particular functionality of computing inner-product values over $\mathbb{Z}_q$. We denote private-key inner-product value encryption by Priv-IPVE.

Definition 3 (Private-Key Inner-Product Value Encryption). *A Priv-IPVE scheme Π consists of four PPT algorithms* Setup, KeyGen, Enc *and* Dec:

- Setup($1^\lambda, n$): *The setup algorithm takes as input a security parameter 1^λ and a vector length parameter n (a positive integer that is polynomial in λ). Then it outputs a public parameter* pp *and a master secret key* msk.
- KeyGen(pp, msk, $\vec{y}$): *The key generation algorithm takes as input a public parameter* pp, *a master secret key* msk *and a key vector $\vec{y} \in \mathbb{Z}_q^n$. Then it outputs a corresponding secret key* $\mathsf{sk}_{\vec{y}}$.
- Enc(pp, msk, $\vec{x}$): *The encryption algorithm takes as input a public parameter* pp, *a master secret key* msk *and a message vector $\vec{x} \in \mathbb{Z}_q^n$. Then it outputs a ciphertext* $\mathsf{ct}_{\vec{x}}$.
- Dec(pp, $\mathsf{sk}_{\vec{y}}$, $\mathsf{ct}_{\vec{x}}$): *The decryption algorithm takes as input a public parameter* pp, *a secret key* $\mathsf{sk}_{\vec{y}}$ *and a ciphertext* $\mathsf{ct}_{\vec{x}}$. *Then it outputs either a value $m \in \mathbb{Z}_q$ or a symbol $\perp$.*

We assume the following property for correctness: for all $(\mathsf{pp}, \mathsf{msk}) \xleftarrow{\mathsf{R}} \mathsf{Setup}$ $(1^\lambda, n)$, *all* $\mathsf{sk}_{\vec{y}} \xleftarrow{\mathsf{R}} \mathsf{KeyGen}(\mathsf{pp}, \mathsf{msk}, \vec{y})$ *and all* $\mathsf{ct}_{\vec{x}} \xleftarrow{\mathsf{R}} \mathsf{Enc}(\mathsf{pp}, \mathsf{msk}, \vec{x})$, Dec $(\mathsf{pp}, \mathsf{sk}_{\vec{y}}, \mathsf{ct}_{\vec{x}})$ *must output $m = \vec{x} \cdot \vec{y}$ without negligible probability.*

Definition 4 (Full-Hiding Security). *An Priv-IPVE scheme Π has full-hiding security if, for any PPT adversaries $\mathcal{A}$, the advantage in the following game is negligible in λ:*

1. *The challenger runs* Setup($1^\lambda, n$) *to generate* pp *and* msk, *and gives* pp *to $\mathcal{A}$. It also chooses a random bit b.*
2. *$\mathcal{A}$ may adaptively make a polynomial number of queries of the following two types:*
    - *Secret key query: For the j-th query, $\mathcal{A}$ submits a pair of vectors $(\vec{y}_0^{(j)}, \vec{y}_1^{(j)})$ and the challenger replies to $\mathcal{A}$ with* $\mathsf{sk}_{\vec{y}_b^{(j)}} \xleftarrow{\mathsf{R}} \mathsf{KeyGen}(\mathsf{pp}, \mathsf{msk}, \vec{y}_b^{(j)})$.
    - *Ciphertext query: For the ℓ-th query, $\mathcal{A}$ submits a pair of vectors $(\vec{x}_0^{(\ell)}, \vec{x}_1^{(\ell)})$ and the challenger replies to $\mathcal{A}$ with* $\mathsf{ct}_{\vec{x}_b^{(\ell)}} \xleftarrow{\mathsf{R}} \mathsf{Enc}(\mathsf{pp}, \mathsf{msk}, \vec{x}_b^{(\ell)})$.
   There is a restriction for queries that $\mathcal{A}$ can make such that all queried vectors must suffice $\vec{x}_0^{(\ell)} \cdot \vec{y}_0^{(j)} = \vec{x}_1^{(\ell)} \cdot \vec{y}_1^{(j)}$ for all j and ℓ.
3. *$\mathcal{A}$ outputs a bit b' as a conjecture of b.*

The advantage of $\mathcal{A}$ in this game is defined as

$$\mathsf{Adv}_{\mathcal{A}}^{\mathsf{FHS},\Pi}(\lambda) := \left| \Pr[\mathsf{Exp}_{\mathcal{A}}^{\mathsf{Game}\,0}(\lambda) \to 1] - \Pr[\mathsf{Exp}_{\mathcal{A}}^{\mathsf{Game}\,1}(\lambda) \to 1] \right|$$

where this game is defined as Game 0 if $b = 0$ and as Game 1 if $b = 1$.

3   Construction

In this section, we present our private-key IPVE scheme with full-hiding security.

$\mathsf{Setup}(1^\lambda, n)$: The setup algorithm selects $(\mathbb{B}, \mathbb{B}^*, \mathsf{param}_\mathbb{V}) \xleftarrow{\mathsf{R}} \mathcal{G}_{\mathsf{ob}}(1^\lambda, 2n + 5)$ and outputs $(\mathsf{pp}, \mathsf{msk})$, where

$$\widehat{\mathbb{B}} := (\boldsymbol{b}_1, \cdots, \boldsymbol{b}_n, \boldsymbol{b}_{2n+1}, \boldsymbol{b}_{2n+2}), \quad \widehat{\mathbb{B}}^* := (\boldsymbol{b}_1^*, \cdots, \boldsymbol{b}_n^*, \boldsymbol{b}_{2n+3}^*, \boldsymbol{b}_{2n+4}^*),$$

$$\mathsf{pp} := (1^\lambda, \mathsf{param}_\mathbb{V}), \quad \mathsf{msk} := \widehat{\mathbb{B}}, \widehat{\mathbb{B}}^*.$$

$\mathsf{KeyGen}(\mathsf{pp}, \mathsf{msk}, \vec{y})$: The key generation algorithm computes and outputs a secret key $\mathsf{sk}_{\vec{y}}$ as

$$\beta, \theta \xleftarrow{\mathsf{U}} \mathbb{Z}_q, \quad \mathsf{sk}_{\vec{y}} := (\vec{y}, 0^n, 0, 0, \beta, \theta, 0)_{\mathbb{B}^*}.$$

$\mathsf{Enc}(\mathsf{pp}, \mathsf{msk}, \vec{x})$: The encryption algorithm computes and outputs a ciphertext $\mathsf{ct}_{\vec{x}}$ as

$$\alpha, \phi \xleftarrow{\mathsf{U}} \mathbb{Z}_q, \quad \mathsf{ct}_{\vec{x}} := (\vec{x}, 0^n, \alpha, \phi, 0, 0, 0)_{\mathbb{B}}.$$

$\mathsf{Dec}(\mathsf{pp}, \mathsf{sk}_{\vec{y}}, \mathsf{ct}_{\vec{x}})$: The decryption algorithm computes the pairing of a secret key and a ciphertext as $d := \tilde{e}(\mathsf{ct}_{\vec{x}}, \mathsf{sk}_{\vec{y}})$. Then it computes m such that $g_T^m = d$ in the polynomial range fixed in advance. If it finds m that satisfies $g_T^m = d$, outputs m. Otherwise, outputs $\perp$.

Correctness: Observe that $d := \tilde{e}(\mathsf{ct}_{\vec{x}}, \mathsf{sk}_{\vec{y}}) = g_T^{\vec{x} \cdot \vec{y}}$. Therefore, m that the decryption algorithm outputs is $\vec{x} \cdot \vec{y}$.

Remark 2. In the schemes of [6,12], we employ two pairs of dual bases, one is used for encrypting vectors and the other is used to encode the same scalars that randomize the corresponding vectors for secret keys and ciphertexts. Because of this construction, it looks like difficult to create another ciphertext from some ciphertexts, while it is easy to do that in our scheme. In other words, our scheme is malleable. However the schemes of [6,12] have not been proven that they are non-malleable. Consequently, our scheme can achieve the same security as [12].

4   Security

4.1   Lemmas for the Security Proof

We consider the following problems and use them to prove the security of our scheme. As mentioned in Sect. 2, we consider type 3 groups.

Definition 5 (Problem 0). *Problem 0 is to guess a bit b, given* $(\mathsf{param}_{\mathsf{P0}},$ $\mathbb{B}, \widehat{\mathbb{B}}^*, \boldsymbol{y}_b, \kappa G_1, \xi G_2)$*, where*

$$\mathsf{param}_{\mathbb{G}} \xleftarrow{\mathsf{R}} \mathcal{G}_{\mathsf{abpg}}(1^\lambda),$$

$$\boldsymbol{B} := (b_{i,j})_{1\le i,j\le 3} \xleftarrow{\mathsf{U}} GL(3, \mathbb{Z}_q), \quad (b^*_{i,j})_{1\le i,j\le 3} := (\boldsymbol{B}^{\mathsf{T}})^{-1},$$

$$\kappa, \xi \xleftarrow{\mathsf{U}} \mathbb{Z}_q^\times, \quad \boldsymbol{b}_i := \kappa \sum_{j=1}^3 b_{i,j}\boldsymbol{a}_j, \quad \boldsymbol{b}^*_i := \xi \sum_{j=1}^3 b^*_{i,j}\boldsymbol{a}^*_j \quad \text{for } i = 1, 2, 3,$$

$$\mathbb{B} := (\boldsymbol{b}_1, \boldsymbol{b}_2, \boldsymbol{b}_3), \quad \widehat{\mathbb{B}}^* := (\boldsymbol{b}^*_1, \boldsymbol{b}^*_3),$$

$$g_T := e(G_1, G_2)^{\kappa\xi}, \quad \mathsf{param}_{\mathsf{P0}} := (\mathsf{param}_{\mathbb{G}}, g_T),$$

$$\delta, \sigma \xleftarrow{\mathsf{U}} \mathbb{Z}_q, \quad \rho \xleftarrow{\mathsf{U}} \mathbb{Z}_q^\times, \quad \boldsymbol{y}_0 := (\delta, 0, \sigma)_\mathbb{B}, \quad \boldsymbol{y}_1 := (\delta, \rho, \sigma)_\mathbb{B}.$$

Definition 6 (Problem 1). *Problem 1 is to guess a bit b, given* $(\mathsf{param}_\mathbb{V},$ $\widehat{\mathbb{B}}, \widehat{\mathbb{B}}^*, \boldsymbol{g}_b)$*, where*

$$(\mathbb{B}, \mathbb{B}^*, \mathsf{param}_\mathbb{V}) \xleftarrow{\mathsf{R}} \mathcal{G}_{\mathsf{ob}}(1^\lambda, 2n + 5),$$

$$\widehat{\mathbb{B}} := (\boldsymbol{b}_1, \cdots, \boldsymbol{b}_{2n}, \boldsymbol{b}_{2n+1}, \boldsymbol{b}_{2n+2}), \quad \widehat{\mathbb{B}}^* := (\boldsymbol{b}^*_1, \cdots, \boldsymbol{b}^*_{2n}, \boldsymbol{b}^*_{2n+3}, \boldsymbol{b}^*_{2n+4}),$$

$$\alpha, \phi \xleftarrow{\mathsf{U}} \mathbb{Z}_q, \quad \tau \xleftarrow{\mathsf{U}} \mathbb{Z}_q^\times,$$

$$\boldsymbol{g}_0 := (0^{2n}, \alpha, \phi, 0, 0, 0)_\mathbb{B}, \quad \boldsymbol{g}_1 := (0^{2n}, \alpha, \phi, 0, 0, \tau)_\mathbb{B}.$$

Definition 7 (Problem 2). *Problem 2 is to guess a bit b, given* $(\mathsf{param}_\mathbb{V},$ $\widehat{\mathbb{B}}, \widehat{\mathbb{B}}^*, \boldsymbol{g}^*_b)$*, where*

$$(\mathbb{B}, \mathbb{B}^*, \mathsf{param}_\mathbb{V}) \xleftarrow{\mathsf{R}} \mathcal{G}_{\mathsf{ob}}(1^\lambda, 2n + 5),$$

$$\widehat{\mathbb{B}} := (\boldsymbol{b}_1, \cdots, \boldsymbol{b}_{2n}, \boldsymbol{b}_{2n+1}, \boldsymbol{b}_{2n+2}), \quad \widehat{\mathbb{B}}^* := (\boldsymbol{b}^*_1, \cdots, \boldsymbol{b}^*_{2n}, \boldsymbol{b}^*_{2n+3}, \boldsymbol{b}^*_{2n+4}),$$

$$\beta, \theta \xleftarrow{\mathsf{U}} \mathbb{Z}_q, \quad \eta \xleftarrow{\mathsf{U}} \mathbb{Z}_q^\times,$$

$$\boldsymbol{g}^*_0 := (0^{2n}, 0, 0, \beta, \theta, 0)_{\mathbb{B}^*}, \quad \boldsymbol{g}^*_1 := (0^{2n}, 0, 0, \beta, \theta, \eta)_{\mathbb{B}^*}.$$

For a PPT algorithm $\mathcal{A}$, the advantage for Problem n ($n = 0, 1, 2$) is defined as

$$\mathsf{Adv}^{\mathsf{P}n}_{\mathcal{A}}(\lambda) := \left| \Pr[\mathcal{A}(1^\lambda, P_0) \to 1] - \Pr[\mathcal{A}(1^\lambda, P_1) \to 1] \right|$$

where P_b is an instance of the Problem n defined above. Then following three lemmas hold.

Lemma 1. *For any PPT adversary $\mathcal{B}$ for Problem 0, there exists a PPT adversary $\mathcal{A}$ for the XDLIN problem such that* $\mathsf{Adv}^{\mathsf{P0}}_{\mathcal{B}}(\lambda) \le \mathsf{Adv}^{\mathsf{XDLIN}}_{\mathcal{A}}(\lambda) + 5/q$.

Lemma 2. $\forall \mathcal{B}, \exists \mathcal{A}, \mathsf{Adv}^{\mathsf{P1}}_{\mathcal{B}}(\lambda) \le \mathsf{Adv}^{\mathsf{P0}}_{\mathcal{A}}(\lambda)$.

Lemma 3. $\forall \mathcal{B}, \exists \mathcal{A}, \mathsf{Adv}^{\mathsf{P2}}_{\mathcal{B}}(\lambda) \le \mathsf{Adv}^{\mathsf{P0}}_{\mathcal{A}}(\lambda)$.

We perform a random linear transformation on $\mathbb{V}$ in the proofs of lemmas. The definition of this operation is the same as Lemma 14 in [19].

Proof (Lemma 1). We show that we can construct a PPT adversary $\mathcal{A}$ for the XDLIN problem from any PPT adversary $\mathcal{B}$ for Problem 0. $\mathcal{A}$ sets $x = 1$ and is given an instance of XDLIN problem. $\mathcal{A}$ sets

$$g_T := e(\kappa G_1, \xi G_2), \quad \mathsf{param}_{\mathsf{P0}} := (\mathsf{param}_{\mathbb{G}}, g_T)$$

$$\boldsymbol{u}_1 := (\xi, 0, 1)_{\mathbb{A}}, \quad \boldsymbol{u}_2 := (0, 0, 1)_{\mathbb{A}}, \quad \boldsymbol{u}_1 := (0, \kappa, 1)_{\mathbb{A}},$$

$$\boldsymbol{u}_1^* := (\kappa, 0, 0)_{\mathbb{A}^*}, \quad \boldsymbol{u}_2^* := (-\kappa, -\xi, \kappa\xi)_{\mathbb{A}^*}, \quad \boldsymbol{u}_1^* := (0, \xi, 0)_{\mathbb{A}^*},$$

$$\boldsymbol{w}_b := (\delta\xi G_1, \sigma\kappa G_1, Y_b).$$

$\mathcal{A}$ can compute $\boldsymbol{u}_1, \boldsymbol{u}_2, \boldsymbol{u}_3, \boldsymbol{u}_1^*, \boldsymbol{u}_3^*$. Then $\mathcal{A}$ generates a random linear transformation W on $\mathbb{G}^3$ and sets

$$\boldsymbol{b}_i := W(\boldsymbol{u}_i) \quad \text{for } i = 1, 2, 3, \quad \boldsymbol{b}_i^* := (W^{-1})^{\mathrm{T}}(\boldsymbol{u}_i^*) \quad \text{for } i = 1, 3,$$

$$\mathbb{B} := (\boldsymbol{b}_1, \boldsymbol{b}_2, \boldsymbol{b}_3), \quad \widehat{\mathbb{B}}^* := (\boldsymbol{b}_1^*, \boldsymbol{b}_3^*), \quad \boldsymbol{y}_b := W(\boldsymbol{w}_b).$$

Then $\mathcal{A}$ gives $(\mathsf{param}_{\mathsf{P0}}, \mathbb{B}, \widehat{\mathbb{B}}^*, \boldsymbol{y}_b, \kappa G_1, \xi G_2)$ to $\mathcal{B}$, and outputs b' if $\mathcal{B}$ outputs b'. If $b = 0$, observe that $\boldsymbol{y}_b = (\delta, 0, \sigma)_{\mathbb{B}}$ when $\kappa, \xi \neq 0$, i.e., except with probability $2/q$. If $b = 1$, $\boldsymbol{y}_b = (\delta, \rho, \sigma)_{\mathbb{B}}$ when $\kappa, \xi, \rho \neq 0$, i.e., except with probability $3/q$. It is the same as an instance of Problem 0.   □

Proof (Lemma 2). We show that we can construct a PPT adversary $\mathcal{A}$ for Problem 0 from any PPT adversary $\mathcal{B}$ for Problem 1. $\mathcal{A}$ is given an instance of Problem 0 $(\mathsf{param}_{\mathsf{P0}}, \mathbb{B}, \widehat{\mathbb{B}}^*, \boldsymbol{y}_b, \kappa G_1, \xi G_2)$. Then $\mathcal{A}$ generates a random linear transformation W on $\mathbb{G}^{2n+5}$, and sets

$\mathsf{param}_{\mathbb{V}} := \mathsf{param}_{\mathsf{P0}},$

$\boldsymbol{d}_i := W(0^{i+2}, \kappa G_1, 0^{2n+2-i}) \quad \text{for } i = 1, \cdots, 2n,$

$\boldsymbol{d}_i := W(0^i, \kappa G_1, 0^{2n+4-i}) \quad \text{for } i = 2n+3, 2n+4,$

$\boldsymbol{d}_{2n+1} := W(\boldsymbol{b}_1, 0^{2n+2}), \quad \boldsymbol{d}_{2n+2} := W(\boldsymbol{b}_3, 0^{2n+2}), \quad \boldsymbol{d}_{2n+5} := W(\boldsymbol{b}_2, 0^{2n+2}),$

$\boldsymbol{d}_i^* := (W^{-1})^{\mathrm{T}}(0^{i+2}, \xi G_2, 0^{2n+2-i}) \quad \text{for } i = 1, \cdots, 2n,$

$\boldsymbol{d}_i^* := (W^{-1})^{\mathrm{T}}(0^i, \xi G_2, 0^{2n+4-i}) \quad \text{for } i = 2n+3, 2n+4,$

$\boldsymbol{d}_{2n+1}^* := (W^{-1})^{\mathrm{T}}(\boldsymbol{b}_1^*, 0^{2n+2}), \quad \boldsymbol{d}_{2n+2}^* := (W^{-1})^{\mathrm{T}}(\boldsymbol{b}_3^*, 0^{2n+2}),$

$\boldsymbol{d}_{2n+5}^* := (W^{-1})^{\mathrm{T}}(\boldsymbol{b}_2^*, 0^{2n+2}), \quad \boldsymbol{h}_b := W(\boldsymbol{y}_b, 0^{2n+2}),$

$\mathbb{D} := (\boldsymbol{d}_1, \cdots, \boldsymbol{d}_{2n+5}), \quad \mathbb{D}^* := (\boldsymbol{d}_1^*, \cdots, \boldsymbol{d}_{2n+5}^*).$

We can see that $(\mathbb{D}, \mathbb{D}^*)$ are dual orthonormal bases. $\mathcal{A}$ does not have $\boldsymbol{b}_2^*$ but it can compute

$$\widehat{\mathbb{D}} := (\boldsymbol{d}_1, \cdots, \boldsymbol{d}_n, \boldsymbol{d}_{2n+1}, \boldsymbol{d}_{2n+2}), \quad \widehat{\mathbb{D}}^* := (\boldsymbol{d}_1^*, \cdots, \boldsymbol{d}_n^*, \boldsymbol{d}_{2n+3}^*, \boldsymbol{d}_{2n+4}^*).$$

Then $\mathcal{A}$ gives $(\mathsf{param}_{\mathbb{V}}, \widehat{\mathbb{D}}, \widehat{\mathbb{D}}^*, \boldsymbol{h}_b)$ to $\mathcal{B}$, and outputs b' if $\mathcal{B}$ outputs b'. We can see that $\boldsymbol{h}_0 := (0^{2n}, \alpha', \phi', 0, 0, 0)_{\mathbb{D}}, \boldsymbol{h}_1 := (0^{2n}, \alpha', \phi', 0, 0, \tau')_{\mathbb{D}}$, where $\alpha' := \delta$, $\phi' := \sigma$ and $\tau' := \rho$. It is the same as an instance of Problem 1.   □

Proof (Lemma 3). The proof of Lemma 3 is similar to that of Lemma 2.   □

4.2   Security Proof of Our Scheme

The proposed scheme has full-hiding security, then the following theorem holds.

Theorem 1. *The proposed Priv-IPVE scheme Π has full-hiding security under the XDLIN assumption. For any PPT adversary $\mathcal{B}$, there exists a PPT adversary $\mathcal{A}$ for the XDLIN problem such that*

$$\mathsf{Adv}_{\mathcal{B}}^{\mathsf{FHS},\Pi}(\lambda) \leq (2q_1 + 4q_2)(\mathsf{Adv}_{\mathcal{A}}^{\mathsf{XDLIN}}(\lambda) + 5/q),$$

where q_1 is a number of $\mathcal{B}$'s secret key queries and q_2 is a number of $\mathcal{B}$'s ciphertext queries.

Proof (Theorem 1). For the proof of Theorem 1, we use a hybrid argument over a series of games that differ in the construction of the challenge ciphertexts and secret keys. The game sequence proceeds as Table 2. It also shows the structure of ciphertexts and secret keys in the end of each game. In the Game 1 sequence, we make $\vec{x}_1$ appear in the domain of $n+1$ to $2n$-th dimensions of each ciphertext one by one. In the Game 2 sequence, for each secret key, we gradually change $\vec{y}_0$ in the domain of 1 to n-th dimensions, into $\vec{y}_1$ in the domain of $n+1$ to $2n$-th dimensions. Now we explain the sequence of the games. We frame a coefficient by a box that was changed from a previous game.

Table 2. Game sequence with structure of ciphertexts and secret keys.

Game	Ciphertexts	Secret keys
Game 0	$(\vec{x}_0, 0^n, \alpha, \phi, 0, 0, 0)_{\mathbb{B}}$	$(\vec{y}_0, 0^n, 0, 0, \beta, \theta, 0)_{\mathbb{B}^*}$
Game 1-1-1		
$\vdots$	$\vdots$	$\vdots$
Game 1-q_2-3	$(\vec{x}_0, \vec{x}_1, \alpha, \phi, 0, 0, 0)_{\mathbb{B}}$	$(\vec{y}_0, 0^n, 0, 0, \beta, \theta, 0)_{\mathbb{B}^*}$
Game 2-1-1		
$\vdots$	$\vdots$	$\vdots$
Game 2-q_1-3	$(\vec{x}_0, \vec{x}_1, \alpha, \phi, 0, 0, 0)_{\mathbb{B}}$	$(0^n, \vec{y}_1, 0, 0, \beta, \theta, 0)_{\mathbb{B}^*}$
Game 3	$(\vec{x}_1, \vec{x}_0, \alpha, \phi, 0, 0, 0)_{\mathbb{B}}$	$(\vec{y}_1, 0^n, 0, 0, \beta, \theta, 0)_{\mathbb{B}^*}$
Game 4	$(\vec{x}_1, 0^n, \alpha, \phi, 0, 0, 0)_{\mathbb{B}}$	$(\vec{y}_1, 0^n, 0, 0, \beta, \theta, 0)_{\mathbb{B}^*}$

Game 0: This game is a original one where the challenger selects 0 as a random bit. Namely, for all $j = 1, \cdots, q_1$ and all $\ell = 1, \cdots, q_2$, the replies to the j-th secret key query for $(\vec{y}_0^{(j)}, \vec{y}_1^{(j)})$ and the ℓ-th ciphertext query for $(\vec{x}_0^{(\ell)}, \vec{x}_1^{(\ell)})$ are

$$\mathsf{sk}_{\vec{y}}^{(j)} := (\vec{y}_0^{(j)}, 0^n, 0, 0, \beta^{(j)}, \theta^{(j)}, 0)_{\mathbb{B}^*},$$
$$\mathsf{ct}_{\vec{x}}^{(\ell)} := (\vec{x}_0^{(\ell)}, 0^n, \alpha^{(\ell)}, \phi^{(\ell)}, 0, 0, 0)_{\mathbb{B}},$$

where $\beta^{(j)}, \theta^{(j)}, \alpha^{(\ell)}, \phi^{(\ell)} \xleftarrow{\mathsf{U}} \mathbb{Z}_q$.

Game 1-μ-1($\mu = 1, \cdots, q_2$): Game 1-0-3 is identical with Game 0. This game is the same as Game 1-(μ-1)-3 except that the reply to the μ-th ciphertext query for $(\vec{x}_0^{(\mu)}, \vec{x}_1^{(\mu)})$ is

$$\mathsf{ct}_{\vec{x}}^{(\mu)} := (\vec{x}_0^{(\mu)}, 0^n, \alpha^{(\mu)}, \phi^{(\mu)}, 0, 0, \boxed{\tau})_{\mathbb{B}},$$

where $\alpha^{(\mu)}, \phi^{(\mu)} \xleftarrow{\mathsf{U}} \mathbb{Z}_q$ and $\tau \xleftarrow{\mathsf{U}} \mathbb{Z}_q^\times$.

Game 1-μ-2($\mu = 1, \cdots, q_2$): This game is the same as Game 1-μ-1 except that the reply to the μ-th ciphertext query for $(\vec{x}_0^{(\mu)}, \vec{x}_1^{(\mu)})$ is

$$\mathsf{ct}_{\vec{x}}^{(\mu)} := (\vec{x}_0^{(\mu)}, \boxed{\vec{x}_1^{(\mu)}}, \alpha^{(\mu)}, \phi^{(\mu)}, 0, 0, \tau)_{\mathbb{B}},$$

where $\alpha^{(\mu)}, \phi^{(\mu)} \xleftarrow{\mathsf{U}} \mathbb{Z}_q$ and $\tau \xleftarrow{\mathsf{U}} \mathbb{Z}_q^\times$.

Game 1-μ-3($\mu = 1, \cdots, q_2$): This game is the same as Game 1-μ-2 except that the reply to the μ-th ciphertext query for $(\vec{x}_0^{(\mu)}, \vec{x}_1^{(\mu)})$ is

$$\mathsf{ct}_{\vec{x}}^{(\mu)} := (\vec{x}_0^{(\mu)}, \vec{x}_1^{(\mu)}, \alpha^{(\mu)}, \phi^{(\mu)}, 0, 0, \boxed{0})_{\mathbb{B}},$$

where $\alpha^{(\mu)}, \phi^{(\mu)} \xleftarrow{\mathsf{U}} \mathbb{Z}_q$.

Game 2-ν-1($\nu = 1, \cdots, q_1$): Game 2-0-3 is identical with Game 1-q_2-3. This game is the same as Game 2-(ν-1)-3 except that the reply to the ν-th secret key query for $(\vec{y}_0^{(\nu)}, \vec{y}_1^{(\nu)})$ is

$$\mathsf{sk}_{\vec{y}}^{(\nu)} := (\vec{y}_0^{(\nu)}, 0^n, 0, 0, \beta^{(\nu)}, \theta^{(\nu)}, \boxed{\eta})_{\mathbb{B}^*},$$

where $\beta^{(\nu)}, \theta^{(\nu)} \xleftarrow{\mathsf{U}} \mathbb{Z}_q$ and $\eta \xleftarrow{\mathsf{U}} \mathbb{Z}_q^\times$.

Game 2-ν-2($\nu = 1, \cdots, q_1$): This game is the same as Game 2-ν-1 except that the reply to the ν-th secret key query for $(\vec{y}_0^{(\nu)}, \vec{y}_1^{(\nu)})$ is

$$\mathsf{sk}_{\vec{y}}^{(\nu)} := (\boxed{0^n, \vec{y}_1^{(\nu)}}, 0, 0, \beta^{(\nu)}, \theta^{(\nu)}, \eta)_{\mathbb{B}^*},$$

where $\beta^{(\nu)}, \theta^{(\nu)} \xleftarrow{\mathsf{U}} \mathbb{Z}_q$ and $\eta \xleftarrow{\mathsf{U}} \mathbb{Z}_q^\times$.

Game 2-ν-3($\nu = 1, \cdots, q_1$): This game is the same as Game 2-ν-2 except that the reply to the ν-th secret key query for $(\vec{y}_0^{(\nu)}, \vec{y}_1^{(\nu)})$ is

$$\mathsf{sk}_{\vec{y}}^{(\nu)} := (0^n, \vec{y}_1^{(\nu)}, 0, 0, \beta^{(\nu)}, \theta^{(\nu)}, \boxed{0})_{\mathbb{B}^*},$$

where $\beta^{(\nu)}, \theta^{(\nu)} \xleftarrow{\mathsf{U}} \mathbb{Z}_q$.

Game 3: This game is the same as Game 2-q_2-3 except that, for all $j = 1, \cdots, q_1$ and all $\ell = 1, \cdots, q_2$, the replies to the j-th secret key query for $(\vec{y}_0^{(j)}, \vec{y}_1^{(j)})$ and the ℓ-th ciphertext query for $(\vec{x}_0^{(\ell)}, \vec{x}_1^{(\ell)})$ are

$$\mathsf{sk}_{\vec{y}}^{(j)} := (\boxed{\vec{y}_1^{(j)}, 0^n}, 0, 0, \beta^{(j)}, \theta^{(j)}, 0)_{\mathbb{B}^*},$$

$$\mathsf{ct}_{\vec{x}}^{(\ell)} := (\boxed{\vec{x}_1^{(\ell)}, \vec{x}_0^{(\ell)}}, \alpha^{(\ell)}, \phi^{(\ell)}, 0, 0, 0)_{\mathbb{B}},$$

where $\beta^{(j)}, \theta^{(j)}, \alpha^{(\ell)}, \phi^{(\ell)} \xleftarrow{\mathsf{U}} \mathbb{Z}_q$.

Game 4: This game is the same as Game 3 except that, for all $\ell = 1, \cdots, q_2$, the replies to the ℓ-th ciphertext query for $(\vec{x}_0^{(\ell)}, \vec{x}_1^{(\ell)})$ are

$$\mathsf{ct}_{\vec{x}}^{(\ell)} := (\vec{x}_1^{(\ell)}, \boxed{0^n}, \alpha^{(\ell)}, \phi^{(\ell)}, 0, 0, 0)_{\mathbb{B}},$$

where $\alpha^{(\ell)}, \phi^{(\ell)} \xleftarrow{\mathsf{U}} \mathbb{Z}_q$. Note that this game is a original one where the challenger selects 1 as a random bit.

Claim 1. *For any PPT distinguisher $\mathcal{B}$ between Game 1-(μ-1)-3 and Game 1-μ-1, there exists a PPT algorithm $\mathcal{A}$ for Problem 1, such that for any security parameter λ,*

$$\left| \Pr[\mathsf{Exp}_{\mathcal{B}}^{\mathsf{Game1}-(\mu-1)-3}(\lambda) \to 1] - \Pr[\mathsf{Exp}_{\mathcal{B}}^{\mathsf{Game1}-\mu-1}(\lambda) \to 1] \right| \leq \mathsf{Adv}_{\mathcal{A}}^{\mathsf{P1}}(\lambda).$$

Proof. We demonstrate that it is possible to construct a PPT algorithm $\mathcal{A}$ for Problem 1 using any PPT distinguisher $\mathcal{B}$ between Game 1-(μ-1)-3 and Game 1-μ-1 as a blackbox. $\mathcal{A}$ takes a role to $\mathcal{B}$ as a challenger of the security game.

1. $\mathcal{A}$ is given a Problem 1 instance $(\mathsf{param}_{\mathbb{V}}, \widehat{\mathbb{B}}, \widehat{\mathbb{B}}^*, \boldsymbol{g}_b)$.
2. $\mathcal{A}$ gives $(1^\lambda, \mathsf{param}_{\mathbb{V}})$ to $\mathcal{B}$ as pp.
3. $\mathcal{A}$ computes $\mathsf{sk}_{\vec{y}}^{(j)}$ using $\widehat{\mathbb{B}}^*$ when $\mathcal{B}$ queries $(\vec{y}_0^{(j)}, \vec{y}_1^{(j)})$, and give it to $\mathcal{B}$.
4. $\mathcal{A}$ computes $\mathsf{ct}_{\vec{x}}^{(\ell)}$ when $\mathcal{B}$ queries $(\vec{x}_0^{(\ell)}, \vec{x}_1^{(\ell)})$ as

$$\mathsf{ct}_{\vec{x}}^{(\ell)} := \sum_{i=1}^{n} x_{0,i}^{(\ell)} \boldsymbol{b}_i + \sum_{i=n+1}^{2n} x_{1,i-n}^{(\ell)} \boldsymbol{b}_i + \alpha^{(\ell)} \boldsymbol{b}_{2n+1} + \phi^{(\ell)} \boldsymbol{b}_{2n+2} \quad \text{if } \ell < \mu,$$

$$\mathsf{ct}_{\vec{x}}^{(\ell)} := \sum_{i=1}^{n} x_{0,i}^{(\ell)} \boldsymbol{b}_i + \boldsymbol{g}_b \quad \text{if } \ell = \mu,$$

$$\mathsf{ct}_{\vec{x}}^{(\ell)} := \sum_{i=1}^{n} x_{0,i}^{(\ell)} \boldsymbol{b}_i + \alpha^{(\ell)} \boldsymbol{b}_{2n+1} + \phi^{(\ell)} \boldsymbol{b}_{2n+2} \quad \text{if } \ell > \mu,$$

where $\alpha^{(\ell)}, \phi^{(\ell)} \xleftarrow{\mathsf{U}} \mathbb{Z}_q$, and give it to $\mathcal{B}$.
5. If $\mathcal{B}$ outputs b', $\mathcal{A}$ outputs b' as it is.

It can be seen that if $b = 0$, $\mathcal{B}$'s view is the same as that in Game 1-(μ-1)-3, and if $b = 1$, $\mathcal{B}$'s view is the same as that in Game 1-μ-1. $\square$

Claim 2. *For any PPT distinguisher $\mathcal{B}$ between Game 1-μ-1 and Game 1-μ-2,*

$$\Pr[\mathsf{Exp}_{\mathcal{B}}^{\mathsf{Game1}-\mu-1}(\lambda) \to 1] = \Pr[\mathsf{Exp}_{\mathcal{B}}^{\mathsf{Game1}-\mu-2}(\lambda) \to 1].$$

Proof. We demonstrate that $\mathcal{B}$'s view in Game 1-μ-1 is the same as that in Game 1-μ-2. For that purpose, we define new bases $(\mathbb{F}, \mathbb{F}^*)$ on $\mathbb{G}^{2n+5}$ such that

$$f_{2n+5} := b_{2n+5} - \sum_{i=n+1}^{2n} \frac{x_{1,i-n}^{(\mu)}}{\tau} b_i,$$

$$f_i^* := b_i^* + \frac{x_{1,i-n}^{(\mu)}}{\tau} b_{2n+5}^* \quad \text{for } i = n+1, \cdots, 2n,$$

$$\mathbb{F} := (b_1, \cdots, b_{2n+4}, f_{2n+5}), \quad \mathbb{F}^* := (b_1^*, \cdots, b_n^*, f_{n+1}^*, \cdots, f_{2n}^*, b_{2n+1}^*, \cdots, b_{2n+5}^*).$$

Observe that $(\mathbb{F}, \mathbb{F}^*)$ are dual orthonormal bases and they are distributed completely at random. Then, the secret keys are

$$\mathsf{sk}_{\vec{y}}^{(j)} := (\vec{y}_0^{(j)}, 0^n, 0, 0, \beta^{(j)}, \theta^{(j)}, 0)_{\mathbb{B}^*} = (\vec{y}_0^{(j)}, 0^n, 0, 0, \beta^{(j)}, \theta^{(j)}, 0)_{\mathbb{F}^*}.$$

On the other hand, the ciphertexts are

$$\mathsf{ct}_{\vec{x}}^{(\mu)} = (\vec{x}_0^{(\mu)}, 0^n, \alpha^{(\mu)}, \phi^{(\mu)}, 0, 0, \tau)_{\mathbb{B}},$$

$$= \sum_{i=1}^{n} x_{0,i}^{(\mu)} b_i + \alpha^{(\mu)} b_{2n+1} + \phi^{(\mu)} b_{2n+2} + \tau b_{2n+5}$$

$$= \sum_{i=1}^{n} x_{0,i}^{(\mu)} b_i + \alpha^{(\mu)} b_{2n+1} + \phi^{(\mu)} b_{2n+2} + \tau \left(f_{2n+5} + \sum_{i=n+1}^{2n} \frac{x_{1,i-n}^{(\mu)}}{\tau} b_i \right)$$

$$= \sum_{i=1}^{n} x_{0,i}^{(\mu)} b_i + \sum_{i=n+1}^{2n} x_{1,i-n}^{(\mu)} b_i + \alpha^{(\mu)} b_{2n+1} + \phi^{(\mu)} b_{2n+2} + \tau f_{2n+5}$$

$$= (\vec{x}_0^{(\mu)}, \vec{x}_1^{(\mu)}, \alpha^{(\mu)}, \phi^{(\mu)}, 0, 0, \tau)_{\mathbb{F}},$$

$$\mathsf{ct}_{\vec{x}}^{(\ell)} = (\vec{x}_0^{(\ell)}, \vec{x}_1^{(\ell)}, \alpha^{(\ell)}, \phi^{(\ell)}, 0, 0, 0)_{\mathbb{B}} = (\vec{x}_0^{(\ell)}, \vec{x}_1^{(\ell)}, \alpha^{(\ell)}, \phi^{(\ell)}, 0, 0, 0)_{\mathbb{F}} \quad \text{if } \ell < \mu,$$

$$\mathsf{ct}_{\vec{x}}^{(\ell)} = (\vec{x}_0^{(\ell)}, 0^n, \alpha^{(\ell)}, \phi^{(\ell)}, 0, 0, 0)_{\mathbb{B}} = (\vec{x}_0^{(\ell)}, 0^n, \alpha^{(\ell)}, \phi^{(\ell)}, 0, 0, 0)_{\mathbb{F}} \quad \text{if } \ell > \mu.$$

Therefore, $\mathcal{B}$'s view in both game is information-theoretically identical. $\square$

Claim 3. *For any PPT distinguisher $\mathcal{B}$ between Game 1-μ-2 and Game 1-μ-3, there exists a PPT algorithm $\mathcal{A}$ for Problem 1, such that for any security parameter λ,*

$$\left| \Pr[\mathsf{Exp}_{\mathcal{B}}^{\mathsf{Game1}-\mu-2}(\lambda) \to 1] - \Pr[\mathsf{Exp}_{\mathcal{B}}^{\mathsf{Game1}-\mu-3}(\lambda) \to 1] \right| \leq \mathsf{Adv}_{\mathcal{A}}^{\mathsf{P1}}(\lambda).$$

Proof. The proof of Claim 3 is the same as that of Claim 1 except that the second equation in Step 4 is $\mathsf{ct}_{\vec{x}}^{(\ell)} := \sum_{i=1}^{n} x_{0,i}^{(\ell)} b_i + \sum_{i=n+1}^{2n} x_{1,i-n}^{(\ell)} b_i + g_b$ if $\ell = \mu$. $\square$

Claim 4. *For any PPT distinguisher $\mathcal{B}$ between Game 2-(ν-1)-3 and Game 2-ν-1, there exists a PPT algorithm $\mathcal{A}$ for Problem 2, such that for any security parameter λ,*

$$\left| \Pr[\mathsf{Exp}_{\mathcal{B}}^{\mathsf{Game2}-(\nu-1)-3}(\lambda) \to 1] - \Pr[\mathsf{Exp}_{\mathcal{B}}^{\mathsf{Game2}-\nu-1}(\lambda) \to 1] \right| \le \mathsf{Adv}_{\mathcal{A}}^{\mathsf{P2}}(\lambda).$$

Proof. We demonstrate that it is possible to construct a PPT algorithm $\mathcal{A}$ for Problem 2 using any PPT distinguisher $\mathcal{B}$ between Game 2-(ν-1)-3 and Game 2-ν-1 as a blackbox. $\mathcal{A}$ takes a role to $\mathcal{B}$ as a challenger of the security game.

1. $\mathcal{A}$ is given a Problem 2 instance $(\mathsf{param}_{\mathbb{V}}, \widehat{\mathbb{B}}, \widehat{\mathbb{B}}^*, \boldsymbol{g}_b^*)$.
2. $\mathcal{A}$ gives $(1^\lambda, \mathsf{param}_{\mathbb{V}})$ to $\mathcal{B}$ as pp.
3. $\mathcal{A}$ computes $\mathsf{sk}_{\vec{y}}^{(j)}$ when $\mathcal{B}$ queries $(\vec{y}_0^{(j)}, \vec{y}_1^{(j)})$ as

$$\mathsf{sk}_{\vec{y}}^{(j)} := \sum_{i=n+1}^{2n} y_{1,i-n}^{(j)} \boldsymbol{b}_i^* + \beta^{(j)} \boldsymbol{b}_{2n+3}^* + \theta^{(j)} \boldsymbol{b}_{2n+4}^* \text{ if } j < \nu,$$

$$\mathsf{sk}_{\vec{y}}^{(j)} := \sum_{i=1}^{n} y_{0,i}^{(j)} \boldsymbol{b}_i^* + \boldsymbol{g}_b^* \text{ if } j = \nu,$$

$$\mathsf{sk}_{\vec{y}}^{(j)} := \sum_{i=1}^{n} y_{0,i}^{(j)} \boldsymbol{b}_i^* + \beta^{(j)} \boldsymbol{b}_{2n+3}^* + \theta^{(j)} \boldsymbol{b}_{2n+4}^* \text{ if } j > \nu,$$

where $\beta^{(j)}, \theta^{(j)} \xleftarrow{\mathsf{U}} \mathbb{Z}_q$, and give it to $\mathcal{B}$.
4. $\mathcal{A}$ computes $\mathsf{ct}_{\vec{x}}^{(\ell)}$ of the Game 2 form using $\widehat{\mathbb{B}}$ when $\mathcal{B}$ queries $(\vec{x}_0^{(\ell)}, \vec{x}_1^{(\ell)})$, and give it to $\mathcal{B}$.
5. If $\mathcal{B}$ outputs b', $\mathcal{A}$ outputs b' as it is.

It can be seen that if $b = 0$, $\mathcal{B}$'s view is the same as that in Game 2-(ν-1)-3, and if $b = 1$, $\mathcal{B}$'s view is the same as that in Game 2-ν-1. $\qquad\qquad\square$

Claim 5. *For any PPT distinguisher $\mathcal{B}$ between Game 2-ν-1 and Game 2-ν-2,*

$$\Pr[\mathsf{Exp}_{\mathcal{B}}^{\mathsf{Game2}-\nu-1}(\lambda) \to 1] = \Pr[\mathsf{Exp}_{\mathcal{B}}^{\mathsf{Game2}-\nu-2}(\lambda) \to 1].$$

Proof. We demonstrate that $\mathcal{B}$'s view in Game 2-ν-1 is the same as that in Game 2-ν-2. For that purpose, we define new bases $(\mathbb{F}, \mathbb{F}^*)$ on $\mathbb{G}^{2n+5}$ such that

$$\boldsymbol{f}_i := \boldsymbol{b}_i - \frac{y_{0,i}^{(\nu)}}{\eta} \boldsymbol{b}_{2n+5} \text{ for } i = 1, \cdots, n,$$

$$\boldsymbol{f}_i := \boldsymbol{b}_i + \frac{y_{1,i-n}^{(\nu)}}{\eta} \boldsymbol{b}_{2n+5} \text{ for } i = n+1, \cdots, 2n,$$

$$\boldsymbol{f}_{2n+5}^* := \boldsymbol{b}_{2n+5}^* + \sum_{i=1}^{n} \frac{y_{0,i}^{(\nu)}}{\eta} \boldsymbol{b}_i^* - \sum_{i=n+1}^{2n} \frac{y_{1,i-n}^{(\nu)}}{\eta} \boldsymbol{b}_i^*,$$

$$\mathbb{F} := (\boldsymbol{f}_1, \cdots, \boldsymbol{f}_{2n}, \boldsymbol{b}_{2n+1}, \cdots, \boldsymbol{b}_{2n+5}), \quad \mathbb{F}^* := (\boldsymbol{b}_1^*, \cdots, \boldsymbol{b}_{2n+4}^*, \boldsymbol{f}_{2n+5}^*).$$

Observe that $(\mathbb{F}, \mathbb{F}^*)$ are dual orthonormal bases and they are distributed completely at random. Then, the secret keys are

$$\mathsf{sk}_{\vec{y}}^{(\nu)} = (\vec{y}_0^{(\nu)}, 0^n, 0, 0, \beta^{(\nu)}, \theta^{(\nu)}, \eta)_{\mathbb{B}^*},$$

$$= \sum_{i=1}^{n} y_{0,i}^{(\nu)} \boldsymbol{b}_i^* + \beta^{(\nu)} \boldsymbol{b}_{2n+3}^* + \theta^{(\nu)} \boldsymbol{b}_{2n+4}^* + \eta \boldsymbol{b}_{2n+5}^*$$

$$= \sum_{i=1}^{n} y_{0,i}^{(\nu)} \boldsymbol{b}_i^* + \beta^{(\nu)} \boldsymbol{b}_{2n+3}^* + \theta^{(\nu)} \boldsymbol{b}_{2n+4}^* + \eta \left(\boldsymbol{f}_{2n+5}^* - \sum_{i=1}^{n} \frac{y_{0,i}^{(\nu)}}{\eta} \boldsymbol{b}_i^* + \sum_{i=n+1}^{2n} \frac{y_{1,i-n}^{(\nu)}}{\eta} \boldsymbol{b}_i^* \right)$$

$$= \sum_{i=n+1}^{2n} y_{1,i-n}^{(\nu)} \boldsymbol{b}_i^* + \beta^{(\nu)} \boldsymbol{b}_{2n+3}^* + \theta^{(\nu)} \boldsymbol{b}_{2n+4}^* + \eta \boldsymbol{f}_{2n+5}^*$$

$$= (0^n, \vec{y}_1^{(\nu)}, 0, 0, \beta^{(\nu)}, \theta^{(\nu)}, \eta)_{\mathbb{F}^*},$$

$$\mathsf{sk}_{\vec{y}}^{(j)} = (0^n, \vec{y}_1^{(j)}, 0, 0, \beta^{(j)}, \theta^{(j)}, 0)_{\mathbb{B}^*} = (0^n, \vec{y}_1^{(j)}, 0, 0, \beta^{(j)}, \theta^{(j)}, 0)_{\mathbb{F}^*} \text{ if } j < \nu,$$

$$\mathsf{sk}_{\vec{y}}^{(j)} = (\vec{y}_0^{(j)}, 0^n, 0, 0, \beta^{(j)}, \theta^{(j)}, 0)_{\mathbb{B}^*} = (\vec{y}_0^{(j)}, 0^n, 0, 0, \beta^{(j)}, \theta^{(j)}, 0)_{\mathbb{F}^*} \text{ if } j > \nu.$$

On the other hand, the ciphertexts are

$$\mathsf{ct}_{\vec{x}}^{(\ell)} = (\vec{x}_0^{(\ell)}, \vec{x}_1^{(\ell)}, \alpha^{(\ell)}, \phi^{(\ell)}, 0, 0, 0)_{\mathbb{B}}$$

$$= \sum_{i=1}^{n} x_{0,i}^{(\ell)} \boldsymbol{b}_i + \sum_{i=n+1}^{2n} x_{1,i-n}^{(\ell)} \boldsymbol{b}_i + \alpha^{(\ell)} \boldsymbol{b}_{2n+1} + \phi^{(\ell)} \boldsymbol{b}_{2n+2}$$

$$= \sum_{i=1}^{n} x_{0,i}^{(\ell)} \left(\boldsymbol{f}_i + \frac{y_{0,i}^{(\nu)}}{\eta} \boldsymbol{b}_{2n+5} \right) + \sum_{i=n+1}^{2n} x_{1,i-n}^{(\ell)} \left(\boldsymbol{f}_i - \frac{y_{1,i-n}^{(\nu)}}{\eta} \boldsymbol{b}_{2n+5} \right) + \alpha^{(\ell)} \boldsymbol{b}_{2n+1} + \phi^{(\ell)} \boldsymbol{b}_{2n+2}$$

$$= \sum_{i=1}^{n} x_{0,i}^{(\ell)} \boldsymbol{f}_i + \sum_{i=n+1}^{2n} x_{1,i-n}^{(\ell)} \boldsymbol{f}_i + \frac{\vec{x}_0^{(\ell)} \cdot \vec{y}_0^{(\nu)} - \vec{x}_1^{(\ell)} \cdot \vec{y}_1^{(\nu)}}{\eta} \boldsymbol{b}_{2n+5} + \alpha^{(\ell)} \boldsymbol{b}_{2n+1} + \phi^{(\ell)} \boldsymbol{b}_{2n+2}$$

$$= \sum_{i=1}^{n} x_{0,i}^{(\ell)} \boldsymbol{f}_i + \sum_{i=n+1}^{2n} x_{1,i-n}^{(\ell)} \boldsymbol{f}_i + \alpha^{(\ell)} \boldsymbol{b}_{2n+1} + \phi^{(\ell)} \boldsymbol{b}_{2n+2}$$

$$= (\vec{x}_0^{(\ell)}, \vec{x}_1^{(\ell)}, \alpha^{(\ell)}, \phi^{(\ell)}, 0, 0, 0)_{\mathbb{F}},$$

because $\vec{x}_0^{(\ell)} \cdot \vec{y}_0^{(j)} = \vec{x}_1^{(\ell)} \cdot \vec{y}_1^{(j)}$ for all j and ℓ. Therefore, $\mathcal{B}$'s view in both game is information-theoretically identical. $\square$

Claim 6. *For any PPT distinguisher $\mathcal{B}$ between Game 2-ν-2 and Game 2-ν-3, there exists a PPT algorithm $\mathcal{A}$ for Problem 2, such that for any security parameter λ,*

$$\left| \Pr[\mathsf{Exp}_{\mathcal{B}}^{\mathsf{Game2}-\nu-2}(\lambda) \to 1] - \Pr[\mathsf{Exp}_{\mathcal{B}}^{\mathsf{Game2}-\nu-3}(\lambda) \to 1] \right| \leq \mathsf{Adv}_{\mathcal{A}}^{\mathsf{P2}}(\lambda).$$

Proof. The proof of Claim 6 is the same as that of Claim 4 except that the second equation in Step 3 is $\mathsf{sk}_{\vec{y}}^{(j)} := \sum_{i=n+1}^{2n} y_{1,i-1}^{(j)} \boldsymbol{b}_i^* + \boldsymbol{g}_b^*$ if $j = \nu$. $\square$

Claim 7. *For any PPT distinguisher $\mathcal{B}$ between Game 2-q_1-3 and Game 3,*

$$\Pr[\mathsf{Exp}_{\mathcal{B}}^{\mathsf{Game2}-q_1-1}(\lambda) \to 1] = \Pr[\mathsf{Exp}_{\mathcal{B}}^{\mathsf{Game3}}(\lambda) \to 1].$$

Proof. We define new bases $(\mathbb{F}, \mathbb{F}^*)$ on $\mathbb{G}^{2n+5}$ as

$$\boldsymbol{f}_i := \boldsymbol{b}_{n+i}, \quad \boldsymbol{f}_i^* := \boldsymbol{b}_{n+i}^*, \quad \boldsymbol{f}_{n+i} := \boldsymbol{b}_i, \quad \boldsymbol{f}_{n+i}^* := \boldsymbol{b}_i^* \quad \text{for } i = 1, \cdots, n,$$

$$\mathbb{F} := (\boldsymbol{f}_1, \cdots, \boldsymbol{f}_{2n}, \boldsymbol{b}_{2n+1}, \cdots, \boldsymbol{b}_{2n+5}), \quad \mathbb{F}^* := (\boldsymbol{f}_1^*, \cdots, \boldsymbol{f}_{2n}^*, \boldsymbol{b}_{2n+1}^*, \cdots, \boldsymbol{b}_{2n+5}^*).$$

These bases are dual orthonormal. Then we can easily see that

$$\mathsf{sk}_{\vec{y}}^{(j)} = (0^n, \vec{y}_1^{(j)}, 0, 0, \beta^{(j)}, \theta^{(j)}, 0)_{\mathbb{B}^*} = (\vec{y}_1^{(j)}, 0^n, 0, 0, \beta^{(j)}, \theta^{(j)}, 0)_{\mathbb{F}^*},$$

$$\mathsf{ct}_{\vec{x}}^{(\ell)} = (\vec{x}_0^{(\ell)}, \vec{x}_1^{(\ell)}, \alpha^{(\ell)}, \phi^{(\ell)}, 0, 0, 0)_{\mathbb{B}} = (\vec{x}_1^{(\ell)}, \vec{x}_0^{(\ell)}, \alpha^{(\ell)}, \phi^{(\ell)}, 0, 0, 0)_{\mathbb{F}}.$$

This is just a conceptual change and $\mathcal{B}$'s view in the both game is the same. $\square$

Claim 8. *For any PPT distinguisher $\mathcal{B}$ between Game 3 and Game 4,*

$$\left| \Pr[\mathsf{Exp}_{\mathcal{B}}^{\mathsf{Game3}}(\lambda) \to 1] - \Pr[\mathsf{Exp}_{\mathcal{B}}^{\mathsf{Game4}}(\lambda) \to 1] \right|$$

$$= \left| \Pr[\mathsf{Exp}_{\mathcal{B}}^{\mathsf{Game0}}(\lambda) \to 1] - \Pr[\mathsf{Exp}_{\mathcal{B}}^{\mathsf{Game1-q_2-3}}(\lambda) \to 1] \right|$$

Proof. The difference between Game 3 and Game 4 is the same as that between Game 0 and Game 1-q_2-3, just switching $\vec{x}_0^{(\ell)}$ with $\vec{x}_1^{(\ell)}$, and $\vec{y}_0^{(j)}$ with $\vec{y}_1^{(j)}$. $\square$

From Claims 1, $\cdots$, 8 and Lemmas 1, 2 and 3, Theorem 1 holds. $\square$

References

1. Abdalla, M., Bourse, F., De Caro, A., Pointcheval, D.: Simple functional encryption schemes for inner products. In: Katz, J. (ed.) PKC 2015. LNCS, vol. 9020, pp. 733–751. Springer, Heidelberg (2015)
2. Abdalla, M., Bourse, F., Caro, A.D., Pointcheval, D.: Better security for functional encryption for inner product evaluations. Cryptology ePrint Archive, Report 2016/11 (2016)
3. Abe, M., Chase, M., David, B., Kohlweiss, M., Nishimaki, R., Ohkubo, M.: Constant-size structure-preserving signatures: generic constructions and simple assumptions. In: Wang, X., Sako, K. (eds.) ASIACRYPT 2012. LNCS, vol. 7658, pp. 4–24. Springer, Heidelberg (2012)
4. Agrawal, S., Agrawal, S., Badrinarayanan, S., Kumarasubramanian, A., Prabhakaran, M., Sahai, A.: Function private functional encryption and property preserving encryption: new definitions and positive results. Cryptology ePrint Archive, Report 2013/744 (2013)
5. Agrawal, S., Libert, B., Stehle, D.: Fully secure functional encryption for inner products, from standard assumptions. Cryptology ePrint Archive, Report 2015/608 (2015)
6. Bishop, A., Jain, A., Kowalczyk, L.: Function-hiding inner product encryption. Cryptology ePrint Archive, Report 2015/672 (2015)
7. Boneh, D., Franklin, M.: Identity-based encryption from the weil pairing. In: Kilian, J. (ed.) CRYPTO 2001. LNCS, vol. 2139, pp. 213–229. Springer, Heidelberg (2001)
8. Boneh, D., Sahai, A., Waters, B.: Functional encryption: definitions and challenges. In: Ishai, Y. (ed.) TCC 2011. LNCS, vol. 6597, pp. 253–273. Springer, Heidelberg (2011)

9. Brakerski, Z., Segev, G.: Function-private functional encryption in the private-key setting. In: Dodis, Y., Nielsen, J.B. (eds.) TCC 2015, Part II. LNCS, vol. 9015, pp. 306–324. Springer, Heidelberg (2015)

10. Boneh, D., Raghunathan, A., Segev, G.: Function-private identity-based encryption: hiding the function in functional encryption. In: Canetti, R., Garay, J.A. (eds.) CRYPTO 2013, Part II. LNCS, vol. 8043, pp. 461–478. Springer, Heidelberg (2013)

11. Boneh, D., Raghunathan, A., Segev, G.: Function-private subspace-membership encryption and its applications. In: Sako, K., Sarkar, P. (eds.) ASIACRYPT 2013, Part I. LNCS, vol. 8269, pp. 255–275. Springer, Heidelberg (2013)

12. Datta, P., Dutta, R., Mukhopadhyay, S.: Functional encryption for inner product with full function privacy. Cryptology ePrint Archive, Report 2015/1255 (2015)

13. Garg, S., Gentry, C., Halevi, S., Raykova, M., Sahai, A., Waters, B.: Candidate indistinguishability obfuscation and functional encryption for all circuits. In: FOCS, pp. 40–49. IEEE (2013)

14. Garg, S., Gentry, C., Halevi, S., Zhandry, M.: Fully secure functional encryption without obfuscation. Cryptology ePrint Archive, Report 2014/666 (2014)

15. Galbraith, S.D., Paterson, K.G., Smart, N.P.: Pairings for cryptographers. Discrete Appl. Math. **156**(16), 3113–3121 (2008)

16. Katz, J., Sahai, A., Waters, B.: Predicate encryption supporting disjunctions, polynomial equations, and inner products. In: Smart, N.P. (ed.) EUROCRYPT 2008. LNCS, vol. 4965, pp. 146–162. Springer, Heidelberg (2008)

17. Lewko, A., Okamoto, T., Sahai, A., Takashima, K., Waters, B.: Fully secure functional encryption: attribute-based encryption and (hierarchical) inner product encryption. In: Gilbert, H. (ed.) EUROCRYPT 2010. LNCS, vol. 6110, pp. 62–91. Springer, Heidelberg (2010)

18. Okamoto, T., Takashima, K.: Hierarchical predicate encryption for inner-products. In: Matsui, M. (ed.) ASIACRYPT 2009. LNCS, vol. 5912, pp. 214–231. Springer, Heidelberg (2009)

19. Okamoto, T., Takashima, K.: Fully secure functional encryption with general relations from the decisional linear assumption. In: Rabin, T. (ed.) CRYPTO 2010. LNCS, vol. 6223, pp. 191–208. Springer, Heidelberg (2010)

20. Okamoto, T., Takashima, K.: Adaptively attribute-hiding (hierarchical) inner product encryption. In: Pointcheval, D., Johansson, T. (eds.) EUROCRYPT 2012. LNCS, vol. 7237, pp. 591–608. Springer, Heidelberg (2012)

21. ONeill, A.: Definitional issues in functional encryption. Cryptology ePrint Archive, Report 2010/556 (2010)

22. Ostrovsky, R., Sahai, A., Waters, B.: Attribute-based encryption with non-monotonic access structures. In: ACM Conference on Computer and Communications Security, pp. 195–203. ACM (2007)

MQSAS - A Multivariate Sequential Aggregate Signature Scheme

Rachid El Bansarkhani[1], Mohamed Saied Emam Mohamed[1(✉)],
and Albrecht Petzoldt[2]

[1] Technische Universität Darmstadt, Darmstadt, Germany
{elbansarkhani,mohamed}@cdc.informatik.tu-darmstadt.de
[2] Kyushu University, Fukuoka, Japan
petzoldt@imi.kyushu-u.ac.jp

Abstract. (Sequential) Aggregate signature schemes enable a group of users $u_1, \ldots, u_k$ with messages $m_1, \ldots, m_k$ to produce a single signature Σ which states the integrity and authenticity of all the messages $m_1, \ldots, m_k$. The length of the signature Σ is thereby significantly shorter than a concatenation of individual signatures. Therefore, aggregate signatures can improve the efficiency of numerous applications, e.g. the BGPsec protocol of Internet routing and the development of new efficient aggregate signature schemes is an important task for cryptographic research. On the other hand, most of the existing schemes for aggregate signatures are based on number theoretic problems and therefore become insecure as soon as large enough quantum computers come into existence. In this paper, we propose a technique to extend multivariate signature schemes such as HFEv- to sequential aggregate signature schemes. By doing so, we create the first multivariate signature scheme of this kind, which is, at the same time, also one of the first post-quantum aggregate signature schemes. Our scheme is very efficient and offers compression rates that outperform current lattice-based constructions for practical parameters.

Keywords: Sequential aggregate signatures · Multivariate cryptography · HFEv-

1 Introduction

(Sequential) aggregate signature schemes enable a group of users $U = \{u_1, \ldots, u_k\}$, each of them having a message m_i to be signed, to generate a single signature Σ which proofs the integrity and authenticity of all the messages $m_1, \ldots, m_k$. The key point hereby is that the size of the aggregate signature Σ is much smaller than a concatenation of the individual signatures. Therefore, (sequential) aggregate signature schemes have a great deal of applications such as the BGPsec protocol [7], which plays an important role in securing the global Internet routing system. In this protocol, each node on a certain path of n hops receives n certificates and the same amount of signatures. It then verifies

© Springer International Publishing Switzerland 2016
M. Bishop and A.C.A. Nascimento (Eds.): ISC 2016, LNCS 9866, pp. 426–439, 2016.
DOI: 10.1007/978-3-319-45871-7_25

the signatures, creates its own signature attesting for this path and sends this result together with the previous signatures to the next hop. As a consequence, the number of certificates and signatures increases linearly with the number of nodes on this path. This amount of bandwidth cost can be reduced drastically by the use of a sequential aggregate signature scheme. Similar ideas can in general be applied to public key infrastructures of any depth requiring chains of certificates and signatures in order to authenticate public keys at the leafs. Such schemes come always into use, when chains and paths need to be authenticated as a condition for the protocol to work.

The currently available solutions for (sequential) aggregate signature schemes are mainly based on classical cryptosystems such as RSA and ECC. However, these schemes will become insecure as soon as large quantum computers arrive. The reason for this is Shor's algorithm [11], which solves number theoretic problems like integer factorization and discrete logarithm in polynomial time on a quantum computer. It is therefore an important task to develop (sequential) aggregate signature schemes whose security is based on hard mathematical problems not affected by quantum computer attacks (so called post-quantum cryptosystems). Besides lattice, code and hash based cryptosystems, multivariate cryptography is one of the main candidates for this [1].

In this paper we show how to extend multivariate signature schemes to sequential aggregate signature schemes. While our technique can be applied to arbitrary multivariate schemes, we mainly concentrate on the HFEv- signature scheme, which is one of the best known and most studied multivariate schemes. Our scheme is the first multivariate and one of the first post-quantum (sequential) aggregate signature schemes and enables high compression rates and therefore very short sizes of the aggregate signature. Furthermore, with regard to its performance, our scheme outperforms current lattice-based constructions [3].

The rest of this paper is organized as follows. In Sect. 2 we repeat the basic concepts of (sequential) aggregate signatures. Section 3 gives an overview of the area of multivariate cryptography and introduces the HFEv- signature scheme, which is the basis of our construction. In Sect. 4 we then present our technique to extend HFEv- to a multivariate sequential aggregate signature scheme. whereas Sect. 5 deals with the security of our construction. Section 6 gives concrete parameter sets for our scheme and compares it with other existing (sequential) aggregate signature schemes. Finally, Sect. 7 concludes the paper.

2   Sequential Aggregate Signatures

In this section we describe the concept of (sequential) aggregate signature schemes. Let $U = \{u_1, \ldots, u_k\}$ be a set of users participating in the protocol, each of them having a message m_i to be signed as well as a key pair $(\mathsf{sk}_i, \mathsf{pk}_i)$ of a digital signature scheme. Each user u_i applies his private key sk_i to generate a signature σ_i for a message m_i (the messages m_i are not necessarily distinct). Let us assume that the k users u_i $(i = 1, \ldots, k)$ desire to prove to a single verifier V that each user u_i indeed signed his message m_i. This could be accomplished by sending all the messages m_i $(i = 1, \ldots, k)$ and a signature $\tilde{\sigma} = (\sigma_1, \ldots, \sigma_k)$ to the verifier V. However, by doing so, the length of $\tilde{\sigma}$ to be transmitted becomes very large if the number k of users increases.

An alternative way of proving the integrity and authenticity of the messages $m_1, \ldots, m_k$ is to combine all the signatures σ_i to a single signature Σ with $|\Sigma| \ll |\tilde{\sigma}|$. To achieve this, one can use a (sequential) aggregate signature scheme.

The signature generation process of a sequential aggregate signature scheme is an iterative process between the users $u_1, \ldots, u_k$ (see Fig. 1).

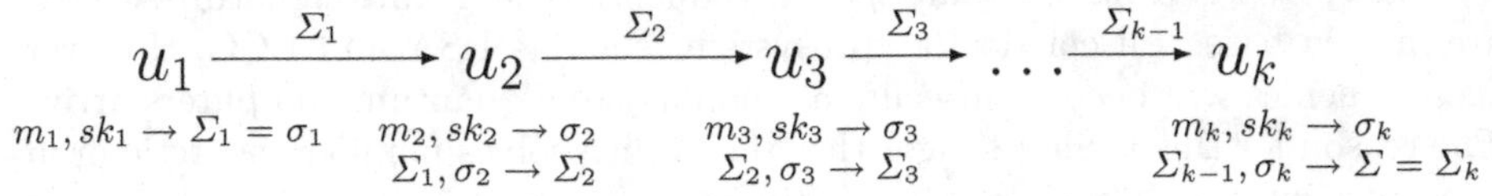

$$u_1 \xrightarrow{\Sigma_1} u_2 \xrightarrow{\Sigma_2} u_3 \xrightarrow{\Sigma_3} \ldots \xrightarrow{\Sigma_{k-1}} u_k$$

$$\begin{array}{llll} m_1, sk_1 \to \Sigma_1 = \sigma_1 & m_2, sk_2 \to \sigma_2 & m_3, sk_3 \to \sigma_3 & m_k, sk_k \to \sigma_k \\ & \Sigma_1, \sigma_2 \to \Sigma_2 & \Sigma_2, \sigma_3 \to \Sigma_3 & \Sigma_{k-1}, \sigma_k \to \Sigma = \Sigma_k \end{array}$$

Fig. 1. Generation of a sequential aggregate signature

The first signer generates a standard signature σ_1 for his message m_1, while the second signer generates a signature on the combination of his message m_2 and σ_1 to obtain an aggregate signature Σ_2 for both the messages m_1 and m_2. Hereby, the goal is to hide a big portion of σ_1 in Σ_2 in such a way that σ_1 can be recovered during the verification. By doing so, we ensure that it is still possible to validate the message m_1. This step is repeated for the signers $u_3, \ldots, u_k$. The last signer produces the final signature $\Sigma = \Sigma_k$, which is now an aggregate signature for all the messages $m_1, \ldots, m_k$.

Compression Rate. Let $|\sigma_i|$ be the size of an individual signature σ_i $(i = 1, \ldots, k)$ and $|\Sigma|$ be the size of the (sequential) aggregate signature Σ. Following [3], we define the compression rate of the aggregate signature scheme by

$$\tau(k) = 1 - \frac{|\Sigma|}{\sum_{i=1}^{k} |\sigma_i|}. \tag{1}$$

The size ratio τ expresses therefore the amount of memory that has been saved due to the use of the aggregate signature scheme. A value of $\tau = 0$ corresponds to an aggregate signature Σ which is as long as the concatenation of all the individual signatures (i.e. no compression at all). A value of $\tau = 1 - \frac{1}{k}$ expresses that the aggregate signature Σ has the size of an individual signature, which corresponds to an optimal aggregate signature scheme.

3   The HFEv- Signature Scheme

In this section we review the HFEv- signature scheme, which is the basis of our construction. Before we give a detailed description of the scheme itself, we start with a short overview of the basic concepts of multivariate cryptography.

3.1   Multivariate Cryptography

The basic objects of multivariate cryptography are systems of multivariate quadratic polynomials (see Eq. (2)).

$$p^{(1)}(x_1,\ldots,x_n) = \sum_{i=1}^{n}\sum_{j=i}^{n} p_{ij}^{(1)} \cdot x_i x_j + \sum_{i=1}^{n} p_i^{(1)} \cdot x_i + p_0^{(1)}$$

$$p^{(2)}(x_1,\ldots,x_n) = \sum_{i=1}^{n}\sum_{j=i}^{n} p_{ij}^{(2)} \cdot x_i x_j + \sum_{i=1}^{n} p_i^{(2)} \cdot x_i + p_0^{(2)}$$

$$\vdots$$

$$p^{(m)}(x_1,\ldots,x_n) = \sum_{i=1}^{n}\sum_{j=i}^{n} p_{ij}^{(m)} \cdot x_i x_j + \sum_{i=1}^{n} p_i^{(m)} \cdot x_i + p_0^{(m)} \tag{2}$$

The security of multivariate schemes is based on the MQ problem.

Definition 1 (MQ Problem). *Given m multivariate quadratic polynomials $p^{(1)}(\mathbf{x}),\ldots,p^{(m)}(\mathbf{x})$ in n variables $x_1,\ldots,x_n$ as shown in Eq. (2), find a vector $\bar{\mathbf{x}} = (\bar{x}_1,\ldots,\bar{x}_n)$ such that $p^{(1)}(\bar{\mathbf{x}}) = \ldots = p^{(m)}(\bar{\mathbf{x}}) = 0$.*

The MQ problem (for $m \approx n$) is proven to be NP-hard even for quadratic polynomials over the field GF(2) [6].

To build a public key cryptosystem based on the MQ problem, one starts with an easily invertible quadratic map $\mathcal{F} : \mathbb{F}^n \to \mathbb{F}^m$ (central map). To hide the structure of $\mathcal{F}$ in the public key, one composes it with two invertible affine (or linear) maps $\mathcal{S} : \mathbb{F}^m \to \mathbb{F}^m$ and $\mathcal{T} : \mathbb{F}^n \to \mathbb{F}^n$. The *public key* is therefore given by $\mathcal{P} = \mathcal{S} \circ \mathcal{F} \circ \mathcal{T}$. The *private key* consists of $\mathcal{S}$, $\mathcal{F}$ and $\mathcal{T}$ and therefore allows to invert the public key. We note that, due to the above construction, the security of multivariate schemes is not only based on the MQ-Problem but also on the EIP-Problem ("Extended Isomorphism of Polynomials") of finding the composition of $\mathcal{P}$.

In this paper we focus on multivariate signature schemes of the BigField family. For this type of multivariate schemes, the map $\mathcal{F}$ is a specially chosen and easily invertible map over a degree n extension field $\mathbb{E}$ of $\mathbb{F}$. One uses an isomorphism $\Phi : \mathbb{F}^n \to \mathbb{E}$ to transform $\mathcal{F}$ into a quadratic map

$$\bar{\mathcal{F}} = \Phi^{-1} \circ \mathcal{F} \circ \Phi \tag{3}$$

from $\mathbb{F}^n$ to itself. The public key of the scheme is therefore given by

$$\mathcal{P} = \mathcal{S} \circ \bar{\mathcal{F}} \circ \mathcal{T} = \mathcal{S} \circ \Phi^{-1} \circ \mathcal{F} \circ \Phi \circ \mathcal{T} : \mathbb{F}^n \to \mathbb{F}^n. \tag{4}$$

with two invertible affine maps $\mathcal{S}, \mathcal{T} : \mathbb{F}^n \to \mathbb{F}^n$.

The standard signature generation and verification process of a multivariate BigField scheme works as shown in Fig. 2.

Signature Generation: To generate a signature for a message $\mathbf{h} \in \mathbb{F}^n$, one computes recursively $\mathbf{x} = \mathcal{S}^{-1}(\mathbf{h}) \in \mathbb{F}^n$, $X = \Phi(\mathbf{x}) \in \mathbb{E}$, $Y = \mathcal{F}^{-1}(X) \in \mathbb{E}$,

Signature Generation

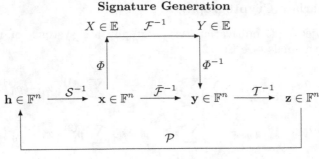

Signature Verification

Fig. 2. General workflow of multivariate BigField signature schemes

$\mathbf{y} = \Phi^{-1}(Y) \in \mathbb{F}^n$ and $\mathbf{z} = \mathcal{T}^{-1}(\mathbf{y})$. The signature of the message $\mathbf{h}$ is given by $\mathbf{z} \in \mathbb{F}^n$.

Verification: To check the authenticity of a signature $\mathbf{z} \in \mathbb{F}^n$, one simply computes $\mathbf{h}' = \mathcal{P}(\mathbf{z}) \in \mathbb{F}^n$. If $\mathbf{h}' = \mathbf{h}$ holds, the signature is accepted, otherwise rejected.

3.2   HFEv-

A famous example for a multivariate signature scheme from the BigField family is the HFEv- signature scheme [8] which uses, additionally to the BigField structure, the Minus and the Vinegar modification for multivariate schemes. The scheme can be described as follows. Let $\mathbb{F} = \mathbb{F}_q$ be a finite field with q elements and $\mathbb{E}$ be a degree n extension field of $\mathbb{F}$. Furthermore, we choose integers D, a and v. Let Φ be the canonical isomorphism between $\mathbb{F}^n$ and $\mathbb{E}$, i.e.

$$\Phi(x_1, \ldots, x_n) = \sum_{i=1}^{n} x_i \cdot X^{i-1}. \tag{5}$$

The central map $\mathcal{F}$ of the HFEv- scheme is a map from $\mathbb{E} \times \mathbb{F}^v$ to $\mathbb{E}$ of the form

$$\mathcal{F}(X) = \sum_{\substack{0 \leq i \leq j \\ q^i + q^j \leq D}} \alpha_{ij} \cdot X^{q^i + q^j} + \sum_{i=0}^{q^i \leq D} \beta_i(v_1, \ldots, v_v) \cdot X^{q^i} + \gamma(v_1, \ldots, v_v), \tag{6}$$

with $\alpha_{ij} \in \mathbb{E}$, $\beta_i : \mathbb{F}^v \to \mathbb{E}$ being linear and $\gamma : \mathbb{F}^v \to \mathbb{E}$ being a quadratic function.

Due to the special form of $\mathcal{F}$, the map $\bar{\mathcal{F}} = \Phi^{-1} \circ \mathcal{F} \circ (\Phi \times \mathrm{id}_v)$ is a quadratic polynomial map from $\mathbb{F}^{n+v}$ to $\mathbb{F}^n$. To hide the structure of $\bar{\mathcal{F}}$ in the public key, one composes it with two affine (or linear) maps $\mathcal{S} : \mathbb{F}^n \to \mathbb{F}^{n-a}$ and $\mathcal{T} : \mathbb{F}^{n+v} \to \mathbb{F}^{n+v}$ of maximal rank.

The *public key* of the scheme is the composed map $\mathcal{P} = \mathcal{S} \circ \bar{\mathcal{F}} \circ \mathcal{T} : \mathbb{F}^{n+v} \to \mathbb{F}^{n-a}$, the *private key* consists of $\mathcal{S}$, $\mathcal{F}$ and $\mathcal{T}$.

Signature Generation: To generate a signature for a message $\mathbf{h} \in \mathbb{F}^{n-a}$, the signer performs the following three steps.

1. Compute a pre-image $\mathbf{x} \in \mathbb{F}^n$ of $\mathbf{h}$ under the affine map $\mathcal{S}$.
2. Lift $\mathbf{x}$ to the extension field $\mathbb{E}$ (using the isomorphism $\varPhi$). Denote the result by X.
   Choose random values for the vinegar variables $v_1, \ldots, v_v \in \mathbb{F}$ and compute $\mathcal{F}_V = \mathcal{F}(v_1, \ldots, v_v)$.
   Solve the univariate polynomial equation $\mathcal{F}_V(Y) = X$ by Berlekamp's algorithm and compute $\mathbf{y}' = \varPhi^{-1}(Y) \in \mathbb{F}^n$.
   Set $\mathbf{y} = (\mathbf{y}'||v_1|| \ldots ||v_v)$.
3. Compute the signature $\mathbf{z} \in \mathbb{F}^{n+v}$ by $\mathbf{z} = \mathcal{T}^{-1}(\mathbf{y})$.

Signature Verification: To check, if $\mathbf{z} \in \mathbb{F}^{n+v}$ is indeed a valid signature for a message $\mathbf{h} \in \mathbb{F}^{n-a}$, one simply computes $\mathbf{h}' = \mathcal{P}(\mathbf{z}) \in \mathbb{F}^{n-a}$. If $\mathbf{h}' = \mathbf{h}$ holds, the signature is accepted, otherwise rejected.

3.3   Gui

Recently, Petzoldt et al. proposed the multivariate signature scheme Gui [9], which is based on the concept of HFEv-. In fact, the private and public keys of Gui are just HFEv- keys over the field GF(2) with specially chosen parameters n, D, a and v. Since the number of equations in the public key and therefore the input size of Gui is only 90 bits, it would be possible for an attacker to find two messages m_1 and m_2 whose hash values collide in these first 90 bits. To overcome this problem, the authors of [9] developed a specially designed signature generation process. For this, they used a special parameter l (denoted as repetition factor). The signature generation process of Gui works as shown in Algorithm 1. Roughly spoken, one computes HFEv- signatures for l different hash values of the message $\mathbf{d}$ and combines them to a single signature of size $(n-a)+l \cdot (a+v)$. Similarly, the verification algorithm (see Algorithm 2) evaluates the public key l times.

4   Our Sequential Aggregate Signature Scheme

In the design of our multivariate sequential aggregate signature scheme we apply the HFEv-/Gui [9] trapdoor functions, which results in a scheme resembling these basic components. However, while in Gui all the partial signatures are computed using the same private key, we use here for every l-th partial signature another key.

Algorithm 1. Signature Generation Process of Gui	**Algorithm 2.** Signature Verification Process of Gui
Input: HFEv- private key $(\mathcal{S}, \mathcal{F}, \mathcal{T})$ message $\mathbf{d}$, repetition factor l **Output:** signature $\sigma \in \mathbb{F}_2^{(n-a)+l\cdot(a+v)}$ $\quad \mathbf{h} \leftarrow \text{SHA-256}(\mathbf{d})$ $\quad S_0 \leftarrow \mathbf{0} \in \mathbb{F}_2^{n-a}$ $\quad \textbf{for } i = 1 \textbf{ to } l \textbf{ do}$ $\qquad D_i \leftarrow \text{first } n - a \text{ bits of } \mathbf{h}$ $\qquad (S_i, X_i) \leftarrow \text{HFEv-}^{-1}(D_i \oplus S_{i-1})$ $\qquad \mathbf{h} \leftarrow \text{SHA-256}(\mathbf{h})$ $\quad \textbf{end for}$ $\quad \sigma \leftarrow (S_l \| X_l \| \dots \| X_1)$ $\quad \textbf{return } \sigma$	**Input:** HFEv- public key $\mathcal{P}$, message $\mathbf{d}$, repetition factor l, signature $\sigma \in \mathbb{F}_2^{(n-a)+l\cdot(a+v)}$ **Output: TRUE** or **FALSE** $\quad \mathbf{h} \leftarrow \text{SHA-256}(\mathbf{d})$ $\quad (S_l, X_l, \dots, X_1) \leftarrow \sigma$ $\quad \textbf{for } i = 1 \textbf{ to } l \textbf{ do}$ $\qquad D_i \leftarrow \text{first } n - a \text{ bits of } \mathbf{h}$ $\qquad \mathbf{h} \leftarrow \text{SHA-256}(\mathbf{h})$ $\quad \textbf{end for}$ $\quad \textbf{for } i = l - 1 \textbf{ to } 0 \textbf{ do}$ $\qquad S_i \leftarrow \mathcal{P}(S_{i+1} \| X_{i+1}) \oplus D_{i+1}$ $\quad \textbf{end for}$ $\quad \textbf{if } S_0 = 0 \textbf{ then}$ $\qquad \textbf{return TRUE}$ $\quad \textbf{else}$ $\qquad \textbf{return FALSE}$ $\quad \textbf{end if}$

4.1 Key Generation

Let $\mathbb{F} = \mathbb{F}_q$ be a finite field with q elements, $n, D, a, v, l \in \mathbb{N}$ be public parameters and $U = \{u_1, \dots, u_k\}$ be a set of users. Every user $u_i \in U$ generates an HFEv-key pair $((\mathcal{S}_i, \mathcal{F}_i, \mathcal{T}_i), \mathcal{P}_i)$ according to the given parameter set. Additionally, he computes a public key identity $\text{id}_i = \mathcal{H}(\mathcal{P}_i)$ using a hash function $\mathcal{H}$ modeled as a random oracle. He publishes his public key $\mathcal{P}_i$ and his public key identity id_i while keeping $\mathcal{S}_i, \mathcal{F}_i$ and $\mathcal{T}_i$ secret.

The reason for introducing public key identities in our scheme is the fact, that the public keys serve as input to the hash function multiple times. For large public keys it is therefore more efficient to utilize public identities instead of the public keys itself, by which we can reduce the input size of the hash functions. This results in faster signature generation engines.

4.2 Signature Generation

Assume that each user u_i has a message m_i to be signed. To generate an aggregate signature Σ for the messages $m_1, \dots, m_k$, every signer u_i ($i = 1, \dots, k$) performs successively Algorithm 3 (using the output of signer u_{i-1} as his input). The final aggregate signature Σ is given by the output of the signer u_k.

The first signer u_1 just computes a standard HFEv- (Gui) signature for the message m_1 (using Algorithm 1) and returns it as the first aggregate signature Σ_1.

In addition to his own private key and message, each signer u_i ($i \in \{2, \dots, k\}$) requires as input lists of the public keys $\{\text{pk}_1, \dots, \text{pk}_{i-1}\}$, the public key identities $\{\text{id}_1, \dots, \text{id}_{i-1}\}$, and the messages $m_1, \dots, m_{i-1}$ of the previous signers

$u_1, \ldots, u_{i-1}$ and the $(i-1)$-th sequential aggregate signature Σ_{i-1}. Before computing his own signature σ_i and combining it with Σ_{i-1}, u_i checks the correctness of Σ_{i-1} via the verification algorithm (see Algorithm 4).

In order to generate the aggregate signature Σ_i, the signer u_i ($i = 2, \ldots, k$) splits up the input aggregate signature Σ_{i-1} into two blocks $\tilde{S}, \tilde{X}$, where $\tilde{S}$ has a length of $n - a$ bits. He uses his own private key sk_i to compute a standard Gui signature $\sigma_i = (S_l, X_l, \ldots, X_1)$ for the hash value $\mathbf{h} = \mathcal{H}(m_1, \ldots, m_i, \mathsf{id}_1, \ldots, \mathsf{id}_i)$ by running Algorithm 1. After that, he combines his signature $(S_l, X_l, \ldots, X_1)$ with the previous aggregate signature (Σ_{i-1}) to generate the new aggregate signature Σ_i. Algorithm 3 shows this process in algorithmic form.

Algorithm 3. Signature Generation Process of MQSAS for each user $i \in 1, \cdots, k$

Input: private key sk_i, message m_i , public keys $\mathsf{pk}_1, \ldots, \mathsf{pk}_{i-1}$, public key identities $\mathsf{id}_1, \ldots, \mathsf{id}_{i-1}$, messages $m_1, \cdots, m_{i-1}$, aggregate signature Σ_{i-1}, where $\Sigma_0 = \emptyset$, repetition factor l

Output: aggregate signature Σ_i

1: **if** i = 1 **then**
2:     $\tilde{S} = \mathbf{0}^{n-a}$
3:     $\tilde{X} \leftarrow \emptyset$
4: **else if** $\mathsf{AggVerify}(i - 1, \Sigma_{i-1}, \mathsf{pk}_1, \cdots, \mathsf{pk}_{i-1}, m_1, \cdots, m_{i-1}) = \mathbf{TRUE}$ **then**
5:     $(\tilde{S}, \tilde{X}) \leftarrow \mathrm{split}(\Sigma_{i-1})$
6: **else**
7:     $print("Incorrect Signature")$
8:     **return**
9: **end if**
10: $\mathbf{h} \leftarrow \mathcal{H}(m_1, \ldots, m_i, id_1, \cdots, id_i)$
11: **for** $j = 1$ to l **do**
12:     $D_j \leftarrow$ first $n - a$ bits of $\mathbf{h}$
13:     $(S_j, X_j) \leftarrow \mathrm{HFEV}-^{-1}(D_j \oplus S_{j-1})$
14:     $\mathbf{h} \leftarrow \mathcal{H}(\mathbf{h})$
15: **end for**
16: $\tilde{S} \leftarrow \tilde{S} \oplus S_l$
17: $\tilde{X} \leftarrow (X_l || \ldots || X_1 || \tilde{X})$
18: $\Sigma_i \leftarrow (\tilde{S}, \tilde{X})$
19: **return** Σ_i

4.3   Signature Verification

To check the authenticity of an aggregate signature Σ_i (Algorithm 4), we parse Σ_i into the sequence of blocks $\tilde{S}, X_{i \cdot l}, \ldots, X_{(i-1) \cdot l+1}, X_{(i-1) \cdot l}, \ldots, X_1$. Here, the length of the block $\tilde{S}$ is $n - a$, while all the other blocks are of length $a + v$. After this, the verification of the aggregate signature Σ_i works very similar to the verification of a Gui signature.

In the j-th iteration, the algorithm first reconstructs the hash values $D_1, \ldots, D_l$ used during the generation of the j-th partial signature. Just as in Algorithm 2 it then evaluates the public key pk_j l times to compute the new value of $\tilde{S}$. At termination, the aggregate signature Σ_i is accepted, if and only if $\tilde{S} = 0$ holds.

Algorithm 4. Verification Process of MQSAS

Input: public keys $\mathsf{pk}_1, \ldots, \mathsf{pk}_i$, public key identities $\mathsf{id}_1, \ldots, \mathsf{id}_i$, messages $m_1, \ldots, m_i$, repetition factor l and aggregate signature Σ_i

Output: boolean value **TRUE** or **FALSE**

1: $(\tilde{S}, X_{i \cdot l}, \ldots, X_{(i-1) \cdot l+1}, X_{(i-1) \cdot l}, \ldots, X_1) \leftarrow \mathrm{split}(\Sigma_i)$
2: **for** $j = i$ to 1 **do**
3:    $\mathbf{h} \leftarrow \mathcal{H}(m_1, \ldots, m_j, \mathsf{id}_1, \cdots, \mathsf{id}_j)$
4:    **for** $k = 1$ to l **do**
5:        $D_k \leftarrow$ first $n - a$ bits of $\mathbf{h}$
6:        $\mathbf{h} \leftarrow \mathcal{H}(\mathbf{h})$
7:    **end for**
8:    **for** $k = l - 1$ to 0 **do**
9:        $\tilde{S} \leftarrow \mathsf{pk}_j(\tilde{S} \| X_{(j-1) \cdot l+k+1}) \oplus D_{k+1}$
10:    **end for**
11: **end for**
12: **if** $\tilde{S} = \mathbf{0}$ **then**
13:    **return TRUE**
14: **else**
15:    **return FALSE**
16: **end if**

The Algorithms 3 and 4 show how to instantiate the MQSAS signature scheme with HFEv-/Gui. However we note that our multivariate sequential aggregate signature scheme can also be instantiated on the basis of every other multivariate signature scheme such as Rainbow. Nevertheless, since HFEv-/Gui leads to optimal compression rates, we restrict here to initializing our scheme with HFEv-/Gui.

5   Security

The security analysis of our scheme is done in two steps. First we show that an attack against our sequential aggregate signature scheme penetrates the (trapdoor-) onewayness of the underlying HFEv- scheme. In the second step we then analyze the practical security of this scheme. However, due to lack of space, we can not present all the details of this analysis here and therefore refer for this to the extended version of our paper [2].

Theorem 1. *Assuming the (t', ϵ')-one-wayness of the public key of HFEv-, the sequential aggregate signature scheme presented in Sect. 4 is $(t, q_H, q_S, n, \epsilon)$-secure against existential forgery under adaptive sequential aggregate chosen-message attack such that*

$$(q_S + q_H + 1) \cdot \epsilon' \geq \epsilon \ and \ t \leq t' - (4kq_H + 4kq_S + 7k - 1)$$

for all t and ϵ.

Proof. see [2].

The practical security of HFEv- has been studied in several papers [5,9]. Additionally to this, we performed a number of experiments with the direct attack against HFEv- instances which showed that the complexity of this attack against the MQSAS instances proposed in the next section is beyond the claimed levels of security. The details of these experiments can be found in the extended version of this paper [2].

6   Parameters and Comparison

In this section we propose concrete parameter sets for the MQSAS scheme and compare its compression capabilities and performance to that of other (sequential) aggregate signature schemes. In particular, we propose 5 parameter sets for 80-bit security and 2 parameter sets for 120-bit security, allowing a trade off between compression rate and performance (see Table 1). The parameters shown in the table are chosen in such a way that the complexity of a direct attack against the scheme is beyond the claimed level of security.

Table 1. Proposed parameters and resulting key sizes for the MQSAS scheme

Security level (bit)	MQSAS $(\mathbb{F}, n, D, a, v, l)$	Public key size (kB)	Private key size (kB)	$|\sigma|$ (bit) (20 signers)	Compression factor $\tau(20)$
80	(GF(2),96,5,6,6,2)	57.7	2.4	570	0.75
	(GF(2),95,9,5,5,2)	55.5	2.3	490	0.78
	(GF(2),94,17,4,4,2)	53.3	2.3	410	0.81
	(GF(2),96,65,2,2,2)	55.7	2.3	254	0.88
	(GF(7),62,8,2,2,1)	47.1	2.9	420	0.89
120	(GF(2),127,9,4,6,2)	133.8	4.1	523	0.81
	(GF(7),93,8,3,3,1)	156.7	6.4	630	0.90

When instantiating our scheme over GF(2), we need, due to the threat of birthday attacks, to choose the repetition factor l of the Gui scheme to be ≥ 2. However, this has negative consequences for the compression capabilities of our scheme. To avoid this, we also propose parameters for the MQSAS scheme over GF(7). We can efficiently store 14 bits in 5 GF(7)-elements, while an element of GF(7) is stored in 3 bits. By doing so, we need 60 GF(7)-elements to store a hash value of 160 bits (80-bit security), while 90 GF(7) elements are needed to store a hash value of 240 bits (120-bit security).

Note that the key sizes listed in Table 1 are those of a single signer u_i. The verifyer of the aggregate signature $\Sigma = \Sigma_k$ is faced with a public key of size k times the value listed in the table.

The size of a sequential aggregate signature of our scheme is given by

$$|\Sigma| = (n - a) + k \cdot l \cdot (a + v), \tag{7}$$

the compression rate τ is given by

$$\tau = 1 - \frac{|\Sigma|}{k \cdot |\sigma|} = 1 - \frac{1}{k} \cdot \left(1 + \frac{(k - 1) \cdot l \cdot (a + v)}{(n - a) + l \cdot (a + v)}\right). \tag{8}$$

Table 2 and Fig. 3 show the signature sizes and the compression rates τ for the parameter sets listed in Table 1 and different numbers of users.

6.1   Implementation

To estimate the performance of MQSAS, we created an implementation of our scheme in C. For the implementation of the underlying HFEv- scheme we thereby adapted the implementation of Gui [9] to our setting. Table 3 shows the computation times needed to generate and verify an aggregate signature for different number of users and parameter sets. The experiments were run on a PC with a Core-i5 3750k processor (Ivy Bridge) at 2.4 GHz and with 16 GB of RAM. The timings in the table are the average values of 500 signature generation/verification processes.

6.2   Discussion

Table 1 indicates that we achieve very short aggregate signatures and high compression rates at security levels of both 80-bit and 120-bit. For example, for 80-bit security, our scheme allows to generate an aggregate signature for 20 signers of

Table 2. Signature sizes and compression rates of the MQ-SAS scheme

MQ-SAS	5 signers		10 signers		20 signers		50 signers		100 signers	
$(\mathbb{F}, n, D, a, v, l)$	$\vert\Sigma\vert$ (bit)	τ	$\vert\Sigma\vert$ (bit)	τ	$\vert\Sigma\vert$ (bit)	τ	$\vert\Sigma\vert$ (bit)	τ	$\vert\Sigma\vert$ (bit)	τ
(GF(2),96,5,6,6,2)	210	0.63	330	0.71	570	0.75	1,290	0.77	2,490	0.78
(GF(2),95,9,5,5,2)	190	0.65	290	0.74	490	0.78	1,090	0.80	2,090	0.81
(GF(2),94,17,4,4,2)	170	0.68	250	0.76	410	0.81	890	0.83	1,690	0.84
(GF(2),96,65,2,2,2)	134	0.73	174	0.83	254	0.88	494	0.90	894	0.91
(GF(7),62,8,2,2,1)	240	0.75	300	0.84	420	0.89	780	0.92	1,380	0.93
(GF(2),127,9,4,6,2)	223	0.69	323	0.77	523	0.82	1,123	0.84	2,123	0.85
(GF(7),93,8,3,3,1)	360	0.75	450	0.84	630	0.89	1,170	0.92	2,070	0.93

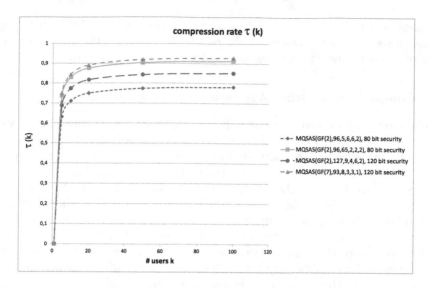

Fig. 3. Compression rate τ of the MQ-SAS scheme

Table 3. Signature generation/verification time of MQSAS

MQ-SAS	Signature generation/verification time (ms)				
$(\mathbb{F}, n, D, a, v, l)$	5 signers	10 signers	20 signers	50 signers	100 signers
(GF(2),96,5,6,6,2)	1.24/0.127	3.13/0.206	6.11/0.339	27.86/0.644	82.91/1.292
(GF(2),95,9,5,5,2)	2.94/0.119	6.75/0.182	8.52/0.228	36.31/0.672	94.25/1.328
(GF(2),94,17,4,4,2)	9.30/0.086	17.19/0.137	32.32/0.242	68.18/0.638	158.01/1.267
(GF(2),127,9,4,6,2)	3.42/0.169	5.23/0.214	11.61/0.439	45.3/0.983	141.4/1.969
(GF(2),96,65,2,2,2)	150.86/0.051	302.23/0.117	599.24/0.231	1,509.69/0.590	3,053.25/1.185

length only 254 bits, which is less than one quarter of a single RSA signature at the same security level. Remarkably, the compression rates for 120-bit security are even higher than in the 80-bit case. On the other hand, the key sizes are considerably larger for the corresponding parameter sets.

Table 2 shows the aggregate signature sizes and associated compression rates for different parameter sets and number of users. The highest compression rates can hereby be achieved by the two schemes MQSAS(GF(7), 62, 8, 2, 2, 1) and MQSAS(GF(7), 93, 8, 3, 3, 1). The reason for this is that, for these parameter sets, we can choose the repetition factor of Gui to be 1. As the table shows, these parameter sets allow compression rates of up to 93 %, which means that we need only the size of 7 individual signatures to prove the validity of an aggregate signature for 100 signers (see also Fig. 3).

In Table 3 we provide the timings for the signing and verification engine of our scheme. In fact, we note, that each signing step by construction invokes the verification engine in order to check the validity of the previous aggregate signature, before the signer is able to proceed. The timings indicate that the parameter set $(GF(2), 96, 65, 2, 2, 2)$, due to the high degree of the HFE polynomial in use, is much slower than the remaining ones. We therefore observe a trade off between the compression rate and the performance of the scheme.

6.3   Comparison to Other Aggregate Signature Schemes

In this subsection we compare our sequential aggregate signature scheme with other constructions. In fact, we observe that multivariate-based sequential aggregate signature schemes are more suitable for practice than their counterparts from classical and lattice-based cryptography. In terms of signature size and performance, HFEv- has been shown to be far more efficient than the other schemes. The size of an individual signature is only slightly more than one hundred bits, whereas the underlying signature schemes of the other sequential aggregate signature schemes produce signatures of size more than 1,000 bits.

With regard to the performance, the timings of our scheme for signing and verification are also significantly better than those of the other schemes. We have shown that the overhead, that our sequential aggregate signature scheme entails, can be at least as low as 7 bits per signature for a reasonable level of security. Hence, almost the whole signature of a signer is concealed within the signature of its successor. Furthermore, the arithmetic operations of our scheme are mainly performed over the field $GF(2)$, which is well studied and thus allows to carry out fast operations which furthermore can be accelerated by the use of special processor instructions such as PCLMULQDQ [9]. On the other hand, lattice-based systems work over $\mathbb{Z}_q^n$, which implies to carry out more complex arithmetic operations such as reductions modulo q for at least $n \geq 256$ components.

Compared to RSA-based sequential aggregate signature schemes, our scheme is much easier to instantiate. The reason for this is that, in the case of RSA, the domains of the participating signers differ (Since every signer has a different public modulus N, the possible range of the hash values can be quite different). Therefore, it is important to agree on how to choose the hash functions beforehand, which leads to a significant communication overhead. Furthermore, the operations of RSA are more complex and hence lead to a less efficient scheme. This has also been observed in [3].

Due to the short signatures and simple operations, our scheme is also a much better candidate than RSA for an aggregate signature scheme on restricted devices (e.g. sensor networks).

7   Conclusion

In this paper we proposed a multivariate sequential aggregate signature scheme on the basis of the HFEv- signature scheme, which is the first multivariate and one of the first post-quantum schemes of this kind. Due to the use of HFEv-/ Gui as the underlying signature scheme, the resulting signatures are very short (less than 1 kbit for 100 signers at 80 bits of security) and we achieve high compression rates (up to 93 % for $k = 100$ signers). Furthermore, due to the efficiency of arithmetic operations over GF(2), our scheme outperforms all current (sequential) aggregate signature schemes in terms of performance.

Acknowledgments. We thank the anonymous reviewers of ISC for their comments which helped to improve this paper. The third author is supported by JSPS KAKENHI 15F15350.

References

1. Bernstein, D.J., Buchmann, J., Dahmen, E. (eds.): Post Quantum Cryptography. Springer, Heidelberg (2009)
2. El Bansarkhani, R., Mohamed, M.S.E., Petzoldt, A.: MQSAS - a multivariate sequential aggregate signature scheme - Extended Versions. IACR eprint 2016/503 (2016)
3. El Bansarkhani, R., Buchmann, J.: Towards lattice based aggregate signatures. In: Pointcheval, D., Vergnaud, D. (eds.) AFRICACRYPT. LNCS, vol. 8469, pp. 336–355. Springer, Heidelberg (2014)
4. Ding, J., Gower, J.E., Schmidt, D.S.: Multivariate Public Key Cryptosystems. Springer, Heidelberg (2006)
5. Ding, J., Yang, B.-Y.: Degree of regularity for HFEv and HFEv-. In: Gaborit, P. (ed.) PQCrypto 2013. LNCS, vol. 7932, pp. 52–66. Springer, Heidelberg (2013)
6. Garey, M.R., Johnson, D.S.: Computers and Intractability: A Guide to the Theory of NP-Completeness. W.H. Freeman and Company, Paris (1979)
7. Network Working Group: A Border Gateway Protocol (BGP-4). RFC 4271. https://tools.ietf.org/html/rfc4271
8. Patarin, J., Courtois, N.T., Goubin, L.: QUARTZ, 128-bit long digital signatures. In: Naccache, D. (ed.) CT-RSA 2001. LNCS, vol. 2020, p. 282. Springer, Heidelberg (2001)
9. Petzoldt, A., Chen, M.-S., Yang, B.-Y., Tao, C., Ding, J.: Design principles for HFEv- based multivariate signature schemes. In: Iwata, T., Cheon, J.H. (eds.) ASIACRYPT 2015. LNCS, vol. 9452, pp. 311–334. Springer, Heidelberg (2015). doi:10.1007/978-3-662-48797-6_14
10. Rivest, R.L., Shamir, A., Adleman, L.: A method for obtaining digital signatures and public-key cryptosystems. Commun. ACM **21**(2), 120–126 (1978)
11. Shor, P.: Polynomial-time algorithms for prime factorization and discrete logarithms on a quantum computer. SIAM J. Comput. **26**(5), 1484–1509 (1997)

Cryptanalysis of Multi-Prime Φ-Hiding Assumption

Jun Xu[1,2], Lei Hu[1,2(✉)], Santanu Sarkar[3], Xiaona Zhang[1,2],
Zhangjie Huang[1,2], and Liqiang Peng[1,2]

[1] State Key Laboratory of Information Security, Institute of Information
Engineering, Chinese Academy of Sciences, Beijing 100093, China
{xujun,hulei,zhangxiaona,huangzhangjie,pengliqiang}@iie.ac.cn
[2] Data Assurance and Communications Security Research Center,
Chinese Academy of Sciences, Beijing 100093, China
[3] Indian Institute of Technology, Sardar Patel Road, Chennai 600 036, India
sarkar.santanu.bir@gmail.com

Abstract. In Crypto 2010, Kiltz, O'Neill and Smith used m-prime RSA modulus N with $m \geq 3$ for constructing lossy RSA. The security of the proposal is based on the Multi-Prime Φ-Hiding Assumption. In this paper, we propose a heuristic algorithm based on the Herrmann-May lattice method (Asiacrypt 2008) to solve the Multi-Prime Φ-Hiding Problem when prime $e > N^{\frac{2}{3m}}$. Further, by combining with mixed lattice techniques, we give an improved heuristic algorithm to solve this problem when prime $e > N^{\frac{2}{3m} - \frac{1}{4m^2}}$. These two results are verified by our experiments. Our bounds are better than the existing works.

Keywords: Multi-Prime Φ-Hiding Assumption · Multi-Prime Φ-Hiding Problem · Lattice · LLL algorithm · Coppersmith's technique · Gauss algorithm

1 Introduction

1.1 Background

The Φ-Hiding Assumption [1] firstly introduced by Cachin, Micali and Stadler in Eurocrypt 1999 was used for building a practical private information retrieval scheme. Based on this assumption, many cryptographic schemes have been designed, such as [3,4,6,12]. This assumption is roughly stated as follows:

"For a given integer N with unknown factorization, it is hard to decide whether a given prime e divides $\Phi(N)$, where Φ is the Euler function."

Obviously, the Φ-Hiding Assumption holds with some requirements on the size of e since it is not true for $e \geq N$. The Φ-Hiding Assumption with RSA modulus $N = pq^{2k}$ has been analyzed in Asiacrypt 2008 [16]. The corresponding result is that a case of this variant fails with a good probability for any prime e.

© Springer International Publishing Switzerland 2016
M. Bishop and A.C.A. Nascimento (Eds.): ISC 2016, LNCS 9866, pp. 440–453, 2016.
DOI: 10.1007/978-3-319-45871-7_26

For cryptographic applications, one would like e to be as large as possible, but from a security point of view, if e divides $\Phi(N)$ and is sufficiently large, then one can recover the factorization of N by using the idea of Coppersmith [2,9,14]. Thus, it is interesting to know the minimal size of e that allows for efficient factoring attacks.

It is well known that one can utilize Coppersmith's method to factorize the standard RSA modulus $N = pq$ when prime $e > N^{\frac{1}{4}}$ divides $\Phi(N) = (p-1)(q-1)$. In Asiacrypt 2012, Kakvi, Kiltz and May proposed a lattice algorithm for obtaining a non-trivial factor of general N under the above condition [10].

In Crypto 2010, Kiltz *et al.* [12] showed that the RSA function $f : x \to x^e$ mod N is a $\log e$ lossy trapdoor permutation (LTDP) under the Φ-Hiding Assumption with $N = pq$. They also showed that the RSA-OAEP is indistinguishable against chosen plaintext attack (IND-CPA) in the standard model under this assumption, which is a long time open problem. Furthermore, they generalized this assumption to the multi-prime situation in order to obtain a more efficient LTDP such that RSA-OAEP can securely encrypt longer plaintext. To be specific, this multi-prime situation is described as follows:

"For a given RSA modulus $N = p_1 \cdots p_m$ where bit-length of the p_i are equal for all $1 \le i \le m$, it is hard to decide whether a given prime e divides $p_i - 1$ for all p_i except one prime factor of N."

The condition that e divides $p_i - 1$ for all p_i except one prime factor of N implies that e^{m-1} divides $\Phi(N) = (p_1 - 1) \cdots (p_m - 1)$. So, this is a special case of e divides $\Phi(N)$. Therefore, it is a variant of the Φ-Hiding Assumption. For the sake of terminology, it is called as the Multi-Prime Φ-Hiding Assumption.

Now when $e|(p_i - 1)$ for $i \in [1, m-1]$, there are integers x_i such that $ex_i = p_i - 1$. So if one obtains the integer root x_i of equation $ex_i = p_i - 1$ for any $i \in [1, m-1]$, factorization of N is easily possible as $\gcd(ex_i + 1, N) = p_i$. Lattice method like Coppersmith's technique can be used to find x_i in polynomial time. So the research goal is to maximize the bound up to which x_i can be computed efficiently. Since the prime p_i are of the same bit-length, one fully breaks the Multi-Prime Φ-Hiding Assumption when the bound of x_i reaches $N^{\frac{1}{m}}$.

Originally, the bound $N^{\frac{1}{m^2}}$ of x_i was received by the Howgrave-Graham method in [9]. Later, this bound was improved up to $N^{O(\frac{1}{m^c})}$ for some $1 < c \le 2$ in [7,12]. Eventually, the bounds $N^{O(\frac{1}{m \log m})}$ were acquired in [15,18,19]. However, it is open whether a bound $N^{O(\frac{1}{m})}$, i.e., the exponent being linear in $\frac{1}{m}$, could be achieved.

1.2 Previous Works

In this subsection, we recall some known attacks on the Multi-Prime Φ-Hiding Problem. Note that if e divides $p_i - 1$ for all $1 \le i \le m$, then $N \equiv 1 \bmod e$. It gives a polynomial time distinguisher. To decide if e is Multi-Prime Φ-Hidden in N, consider the system of equations

$$ex_1 + 1 \equiv 0 \bmod p_1, \quad ex_2 + 1 \equiv 0 \bmod p_2, \quad \ldots, \quad ex_{m-1} + 1 \equiv 0 \bmod p_{m-1}.$$

Let $x_1 = N^\delta$. Here all p_i are of sizes of the same magnitude for $1 \leq i \leq m - 1$. Usually we have

$$x_2 \approx \cdots \approx x_{m-1} \approx N^\delta.$$

The Howgrave-Graham method [9] can be used to find the desired small solutions of a modular linear equation

$$ex_i + 1 = 0 \bmod p_i \text{ for some } i \in \{1, \cdots, m - 1\}.$$

Using Howgrave-Graham's method, one can solve the Multi-Prime Φ-Hiding Problem in polynomial time if

$$\delta < \frac{1}{m^2}.$$

In Crypto 2010, Kiltz et al. [12] constructed a polynomial equation

$$e^{m-1} \left(\prod_{i=1}^{m-1} x_i \right) + \cdots + e \left(\sum_{i=1}^{m-1} x_i \right) + 1 \equiv 0 \bmod \prod_{i=1}^{m-1} p_i$$

by multiplying all given equations. Then they linearized the polynomial and solved it by the Herrmann-May theorem [8]. They showed that one can solve the Multi-Prime Φ-Hiding Problem in polynomial time if[1]

$$\delta < \frac{2}{m} \left(\frac{1}{m} \right)^{\frac{m}{m-1}}.$$

Later in Africacrypt 2011, Herrmann [7] improved the work of Kiltz et al. He used the Herrmann-May theorem to find the desired root (x, y) in equation

$$e^2 x + ey + 1 = 0 \bmod \prod_{i=1}^{m-1} p_i,$$

where $x = e^{m-3} \prod_{i=1}^{m-1} x_i + \cdots + \sum_{j>i} x_i x_j, y = \sum_{i=1}^{m-1} x_i$. He solved the Multi-Prime Φ-Hiding Problem in polynomial time if

$$\delta < \frac{2}{3} \left(\frac{1}{m} \right)^{\frac{3}{2}}.$$

In ACISP 2012, Tosu and Kunihiro [19] generalized the method of Herrmann. Instead of taking two variables, they considered linear polynomials of k variables for $k \in [1, m - 1]$. They proved that one can solve the Multi-Prime Φ-Hiding Problem in polynomial time if

$$\delta < \max_{1 \leq k \leq m-1} \left\{ \frac{2}{k+1} \left(\frac{1}{m} \right)^{\frac{k+1}{k}} \right\}.$$

---

[1] There is a minor mistake in proceedings version of Crypto 2010 as reported in [7, Page 97].

For large m, Tosu and Kunihiro further optimized k and got

$$\delta < \frac{2}{em(\ln m + 1)}$$

where e is the base of the natural logarithm. Thus, asymptotically bound of δ is

$$\frac{2}{em \ln m} = O(\frac{1}{m \log m}).$$

In SPACE 2012, Sarkar [15] observed that the sizes of two components of the desired root (x, y) in the analysis of Herrmann are not balanced. Based on this observation, he obtained better bound on δ than the work of Herrmann.

Takayasu and Kunihiro generalized the work of Herrmann and May [8] in [17, 18]. Their bounds are better when components of the desired root are of different size. Since there is a big difference between the sizes of x and y in Φ-Hiding Polynomial of [7], one can get a better bound on δ than the work of Herrmann. The bound of δ in the work of [17] is very close to [15], however, the work of [17] is more flexible and it can deal with modular equations with more variables than [15].

1.3   Our Contribution

In this paper, we show that the Multi-Prime Φ-Hiding Assumption does not hold when $\delta < \frac{1}{3m}$. For the first time, we obtain such a bound of δ which is linear in $\frac{1}{m}$. Thus we can solve the Multi-Prime Φ-Hiding Problem in polynomial time if

$$e > N^{\frac{1}{m} - \frac{1}{3m}} = N^{\frac{2}{3m}}.$$

Further, we improve the bound of δ up to $\frac{1}{3m} + \frac{1}{4m^2}$. This improvement is enormous for small values of m. Hence Multi-Prime Φ-Hiding Problem can be solved in polynomial time if

$$e > N^{\frac{1}{m} - (\frac{1}{3m} + \frac{1}{4m^2})} = N^{\frac{2}{3m} - \frac{1}{4m^2}}.$$

1.4   Organization of the Paper

We organize our paper as follow. In Sect. 2, we recall some preliminaries. In Sect. 3, we propose an algorithm using lattice technique. We give an improved algorithm using mixed lattice methods in Sect. 4. In Sect. 5, we give the comparison of our work with the existing results. We present our experiment results in Sect. 6. Section 7 concludes the paper.

2   Preliminaries

2.1   Lattice

A lattice $\mathcal{L}$ is a discrete subgroup of $\mathbb{R}^n$. An alternative equivalent definition of an integer lattice can be given using a basis.

Let $\mathbf{b_1}, \cdots, \mathbf{b_m}$ be linear independent row vectors in $\mathbb{R}^n$, a lattice $\mathcal{L}$ spanned by them is

$$\mathcal{L} = \left\{ \sum_{i=1}^{m} k_i \mathbf{b_i} \mid k_i \in \mathbb{Z} \right\}.$$

The set $\{\mathbf{b_1}, \cdots, \mathbf{b_m}\}$ is called a basis of $\mathcal{L}$ and $B = [\mathbf{b_1}^T, \cdots, \mathbf{b_m}^T]^T$ is the corresponding basis matrix. The dimension and determinant of $\mathcal{L}$ are respectively

$$\dim(\mathcal{L}) = m, \det(\mathcal{L}) = \sqrt{\det(BB^T)}.$$

When $m = n$, lattice is called full rank. In case of a full rank lattice, $\det(\mathcal{L}) = |\det(B)|$. From Hadamard's inequality, it is known that $\det(B) \le \prod_{i=1}^{n} \|\mathbf{b_i}\|$, where $\|\mathbf{b}\|$ denotes Euclidean ℓ_2 norm of a vector $\mathbf{b}$.

For any two-dimensional lattice $\mathcal{L}$, the Gauss algorithm can find out the reduced basis vectors $\mathbf{v_1}$ and $\mathbf{v_2}$ satisfying

$$\|\mathbf{v_1}\| \le \|\mathbf{v_2}\| \le \|\mathbf{v_1} \pm \mathbf{v_2}\|$$

in time $O\left(\log^2(\max\{\|\mathbf{v_1}\|, \|\mathbf{v_2}\|\})\right)$. Here $\mathbf{v_1}$ is the shortest nonzero vector in $\mathcal{L}$ and $\mathbf{v_2}$ is the shortest vector in $\mathcal{L} \setminus \{k\mathbf{v_1} \mid k \in \mathbb{Z}\}$. A shortest vector $\mathbf{v}$ of an n dimensional lattice satisfies the Minkowski bound $\|\mathbf{v}\| \le \sqrt{n}(\det(\mathcal{L}))^{\frac{1}{n}}$. The following result will be used in Sect. 4.

Lemma 1. (See, e.g., [5]). *Let $\mathbf{v_1}$ and $\mathbf{v_2}$ be the reduced basis vectors of $\mathcal{L}$ by the Gauss algorithm and $\mathbf{x} \in \mathcal{L}$. For the unique pair of integers (α, β) that satisfies $\mathbf{x} = \alpha\mathbf{v_1} + \beta\mathbf{v_2}$, we have*

$$\|\alpha\mathbf{v_1}\| \le \frac{2}{\sqrt{3}}\|\mathbf{x}\|, \ \|\beta\mathbf{v_2}\| \le \frac{2}{\sqrt{3}}\|\mathbf{x}\|.$$

2.2 Finding Small Roots

Coppersmith gave rigorous methods for extracting small roots of modular univariate polynomials and bivariate integer polynomials. These methods can extend to multivariate cases under the following assumption.

Assumption 1. *Let $h_1, \cdots, h_n \in \mathbb{Z}[x_1, \cdots, x_n]$ be the polynomials that are found by Coppersmith's algorithm. Then the ideal generated by the polynomial equations $h_1(x_1, \cdots, x_n) = 0, \cdots, h_n(x_1, \cdots, x_n) = 0$ has dimension zero.*

Herrmann and May used the idea of Coppersmith's technique to analyze modular linear polynomials and got the following result for bivariate linear polynomials.

Theorem 1 ([8]). *Let $\epsilon > 0$ and N be a large integer with a divisor $p \geq N^\beta$. Let $f(x_1, x_2) \in \mathbb{Z}[x_1, x_2]$ be a linear polynomial. Under Assumption 1, one can find all solutions (x_1, x_2) of the equation $f(x_1, x_2) = 0 \mod p$ with $|x_1| \leq N^{\gamma_1}, |x_2| \leq N^{\gamma_2}$ in polynomial time if*

$$\gamma_1 + \gamma_2 \leq 3\beta - 2 + 2(1-\beta)^{\frac{3}{2}} - \epsilon.$$

In our analyses, we consider the asymptotic case and ignore the low order term.

2.3 Multi-Prime Φ-Hiding Assumption

We briefly introduce the Multi-Prime Φ-Hiding Assumption and the corresponding problem. Please refer to [7,12,15,19] for more details.

Definition 1 (Multi-Prime Φ-Hiding Problem). *Let $N = p_1 \cdots p_m$ be a Multi-Prime RSA modulus where the p_i are of the same bit length for $1 \leq i \leq m$. Let e be a given prime of the size $N^{\frac{1}{m} - \delta}$. The problem is to decide whether*

$$e \mid (p_1 - 1), \cdots, e \mid (p_{m-1} - 1), e \nmid (p_m - 1).$$

Definition 2 (Multi-Prime Φ-Hiding Assumption). *There is no polynomial time algorithm that solves the Multi-Prime Φ-Hiding Problem with a non-negligible probability of success.*

3 Algorithm Using Lattice Technique

In this section we give an algorithm for solving the Multi-Prime Φ-Hiding Problem. Our algorithm can be derived from the following theorem.

Theorem 2. *Let $N = p_1 \cdots p_m$ be a Multi-Prime RSA modulus where the p_i are of same bit length for $1 \leq i \leq m$. Let e be a prime of the size $N^{\frac{1}{m} - \delta}$. Under Assumption 1, we can solve the Multi-Prime Φ-Hiding Problem in polynomial time when*

$$\delta < \frac{1}{3m}.$$

Proof. Let $r = N \mod e$ and $s = (\frac{N-r}{e}) \mod e$. If $e \mid (p_1 - 1), \cdots, e \mid (p_{m-1} - 1)$ and $e \nmid (p_m - 1)$, there exist unknown integers $x_1, \cdots, x_{m-1}$ such that

$$ex_1 + 1 = p_1, \cdots, ex_{m-1} + 1 = p_{m-1}.$$

Since $N = p_1 \cdots p_m$, we have $(ex_1 + 1) \cdots (ex_{m-1} + 1) \cdot p_m = N$. Then taking modulo e on both sides we get

$$p_m \mod e = N \mod e = r.$$

Thus, there is an equation $ex_m + r = p_m$ with unknown x_m. We multiply all equations together to get $(ex_1 + 1) \cdots (ex_{m-1} + 1)(ex_m + r) = N$. So we have

$$(e^{m-1} \prod_{i=1}^{m-1} x_i + \cdots + e^2 \sum_{1 \le i < j \le m-1} x_i x_j + e \sum_{1 \le i \le m-1} x_i + 1)(ex_m + r) = N$$

$$\Rightarrow (e^{m-1} \prod_{i=1}^{m-1} x_i + \cdots + e^2 \sum_{1 \le i < j \le m-1} x_i x_j + e \sum_{1 \le i \le m-1} x_i + 1)(ex_m + r) = N$$

$$\Rightarrow (e \sum_{1 \le i \le m-1} x_i + 1)(ex_m + r) \equiv N \bmod e^2$$

$$\Rightarrow ex_m + er \sum_{1 \le i \le m-1} x_i + r \equiv N \bmod e^2$$

$$\Rightarrow ex_m + er \sum_{1 \le i \le m-1} x_i \equiv es \bmod e^2$$

$$\Rightarrow x_m + r \sum_{1 \le i \le m-1} x_i - s \equiv 0 \bmod e.$$

Let $y_1 = x_m$, $y_2 = x_1 + \cdots + x_{m-1}$. Consider the bivariate modular linear equation

$$f(y_1, y_2) = y_1 + ry_2 - s. \tag{1}$$

The Eq. (1) has root $\mathbf{y} := (x_m, x_1 + \cdots + x_{m-1})$ in $\mathbb{Z}_e$ as $f(y_1, y_2) \equiv 0 \bmod e$.

First, let us bound the size of $\mathbf{y}$. Since $0 < x_i = \frac{p_i - 1}{e} < \frac{N^{\frac{1}{m}}}{N^{\frac{1}{m} - \delta}} = N^\delta$ for $i = 1, \ldots, m$, we have

$$0 < x_1 + \cdots + x_{m-1} < (m-1)N^\delta = e^{\log_e(m-1) + \frac{\delta}{\frac{1}{m} - \delta}}.$$

Next, we use Theorem 1 for solving Eq. (1). Since modulus e is known, we take $\beta = 1$. Here $\gamma_1 = \frac{\delta}{\frac{1}{m} - \delta}$ and $\gamma_2 = \log_e(m-1) + \frac{\delta}{\frac{1}{m} - \delta}$. Under Assumption 1, we can find all solution (y_1, y_2) in polynomial time when

$$\gamma_1 + \gamma_2 = \frac{2\delta}{\frac{1}{m} - \delta} + \log_e(m-1) \le 1 - \epsilon.$$

Considering the asymptotic case and ignoring the lower order terms, the above condition is simplified to

$$\delta < \frac{1}{3m}.$$

Further, we check whether $\gcd(ey_1 + r, N)$ gives a nontrivial factor of N for every candidate. Thus, we can find out the desired root $\mathbf{y}$ and recover p_m. Conversely, if we cannot get a non-trivial factor of N under Assumption 1, then relation $e \mid (p_1 - 1), \cdots, e \mid (p_{m-1} - 1), e \nmid (p_m - 1)$ in the Multi-Prime Φ-Hiding Problem does not hold.

Based on the Theorem 2, we have the Algorithm 1 to solve the Multi-Prime Φ-Hiding Problem.

Algorithm 1. Solving Multi-Prime Φ-Hiding Problem

Input: Public key (N, e) and m is the number of prime factors of N.
Output: Decide whether $e \mid (p_1 - 1), \cdots, e \mid (p_{m-1} - 1), e \nmid (p_m - 1)$.
1: Compute $r = N \bmod e$ and $s = (\frac{N-r}{e}) \bmod e$.
2: Solve equation $y_1 + ry_2 - s \equiv 0 \bmod e$ using Theorem 1.
3: If $\gcd(ey_1 + r, N)$ for all solutions (y_1, y_2) are trivial factors of N, output no. Else, output yes.

---

4   Improved Algorithm Using Mixed Lattice Methods

In this section we present an improved algorithm in order to improve the bound $\delta < \frac{1}{3m}$. This algorithm is obtained by dealing with Eq. (1) with mixed lattice methods in the following theorem.

Theorem 3. *Let $N = p_1 \cdots p_m$ be a Multi-Prime RSA modulus where the p_i are of same bit length for $1 \le i \le m$. Let e be a prime of the size $N^{\frac{1}{m} - \delta}$. Under Assumption 1, we can solve the Multi-Prime Φ-Hiding Problem in polynomial time when*

$$\delta < \frac{4}{3m} - \frac{2}{3} + \frac{2}{3}\left(1 - \frac{1}{m}\right)^{3/2}.$$

Proof. If $e \mid (p_1 - 1), \cdots, e \mid (p_{m-1} - 1)$ and $e \nmid (p_m - 1)$, we know

$$y_1 + ry_2 \equiv s \bmod e \qquad (2)$$

has integer root $\mathbf{y} := (x_m, x_1 + \cdots + x_{m-1})$, where

$$\|\mathbf{y}\| = \sqrt{(x_1 + \cdots + x_{m-1})^2 + x_m^2} < m \cdot N^\delta.$$

The set of solutions

$$\mathcal{L} = \left\{ (y_1, y_2) \in \mathbb{Z}^2 \mid y_1 + ry_2 \equiv 0 \bmod e \right\}$$

forms an additive discrete subgroup of $\mathbb{Z}^2$. Thus, $\mathcal{L}$ is a 2-dimensional integer lattice. Lattice $\mathcal{L}$ is spanned by the row vectors of the basis matrix

$$B = \begin{bmatrix} -r & 1 \\ e & 0 \end{bmatrix}.$$

Let us briefly check integer span of B, denoted by $\text{span}(B)$ is indeed equal to $\mathcal{L}$. First both $(-r, 1)$ and $(e, 0)$ are solutions of $y_1 + ry_2 \equiv 0 \bmod e$. Thus $\text{span}(B) \subseteq \mathcal{L}$. Conversely, let $(y_1, y_2) \in \mathcal{L}$. So we have $y_1 + ry_2 = ke$ for some $k \in \mathbb{Z}$. Then $(y_2, k)B = (y_1, y_2) \in \text{span}(B)$. Thus $\mathcal{L} \subseteq \text{span}(B)$.

Consider the set

$$\mathcal{L}' = \left\{ (s + y_1, y_2) \mid (y_1, y_2) \in \mathcal{L} \right\}.$$

It is clear that for any $(x, y) \in \mathcal{L}'$, (x, y) will satisfy the Eq. (2).

Let $\mathbf{u} := (u_1, u_2)$ be the smallest length vector in $\mathcal{L}'$, which can be obtained by the closest vector algorithm on the lattice $\mathcal{L}$ from the point $(-s, 0)$ in polynomial time (see, e.g., [11]). Obviously, $\|\mathbf{u}\| \le \|\mathbf{y}\| < m \cdot N^\delta$.

Let $\mathbf{v_1} := (v_{11}, v_{12}), \mathbf{v_2} := (v_{21}, v_{22})$ be Gauss-reduced basis vectors of $\mathcal{L}$. Since $\mathbf{y} - \mathbf{u}$ belongs to $\mathcal{L}$, there exist integer coefficients α_1, α_2 such that

$$\mathbf{y} - \mathbf{u} = \alpha_1 \mathbf{v_1} + \alpha_2 \mathbf{v_2}. \tag{3}$$

Observing that the first component of $\mathbf{y}$ is equal to $\frac{p_m - r}{e}$ and rearranging Eq. (3), we get $ev_{11}\alpha_1 + ev_{21}\alpha_2 + eu_1 + r = p_m$. In Appendix A, we prove that $|v_{11}| \le \sqrt{2e}$ and $v_{11} \ne 0$. Thus, the bivariate modular linear equation

$$(ev_{11})x_1 + (ev_{21})x_2 + (eu_1 + r) \equiv 0 \bmod p_m \tag{4}$$

has an integer root (α_1, α_2).

First, let us bound the sizes of unknown α_1 and α_2. From (3), according to Lemma 1, we obtain

$$|\alpha_1| \le \frac{2\|\mathbf{y} - \mathbf{u}\|}{\sqrt{3}\|\mathbf{v_1}\|} \le \frac{2(\|\mathbf{y}\| + |\mathbf{u}|)}{\sqrt{3}\|\mathbf{v_1}\|} < \frac{4mN^\delta}{\sqrt{3}\|\mathbf{v_1}\|},$$

$$|\alpha_2| \le \frac{2\|\mathbf{y} - \mathbf{u}\|}{\sqrt{3}\|\mathbf{v_2}\|} \le \frac{2(\|\mathbf{y}\| + |\mathbf{u}|)}{\sqrt{3}\|\mathbf{v_2}\|} < \frac{4mN^\delta}{\sqrt{3}\|\mathbf{v_2}\|}.$$

So $|\alpha_1\alpha_2| < \frac{16m^2N^{2\delta}}{3\|\mathbf{v_1}\|\|\mathbf{v_2}\|}$. Notice that $e = \det(\mathcal{L}) \le \|\mathbf{v_1}\|\|\mathbf{v_2}\|$. Thus we have

$$|\alpha_1\alpha_2| < \frac{16m^2N^{2\delta}}{3\|\mathbf{v_1}\|\|\mathbf{v_2}\|} = N^{3\delta - \frac{1}{m} + \log_N \frac{16m^2}{3}}, \text{ as } e = N^{\frac{1}{m} - \delta}.$$

Next, we use Theorem 1 to solve the Eq. (4), where the size of unknown modulus p_m is $N^{\frac{1}{m}}$. So we take $\beta = \frac{1}{m}$. Under Assumption 1, we can find all roots (x_1, x_2) of the Eq. (4) in polynomial time when

$$3\delta - \frac{1}{m} + \log_N \frac{16m^2}{3} \le \frac{3}{m} - 2 + 2\left(1 - \frac{1}{m}\right)^{\frac{3}{2}} - \epsilon.$$

Ignoring the term $\log_N \frac{16m^2}{3}$ as $m \ll N$, we get

$$\delta < \frac{4}{3m} - \frac{2}{3} + \frac{2}{3}\left(1 - \frac{1}{m}\right)^{3/2}.$$

Furthermore, we check whether $\gcd(ev_{11}x_1 + ev_{21}x_2 + eu_1 + r, N)$ gives a nontrivial factor of N for every candidate. Thus, we can obtain the desired root (α_1, α_2) and recover the factor p_m of N. □

Since $(1 - \frac{1}{m})^{\frac{3}{2}} = 1 - \frac{3}{2m} + \frac{3}{8m^2} + o(\frac{1}{m^2})$, we have $\frac{4}{3m} - \frac{2}{3} + \frac{2}{3}\left(1 - \frac{1}{m}\right)^{3/2} \approx \frac{1}{3m} + \frac{1}{4m^2}$. Thus the simplified condition is

$$\delta < \frac{1}{3m} + \frac{1}{4m^2}.$$

So when $e > N^{\frac{1}{m} - \frac{1}{3m} - \frac{1}{4m^2}} = N^{\frac{2m}{3} - \frac{1}{4m^2}}$, one can solve Multi-Prime Φ-Hiding Problem in polynomial time.

Since $\frac{4}{3m} - \frac{2}{3} + \frac{2}{3}\left(1 - \frac{1}{m}\right)^{3/2} > \frac{1}{3m}$, bound of δ in Theorem 3 is better than that of Theorem 2. In Fig. 1, we present the two bounds pictorially.

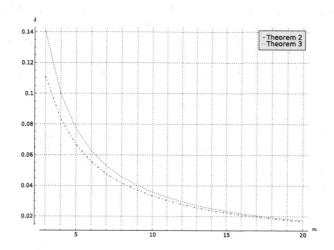

Fig. 1. Comparison between the bounds δ for Theorem 2 and Theorem 3 when $3 \leq m \leq 20$.

Based on the Theorem 3, we have the Algorithm 2 to solve the Multi-Prime Φ-Hiding Problem.

Algorithm 2. Further Solving Multi-Prime Φ-Hiding Problem

Input: Public key (N, e) and m is the number of prime factors of N.
Output: Decide whether $e \mid (p_1 - 1), \cdots, e \mid (p_{m-1} - 1), e \nmid (p_m - 1)$.
1: Compute $r = N \bmod e$ and $s = \left(\frac{N-r}{e}\right) \bmod e$.
2: Find the smallest Euclidean length root (u_1, u_2) of equation $y_1 + ry_2 \equiv s \bmod e$ using the closest vector algorithm.
3: Generate lattice $\mathcal{L}$ spanned by the row vectors of the matrix

$$\begin{bmatrix} -r & 1 \\ e & 0 \end{bmatrix}.$$

4: Compute Gauss-reduced basis vectors (v_{11}, v_{12}) and (v_{21}, v_{22}) of lattice $\mathcal{L}$.
5: Solve equation $(ev_{11})x_1 + (ev_{21})x_2 + (eu_1 + r) \equiv 0 \bmod p_m$ using Theorem 1.
6: If $\gcd(ev_{11}x_1 + ev_{21}x_2 + eu_1 + r, N)$ for all solutions (x_1, x_2) are trivial factors of N, output no. Else, output yes.

---

5   Comparison with the Existing Works

In this section, we compare our results with previous works.

In Fig. 2, we compare our results with the existing works pictorially. We observe that the curve of [18] is almost identical with that of [15]. So we do not plot it explicitly. It is clear from the figure that our bound is much better than the existing bounds. Thus our new attack solves the Multi-Prime Φ-Hiding Problem for more values of e than the existing works. One can see that existing curves [7,15,19] are very close to each other. On the other hand, we are achieving much improved curve. More importantly, for small values of m, these differences are more prominent. For example when $m = 4$, new bound of δ becomes 0.09968 whereas existing was

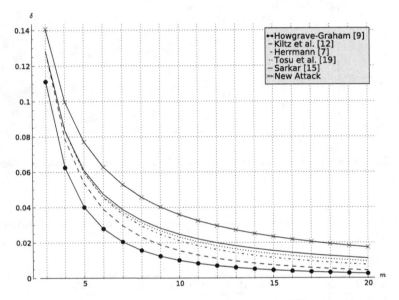

Fig. 2. Comparison of our bound $\delta < \frac{4}{3m} - \frac{2}{3} + \frac{2}{3}\left(1 - \frac{1}{m}\right)^{3/2}$ with the existing works for $3 \le m \le 20$.

Table 1. Comparison of bit lengths of the minimum e with 2048-bit N

m Results	3	4	5	6	7	8	9	10
Kiltz *et al.* [12]	420	351	301	262	233	209	190	174
Herrmann [7]	420	342	288	249	219	196	177	162
Tosu *et al.* [19]	420	342	288	248	217	192	173	158
Sarkar [15]	420	341	286	245	214	190	170	155
Takayasu *et al.* [18]	421	341	286	245	214	190	170	154
$\left(\frac{1}{m} - \frac{1}{3m}\right) \cdot 2048$	456	342	274	228	196	171	152	137
$\left(\frac{1}{m} - \left(\frac{4}{3m} - \frac{2}{3} + \frac{2}{3}(1 - \frac{1}{m})^{3/2}\right)\right) \cdot 2048$	395	308	252	213	185	163	146	132

0.08358 in [18]. Thus the improvement is significant for small values of m. Also m cannot be large as in that case Elliptic Curve Factorization [13] will be efficient.

In Table 1, we present the minimum bit lengths of e for which Φ-Hidding Problem is polynomial time solvable for different values of m. Here we take 2048-bit N. From the table, it is clear that all existing bounds are almost same for $m = 3$. Though early works improve the work of [12] for values $m > 3$, this work improves the bound on δ for $m > 3$ as well as $m = 3$. So one can solve Φ-Hidding Problem in polynomial time for much smaller values of e.

6 Experiment Results

We implement the above attacks with LLL algorithm in Magma on a PC with Intel(R) Core(TM) Quad CPU (2.83 GHz, 3.25 GB RAM, Windows XP). In our experiments, Assumption 1 is always verified. We present our experimental results in Table 2. As we can see that the experimental results and theoretical upper bounds on δ are perfectly match.

For Theorem 2, we take $3 \le m \le 10$. We use Theorem 1 to solve equation $y_1 + ry_2 - s = 0 \bmod e$. For a positive integer t, we generate polynomials

$$g_{k,i}(y_1, y_2) := y_2^i (y_1 + ry_2 - s)^k e^{t-k}$$

which share the common root $\mathbf{y}$ modulo e^t, where $k = 0, \cdots, t; i = 0, \ldots, t - k$. In our experiments, we choose $t = 8$. The dimensions of the involved lattices are $\frac{1}{2}(t^2 + 3t + 2) = 45$. Then the desired root can be obtained by lattice reduction. Hence the factor p_m of the modulus N can be recovered when e is Multi-Prime Φ-Hidden in N and the corresponding δ satisfies the experimental value.

For Theorem 3, we present the situations of $3 \le m \le 5$. We neglect running times of the closest vector algorithm and the Gauss algorithm as they are negligible since the corresponding lattices are only two-dimensional. In order to use

Table 2. Experiment results for different values of m with 2048 bit N

Analyses	m	δ (theoretical)	δ (experimental)	LLL (seconds)	Gröbner (seconds)
Theorem 2	3	0.1111	0.1100	60.497	0.842
	4	0.0833	0.0820	32.854	0.484
	5	0.0667	0.0657	19.859	0.421
	6	0.0556	0.0548	15.241	0.296
	7	0.0476	0.0469	11.778	0.287
	8	0.0417	0.0409	8.299	0.187
	9	0.0370	0.0355	6.349	0.125
	10	0.0333	0.0315	5.444	0.078
Theorem 3	3	0.1407	0.1320	3975.826	1120.540
	4	0.0997	0.0891	2059.156	121.734
	5	0.0770	0.0683	1866.188	109.938

Theorem 1, we first multiply the equation $(ev_{11})x_1 + (ev_{21})x_2 + (eu_1 + r) \equiv 0 \bmod p_m$ by $(ev_{11})^{-1}$ modulo N and get a monic equation $f(x_1, x_2) \equiv 0 \bmod p_m$. Then, we collect the polynomials which share a common root (α_1, α_2) modulo N^l

$$h_{k,i}(x_1, x_2) := x_2^i f^k(x_1, x_2) N^{\max\{l-k, 0\}}$$

for $k = 0, \cdots, t; i = 0, \ldots, t-k$ and $l = \left\lfloor \left(1 - \sqrt{\frac{m-1}{m}}\right) t \right\rfloor$. In our experiments, we take $t = 12$. The dimensions of the corresponding lattices are $\frac{1}{2}(t^2 + 3t + 2) = 91$. Finally, we obtain the desired (α_1, α_2).

7   Conclusion

In this paper, we have reduced the Multi-Prime Φ-Hiding Problem to the problem of finding small root of a bivariate modular linear equation. Based on this, we have proposed two algorithms using lattice techniques to solve the problem. We have obtained better bounds than the existing works.

Acknowledgements. The authors would like to thank anonymous reviewers for their helpful comments and suggestions. The work of this paper was supported by the National Key Basic Research Program of China (Grants 2013CB834203), the National Natural Science Foundation of China (Grants 61472417, 61472415 and 61502488), the Strategic Priority Research Program of Chinese Academy of Sciences under Grant XDA06010702, and the State Key Laboratory of Information Security, Chinese Academy of Sciences.

A   Proof on $|v_{11}| \leq \sqrt{2e}$ and $v_{11} \neq 0$

Proof. Note that $\mathbf{v_1} = (v_{11}, v_{12})$ is the shortest nonzero vector in lattice $\mathcal{L}$. According to Minkowski bound, we know that

$$\|\mathbf{v_1}\| \leq \sqrt{2 \det(\mathcal{L})} = \sqrt{2e}.$$

Since v_{11} is a component of $\mathbf{v_1}$, we have $|v_{11}| \leq \sqrt{2e}$. Now, we prove that $v_{11} \neq 0$. Since $v_1 \in \mathcal{L}$, there exists some integer c_1 such that

$$v_{11} + rv_{12} = c_1 e.$$

If $v_{11} = 0$, we get $rv_{12} = c_1 e$. Since e is a prime and $0 < r < e$, e divides v_{12}. Thus e divides $\|\mathbf{v_1}\|$. So $\|\mathbf{v_1}\| \geq e$. However, it is impossible since $\|\mathbf{v_1}\| \leq \sqrt{2e}$. Therefore, $v_{11} \neq 0$. $\square$

References

1. Cachin, C., Micali, S., Stadler, M.A.: Computationally private information retrieval with polylogarithmic communication. In: Stern, J. (ed.) EUROCRYPT 1999. LNCS, vol. 1592, p. 402. Springer, Heidelberg (1999)

2. Coppersmith, D.: Small solutions to polynomial equations, and low exponent RSA vulnerabilities. J. Cryptol. **10**(4), 233–260 (1997)

3. Gentry, C., Mackenzie, P., Ramzan, Z.: Password authenticated key exchange using hidden smooth subgroups. In: Proceedings of the 12th ACM Conference on Computer and Communications Security CCS 2005, pp. 299–309. ACM, New York (2005)

4. Gentry, C., Ramzan, Z.: Single-database private information retrieval with constant communication rate. In: Caires, L., Italiano, G.F., Monteiro, L., Palamidessi, C., Yung, M. (eds.) ICALP 2005. LNCS, vol. 3580, pp. 803–815. Springer, Heidelberg (2005)

5. Gomez, D., Gutierrez, J., Ibeas, A.: Attacking the pollard generator. IEEE Trans. Inf. Theor. **52**(12), 5518–5523 (2006)

6. Hemenway, B., Ostrovsky, R.: Public-key locally-decodable codes. In: Wagner, D. (ed.) CRYPTO 2008. LNCS, vol. 5157, pp. 126–143. Springer, Heidelberg (2008)

7. Herrmann, M.: Improved cryptanalysis of the Multi-Prime ϕ - Hiding Assumption. In: Nitaj, A., Pointcheval, D. (eds.) AFRICACRYPT 2011. LNCS, vol. 6737, pp. 92–99. Springer, Heidelberg (2011)

8. Herrmann, M., May, A.: Solving linear equations modulo divisors: on factoring given any bits. In: Pieprzyk, J. (ed.) ASIACRYPT 2008. LNCS, vol. 5350, pp. 406–424. Springer, Heidelberg (2008)

9. Howgrave-Graham, N.: Approximate integer common divisors. In: Silverman, J.H. (ed.) CaLC 2001. LNCS, vol. 2146, p. 51. Springer, Heidelberg (2001)

10. Kakvi, S.A., Kiltz, E., May, A.: Certifying RSA. In: Wang, X., Sako, K. (eds.) ASIACRYPT 2012. LNCS, vol. 7658, pp. 404–414. Springer, Heidelberg (2012)

11. Kannan, R.: Minkowski's convex body theorem and integer programming. Math. Oper. Res. **12**(3), 415–440 (1987)

12. Kiltz, E., O'Neill, A., Smith, A.: Instantiability of RSA-OAEP under chosen-plaintext attack. In: Rabin, T. (ed.) CRYPTO 2010. LNCS, vol. 6223, pp. 295–313. Springer, Heidelberg (2010)

13. Lenstra Jr., H.W.: Factoring integers with elliptic curves. Ann. Math. **126**, 649–673 (1987)

14. May, A.: Using LLL-reduction for solving RSA and factorization problems. In: Nguyen, P.Q., Valle, B. (eds.) The LLL Algorithm. Information Security and Cryptography, pp. 315–348. Springer, Heidelberg (2010)

15. Sarkar, S.: Reduction in lossiness of RSA trapdoor permutation. In: Bogdanov, A., Sanadhya, S. (eds.) SPACE 2012. LNCS, vol. 7644, pp. 144–152. Springer, Heidelberg (2012)

16. Schridde, C., Freisleben, B.: On the validity of the Φ-hiding assumption in cryptographic protocols. In: Pieprzyk, J. (ed.) ASIACRYPT 2008. LNCS, vol. 5350, pp. 344–354. Springer, Heidelberg (2008)

17. Takayasu, A., Kunihiro, N.: Better lattice constructions for solving multivariate linear equations modulo unknown divisors. In: Boyd, C., Simpson, L. (eds.) ACISP. LNCS, vol. 7959, pp. 118–135. Springer, Heidelberg (2013)

18. Takayasu, A., Kunihiro, N.: Better lattice constructions for solving multivariate linear equations modulo unknown divisors. IEICE Trans. **97-A**(6), 1259–1272 (2014)

19. Tosu, K., Kunihiro, N.: Optimal bounds for multi-prime Φ-hiding assumption. In: Susilo, W., Mu, Y., Seberry, J. (eds.) ACISP 2012. LNCS, vol. 7372, pp. 1–14. Springer, Heidelberg (2012)

Author Index